NUMERICAL SOLUTION OF PARTIAL DIFFERENTIAL EQUATIONS—II

SYNSPADE 1970

Proceedings of the Second Symposium on the Numerical Solution of Partial Differential Equations, SYNSPADE 1970, Held at the University of Maryland, College Park, Maryland, May 11-15, 1970

Symposium on the Numerical Solution of Partial Differential Equations.

Numerical Solution of Partial Differential Equations—II

SYNSPADE 1970

EDITED BY
BERT HUBBARD

Institute for Fluid Dynamics
and Applied Mathematics
University of Maryland
College Park, Maryland

ACADEMIC PRESS 1971
New York • London

ACADEMIC PRESS, INC.
111 Fifth Avenue, New York, New York 10003

United Kingdom Edition published by
ACADEMIC PRESS, INC. (LONDON) LTD.
Berkeley Square House, London W1X 6BA

LIBRARY OF CONGRESS CATALOG CARD NUMBER: 73-137624
AMS (MOS) 1970 Subject Classification 65-65

PRINTED IN THE UNITED STATES OF AMERICA

CONTENTS

CONTENTS

CONTRIBUTORS

Numbers in parentheses indicate the pages on which authors' contributions begin.

Jean Pierre Aubin, Université Paris, Paris, France (1)

Ivo Babuška, Institute for Fluid Dynamics and Applied Mathematics, University of Maryland, College Park, Maryland (69)

J. H. Bramble, Cornell University, Ithaca, New York (107)

Hermann G. Burchard, Indiana University, Bloomington, Indiana (1)

David L. Colton, Indiana University, Bloomington, Indiana (273)

Jim Douglas, Jr., University of Chicago, Chicago, Illinois (133)

Todd Dupont, University of Chicago, Chicago, Illinois (133)

Alan J. Faller, University of Maryland, College Park, Maryland (215)

Gaetano Fichera, University of Rome, Rome, Italy (243)

P. R. Garabedian, New York University, New York, New York (253)

Robert P. Gilbert, Indiana University, Bloomington, Indiana (273)

Herbert B. Keller, Firestone Laboratories, California Institute of Technology, Pasadena, California (327)

R. B. Kellogg, University of Maryland, College Park, Maryland (351)

D. G. Korn, New York University, New York, New York (253)

Heinz-Otto Kreiss, Uppsala University, Uppsala, Sweden (401)

CONTRIBUTORS

M. M. Lavrentiev, Computer Center, Siberian Branch of the Academy of Sciences of the U. S. S. R., Novosibirsk, U. S. S. R. (417)

Pedro V. Marcal, Brown University, Providence, Rhode Island (433)

G. I. Marchuk, Computer Center, Siberian Branch of the Academy of Sciences of the U. S. S. R., Novosibirsk, U. S. S. R. (469)

Werner C. Rheinboldt, Computer Science Center, University of Maryland, College Park, Maryland (501)

A. H. Schatz, Cornell University, Ithaca, New York (107)

Gilbert Strang, Massachusetts Institute of Technology, Cambridge, Massachusetts (547)

Vidar Thomée, Department of Mathematics, Chalmers Institute of Technology and The University of Göteborg, Göteborg, Sweden (585)

Richard S. Varga, Kent State University, Kent, Ohio (623)

PREFACE

The Second Symposium on the Numerical Solution of Partial Differential Equations, SYNSPADE 1970, brought together an international audience of over 150 active researchers in this field and related fields of application. Held May 11-15, 1970, it was the second conference of its kind to be held at the University of Maryland, College Park, Maryland. The prior conference took place five years earlier (1965) and both proceedings have been published by Academic Press.

As before, the symposium was composed of invited hourly lectures (18 in number) and a larger number (27) of ten-minute contributed papers. This volume contains, in complete form, the papers given by the invited speakers. As can be seen from the table of contents, they range from talks on mathematical numerical analysis to descriptions of numerical methods applied to problems in fluid dynamics, meteorology, and mechanics.

Certain new themes which have appeared in this field in the intervening period between the two symposia have given this volume a different flavor from the earlier one. For example, the finite element method which has been used for a long time by engineers and was represented by Varga's paper in 1965 now has become a focal point of research in boundary value problems and appears in roughly half of the papers in this volume. The paper by Garabedian and Korn represents a promising advance in the difficult problem of airfoil design. Continuing themes from the last symposium are represented by the papers of Fichera, Kreiss, and Thomée.

SYNSPADE 1970 was sponsored by The Institute for Fluid Dynamics and Applied Mathematics, University of Maryland, and was funded by the National Science Foundation. The generous financial assistance of NSF and the hard work of many members of the University of Maryland faculty and staff contributed immeasurably to the success of our meeting. The symposium committee itself consisted of R. B. Kellogg (chairman), A. K. Aziz, I. Babuška, B. Hubbard, L. Karlovitz, J. Osborn, and J. S. Vandergraft.

SOME ASPECTS OF THE METHOD OF THE HYPERCIRCLE APPLIED TO ELLIPTIC VARIATIONAL PROBLEMS*

Jean Pierre Aubin** and Hermann G. Burchard***

Introduction

In this paper we try to combine techniques from two closely related fields of investigation: recently, methods of the Rayleigh-Ritz-Galerkin type for the approximation of boundary value problems using spline basis functions and Sobolev spaces have received the attention of many writers [4, 6, 7, 10, 20, and others]. While the Rayleigh-Ritz method gives upper bounds for the energy integral, it was Trefftz [22] who in 1926 showed how to find lower bounds by a complementary method, which now bears his name. This method was studied by Friedrichs [12] who showed that Trefftz' method is nothing else but the Rayleigh-Ritz method applied to a variational problem conjugate to the one originally given and obtained from it by a Legendre transformation. The study of Trefftz' method was taken up again in 1947 by Prager and Synge [19] and Diaz and Weinstein [11] and subsequently by many others, e.g. [8, 13,

---

* Sponsored by the Mathematics Research Center, United States Army, Madison, Wisconsin, under Contract No. DA-31-124-ARO-D-462.

** Université Paris IX.

*** Indiana University.

14, 17, 18, 21] and references therein. These authors were intrigued by the fact that the combination of Ritz' and Trefftz' method not only gives upper and lower bounds for the energy integral but can also be made, by Schwarz' inequality, to yield pointwise upper and lower bounds for the solution.

The immediate purpose of this paper is twofold:

a) To obtain "a posteriori" bounds for the error in approximating the solution of boundary value problems. We include the case considered by Weinberger [23], where the underlying quadratic form is not positive.

b) To set up a conjugate problem (equivalent with the originally given boundary value problem) the boundary conditions of which are "natural" whenever the original boundary conditions are "forced" and vice versa. The approximation of the conjugate problem is studied.

We illustrate the procedure in the case of the Dirichlet problem

(I) (i) $-\Delta u + \lambda u = f \quad \text{on} \quad \Omega \subset R^n$

(ii) $\gamma u = u|_{\Gamma} = t \quad \text{on} \quad \Gamma = \partial\Omega .$

For convenience we assume $\lambda > 0$. We can split $-\Delta$:

(II) $$-\Delta = -\text{div grad} = D^*D ,$$

where $Du = \text{grad } u$ and $D^*\vec{u} = -\text{div } \vec{u}$. Also we have Green's formula

$$\text{(III)} \quad \sum_{i=1}^{n} \int_{\Omega} u_i D_i v dx = - \int_{\Omega} \operatorname{div} \vec{u} \cdot v dx + \int_{\Gamma} \beta\vec{u}\, \gamma v d\sigma(x)$$

with $\beta\vec{u} = \vec{u} \cdot \vec{n}$, $\vec{n}$ being the outward normal to Γ, and $\vec{u} = (u_i)_{i=1}^{n}$. Then the Dirichlet problem (I) is equivalent with "conjugate" problem

$$\text{(IV)} \qquad \text{(i)} \quad DD^{*}\vec{u} + \lambda\vec{u} = Df$$

$$\text{(ii)} \quad \gamma\lambda^{-1}(f - D^{*}\vec{u}) = t \; .$$

The two solutions u and $\vec{u}$ are related by

$$\text{(V)} \qquad u = \lambda^{-1}(f-D^{*}\vec{u}) \; , \quad Du = \vec{u} \; .$$

In (I) the boundary conditions are "forced," in (IV) they are "natural." This is an important fact, e.g. for approximation purposes, since natural boundary conditions are easier to deal with. T. Kato [14] has also used the factorization $T^{*}T$ for the purpose of obtaining a posteriori bounds. We remark that this example is treated differently in the general procedure below, with little effort as long as $\lambda > 0$. For the case $\lambda = 0$, see [21]. We have assumed $\lambda > 0$ throughout because of the great convenience in defining a conjugate problem with properties similar to those of the primal. However, e.g. the results of section (2.5) are easily established for any λ.

Furthermore, Green's formula (III) expresses orthogonality relations which lead to a posteriori estimates of the error: Let u^h be any approximant of the solution u

of (I) satisfying $\gamma u^h = t$, and let $\vec{u}^h$ be an approximant of the solution $\vec{u}$ of (IV); $\vec{u}^h$ need not satisfy any boundary conditions. Then the following estimate holds.

$$\text{(VI)} \qquad \int_\Omega |\mathrm{grad}(u-u^h)|^2 dx + \lambda \int_\Omega |u-u^h| dx \leqslant \int_\Omega |\vec{u}^h - \mathrm{grad}\, u^h|^2 dx + \lambda^{-1} \int |f-\lambda u^h + \mathrm{div}\, \vec{u}^h|^2 dx .$$

Obviously the right hand side of this inequality converges to zero if u^h converges to u and $\vec{u}^h$ converges to $\vec{u}$.

In the first section, we study these two problems in the case of boundary value problems for differential operators of order $2m$.

In the second section we generalize these results to "abstract" boundary value problems.

The third and fourth sections are devoted to the construction of approximate finite difference schemes for the conjugate problem. Results are obtained regarding convergence and error estimates.

The extension of the results of this paper to the case of boundary value problems for nonlinear monotone coercive operators and to unilateral boundary value problems (which are equivalent to variational inequalities) is easy; we do not treat them here. Likewise not dealt with by us are pointwise and optimal bounds; for these questions we refer to Golomb and Weinberger [13] and Payne and Weinberger [18].

The paper assumes familiarity with some facts from the theories of Sobolev spaces and approximation of differential equations. The reader may wish to consult the book by Lions and Magenes [16] and also e.g., [1], [2], and [4]. The footnotes at the end of each section hopefully will help in tracing the needed material.

1. Example of Conjugate Problems and a posteriori Estimates

We illustrate by means of an example the construction of the conjugate problem of a boundary value problem. We point out that "forced" boundary conditions become "natural" boundary conditions (section 1-2). We deduce a posteriori estimates from Green's formula.

1.1. Construction of the conjugate problem.

Let $\Omega \subset R^n$ be a smooth bounded open subset, Γ its boundary. Let $a_{pq}(x)$ be functions belonging to $L^\infty(\Omega)$ and Λ the differential operator†

$$(1.1) \qquad \Lambda u = \sum_{|p|,|q|\leq m} (-1)^{|q|} D^q(a_{pq}(x)D^p u) .$$

---

† $H^m(\Omega)$ is the Sobolev space of functions u of $L^2(\Omega)$ such that the weak derivatives $D^p u = \dfrac{\partial^{|p|}}{\partial x_1^{p_1} \cdots \partial x_n^{p_n}} u$ belong to $L^2(\Omega)$ for $|p| = p_1 + \cdots + p_n \leq m$. cf. [16].

$\overset{\circ}{H}{}^m(\Omega)$ is the closure in $H^m(\Omega)$ of the space $\mathcal{D}(\Omega)$ of infinitely differentiable functions with compact support.

$H^{-m}(\Omega)$ is the dual of $\overset{\circ}{H}{}^m(\Omega)$. Sobolev spaces are also defined for fractional indices, cf. [16], e.g. by Fourier transform.

Finally, $H^m(\Omega,\Lambda)$ is the space of functions u of $H^m(\Omega)$ such that Λu belongs to $L^2(\Omega)$. Duality is indicated by primes.

It is the formal operator associated with the bilinear form

$$(1.2)\qquad a(u,v) = \sum_{|p|,|q|\leq m} \int_\Omega a_{pq}(x) D^p u D^q v dx$$

which is continuous on $V \times V$, where $V = H^m(\Omega)$. Let us recall the Green's formula based on the trace theorem for the trace operator $u = (u, \frac{\partial}{\partial n} u, \cdots, \frac{\partial^{m-1}}{\partial n^{m-1}} u)\big|_\Gamma$ † (cf [3] and [16])

$$(1.3)\qquad a(u,v) = (\Lambda u, v) + \sum_{0\leq j\leq m-1} \langle \delta_{2m-j-1} u, \gamma_j v\rangle$$

$$\text{for any } u \in H^m(\Omega,\Lambda) \text{ and } v \in H^m(\Omega) .$$

---

† We recall the general Green's formula (see [3] and [4], section IV-1): let V, H, T be Hilbert spaces and $\gamma \in L(V,T)$ satisfying

i) γ maps V onto T

ii) V and $V_o = \ker\gamma$ are dense subspaces of H.

If $a(u,v)$ is a continuous bilinear form on $V \times V$, we define its formal operator $\Lambda \in L(V,V_o')$ by

$$(\Lambda u, v) = a(u,v) \quad \text{for any } u \in V,\ v \in V_o ,$$

and its domain $V(\Lambda)$ by

$$V(\Lambda) = \{u \in V \text{ such that } \Lambda u \in H\} .$$

Then there exists a unique operator $\delta \in L(V(\Lambda),T')$ such that

$$a(u,v) = (\Lambda u,v) + \langle \delta u,\gamma v\rangle \quad \text{for any } u \in V(\Lambda),\ v \in V .$$

In this example, (1.3), we choose

$$V = H^m(\Omega),\quad V_o = H_o^m(\Omega),\quad H = L^2(\Omega)$$

$$T = \prod_{j=o}^{m-1} H^{m-j-\frac{1}{2}}(\Gamma) .$$

In this case assumptions i) and ii) above are guaranteed by

Let us introduce the following data

$$
(1.4)\quad \begin{cases} \text{i)} & f \in H = L^2(\Omega) \ , \\ \text{ii)} & g_j \in H^{m-j-\frac{1}{2}}(\Gamma) \quad \text{for} \quad 0 \leqslant j \leqslant p-1 \ , \quad 1 \leqslant p \leqslant m \ , \\ \text{iii)} & h_j \in H^{m-j-\frac{1}{2}}(\Gamma) \quad \text{for} \quad m \leqslant j \leqslant 2m-1-p \end{cases}
$$

and let us consider a solution $u \in H^m(\Omega, \Lambda)$ (if any) of the following boundary problem

$$
(1.5)\quad \begin{cases} \text{i)} & \Lambda u + \lambda u = f \quad \text{in} \quad \Omega \ , \quad \lambda \neq 0 \ , \\ \text{ii)} & \gamma_j u = g_j \quad \text{for} \quad 0 \leqslant j \leqslant p-1 \ , \\ \text{iii)} & \delta_j u = h_j \quad \text{for} \quad m \leqslant j \leqslant 2m-1-p \ . \end{cases}
$$

We shall construct another boundary-value problem called

---

the following trace theorem: the operator

$$
\gamma u = (\gamma_o u, \cdots, \gamma_{m-1} u) \qquad \text{where} \quad \gamma_j u = \frac{\partial^j}{\partial n^j}
$$

can be extended to a continuous linear operator mapping V onto T and $H^m_o(\Omega)$ coincides with the kernel of γ (see [16], Chapter 1).

Then there exists $\delta = (\delta_{2m-1}, \cdots, \delta_m)$ mapping $V(\Lambda) = H^m(\Omega,\Lambda)$ into T' where

$$
T' = \prod_{j=0}^{m-1} H^{-m+j+\frac{1}{2}}(\Gamma) = \prod_{k=2m-1}^{m} H^{m-k-\frac{1}{2}}(\Gamma)
$$

such that (1.3) holds.

"conjugate problem" equivalent to (1.5)[†]. For this purpose, let us introduce the space

$$(1.6)\quad \begin{cases} \vec{H} = (L^2(\Omega))^M \text{ of elements } \vec{u} = (u^q)_{|q|\leq m}, \\ \text{where } M \text{ is the number of multi-integers} \\ q \text{ such that } |q| \leq m \end{cases}$$

supplied with the following inner product and norm

$$(1.7)\quad [\vec{u}, \vec{v}] = \sum_{|q|\leq m} (u^q, v^q) , \quad [\vec{u}] = ([\vec{u}, \vec{u}])^{\frac{1}{2}} .$$

Next, we consider the operator $D \in L(V, \vec{H})$ defined by

$$(1.8)\quad Du = (D^q u)_{|q|\leq m} ,$$

and its formal adjoint $D^* \in L(\vec{H}, H^{-m}(\Omega))$ defined by

$$(1.9)\quad (D^*\vec{u}, v) = [\vec{u}, Dv] \text{ for any } v \in H_0^m(\Omega) .$$

Since $H_0^m(\Omega)$ is the closure in $H^m(\Omega)$ of the space of infinitely differentiable functions with compace support in Ω, we have

$$(1.10)\quad D^*\vec{u} = \sum_{|q|\leq m} (-1)^{|q|} D^q u^q .$$

---

† The introduction of this problem can be motivated by considering a kind of Legendre transform of a variational problem in cases where (1.5) is equivalent to such, cf. [12].

Finally, we introduce the domain $\vec{H}(D^*)$ of D^* defined by

$$(1.11) \qquad \vec{H}(D^*) = \{\vec{u} \in H \text{ such that } D^*\vec{u} = \sum_{|q|\leq m} (-1)^{|q|} D^q u^q \in L^2(\Omega)\} .$$

Then, we can prove a Green's formula†:

Lemma 1.1. There exist unique operators $\beta_j \in L(\vec{H}(D^*), H^{m-j-\frac{1}{2}}(\Gamma))$ (for $m \leq j \leq 2m-1$) such that

$$(1.12) \qquad \sum_{|p|\leq m} (u^p, D^p v) = \left(\sum_{|p|\leq m} (-1)^{|p|} D^p u^p, v \right) + \sum_{0\leq j\leq m-1} \langle \beta_{2m-1-j}\vec{u}, \gamma_j v \rangle$$

holds when $\vec{u}$ ranges over $\vec{H}(D^*)$ and v ranges over $V = H^m(\Omega)$.

Finally, let us introduce the following operator $\mathcal{A} \in L(\vec{H}, \vec{H})$

$$(1.13) \qquad (\mathcal{A}\vec{u})^q = \sum_{|p|\leq m} a_{pq}(x) u^p \quad \text{for any } q \text{ such that } |q| \leq m .$$

Then it is clear that we have the following relations

---

† This follows again from the trace theorem for γ, see Theorem 2.1 below and Theorem 1.1 of [3].

$$(1.14)\quad \begin{cases} \text{i)} & \Lambda u = D^{*}\mathcal{A}Du \quad \text{for any} \quad u \in H^{m}(\Omega, \Lambda)\ , \\ \text{ii)} & \delta_{j}u = \beta_{j}\mathcal{A}Du \quad \text{for any} \quad u \in H^{m}(\Omega, \Lambda)\ .^{\dagger} \end{cases}$$

Notice that D maps $H = L^{2}(\Omega)$ into $\vec{\mathcal{D}}'(\Omega) = (\mathcal{D}'(\Omega))^{M}$ and that, by definition, $V = H^{m}(\Omega)$ is the domain of D, i.e. the space of $u \in L^{2}(\Omega)$ such that $Du \in \vec{H} = (L^{2}(\Omega))^{M}$.

For the sake of simplicity, we shall assume that

$$(1.15)\qquad \mathcal{A} \text{ is an isomorphism from } \vec{H} \text{ onto } \vec{H}$$

and we can find functions α_{pq} in $L^{\infty}(\Omega)$ such that

$$(1.16)\quad (\mathcal{A}^{-1}u)^{q} = \sum_{|p|\leq m} \alpha_{pq}(x)u^{p} \quad \text{for any } q \text{ such that } |q| \leq m\ .$$

Now, we can define the "conjugate problem" of the boundary value problem (1.4). We look for $\vec{u} \in \vec{H}(D^{*})$ satisfying the following system of partial differential equations

$$(1.17)\quad \begin{cases} \text{i)} & \sum_{|p|\leq m} (-1)^{|p|} D^{p+q}u^{p} + \lambda \sum_{|p|\leq m} \alpha_{pq}(x)u^{p} = D^{q}f \\ & \qquad \text{for any } |q| \leq m\ , \\ \text{ii)} & \gamma_{j}\Big(f - \sum_{|p|\leq m} (-1)^{|p|} D^{p}u^{p}\Big) = \lambda g_{j} \quad \text{for } 0 \leq j \leq p-1\ , \end{cases}$$

---

† Notice that Green's formula (1.3) follows from Lemma 1.1 since

$$a(u,v) = [\mathcal{A}Du, Dv] = (D^{*}\mathcal{A}Du, v) + \langle \beta\mathcal{A}Du, \gamma v\rangle$$

when $u \in V(\Lambda) = \{u \text{ such that } \mathcal{A}Du \in \vec{H}(D^{*})\}$ and when $v \in V$. Here $\beta = (\beta_{2m-1}, \cdots, \beta_{m})$.

$$\left| \text{iii)} \quad \beta_j \vec{u} = h_j \quad \text{for} \quad m \leq j \leq 2m-1-p \; . \right.$$

We can write (1.17)i) in the shorter form

$$(1.18) \qquad DD^* \vec{u} + \lambda \mathcal{A}^{-1} \vec{u} = Df$$

and equation (1.5)i) in the form

$$(1.19) \qquad D^* \mathcal{A} Du + \lambda u = f \; .$$

Proposition 1.1. The boundary-value problem (1.5) and its conjugate problem (1.17) are equivalent: If $u \in H^m(\Omega,\Lambda)$ is solution of (1.5), then $\vec{u} = \mathcal{A}Du \in \vec{H}(D^*)$ is solution of (1.17). Conversely, if $\vec{u} \in \vec{H}(D^*)$ is a solution of (1.17), then $u = \lambda^{-1}(f - D^*\vec{u}) \in H^m(\Omega,\Lambda)$ is a solution of (1.5).

The first statement is obvious. Let $\vec{u}$ be a solution of (1.17) and $u = \lambda^{-1}(f - D^*\vec{u})$. Since $\vec{u} \in \vec{H}(D^*)$ and since $f \in L^2(\Omega)$, we obtain that $u \in L^2(\Omega)$. Next, we deduce from (1.17)i) that $Du = \mathcal{A}^{-1}\vec{u}$ to $L^2(\Omega)^M$. Therefore, $u \in H^m(\Omega)$ and we can obtain the boundary conditions (1.5)ii). Applying $D^*\mathcal{A}$ to (1.17)i), we deduce that $\Lambda u + \lambda u = f$ and thus $u \in H^m(\Omega,\Lambda)$. Finally, since $\vec{u} = \mathcal{A}Du$, the boundary conditions (1.17)iii) imply the boundary conditions (1.5)iii).

1.2. Variational formulation of the problem and its conjugate.

We know† that the boundary-value problem (1.5) is equivalent to the following variational equation: Find

† See [16], Chapter II.

$u \in H^m(\Omega)$ such that

$$(1.20)\quad \begin{cases} \text{i)} & a(u,v) = (f,v) + \sum_{p\leq j\leq m-1} \langle h_{2m-1-j}, \gamma_j v\rangle \\ & \text{for any } v \in H^m(\Omega) \text{ such that } \gamma_j v = 0 \\ & \text{for } 0 \leq j \leq p-1 , \\ \text{ii)} & \gamma_j u = g_j \quad \text{for } 0 \leq j \leq p-1 . \end{cases}$$

We next show that the conjugate problem (1.17) is also equivalent to a variational equation.

We introduce the bilinear form $\alpha(\vec{u},\vec{v})$ defined by

$$(1.21)\qquad \alpha(\vec{u},\vec{v}) = [\mathcal{A}^{-1}\vec{u},\vec{v}] \quad \text{for any } \vec{u},\vec{v} \in \vec{H} .$$

Let $\vec{u} \in \vec{H}(D^*)$ be a solution of

$$(1.22)\quad \begin{cases} \text{i)} & (D^*\vec{u}, D^*\vec{v}) + \lambda\alpha(\vec{u},\vec{v}) \\ & = (f, D^*\vec{v}) + \lambda \sum_{p\leq j\leq m-1} \langle g_j, \beta_{2m-1-j}\vec{v}\rangle \\ & \text{for any } \vec{v} \in \vec{H}(D^*) \text{ such that } \beta_j\vec{v} = 0 \\ & \text{for } m \leq j \leq 2m-1-p , \\ \text{ii)} & \beta_j\vec{u} = h_j \quad \text{for } m \leq j \leq 2m-1-p . \end{cases}$$

It is easy to deduce from the Green's formula (1.12) (Lemma 1.1) the following proposition:

<u>Proposition 1.2.</u> The conjugate problem (1.17) and the

variational problem (1.22) are equivalent.

We point out the following corollary:

Corollary 1.1. The variational equations (1.20) and (1.22) are equivalent. The boundary conditions $\gamma_j u = g_j$ $(0 \leq j \leq p-1)$ are forced in the variational formulation (1.20) and are natural in the variational formulation (1.22). Conversely, the boundary conditions $\beta_j \vec{u} = \delta_j u = h_j$, $m \leq j \leq 2m-1-p$ are natural in (1.20) and are forced in (1.22).

This is an important fact for approximation purposes. In particular, the boundary conditions of the Dirichlet problem are "natural" boundary conditions in the variational formulation of its conjugate problem.†

1.3. A posteriori estimates of error.

The Green's formula (1.12) for the operator D expresses orthogonality relations we can use in order to obtain a posteriori estimates of error of approximate solutions for the initial and the conjugate problems.

Let $u^h \in H^m(\Omega)$, satisfying the boundary conditions $\gamma_j u^h = g_j$ $(0 \leq j \leq p-1)$, be an approximant of the solution u of (1.5). Let $\vec{u}^h \in \vec{H}(D^*)$ satisfying the boundary conditions $\beta_j \vec{u}^h = h_j$ $(m \leq j \leq 2m-1-p)$ be an approximant of the

† There are other variational formulations for which the Dirichlet conditions are "natural". For instance, let us quote the following one, pointed out to us by Professor J. L. Lions.

Find $u \in H^m(\Omega, \Lambda)$ satisfying

$$(\Lambda u, \Lambda v) + \lambda a_*(u,v) = (f, \Lambda v) + \sum_{j=0}^{m-1} \langle g_j, \delta_{2m-1-j} v \rangle$$

for any $v \in H^m(\Omega, \Lambda)$. Here $a_*(u,v) = \overline{a(v,u)}$.

conjugate boundary value problem (1.17).

We will estimate the error $u - u^h$ in terms of the approximants u^h and $\vec{u}^h$ of the solutions u and $\vec{u}$ of (1.5) and (1.17).

For that purpose, let us assume that $\mathcal{A}$ is $\vec{H}$-elliptic: There exists a positive constant c such that

$$(1.23) \qquad [\mathcal{A}\vec{v},\vec{v}] \geq c[\vec{v}]^2 \quad \text{for any } \vec{v} \in \vec{H} .$$

Then $\mathcal{A}$ is an isomorphism (by the Lax-Milgram theorem) and the form $a(u,v)$ is $H^m(\Omega)$-elliptic since

$$(1.24) \qquad a(v,v) = [\mathcal{A}Dv,Dv] \geq c[Dv]^2 = c\,\|v\|^2_{H^m(\Omega)} .$$

In this case, there exist unique solutions u and $\vec{u}$ of (1.5) and (1.17).

<u>Proposition 1.3</u>. Assume (1.23) and $\lambda > 0$. Let $u \in H^m(\Omega,\Lambda)$ by the solution of (1.5), $\vec{u} \in \vec{H}(D^*)$ be the solution of (1.17), $u^h \in H^m(\Omega)$ and $\vec{u}^h \in \vec{H}(D^*)$ satisfying

$$(1.25) \qquad \gamma_j u^h = g_j \quad \text{for } 0 \leq j \leq p-1 ,$$

$$\beta_j \vec{u}^h = h_j \quad \text{for } m \leq j \leq 2m-1-p .$$

Then the following a posteriori estimates hold:

$$(1.26) \qquad c[D(u - u^h)]^2_{\vec{H}} + \lambda\|\vec{u} - \vec{u}^h\|^2_H$$
$$\leq c^{-1}[\vec{u}^h - \mathcal{A}Du^h]^2_{\vec{H}} + \lambda^{-1}\|f - \lambda u^h - D^*\vec{u}^h\|^2_H$$

and, with a bound C for the operator $\mathcal{A}$,

$$(1.27) \qquad c[\mathcal{A}^{-1}(\vec{u} - \vec{u}^h)]^2_{\vec{H}} + \lambda^{-1}\|D^*(\vec{u} - \vec{u}^h)\|^2_H$$
$$\leqslant c^{-1}C^2[\mathcal{A}^{-1}\vec{u}^h - Du^h]^2_{\vec{H}} + \lambda^{-1}\|f - \lambda u^h - D^*\vec{u}^h\|^2_H .$$

We shall prove this proposition in an abstract framework (see Theorem 2.2, Section 2.2 and Corollary 2.1, Section 2.5).

We observe: If u^h converges to u in $H^m(\Omega)$ and if $\vec{u}^h$ converges to $\vec{u}$ in $\vec{H}(D^*)$, then the right-hand side of (1.26) (respectively (1.27)) converges to 0 .

Therefore, for the effective use of these a posteriori estimates, we shall have to construct approximations of the conjugate problem (1.17), as well as for the initial problem (1.5). (See below Section 2.2.) See also [21].

<u>Remark 1.1</u>. The assumption $\lambda \neq 0$ is motivated by the fact that the variational formulation of the conjugate problem is analogous to the variational formulation of the initial problem. Furthermore, this variational formulation is convenient in order to construct approximate problems (see Section 4 below).

Nevertheless, this assumption is not as restrictive as it may seem: If we assume that one can write $\Lambda = D^*\mathcal{A}D$, where $\mathcal{A}$ is $\vec{H}$-elliptic, then it is possible to choose $\lambda > 0$, sufficiently small in such a way that $\Lambda = D^*\mathcal{A}_0 D + \lambda$, where $\mathcal{A}_0$ is still $\vec{H}$-elliptic.

Remark 1.2. There are several ways of splitting a differential operator Λ into a product $D^*\mathcal{A}D$.

Consider for instance $\Lambda = \Delta^2 + 1$. Then we can choose the above decomposition with $m = 2$. In this case, the matrix $\mathcal{A}$ is not $\vec{H}$-elliptic.

Alternatively, we can choose the trivial splitting, where

$$V = H(\Omega,\Delta) \ , \quad \vec{H} = (L^2(\Omega))^2 \ , \quad D = (1,\Delta) \ .$$

There exist other possibilities. For instance, we can take

$$\vec{H} = (L^2(\Omega))^{n+1} \ , \quad D = (1, D_1^2, \ldots, D_i^2, \ldots, D_n^2)$$

and V will be the completion of $H^2(\Omega)$ for the norm $(\|u\|^2_{L^2(\Omega)} + \sum_{1\leq i\leq n} \|D_i^2 u\|^2_{L^2(\Omega)})^{\frac{1}{2}}$.

2. A posteriori Estimates and Conjugate Problems of Abstract Boundary Value Problems

The example we have studied in the first section is a particular case of a more general problem.

We start with an abstract Green's formula relating an operator D and its formal adjoint:

$$[\vec{u}, Dv] = (D^*\vec{u}, v) + \langle \beta_1\vec{u}, \gamma_1 v\rangle + \langle \beta_2\vec{u}, \gamma_2 v\rangle$$

(Section 2.1) and we consider the class of boundary value problems

$$\text{i)}\quad D^*\mathcal{A}Du + \lambda u = f$$

$$\text{ii)}\quad \gamma_1 u = t_1$$

$$\text{iii)}\quad \beta_2 \mathcal{A}Du = t_2 \ .$$

In Section 2.2, we deduce a posteriori estimates from the Green's formula. These a posteriori estimates motivate the introduction of the conjugate problem (Section 2.3) and the study of the approximation of their solutions (Section 2.4).

We go back in Section 2.5 to the study of a posteriori estimates by using the Fredholm alternative.

2.1. The operator D and its formal adjoint D^*.

We consider a continuous linear operator D mapping a Hilbert space V into a Hilbert space $\vec{H}$.

Furthermore, let us introduce Hilbert spaces H and T and a continuous linear operator γ from V into T.

We shall assume that

(2.1)
- i) D is an isometry from V into $\vec{H}$,
- ii) γ maps V onto T,
- iii) V and $V_0 = \ker\gamma$ are dense subspaces of H,
- iv) H and $\vec{H}$ are identified with their duals.

We use the following notations

(2.2) i) (f,v) denotes the inner product on $H \times H$,

ii) $[\vec{f},\vec{v}]$ denotes the inner product on $\vec{H} \times \vec{H}$,

iii) $<t,s>$ denotes the duality pairing on $T' \times T$.

The formal adjoint $D^* \in L(\vec{H},V_0')$ of D is defined by

$$(2.3) \qquad (D^*\vec{u},v) = [\vec{u},Dv] \text{ for any } \vec{u} \in \vec{H}, \quad v \in V_0,$$

and its domain $\vec{H}(D^*)$ is defined by

$$(2.4) \qquad \vec{H}(D^*) = \{\vec{u} \in \vec{H} \text{ such that } D^*\vec{u} \in H\}$$

supplied with the graph norm $\|\vec{u}\|_{\vec{H}(D^*)} = (\|\vec{u}\|^2_{\vec{H}} + \|D^*u\|^2_H)^{\frac{1}{2}}$ †.

We can now prove the following Green's formula.

<u>Theorem 2.1</u>. We assume (2.1). Then there exists a unique operator $\beta \in L(\vec{H}(D^*), T')$ such that the Green's formula

$$(2.5) \qquad [\vec{u},Dv] = (D^*\vec{u},v) + <\beta\vec{u}, \gamma v>$$

holds, where $\vec{u}$ ranges over $\vec{H}(D^*)$ and v ranges over V.

See [3], Theorem 1.1, and ††.

---

† This definition is meaningful: Since V_0 is dense in H, we can identify $H = H'$ by transposition, with a dense subspace of V_0'.

††We sketch the proof. Assume $\vec{u} \in \vec{H}(D^*)$. Then $D'\vec{u} - D^*\vec{u}$ belongs to the orthogonal $V_0^{\perp}$ of V_0 in V'. Since $V_0 = \ker \gamma$, we see that γ' is an isomorphism from T' onto $V_0^{\perp}$. Thus there exists a unique element $\beta\vec{u}$ in T' such that $D'\vec{u} - D^*\vec{u} = \gamma'\beta\vec{u}$ (i.e. such that (2.5) holds). It is clear that β is linear and continuous.

Now we shall extend D to H and express that $V = H(D)$ is the "maximal domain" of D.

Let $\vec{Y}$ be a reflexive topological vector space such that

$$(2.6) \begin{cases} \text{i)} & \vec{Y} \subset \vec{H}(D^*) \text{ and is dense in } \vec{H}\text{, with continuous injection,} \\ \text{ii)} & \beta u = 0 \text{ for any } \vec{u} \in \vec{Y}. \end{cases}$$

We can extend D to a continuous linear operator from H into $\vec{Y}'$. Indeed, let $D^{**} \in L(H,Y')$ be defined by

$$[D^{**}u, \vec{v}] = (u, D^*\vec{v}) \quad \text{for any} \quad u \in H, \quad \vec{v} \in \vec{Y}.$$

When u ranges over $V \subset H$, formula (2.5) implies that $D^{**}u = Du$ by (2.6)ii). Since V is dense in H and $\vec{H}$ is dense in $\vec{Y}'$ (by (2.6)i)), the map D^{**} is the unique extension of $D \in L(V,\vec{H})$ to $L(H,\vec{Y}')$ and hence, we set $D = D^{**}$.

Finally, we will assume that

$$(2.7) \qquad V = H(D) = \{u \in H \text{ such that } Du \in \vec{H}\}.$$

Actually, we split the Green's formula (2.5) in the following way. Let σ_i be a continuous projector of T, $\sigma_2 = 1 - \sigma_1$ and

$$(2.8) \quad \gamma_i = \sigma_i\gamma, \quad \beta_i = \sigma_i'\beta, \quad T_i = \sigma_i T, \quad T_i' = \sigma_i' T' \qquad (i = 1, 2).$$

Proposition 2.1. We assume (2.1). Then

$$(2.9)\quad \begin{cases} \text{i) } \gamma_2 \text{ maps } V_1 = \{u \in V \text{ such that } \gamma_1 u = 0\} \\ \quad \text{onto } T_2 \\ \text{ii) } \beta_1 \text{ maps } \vec{H}_2(D^*) = \{\vec{u} \in \vec{H}(D^*) \text{ such that} \\ \quad \gamma_2 u = 0\} \text{ onto } T_1' . \end{cases}$$

The first statement follows from (2.1)ii). In order to prove the second statement, consider the variational equation

$$(2.10)\qquad [Du, Dv] = \langle t_1, \gamma_1 v\rangle \text{ for any } v \in V$$

where t_1 is given in T_1'. By (2.1)i), the continuous bilinear form $[Du, Dv]$ is V-elliptic. Hence there exists a unique solution u of (2.10).[†] We deduce from (2.3) that $D^*Du = 0$, hence $\vec{u} = Du \in \vec{H}(D^*)$. The Green's formula shows that $\vec{u} \in H_2(D^*)$ and $\beta_1\vec{u} = t_1$.

2.2. Boundary value problems for D^*AD and a posteriori estimates.

Let $A \in L(\vec{H}, \vec{H})$ be given. We associate with A the space

$$(2.11)\qquad V(\Lambda) = \{u \in V \text{ such that } ADu \in \vec{H}(D^*)\}$$

† By the Lax-Milgram theorem.

and let

$$(2.12)\quad \begin{cases} \text{i)} & \Lambda = D^{*}AD \in L(V(\Lambda),H) , \\ \text{ii)} & \delta_i = \beta_i AD \in L(V(\Lambda),T_i') , \\ & i = 1,2 \ ; \quad \delta = \beta AD \in L(V(\Lambda),T') . \end{cases}$$

Let $f \in H$, $t_1 \in T_1$ and $t_2 \in T_2'$ be given. We consider a solution u of the boundary-value problem

$$(2.13)\quad \begin{cases} \text{i)} & u \in V(\Lambda) , \\ \text{ii)} & \Lambda u + \lambda u = f , \\ \text{iii)} & \gamma_1 u = t_1 , \\ \text{iv)} & \delta_2 u = t_2 . \end{cases}$$

Let us begin by proving the following result (we shall prove other results in section 2.5).

<u>Theorem 2.2</u>. Let us assume (2.1). Furthermore, suppose that $\lambda > 0$ and that $a(u,v) = [ADu,Dv]$ is V_1-elliptic: there exists $c > 0$ such that

$$(2.14)\quad a(v,v) = [ADv,Dv] \geq c \,\|v\|_V^2 = c\,\|Dv\|_{\vec{H}}^2$$

for any $v \in V$ such that $\gamma_1 v = 0$.

Let u be the (unique) solution of (2.13), $u^h \in V$ and $\vec{u}^h \in \vec{H}(D^*)$ be given elements satisfying the boundary condition

$$(2.15) \quad \begin{cases} \text{i)} \quad \gamma_1 u^h = t_1 , \\ \text{ii)} \quad \beta_2 \vec{u}^h = t_2 . \end{cases}$$

Then the following a posteriori estimate holds:

$$(2.16) \quad c \, \|u-u^h\|_V^2 + \lambda \, \|u-u^h\|_H^2 \leqslant c^{-1} \, \|\vec{u}^h - \mathcal{A}Du^h\|_H^2 + \lambda^{-1} \, \|f-D^*\vec{u}^h - \lambda u^h\|_H^2 \; .$$

Indeed, when v ranges over V_1, the Green's formula (2.5) implies that

(2.17)

$$a(u-u^h,v) + \lambda(u-u^h,v) = (D^*\mathcal{A}Du+\lambda u,v) + \langle t_2,\gamma_2 v\rangle - [\mathcal{A}Du^h,Dv] - \lambda(u^h,v) = (f-\lambda u^h,v) - [\mathcal{A}Du^h,Dv] + \langle t_2, \gamma_2 v\rangle \; .$$

On the other hand, if $\vec{u}^h \in \vec{H}(D^*)$ satisfies $\beta_2 \vec{u}^h = t_2$, we deduce from (2.5) that $[\vec{u}^h, Dv] - (D^*\vec{u}^h, v) - \langle t_2, \gamma_2 v\rangle = 0$ for any $v \in V_1$. Thus the following variational equation is obtained for u:

$$(2.18) \quad \text{i)} \quad a(u-u^h,v) + g(u-u^h,v) = (f-\lambda u^h - D^*\vec{u}^h,v) + [\vec{u}^h - \mathcal{A}Du^h,Dv] \quad \text{for any} \quad v \in V_1$$

ii) $u-u^h \in V_1$.

Here $\gamma_1 u^h = \gamma_1 u = t_1$, so we obtain (ii). Thus, replacing v by $u-u^h$ in (2.18), we obtain (2.16) from (2.14).

We remark here that (2.18) is an equivalent variational formulation of the boundary value problem (2.13), as follows from Theorem 2.1 and Proposition 2.1. This implies the existence and uniqueness of the solution u of (2.13), cf. [16], Chapter II.

In (2.16) u^h is an approximant for the solution u of (2.13). For $u^h = u$ the right-hand side of (2.16) converges to 0 if $\vec{u}^h$ converges to an element $\vec{u}$ of $\vec{H}(D^*)$ satisfying

$$(2.19) \qquad \vec{u} = \mathcal{A}Du, \quad f-\lambda u = D^*\vec{u}, \quad \beta_2\vec{u} = t_2, \quad \gamma_1 u = t_1 .$$

If we eliminate u in the relations (2.19), using (2.16), (2.17), we find that

$$(2.20) \qquad \vec{u} = \mathcal{A}\vec{v} ,$$

where $\vec{v}$ is the solution of the problem:

$$(2.21) \qquad \begin{aligned} &\text{i)} \quad \vec{v} \in \vec{H}, \quad \mathcal{A}\vec{v} \in \vec{H}(D^*) , \\ &\text{ii)} \quad DD^*\mathcal{A}\vec{v} + \lambda\vec{v} = Df , \\ &\text{iii)} \quad \gamma_1(f-D^*\mathcal{A}\vec{v}) = \lambda t_1 , \\ &\text{iv)} \quad \beta_2\mathcal{A}\vec{v} = t_2 . \end{aligned}$$

It is clear that if u is solution of (2.13), then $\vec{v} = Du$ is solution of (2.21). Conversely, it is shown below that if $\vec{v}$ satisfies (2.21), then $u = \lambda^{-1}(f - D^{*}\mathcal{A}\vec{v})$ satisfies (2.13).

Hence, in order to make use of the a posteriori estimate (2.16), our next task will be approximate $\vec{u} = \mathcal{A}\vec{v}$, where $\vec{v}$ is solution of (2.21). For that purpose, we will rewrite equations (2.21) in the case where $a(u,v)$ is V-elliptic.

2.3. Conjugate problems of boundary value problems for $D^{*}\mathcal{A}D$.

Let us assume that $a(u,v)$ is V-elliptic: there exists a positive constant c such that

(2.22)

$$a(v,v) = [\mathcal{A}Dv, Dv] \geq c\,\|Dv\|^2_{\vec{H}} = c\,\|v\|^2_V \quad \text{for any} \quad v \in V\,.$$

We set

(2.23) $\quad \mathcal{M} = D(V)\,, \quad$ a closed subspace of $\vec{H}$.

Then (2.22) implies that there exists at least one operator $\mathcal{B} \in L(\vec{H},\vec{H})$ such that

(2.24) $\quad \mathcal{B}\mathcal{A}\vec{v} = \vec{v}$ for any $\vec{v} \in \mathcal{M} = D(V)$.

We shall say that the following problem is the "conjugate problem" of the boundary value problem (2.13):

(2.25)

$$\text{i)}\quad \vec{u} \in \vec{H}(D^*) \cap \mathcal{AM} = \mathcal{D}(\mathcal{B}) \ ,$$

$$\text{ii)}\quad DD^*\vec{u} + \lambda\mathcal{B}u = Df \ ,$$

$$\text{iii)}\quad \gamma_1(f-D^*\vec{u}) = \lambda t_1 \ ,$$

$$\text{iv)}\quad \beta_2\vec{u} = t_2 \ .$$

<u>Proposition 2.2</u>. We assume (2.1), (2.6), (2.7) and (2.22). Then the problems (2.13) and (2.25) are equivalent: if u is solution of (2.13), then $\vec{u} = \mathcal{A}Du$ is solution of (2.25) and conversely, if $\vec{u}$ is solution of (2.25), then $u = \lambda^{-1}(f-D^*\vec{u})$ is solution of (2.13).

The first statement is obvious. To prove the second statement, let us set $u = \lambda^{-1}(f-D^*\vec{u})$. Then $u \in H$ since $f \in H$ and $\vec{u} \in \vec{H}(D^*)$. We now can write (2.25)ii) in the form $Du = \mathcal{B}\vec{u}$. Therefore $u \in H(D) = V$ by (2.7), and also $u \in V(\Lambda)$: since $\vec{u} \in \mathcal{AM}$ we obtain $\vec{u} = \mathcal{A}Du$, using (2.1)i) and (2.24), hence $D^*\mathcal{A}Du = D^*\vec{u} = f - \lambda u$. On the other hand, (2.25)iii) and iv) imply (2.13)iii) and iv).

We now describe the variational formulation of the conjugate problem: find $\vec{u}$ satisfying

(2.26)

$$\text{i)}\quad u \in \mathcal{D}(\mathcal{B}) = H(D^*) \cap \mathcal{AM} \ ,$$

$$\text{ii)}\quad (D^*\vec{u},D^*\vec{v}) + \lambda[\mathcal{B}\vec{u},\vec{v}] = (f,D^*\vec{v}) + \lambda\langle t_1, \beta_1\vec{v}\rangle \quad \text{for any}$$

$$\vec{v} \in \vec{H}_2(D^*) = \{\vec{v} \in H(D^*) \ \text{ such that } \ \beta_2\vec{v} = 0\}$$

iii) $\beta_2\vec{u} = t_2$.

It is clear that, under the assumptions of Proposition 2.2, problems (2.25) and (2.26) are equivalent.

It will be useful to give another variational formulation.

By Proposition 2.1, there exists at least one element $\vec{w} \in H(D^*)$ such that

(2.27) $$\beta_2\vec{w} = t_2 \ .$$

Then $\vec{u}_2 = \vec{u} - \vec{w}$ is solution of the variational equation

(2.28)

i) $\vec{u}_2 \in (\mathcal{D}(\mathcal{B})-\vec{w}) \quad \vec{H}_2(D^*)$ where $\beta_2\vec{w} = t_2$,

ii) $(D^*\vec{u}_2, D^*\vec{v}) + \lambda[\mathcal{B}\vec{u}_2,\vec{v}] = \ell_2(\vec{v})$ for any $\vec{v} \in \vec{H}_2(D^*)$

where $\ell_2(v)$ is the continuous linear form defined by

$$\ell_2(v) = (f,D^*\vec{v}) + \lambda<t_1, \beta_1 v> - (D^*\vec{w},D^*\vec{v}) - \lambda[\mathcal{B}\vec{w},\vec{v}] \ .$$

Notice that

(2.29)

i) we can choose $\vec{w} \in \mathcal{D}(\mathcal{B})$ satisfying $\beta_2\vec{w} = t_2$ (take $\vec{w} = \vec{u}$) ,

ii) the bilinear form $(D^*\vec{u},D^*\vec{v}) + \lambda[\mathcal{B}\vec{u},\vec{v}]$ is only $\mathcal{D}(\mathcal{B})$ - elliptic when $a(u,v)$ is V-elliptic.

2.4. Remarks on the approximation of the conjugate problem.

In the above section, we have shown that if $a(u,v)$ is V-elliptic, the solution $\vec{u}_2 = \vec{u} - \vec{w}$ of (2.28) is the solution of a variational equation.

Thus by using the results of, for instance, [1], [5] or [9], we know how to construct approximate problems of (2.28) whenever we know convergent approximations† (or convergent external approximations)†† of the space $\mathcal{D}(\mathcal{B}) \cap H_2(D^*)$.

Such approximations will depend on the operator $\mathcal{A}$, and we do not know how to construct convergent approximations for all spaces $\mathcal{D}(\mathcal{B}) \cap \vec{H}_2(D^*)$ with $\mathcal{A}$ ranging over $L(\vec{H},\vec{H})$.

Instead, we consider only the situation where it is enough to construct convergent approximations (or convergent

---

†If X is a Hilbert space, we call approximation (X_h, p_h, r_h) of X the data of

i) a "discrete" space X_h (whose dimension is usually finite)

ii) an isomorphism p_h from X_h into X (called prolongation)

iii) a map r_h from X into X_h (called restriction).

A family of approximations is said convergent iff

$$\lim_{h \to 0} \|u - p_h r_h u\|_X = 0 \quad \text{for any} \quad u \in X$$

cf. [1], [9].

††If a Hilbert space X is a closed subspace of a Hilbert space $\bar{X}$, we say that approximation (X_h, p_h, u_h) of $\bar{X}$ are convergent external approximations [5] of X iff

i) $\lim_{h \to 0} \|u - p_h r_h u\|_X = 0$ for any $u \in X$,

ii) if $p_h u_h$ converges weakly to u in $\bar{X}$, then u actually belongs to X.

partial approximations) of the spaces $\vec{H}(D^*)$ and, possibly, of its closed subspaces $\vec{H}_2(D^*)$ defined by homogeneous boundary-conditions.

Therefore, we shall assume that

(2.30)

i) $\mathcal{A} \in L(\vec{H},\vec{H})$ is $\vec{H}$-elliptic,

ii) $\mathcal{A}$, $\mathcal{B} = \mathcal{A}^{-1}$ and $\alpha(\vec{u},\vec{v}) = [\mathcal{B}\vec{u},\vec{v}]$ are given explicitly.

Let us set

(2.31)

i) $b(\vec{u},\vec{v}) = (D^*\vec{u},D^*\vec{v}) + \lambda\alpha(\vec{u},\vec{v})$

ii) $\ell_2(v) = (f,D^*\vec{v}) + \lambda\langle t_1, \beta_1 v\rangle - (D^*\vec{w},D^*\vec{v})$

$- \lambda\alpha(\vec{w},\vec{v})$ with $\beta_2\vec{w} = t_2$.

Then $b(\vec{u},\vec{v})$ is an $\vec{H}(D^*)$-elliptic continuous bilinear form and $\ell_2(\vec{v})$ is a continuous linear form on $\vec{H}(D^*)$.

Let $\vec{u}_2 = \vec{u} - \vec{w}$ be the solution of

(2.32)

i) $\vec{u}_2 \in \vec{H}_2(D^*)$,

ii) $b(\vec{u}_2,\vec{v}) = \ell_2(\vec{v})$ for any $\vec{v} \in \vec{H}_2(D^*)$

(i.e., such that $\beta_2\vec{v} = 0$) .

Proposition 2.3. Let us assume (2.1), (2.6), (2.7) and (2.30). Then the problems (2.13) and (2.32) are equivalent in the following sense. If u is the (unique) solution of (2.13), then $\vec{u}_2 = \mathcal{A}DU - \vec{w}$ is a solution of (2.32). Conversely, if $\vec{u}_2$ is the (unique) solution of (2.32), then $u = \lambda^{-1}(f - D^*\vec{w} - D^*\vec{u}_2)$ is a solution of (2.13).

The proof of this proposition is left to the reader. Then, if we know convergent approximations $(\vec{V}_h, \vec{p}_h, \vec{r}_h)$ of $\vec{H}_2(D^*)$, the solution $\vec{u}$ of the conjugate problem will be approximated by $\vec{u}^h = \vec{w} + \vec{p}_h\vec{u}_h$ where $\vec{u}_h \in \vec{V}_h$ is the solution of

$$(2.33)\quad b(\vec{p}_h\vec{u}_h, \vec{p}_h\vec{v}_h) = \ell_2(\vec{p}_h\vec{v}_h) \quad \text{for any} \quad \vec{v}_h \in \vec{V}_h .$$

(see [1], [2] for instance).

2.5. A posteriori estimates and Fredholm alternative.

In this section we drop the assumption of V_1-ellipticity of $\mathcal{A}$, which we used in the derivation of the a posteriori estimate (2.16). Instead we assume that we can write (changing the meaning of $\mathcal{B}$)

$$(2.34)\qquad \mathcal{A} = \mathcal{A}_\alpha = \mathcal{A}_0(\alpha - \mathcal{B}), \quad 0 \neq \alpha = \lambda + b .$$

Here, $\mathcal{A}_0$ is an hermitian isomorphism of $\vec{H}$ and b a constant.[†] Regarding $\mathcal{B}$ we assume the following conditions:

[†] The assumption that $\mathcal{A}_0$ is $\vec{H}$-elliptic can be replaced by the weaker assumption of DV_1-ellipticity of $\mathcal{A}_0$. If this condition is satisfied, let

(2.35)

i) $\mathcal{B}$ is linear and bounded from $\vec{H}$ to $\vec{H}$ and compact on DV,

ii) The injection $j: V \to H$ is compact.

As we shall see, these assumptions are natural. We describe an Application. Let Ω be a bounded smooth open subset of R^n, $H = L^2(\Omega)$, $V = H^m(\Omega)$, $m \geq 1$, and $\vec{H} = H^M$, cf. (1.6). D is as defined in (1.8). For $\vec{u} = (\vec{u}^q)_{|q| \leq m}$ we let

(2.36)

$$\mathcal{A}_0\vec{u} = \Big(\sum_{|q|\leq m} a^0_{p,q}\vec{u}^q\Big)_{|P|\leq m}, \quad \mathcal{B}\vec{u} = \Big(\sum_{|q|<m} b_{p,q}\vec{u}^q\Big)_{|p|\leq m}.$$

Here the $a^0_{p,q} = a^0_{q,p}$ as well as the $b_{p,q}(|p|, |q| \leq m)$ are functions in $L^\infty(\Omega)$. We assume that for some $c > 0$ and all $(t^q)_{|q|\leq m} \in R^M$ we have

$$(2.37) \qquad \sum_{|p|,|q|\leq m} a^0_{p,q}(x)t^p t^q \geq c \sum_{|q|\leq m} |t^q|^2$$

$$\tilde{\mathcal{A}} = P_1\mathcal{A}_0P_1 + 1 - P_1, \quad \tilde{\mathcal{B}} = \tilde{\mathcal{A}}_0^{-1}\mathcal{A}_0\mathcal{B},$$

where P_1 is the orthogonal projection of $\vec{H}$ onto DV_1. $\tilde{\mathcal{A}}_0$ is $\vec{H}$-elliptic, and for all $v, w \in V_1$

$$[\tilde{\mathcal{A}}_0(\alpha-\tilde{\mathcal{B}})Dv, Dw] = [\mathcal{A}_0(\alpha-\mathcal{B})Dv, Dw].$$

Hence, replacing $\mathcal{A} = \mathcal{A}_0(\alpha-\mathcal{B})$ by $\tilde{\mathcal{A}} = \mathcal{A}_0(\alpha-\tilde{\mathcal{B}})$ we pass to an equivalent problem. However, the computation of $P_1\vec{w}$ involves solving a boundary value problem, and we do not know whether the preceding construction of $\tilde{\mathcal{A}}_0$ can be replaced by a "practical" one (obviously there are many other modifications of $\mathcal{A}_0$ doing the same service).

so that $\mathcal{A}_0$ is H-elliptic.[†] Consistent with (2.36) we require

$$b_{pq} = 0 \quad \text{for} \quad |q| = m \; . \tag{2.38}$$

Proposition 2.4. With the preceding definitions the assumptions (2.35) are satisfied.

Indeed, the injection j from $H^m(\Omega)$ to $L^2(\Omega)$ is compact for $m \geq 1$, by Rellich's theorem and so is the injection J from $H^m(\Omega)$ to $H^{m-1}(\Omega)$. For $\vec{u} = Du \in DV$ let

$$\pi\vec{u} = u \; .$$

Finally, for $u \in H^{m-1}(\Omega)$ let

$$D_0 u = \vec{u} \in \vec{H} \; , \quad \text{with} \quad \vec{u}^q = \begin{cases} D^q u & \text{for} \quad |q| < m \\ 0 & \text{for} \quad |q| = m \end{cases}$$

Then clearly

$$\mathcal{B}\Big|_{DV} = \mathcal{B}D_0 J\pi \; .$$

Since all operators are bounded and J compact we conclude (2.35).

We introduce the Hilbert space

[†]See preceding footnote.

$$\hat{H} = H \times \vec{H} \quad \text{with elements} \quad \{u,\phi\} = \hat{\phi}, \quad u \in H, \quad \phi \in \vec{H}$$

with the equivalent inner products

$$(2.39) \begin{cases} \text{i)} \quad h(\{u,\phi\}, \{v,\psi\}) = (u,v) + [\phi,\psi] , \\ \text{ii)} \quad h_0(\{u,\phi\}, \{v,\psi\}) = (u,v) + [\mathcal{A}_0\phi,\psi] . \end{cases}$$

When referring to norm and inner product in $\hat{H}$ we shall always mean the ones induced by h.

We need several operations on $\hat{H}$:

$$\hat{\mathcal{A}}\{u,\phi\} = \{\lambda u,\mathcal{A}\phi\}, \; \hat{\mathcal{B}}\{u,\phi\} = \{bu,\mathcal{B}\phi\}, \; \hat{A}_0\{u,\phi\} = \{u,\mathcal{A}_0\phi\} .$$

Now equation (2.34) leads to

$$(2.40) \qquad \hat{\mathcal{A}} = \hat{A}_0(\alpha-\hat{\mathcal{B}}) .$$

Furthermore, let

$$\hat{\mathcal{M}}_1 = \{\{u,Du\} : \; u \in V_1\}$$

and let P (and P_0) be the projection in $\hat{H}$ which is orthogonal with respect to h (h_0) and has range $\hat{\mathcal{M}}_1$. Next we formulate the abstract boundary value problem (2.13) in a form convenient for the application of the Fredholm alternative and for obtaining a posteriori bounds. We start from the variational equation (2.18) which we rewrite slightly. We assume given approximate solutions u^h of (2.13) and $\vec{u}^h$ of (2.21) as in (2.18) and we let

$$(2.41) \qquad \vec{u}^h = \mathcal{A}\psi^h \ , \quad v^h = \frac{1}{\lambda}(f - D^*\mathcal{A}\psi^h) \ .$$

I.e. we assume that we can find explicitly a solution ψ^h of the equation $\vec{u}^h = \mathcal{A}\psi^h$. We now can write instead of (2.18) the variational problem: to find $u \in V$ such that

$$(2.42) \begin{cases} \text{i)} \ \lambda(u-u^h,v) + [\mathcal{A}D(u-u^h), Dv] = \lambda(v^h-u^h,v) \\ \qquad + [\mathcal{A}(\psi^h-Du^h),Dv] \quad \text{for all} \quad v \in V_1 \ , \\ \text{ii)} \ u - u^h \in V_1 \ . \end{cases}$$

The equivalence of this with (2.18) and hence (2.13) is trivial. With the notation (2.39) we can write (2.42)i) in the form

$$(2.43) \qquad P\hat{\mathcal{A}}\{u-u^h, D(u-u^h)\} = P\hat{\mathcal{A}}\{v^h-u^h, \psi^h-Du^h\} \ .$$

Instead of P we can also use P_0 and then obtain, on account of $P_0\hat{\mathcal{A}}_0^{-1}\hat{\mathcal{A}} = P_0(\alpha-\hat{\mathcal{B}})$

(2.44)

$$(\alpha-P_0\hat{\mathcal{B}})\{u-u^h, D(u-u^h)\} = P_0(\alpha-\hat{\mathcal{B}})\{v^h-u^h, \psi^h-Du^h\} \ .$$

On the left we have made use of the condition

$$(2.45) \qquad \{u-u^h, D(u-u^h)\} \in \mathcal{M}_1 \ ,$$

which is synonymous with (2.42)ii).

Thus, both (2.43) and (2.44) appear as equations of the form $T\hat{u} = \hat{f}$ for an unknown $\hat{u}$ in $\hat{\mathcal{M}}_1$, with given element $\hat{f} \in \hat{\mathcal{M}}_1$, and an operator T which maps $\hat{\mathcal{M}}_1$ into $\hat{\mathcal{M}}_1$. On account of (2.35) one verifies that $\hat{\mathcal{B}}$ is compact on $\hat{\mathcal{M}}_1$. Thus

$$(2.46) \qquad P_0\hat{\mathcal{B}}\Big|_{\hat{\mathcal{M}}_1} = P_0\hat{\mathcal{B}}P_0' \in L(\hat{\mathcal{M}}_1, \hat{\mathcal{M}}_1)$$

is a compact operator on the Hilbert space $\hat{\mathcal{M}}_1$. We obtain the immediate result:

<u>Theorem 2.3</u>. For any given $\alpha \neq 0$, either equation (2.44) has a unique solution $u \in V$ such that $u-u^h \in V_1$ or α is an eigenvalue of the compact operator (2.46), i.e. there exists an element $v \in V$, such that

$$(2.47) \qquad (\alpha - P_0\hat{\mathcal{B}})\{v, Dv\} = 0, \quad v \neq 0, \quad v \in V_1 .$$

In the first case u is the unique solution of the boundary value problem (2.13) where now

(2.48)

$$\Lambda = \Lambda_\alpha = D^*\mathcal{A}_\alpha D, \quad \mathcal{A}_\alpha = \mathcal{A}_0(\alpha - \mathcal{B}), \quad \lambda = \lambda_\alpha = \alpha + b .$$

In the second case v is a solution of the eigenvalue problem: to find $v \neq 0$, $v \in V$ such that

$$(2.49) \qquad \Lambda_\alpha v + \lambda_\alpha v = \alpha(D^*\mathcal{A}_0 DV + v) - (D^*\mathcal{A}_0\mathcal{B}DV + bv) = 0$$

$$\gamma_1 v = 0, \quad \beta_2\mathcal{A}\, Dv = \beta_2\mathcal{A}_0(\alpha - \mathcal{B})v = 0 .$$

Here, the equivalence of (2.47) and (2.49) follows from Green's formula (2.5) and Proposition 2.1.

Corollary 2.1. If $\alpha \neq 0$ is not an eigenvalue of equation (2.49), then a posteriori estimates of the error in the approximate solutions u^h, ψ^h exist in the form

(2.50)

$$\|u-u^h\|_H^2 + \|D(u-u^h)\|_{\vec{H}}^2 \leq K^2(\|u^h-v^h\|_H^2 + \|Du^h-\psi^h\|_{\vec{H}}^2)$$

(2.51)

$$\|u-v^h\|_H^2 + \|Du-\psi^h\|_{\vec{H}}^2 \leq K^2(\|u^h-v^h\|_H^2 + \|Du^h-\psi^h\|_{\vec{H}}^2) .$$

Here v^h is given by (2.41) and K is a bound such that

(2.52) $$\|Q\|_{\hat{H}} \leq K, \quad Q = \mathcal{R}_\alpha P_0(\alpha-\hat{\mathcal{B}}) ,$$

where $\mathcal{R}_\alpha$ is the resolvent of $P_0\hat{\mathcal{B}}\Big|_{\mathcal{M}_1}$

$$\mathcal{R}_\alpha = (\alpha-P_0\hat{\mathcal{B}}\Big|_{\hat{\mathcal{M}}_1})^{-1} .$$

Indeed, by the preceding theorem if $\alpha \neq 0$ is not an eigenvalue of (2.49), then from (2.44) we obtain

(2.53) $$\{u-u^h, D(u-u^h)\} = Q\{v^h-u^h, \psi^h-Du^h\}$$

with Q given by (2.52) and $\mathcal{R}_\alpha$, hence Q, is bounded. This implies (2.50), provided (2.52) holds. Next it follows from (2.53) that

$$\{u-v^h,\ Du-\psi^h\} = (1-Q)\{u^h-v^h,\ Du^h-\psi^h\} .$$

Thus we obtain (2.51) in case

$$\|1-Q\|_{\hat{H}} \leqslant K .$$

By the Lemma below, this follows from (2.52). It was pointed out to us by Professor T. Kato that Lemma 2.1 was proved by him in [15].

Lemma 2.1. Let Q be an idempotent bounded linear operator of a Hilbert space $\mathcal{H}$ into itself. Then

$$\|Q\|_{\mathcal{H}} = \|1-Q\|_{\mathcal{H}} .$$

For the proof we abbreviate $\mathcal{V}$ = ran Q, $\mathcal{W}$ = ker Q = ran(1-Q). Then every element u of the given Hilbert space $\mathcal{H}$ can be decomposed

$$u = v-w,\quad v \in \mathcal{V},\quad w \in \mathcal{W} .$$

With this we obtain that the following statements (i) - (iv) are mutually equivalent.

i) $\|Q\|_{\mathcal{H}} \leqslant K$

ii) $\|v\|_{\mathcal{H}} \leqslant K\,\|v-w\|_{\mathcal{H}}$ (all $v \in \mathcal{V}$, $w \in \mathcal{W}$)

iii) $\|v\|_{\mathcal{H}} \leqslant K \inf\limits_{w \in \mathcal{W}} \|v-w\|_{\mathcal{H}}$

$$= K \inf_{0 \neq w \in \mathcal{W}} \left\{ \|v-w\|^2_{\mathcal{H}} - \frac{(v,w)^2_{\mathcal{H}}}{\|w\|^2_{\mathcal{H}}} \right\}^{\frac{1}{2}} \quad (v \in \mathcal{V})$$

$$\text{iv)} \quad 1 \leqslant K \inf_{\substack{0 \neq v \in \mathcal{V} \\ 0 \neq w \in \mathcal{W}}} \left\{ 1 - \frac{(v,w)^2_{\mathcal{H}}}{\|v\|^2_{\mathcal{H}} \, \|w\|^2_{\mathcal{H}}} \right\}^{\frac{1}{2}} .$$

In (iii) we have made use of an elementary formula for the distance of v from the one-dimensional subspace $\{\alpha w: \alpha \in R^1\}$. Now, (iv) is entirely symmetric in $\mathcal{V}$ and $\mathcal{W}$. Hence (iv), and (i), are equivalent with $\|1-Q\|_{\mathcal{H}} \leqslant K$, as was to be shown.

Remark 2.1. Formula (iv) above suggests the interesting geometric interpretation that

$$\|Q\| = 1/\sin\theta$$

where θ is the angle between the subspaces $\mathcal{V}$ and $\mathcal{W}$. An interpretation of theorem 2.3 is that in the situation described there $\theta = 0$ implies $\mathcal{V} \cap \mathcal{W}$ is not zero.

Remark 2.2. Bounds for $\|Q\|_{A}$ can be obtained in various ways, one case is the one dealt with in Theorem 2.2: it is not hard to show, using (2.43), that

$$Q = (P\hat{A}|_{\hat{\mathcal{M}}_1})^{-1} P\hat{A} \tag{2.54}$$

is another representation of Q as product of bounded operators if α is not an eigenvalue. In the case of Theorem 2.2 this leads to

(2.55) $$\|Q\|_H \leqslant \max\{\frac{1}{\lambda}, \frac{1}{c}\} \max\{\lambda, \|\mathcal{A}\|_{\vec{H}}\}$$

We omit the proof, since in this case the estimates (2.16) can be derived more directly.

Corollary 2.2. If $\mathcal{A} = \mathcal{A}_0(\alpha-\mathcal{B})$ is hermitian then the a posteriori estimates (2.50), (2.51) hold with

$$K = K_1 K_2 K_3 .$$

Here we assume

(2.56) $$K_1 \geqslant 1/\inf\{|\alpha-\sigma| : \sigma \text{ is in the spectrum of (2.49)}\}$$

(2.57) $$K_2 \geqslant \|\hat{\mathcal{A}}_0\|_{\hat{H}} \, \|\mathcal{A}_0^{-1}\|_{\hat{H}}$$

(2.58) $$K_3 \geqslant \|\alpha-\hat{\mathcal{B}}\|_{\hat{H}} .$$

For the proof one first verifies that $P_0\hat{\mathcal{B}}|_{\mathcal{M}_1}$ is hermitian with respect to the inner product h_0 on $\hat{H}$. Then the right hand side in (2.56) is the well known formula for the norm of the resolvent $\mathcal{R}_\alpha$ with respect to the norm induced by h_0. In the same norm,

$$\|P_0\|_{h_0} = 1 .$$

The corollary then follows from (2.52) and the easily established inequality

$$\|T\|_{\hat{H}} \leqslant (\|\hat{\mathcal{A}}_0\|_H \ \|\hat{\mathcal{A}}_0^{-1}\|_{\hat{H}})^{\frac{1}{2}} \|T\|_{h_0} .$$

Here T is any bounded linear operator on $\hat{H}$.

A related estimate has been obtained by Weinberger in the case of Helmholtz' equation [23].

3. Approximations of the Spaces $\vec{H}^m(\Omega,D^*)$ and $\vec{H}_0^m(\Omega,D^*)$

In order to construct approximate problems of conjugate problems, we have to construct approximations of the domain of the operator D^* (see section 2.4). We restrict our study to the case where D is the operator $Du = (D^q u)_{|q|\leqslant m}$ mapping $H^m(\Omega)$ into $(L^2(\Omega))^M$ and we shall construct convergent "piecewise polynomial" approximations (or "spline" approximations) of the domain $\vec{H}^m(\Omega,D^*)$ of D^*, defined by $D^*u = \sum_{|q|\leqslant m} (-1)^{|q|} D^q u^q$, (section 3.1).

We estimate in section 3.2 the behavior of the error functions of these approximations.

Finally, in section 3.3, we construct and study approximations of the subspace $\vec{H}_0^m(\Omega,D^*)$ of vector functions u satisfying the boundary conditions $\beta_j \vec{u} = 0$.

3.1. Construction of approximations of the space $\vec{H}^m(\Omega,D^*)$.

Let $\Omega \subset R^n$ be a smooth bounded subset. Then assumptions (2.1), (2.6) and (2.7) are satisfied when we choose

(3.1)

i) $V = H^m(\Omega)$, $H = L^2(\Omega)$, $T = \prod_{0 \leqslant j \leqslant m-1} H^{m-j-\frac{1}{2}}(\Gamma)$,

$\vec{H} = (L^2(\Omega))^M$ (where M is the number of multi-integers q such that $|q| \leqslant m$) ,

ii) $Du = (D^q u)_{|q| \leqslant m}$, $\gamma = (\gamma_0, \cdots, \gamma_{m-1})$ where $\gamma_j = \partial^j/\partial n^j$,

iii) $Y = (\mathcal{D}(\Omega))^M$

(see first section).

In this section, we set

(3.2) $$\vec{H}^m(\Omega, D^*) = \vec{H}(D^*) = \{\vec{u} = (u^q)_{|q| \leqslant m} \in \vec{H}$$
$$\text{such that } D^*\vec{u} = \sum_{|q| \leqslant m} (-1)^{|q|} D^q u^q \in L^2(\Omega)\} \;.$$

We shall construct a family of approximations of this space $\vec{H}^m(\Omega, D^*)$.

We begin by recalling some notations. If $u_h = (u_h^k)_{k \in Z^n} \in \ell^2(Z^n)$, if $\theta_{kh}(x) = \theta_{k_1 h_1}(x_1) \cdots \theta_{k_n h_n}(x_n)$[†] is the characteristic function of $(kh, (k+1)h) = \prod_{1 \leqslant i \leqslant n} (k_i h_i, (k_i+1)h_i)$, we set

---

[†] $h = (h_1, \cdots, h_n)$ is a positive vector of R^n, $k = (k_1, \cdots, k_n) \in Z^n$ is a multi-integer with positive or negative components.

(3.3) $p_h^o u_h = \sum_{k \in Z^n} u_h^k \theta_{kh}(x), \; p_h^j u_h = \pi_{j,h} * p_h^o u_h$ †

where

(3.4)

i) $\pi_{j,h}(x) = h^{-1}\pi_j(x/h) = (h_1 \cdots h_n)^{-1}\pi_{j_1}(x_1/h_1)\cdots\pi_{j_n}(x_n/h_n)$,

ii) $\pi_j(x) = \begin{cases} \text{the Dirac measure } \delta \text{ if } j = 0 \\ \theta *^j(x) = (\theta * \cdots * \theta) \text{ if } j > 0^{\dagger\dagger} \end{cases}$ when $n = 1$.

Let us denote by ρ the operator of restriction to Ω (defined by $\rho u = u|_\Omega$) and by $(s) = (s, \cdots, s)$ the multi-integer whose components equal s.

3.1.a. <u>The discrete space</u> $\vec{V}_h = \vec{H}_h^{m,s}(\Omega, D^*)$.

If s is an integer, let

(3.5) $s_q = (s) + q = (s+q_1, \cdots, s+q_n)$

and

(3.6)

i) $\mathcal{R}_h^{s_q}(\Omega) = \{k \in Z^n$ such that support $(\pi_{s_q,h} * \theta_{kh}) \cap \Omega \neq \phi\}$

---

† $p_h^o u_h$ is a step function.

†† Here, $*$ denotes the convolution product. If $n = 1$ and if $j > 0$, $\pi_j(x)$ is a function of support $(0,j)$, whose restriction to each interval $(k,k+1)$ is a polynomial of degree $j-1$. It belongs to $H^{j-1}(R)$.

ii) $\overset{s}{H}{}_h^q(\Omega) = \{u_h = (u_h^k)_{k \in Z^n}$ such that $u_h^k = 0$ for $k \notin \overset{s}{\mathcal{R}}{}_h^q(\Omega)\}$.

Then we define the discrete space $\vec{V}_h$ in the following way:

(3.7) $\vec{V}_h = \vec{H}_h^{m,s}(\Omega, D^*) = \prod_{|q| \leq m} \overset{s}{H}{}_h^q(\Omega)$ is the space of

$$\vec{u}_h = (u_h^q)_{|q| \leq m}, \; u_h^q \in \overset{s}{H}{}_h^q(\Omega) ,$$

supplied with the duality pairing

$$[\vec{u}_h, \vec{v}_h]_h = \sum_{|q| \leq m} (u_h^q, v_h^q)_h \; ;$$

$$(u_h, v_h)_h = h_1 \cdots h_n \sum_{k \in Z^n} u_h^k v_h^k .$$

(It will be clear whether u_h^j denotes a component of $\vec{u}_h = (u_h^j)_{|j| \leq m}$ or the component $\vec{u}_h^j$ of $u_h = (u_h^j)_{j \in Z^n}$) .

3.1.b. <u>The prolongations</u> $p_h = p_{h,}^{(s)}$.

We shall set

$$(3.8) \quad \begin{cases} \text{i)} \quad \overset{s}{p}{}_{h,\Omega}^q = \rho \overset{s}{p}{}_h^q \\ \text{ii)} \quad \vec{p}_h \vec{u}_h = \vec{p}_{h,\Omega}^{(s)} \vec{u}_h = (\overset{s}{p}{}_{h,\Omega}^q u_h^q)_{|q| \leq m} . \end{cases}$$

Then $\vec{p}^{(s)}_{h,\Omega}\vec{u}_h$ belongs to $\vec{H}^m(\Omega,D^*)$ since, by the commutation formulas† $D^q\pi_{q,h}* = \nabla^q_h$ and $\nabla^q_h p^{(s)}_h \nabla^q_h$, we obtain $D^q p^{s}_{h,\Omega}{}^q \nabla^q_h = p^{(s)}_{h,\Omega}\nabla^q_h$ and thus

(3.9)

$$\text{i)}\quad D^*\vec{p}^{(s)}_{h,\Omega}\vec{u}_h = p^{(s)}_h D^*_h\vec{u}_h = p^{(s)}_{h,\Omega}\Big(\sum_{|q|\leqslant m}(-1)^{|q|}\nabla^q_h u^q_h\Big),$$

where

$$\text{ii)}\quad D^*_h\vec{u}_h = \sum_{|q|\leqslant m}(-1)^{|q|}\nabla^q_h u^q_h .$$

3.1.c. <u>Restriction</u> $\hat{r}_h$.

For the restriction r_h we shall choose the optimal restriction $\hat{r}_h$ associated with $\vec{p}_h = \vec{p}^{(s)}_{h,\Omega}$ in $\vec{H}^m(\Omega,D^*)$.††

---

†If $n = 1$, it is clear that $D\theta_h = \frac{1}{h}(\delta-\delta(h))$. Then

$$D^q\pi_{q,h}* = \frac{1}{h^q}\sum_{k\leqslant q}(-1)^{|k|}\binom{q}{k}\delta(kh)* = \nabla^q_h$$

is the finite difference of order q $\left(\text{we set } \binom{q}{k} = \binom{q_1}{k_1}\cdots\binom{q_n}{k_n}\right)$. if $u_h = (u^k_h)_{k\in Z^n}$ we define

$$(\nabla^q_h u_h)^j = \frac{1}{h^q}\sum_{k\leqslant q}(-1)^k\binom{q}{k}u^{j-k}_h .$$

††We define it as follows. Let $[[\vec{u},\vec{v}]] = [\vec{u},\vec{v}] + (D^*\vec{u},D^*\vec{v})$ be the inner product of $\vec{H}^m(\Omega,D^*)$. Then $\hat{r}_h\vec{u}$ is the solution of the variational equation

$$[[\vec{u}-\vec{p}^{(s)}_{h,\Omega}\hat{r}_h\vec{u},\ \vec{p}^{(s)}_{h,\Omega}v_h]] = 0 \quad \text{for any } v_h \in V_h .$$

In other words, $\vec{p}^{(s)}_{h,\Omega}\hat{r}_h$ is the orthogonal projector from $\vec{H}^m(\Omega,D^*)$ onto the subspace $\vec{p}^{(s)}_{h,\Omega}\vec{H}^{m,s}_h(\Omega,D^*)$ of $\vec{H}^m(\Omega,D^*)$.

3.1.d. Convergence of the approximations of $\vec{H}^m(\Omega,D^*)$.

Theorem 3.1. Let us assume Ω smooth.

Let $(\vec{H}_h^{m,s}(\Omega,D^*)$, $\vec{p}_{h,r}^{(s)}$, $\hat{r}_h)$ be the approximations defined by (3.7) and (3.8). These approximations are convergent in $\vec{H}^m(\Omega,D^*)$.

Since the operators $1 - \vec{p}_{h,\Omega}^{(s)}\hat{r}_h$ are bounded by 1 in $L(\vec{H}^m(\Omega,D^*)$, $\vec{H}^m(\Omega,D^*))$ it is sufficient to prove $\vec{u} - \vec{p}_{h,\Omega}^{(s)}\hat{r}_h\vec{u}$ converges to 0 for any $\vec{u}$ in a dense subspace of $\vec{H}^m(\Omega,D^*)$. We begin by proving the following lemma.

Lemma 3.1. If Ω is smooth, then $(\mathcal{E}(\overline{\Omega}))^M$ is dense in $\vec{H}^m(\Omega,D^*)$.†

Let $\ell(\vec{v})$ be a continuous linear form vanishing on $(\mathcal{E}(\Omega))^M$. We have to prove that $\ell(\vec{v}) = 0$ for any $\vec{v} \in \vec{H}^m(\Omega,D^*)$.†† Since $\vec{H}^m(\Omega,D^*)$ is the domain of D^* , we can write the form $\ell(\vec{v})$ in the following way:†††

$$\ell(\vec{v}) = [\vec{f},\vec{v}] + (u,D^*\vec{v}) \quad \text{where} \tag{3.10}$$

$$\vec{f} \in \vec{H} \quad \text{and} \quad u \in H = L^2(\Omega) .$$

† $\mathcal{E}(\bar{\Omega})$ is the space of infinitely differentiable functions on the closure of Ω .

†† By the Hahn-Banach theorem.

††† By identifying $\vec{H}^m(\Omega,D^*)$ with the closed subspace of $\vec{H} \times H$ consisting of pairs $(\vec{u},D^*\vec{u})$. The decomposition (3.10) is not unique.

In particular, $\ell(\vec{v})$ vanishes on $(\mathcal{D}(\Omega))^M$.† Hence

$$\ell(\vec{v}) = [\vec{f},\vec{v}] + [Du,\vec{v}] = [\vec{f}+Du,\vec{v}] = 0 \tag{3.11}$$

$$\text{for any } v \in (\mathcal{D}(\Omega))^M .$$

This implies that $Du = -\vec{f}$ and thus, that $u \in H^m(\Omega)$ since $u \in L^2(\Omega)$ and $Du \in \vec{H}$. Therefore, we can use the Green's formula (1.12) and write

$$\ell(\vec{v}) = \sum_{0\leq j\leq m-1} < \beta_{2m-1-j}\vec{v}, \gamma_j u > = 0 \tag{3.12}$$

$$\text{for any } \vec{v} \in (\mathcal{E}(\bar{\Omega}))^M .$$

By Proposition 2.1, we know that $\beta = (\beta_{2m-1},\cdots,\beta_m)$ maps $\vec{H}^m(\Omega,D^*)$ onto T', where $T = \prod_{0\leq j\leq m-1} H^{m-j-\frac{1}{2}}(\Gamma)$.

We assume that Ω is smooth enough in order that

(3.13)
- i) $(\mathcal{E}(\Gamma))^m$ is dense in T',
- ii) There exists $\vec{v} \in (\mathcal{E}(\bar{\Omega}))^M$ such that $\beta_j v = t_j$ for any $t_j \in \mathcal{E}(\Gamma)$.

Then β maps $(\mathcal{E}(\bar{\Omega}))^M$ onto $(\mathcal{E}(\Gamma))^m$, which is dense in T'; therefore $\gamma_j u = 0$ for $0 \leq j \leq m-1$ by (3.12) and the linear form ℓ vanishes. This proves Lemma 3.1.

---

† $\mathcal{D}(\Omega)$ is the space of functions of $\mathcal{E}(\bar{\Omega})$ with compact support in Ω.

Now, if $\vec{u} \in (\mathcal{E}(\bar{\Omega}))^M$, we know that there exist restrictions $r_{h,\Omega}$ such that†

(3.14)

$$\text{i)}\quad u^q - p^{s_q}_{h,\Omega}\, r_{h,\Omega}\, u^q \text{ converges to 0 in } L^2(\Omega)\ ,$$

$$\text{ii)}\quad D^q u^q - D^q\, p^{s_q}_{h,\Omega}\, r_{h,\Omega}\, u^q \text{ converges to 0 in } L^2(\Omega)\ .$$

Hence, if we set

$$(3.15)\qquad \vec{r}_{h,\Omega}\vec{u} = (r_{h,\Omega}u^q)_{|q|\leq m}\ ,$$

we deduce from (3.14) that

(3.16)

$$\|\vec{u}-\vec{p}^{(s)}_{h,\Omega}\hat{r}_h\vec{u}\|_{\vec{H}^m(\Omega,D^*)} \leq \|\vec{u}-\vec{p}^{(s)}_{h,\Omega}\vec{r}_{h,\Omega}\vec{u}\|_{\vec{H}^m(\Omega,D^*)}$$

converges to 0.

Therefore, the approximations are convergent.

3.2. Estimate of the error-function.

Let us define

---

†We choose for instance

$$(r_{h,\Omega}u)^k = \frac{s}{h_1\cdots h_n}\int_{(kh,(k+1)h)} \omega u(x)\,ds$$

where ω is a continuous right inverse of the operator ρ of restrictions to Ω (see [1], [2]).

(3.17)

i) $\mathcal{V} = \vec{H}^m(\Omega, D^*)$

ii) $\mathcal{U} = \prod_{|q| \leq m} H^{s+1}(\Omega, D^q)$, $\mathcal{U}(R^n) = \prod_{|q| \leq m} H^{s+1}(R^n, D^q)$.

Here $H^{s+1}(\Omega, D^q)$ is the space of functions u of $H^{s+1}(\Omega)$ such that $D^q u \in H^{s+1}(\Omega)$.

We will estimate the error-function

(3.18)

$$e_{\mathcal{U}}^{\mathcal{V}}(\vec{p}_{h,\Omega}^{(s)}) = \sup_{u \in \mathcal{U}} \|\vec{u} - \vec{p}_{h,\Omega}^{(s)} \hat{r}_h \vec{u}\|_{\vec{H}^m(\Omega, D^*)} \Big/ \|\vec{u}\|_{\mathcal{U}} .$$

Let us assume that Ω satisfies the following property

(3.19) The operator ρ of restriction to Ω maps $H^{s+1}(R^n, D^q)$ onto $H^{s+1}(\Omega, D^q)$ for any q such that $|q| \leq m$.

Theorem 3.2. Assume (3.19). Then there exists a constant M independent of h such that

(3.20)

$$e_{\mathcal{U}}^{\mathcal{V}}(\vec{p}_{h,\Omega}^{(s)}) \leq M|h|^{s+1} \quad \text{where} \quad |h| = \max(h_1, \cdots, h_n) .$$

We begin by proving the theorem in the case where $\Omega = R^n$ and the supports of the functions occuring are contained in a fixed compact subset K of R^n.

Let us introduce the restrictions r_h^q, s_h^q and $r_h^{(s)}$ defined by

(3.21)

$$\text{i)}\quad (D^q(u-\overset{s}{p}{}_h^q r_h^q u),\ \overset{s}{p}{}_h^q v_h) = 0 \quad \text{for any} \quad v_h \in \ell^2(Z^n)\ ,$$

$$\text{ii)}\quad (u-\overset{s}{p}{}_h^q\, s_h^q u,\ \overset{s}{p}{}_h^q v_h) = 0 \quad \text{for any} \quad v_h \in \ell^2(Z^n)\ ,$$

$$\text{iii)}\quad (u-p_h^{(s)} r_h^{(s)} u,\ p_h^{(s)} v_h) = 0 \quad \text{for any} \quad v_h \in \ell^2(Z^n)\ .$$

Let

(3.22)

$$\varepsilon_q(h) = \sup_{\substack{\text{support }(u)\subset K\\ u\in H^{s+1}(R^n,D^q)}} \|u-\overset{s}{p}{}_h^q s_h^q u\|_{L^2(R^n)} \Big/ \|D^q u\|_{L^2(R^n)}\ .$$

$$\text{(3.23)}\qquad \vec{r}_h^{(s)}\vec{u} = (r_h^q u^q)_{|q|\le m}\ .$$

Since the functions occuring have their supports contained in K, $\varepsilon_q(h)$ is bounded.

We next deduce from (3.21), (3.22) and the commutation formulas (3.9) that

(3.24)

$$\text{i)}\quad D^*(\vec{u}-\vec{p}_h^{(s)}\vec{r}_h^{(s)}\vec{u}) = (1-p_h^{(s)} r_h^{(s)})D^*\vec{u}\ ,$$

$$\text{ii)}\quad \|u-\overset{s}{p}{}_h^q r_h^q u\|_{L^2(R^n)} \le \varepsilon_q(h)\, \|(1-p_h^{(s)} r_h^{(s)})D^q u\|_{L^2(R^n)}\ .$$

Indeed, we deduce from (3.21)i) and the commutation formulas (3.9) that

$$(3.25) \qquad \begin{cases} (D^q u - p_h^{(s)} \nabla_h^q r_h^q,\ p_h^{(s)} \nabla_h^q v_h) = 0 \\ \quad \text{for any } v_h \in \ell^2(Z^n) . \end{cases}$$

Since the range of ∇_h^q is dense in $\ell^2(Z^n)$, we can replace $\nabla_h^q v_h$ by v_h in (3.25). Hence (3.21)iii) implies that

$$(3.26) \qquad \nabla_h^q r_h^q u = r_h^{(s)} D^q u .$$

Now (3.9), (3.23) and (3.26) imply (3.24)i).

On the other hand, we deduce from the identity $u - p_h^q r_h^q u = (1-p_h^q s_h^q)(u-p_h^q r_h^q u)$, from (3.22) and from the commutation formulas (3.9) and (3.26) the inequalities

$$\begin{aligned} \|u-p_h^q r_h^q u\| &\leqslant \varepsilon_q(h)\, \|D^q(u-p_h^q r_h^q u)\| \\ &= \varepsilon_q(h)\, \|(1-p_h^{(s)} r_h^{(s)}) D^q u\| . \end{aligned}$$

Now, since $\|u-p_h^{(s)} r_h^{(s)} u\|_{L^2(R^n)} \leqslant M|h|^{s+1} \|u\|_{H^{s+1}(R^n)}$ (see [1], [2]), we deduce from (3.24) that

(3.27)

$$\begin{aligned} &\|\vec{u}-\vec{p}_h^{(s)} \hat{r}_h \vec{u}\|_{\vec{H}^m(R^n,D)} \\ &\quad \leqslant \|\vec{u}-\vec{p}_h^{(s)} \vec{r}_h^{(s)} \vec{u}\|_{\vec{H}^m(R^n,D^*)} \leqslant M|h|^{s+1} \|\vec{u}\|_{\mathcal{U}(R^n)} . \end{aligned}$$

Now we use (3.27) to prove the theorem for a smooth bounded subset Ω satisfying (3.19). Indeed, by (3.19), there exist continuous right-inverses $\omega_q \in L(H^{s+1}(\Omega,D^q), H^{s+1}(R^n,D^q))$ of the operator ρ. After multiplying by a smooth function with compact support K, with value 1 on Ω, the supports of the functions $\omega_q u$ are contained in K.

We let $\vec{\vec{\omega u}} = (\omega_q u^q)_{|q|\leqslant m}$ (so that $\vec{u} = \rho\vec{\vec{\omega u}}$ for $\vec{u} \in \mathcal{U}$) and we obtain from (3.27)

$$\|\vec{u}-\vec{p}_{h,}^{(s)}\hat{r}_h\vec{u}\|_{\vec{H}^m(\Omega,D^*)} \leqslant \|(1-\vec{p}_{h,}^{(s)}r_h^{(s)})\vec{\vec{\omega u}}\|_{\vec{H}^m(\Omega,D^*)}$$

$$\leqslant \|(1-\vec{p}_h^{(s)}\vec{r}_h^{(s)})\vec{\vec{\omega u}}\|_{\vec{H}^m(R^n,D^*)} \leqslant M|h|^{s+1}\,\|\vec{\vec{\omega u}}\|_{\mathcal{U}(R^n)}$$

$$\leqslant M|h|^{s+1}\,\|\vec{u}\|_{\mathcal{U}}$$

This completes the proof of the theorem.

3.3. Approximations of the space $\vec{H}_0^m(\Omega,D^*)$.

We denote by $\vec{H}_0^m(\Omega,D^*)$ the kernel of $\beta = (\beta_{2m-1}, \dots, \beta_m)$:

$$\vec{H}_0^m(\Omega,D^*) = \{\vec{v} \in \vec{H}^m(\Omega,D^*) \text{ such that } \beta_j\vec{v} = 0 \text{ for } m \leqslant j \leqslant 2m-1\}. \tag{3.28}$$

We shall use the following characterization of $\vec{H}_0^m(\Omega,D^*)$.

Proposition 3.1. Let us assume that Ω is smooth. Then $\vec{H}_0^m(\Omega,D^*)$ coincides with the closure in $\vec{H}^m(\Omega,D^*)$ of the space $(\mathcal{D}(\Omega))^M$.

First, we know that $(\mathcal{D}(\Omega))^M$ is contained in $H_0^m(\Omega,D^*)$. Hence it suffices to prove that the orthogonal of $(\mathcal{D}(\Omega))^M$ is contained in the orthogonal of $\vec{H}_0^m(\Omega,D^*)$, which is the (closed) range of β since $\vec{H}_0^m(\Omega,D^*)$ is the kernel of the surjective operator β.

Let $\ell(\vec{v})$ be a continuous linear form vanishing on $(\mathcal{D}(\Omega))^M$. We have seen, in the proof of Lemma 3.1, (3.12), that such a form can be written

$$(3.12) \qquad \ell(\vec{v}) = \langle\gamma u, \beta\vec{v}\rangle = \sum_{0\leq j\leq m-1} \langle\gamma_j u, \beta_{2m-1-j}\vec{v}\rangle$$

where u belongs to $H^m(\Omega)$.

Now (3.29) implies that $\ell(\vec{v}) = [\beta'\gamma u, \vec{v}]$ belongs to the range of β' and the proposition is proved.

Next construct a family of approximations of the space $\vec{H}_0^m(\Omega,D^*)$.

3.3.a. <u>The discrete space</u> $\vec{H}_{0,h}^{m,s}(\Omega,D^*)$.

We introduce the following items:

(3.29)

i) $\mathcal{R}_{0,h}^{s_q}(\Omega) = \{k \subset Z^n$ such that support $(\pi_{s_q,h} * \theta_{kh}) \subset \Omega\}$

ii) $H_{0,h}^{s_q}(\Omega) = \{u_h = (u_h^k)_{k\in Z^n}$ such that $u_h^k = 0$ for $k \notin \mathcal{R}_{0,h}^{s_q}(\Omega)\}$.

We choose the discrete space $\vec{V}_h = \vec{H}^{m,s}_{0,h}(\Omega,D^*)$ defined by

(3.30) $\vec{H}^{m,s}_{0,h}(\Omega,D^*) = \prod_{|q|\leq m} \overset{s}{H}{}^{q}_{0,h}(\Omega)$ is the space

of the $\vec{u}_h = (u_h^q)_{|q|\leq m}$.

3.3.b. The prolongations $\vec{p}_h(s)$.

As prolongation we take the operator $\vec{p}_h^{(s)}$ defined by:

$$\vec{p}_h^{(s)}\vec{u}_h = (\overset{s}{p}{}_h^{q} u_h^q)_{|q|\leq m} \tag{3.31}$$

Since the support of $\vec{p}_h^{(s)} u_h$ is contained in Ω (by (3.29) and (3.30)), $\vec{p}_h^{(s)}$ is an isomorphism from $\vec{H}^{m,s}_{0,h}(\Omega,D^*)$ onto its closed range in $\vec{H}^m_0(\Omega,D^{\Xi})$ and satisfies the commutation formulas

(3.32)

i) $D^q \overset{s}{p}{}_h^{q} u_h = p_h^{(s)} \nabla_h^q u_h$,

ii) $D^*\vec{p}_h^{(s)}\vec{u}_h = p_h^{(s)}\left(\sum_{|q|\leq m}(-1)^{|q|}\nabla_h^q u_h^q\right) = p_h^{(s)} D_h^* \vec{u}_h$.

3.3.c. The restriction $\hat{r}_h$.

As restriction r_h we choose the optimal restriction $\hat{r}_h$ associated with $\vec{p}_h^{(s)}$ in $\vec{H}^m_0(\Omega,D^*)$.

3.3.d. Convergence of the approximations of $\vec{H}^m_0(\Omega,D^*)$.

Theorem 3.3. Let us assume that Ω is smooth and satisfies property (3.11) of [1], page 96. Then the approximations $(\vec{H}^{m,s}_{0,h}(\Omega,D^*), \vec{p}_h^{(s)}, \hat{r}_h)$ of $\vec{H}^m_0(\Omega,D^*)$ defined by (3.30) and (3.32) are convergent.

The proof is analogous to the proof of Theorem 3.1: the theorem follows from Proposition 3.1 and from Theorem 3.2 of [1], page 96.

Let

(3.33) i) $\mathcal{V}_0 = \vec{H}^m_0(\Omega,D^*)$,

ii) $\mathcal{U}_0 = \prod_{|q|\leq m} H_0^{s+1}(\Omega,D^q)$.

Here $H_0^{s+1}(\Omega,D^q)$ is the closure in $H^{s+1}(\Omega,D^q)$ of $\mathcal{D}(\Omega)$.

The proof of the following theorem is analogous to the proof of Theorem 3.2.

<u>Theorem 3.4</u>. We make the assumptions of Theorem 3.3. Then the error-function

(3.34)

$$e^{\mathcal{V}_0}_{\mathcal{U}_0}(\vec{p}_h^{(s)}) = \sup \|\vec{u}-\vec{p}_h^{(s)}\hat{r}_h\vec{u}\|_{\vec{H}^m(\Omega,D^*)} \Big/ \|\vec{u}\|_{\mathcal{U}_0}$$

satisfies the following inequality:

(3.35) $$e^{\mathcal{V}_0}_{\mathcal{U}_0}(\vec{p}_h^{(s)}) \leq M|h|^{s+1} .$$

4. <u>Approximation of Conjugate Boundary Value Problems</u>

Under convenient assumptions, we have shown that the solution of the conjugate Dirichlet problem (respectively the conjugate Neumann problem) is equivalent to an $\vec{H}^m(\Omega,D^*)$ - elliptic variational equation (respectively $\vec{H}_0^m(\Omega,D^*)$ - elliptic variational equation).

On the other hand, we have constructed convergent approximations of the spaces $\vec{H}^m(\Omega,D^*)$ and $\vec{H}_0^m(\Omega,D^*)$. Therefore, we can construct approximate problems of the conjugate Dirichlet problems (section 4.1) and conjugate Neumann problems (section 4.3) and we state the results regarding convergence and error-estimates.

In section 4.2, we use external approximations of the space $\vec{H}^m(\Omega,D^*)$ to construct the "natural" system of finite difference equations approximating the conjugate Dirichlet problem.

4.1. Approximation of the conjugate Dirichlet problem

Let $\mathcal{A} = (a_{pq}(x))_{|p|,|q|\leq m}$ be a matrix of functions of $L^\infty(\Omega)$ satisfying

(4.1)

i) $\mathcal{A}$ is $\vec{H}$ = elliptic (where $\vec{H} = (L^2(\Omega))^M$,

ii) $\mathcal{B} = A^{-1} = (\alpha_{pq}(x))_{|p|,|q|\leq m}$.

Let $D \in L(H^m(\Omega),\vec{H})$ and $D^* \in L(\vec{H}^m(\Omega,D^*),\ L^2(\Omega))$ be the operators

(4.2)

$$Du = (D^q u)_{|q|\leq m} \quad \text{and} \quad D^*\vec{u} = \sum_{|q|\leq m} (-1)^{|q|} D^q u^q$$

and Λ be the differential operator defined by

(4.3)

$$\Lambda u = D^*ADu = \sum_{|p|,|q|\leq m} (-1)^{|q|} D^q(a_{pq}(x)D^p u)$$

mapping $H^m(\Omega,\Lambda)$ into $L^2(\Omega)$.

We consider the following data

$$(4.4) \quad f \in L^2(\Omega), \quad t_j \in H^{m-j-\frac{1}{2}}(\Gamma) \quad \text{for} \quad 0 \leq j \leq m-1$$

and the solution u of the Dirichlet problem

(4.5)

$$\Lambda u + \lambda u = 0, \quad \gamma_j u = t_j \quad \text{for} \quad 0 \leq j \leq m-1 \ (\lambda > 0) \ .$$

Let $\vec{u}$ be the solution of the conjugate Dirichlet problem

(4.6)

i) $\vec{u} \in \vec{H}^m(\Omega,D^*)$,

$$\text{ii)} \quad (D^*\vec{u},D^*\vec{v}) + \lambda \sum_{|p|,|q|\leq m} \int_\Omega \alpha_{pq}(x)u^p v^q dx = (f,D^*\vec{v})$$

$$+ \sum_{0\leq j\leq m-1} \langle t_j, \beta_{2m-1-j}\vec{v}\rangle \quad \text{for any} \quad \vec{v} \in \vec{H}^m(\Omega,D^*)$$

related to the solution u of (4.5) by

(4.7)

$$u = \lambda^{-1}(f-D^*\vec{u}), \quad D^q u = (\mathcal{B}u)^q = \sum_{|p|\leq m} \alpha_{pq}(x)u^p) \ .$$

(See sections. 1.2 and 2.4).

We approximate the solution $\vec{u}$ of the conjugate problem (4.6) by using the approximations $(\vec{H}_h^{m,s}(\Omega,D^*)$, $p_{h,\Omega}^{(s)}, \hat{r}_h)$ of $\vec{H}^m(\Omega,D^*)$ which are defined and studied in Section 3.1. Then the internal approximation† of (4.6) is the following system of discrete variational equations

$$
(4.8)\quad \begin{cases}
\text{i)}\ \vec{u}_h = (u_h^q)_{|q|\leq m} \in \vec{H}_h^{m,s}(\Omega,D^*) \\
\qquad = \prod_{|q|\leq m} H_h^{s_q}(\Omega)\ ,\quad s_q = (s) + q\ , \\
\text{ii)}\ (D^*\vec{p}_{h,\Omega}^{(s)}\vec{u}_h,\ D^*\vec{p}_{h,\Omega}^{(s)}\vec{v}_h) \\
\qquad + \lambda \sum_{|p|,|q|\leq m} \int_\Omega \alpha_{pq}(x) p_{h,\Omega}^{s_p} u_h^p p_{h,\Omega}^{s_q} v_h^q dx \\
\qquad = (f, D^*\vec{p}_{h,\Omega}^{(s)}\vec{v}_h) + \lambda \sum_{0\leq j\leq m-1} \langle t_j, \beta_{2m-1-j}\vec{p}_{h,\Omega}^{(s)}\vec{v}_h \rangle \\
\text{for any } \vec{v}_h \in \vec{H}_h^{m,s}(\Omega,D^*)\ .
\end{cases}
$$

† Let $b(u,v)$ be an X-elliptic continuous bilinear form, $\ell(v)$ a continuous linear form on X and (x_h, p_h, u_h) approximations of X . Let $u \in X$ be the solution of the variational equation

$$b(u, v) = \ell(v) \quad \text{for any } v \in X\ .$$

We say that the discrete variational equation

$$b(p_h u_h, p_h v_h) = \ell(p_h v_h) \quad \text{for } v_h \quad X_h$$

is the internal approximation of the above variational equation. Here we take $X = \vec{H}^m(\Omega,D^*)$ and

$$b(\vec{u}, \vec{v}) = (D^*\vec{u}, D^*\vec{v}) + \lambda[A^{-1}\vec{u}, \vec{v}]$$

$$\ell(\vec{v}) = (f, D^*\vec{v}) + \langle t, \beta\vec{v} \rangle$$

$$X_h = \vec{H}_h^{m,s}(\Omega,D^*)\ ,\quad p_h = \vec{p}_{h,\Omega}^{(s)}\ .$$

We will make explicit this system of discrete equations at the end of this section. First, we state the results regarding convergence and error estimates.

Theorem 4.1. Let us assume (4.1) and that Ω is smooth. Let $f \in L^2(\Omega)$, $t_j \in H^{m-j-\frac{1}{2}}(\Gamma)$ $(0 \leq j \leq m-1)$ be given, let $u \in H^m(\Omega,\Lambda)$ be the solution of (4.5), $\vec{u} \in \vec{H}^m(\Omega,D^*)$ be the solution of (4.6) and $\vec{u}_h \in \vec{H}_h^{m,s}(\Omega,D^*)$ be the solution of (4.8).

Then

(4.9) $\quad \vec{p}^{(s)}_{h,\Omega}\vec{u}_h$ converges to $\vec{u}$ in $\vec{H}^m(\Omega,D^*)$.

In other words, we have

(4.10)

i) $\lambda^{-1}(f-p^{(s)}_{h,\Omega}D^*_h\vec{u}_h)$ converges to u in $L^2(\Omega)$,

ii) $\sum\limits_{|p|\leq m} \alpha_{pq}(x)p^{s\;p}_{h,\Omega}u^p_h$ converges to $D^q u$ in $L^2(\Omega)$.

This theorem follows from Theorem 3.1 of [2] and from Theorem 3.1 of section 3.1.

Now, let us introduce the following spaces in which we will estimate the error.

(4.11)

i) $H_0^{s+1}(\Omega,D^q)$ is the closure of $\mathcal{D}(\Omega)$ in $\vec{H}^{s+1}(\Omega,D^q)$,

ii) $H^{-s-1}(\Omega,D^q)$ is the dual of $H_0^{s+1}(\Omega,D^q)$.

We observe: with $\vec{v}$ ranging over $\vec{H}^m(\Omega,D^*)$

(4.12) $$D^q D^* \vec{v} + \lambda \sum_{|p|\le m} \alpha_{pq}(x) v^p \text{ belongs to } H^{-s-1}(\Omega,D^q) \text{ for } |q|\le m .$$

Theorem 4.2. We make the assumptions of Theorem 4.1. Then the following estimates hold:

(4.13) $$\Big\| D^q D^*(\vec{u} - p^{(s)}_{h,\Omega}\vec{u}_h) + \sum_{|p|\le m} \alpha_{pq}(x)(u^p - p^{s\,p}_{h,\Omega} u^p_h) \Big\|_{H^{-s-1}(\Omega,D^q)} \le M|h|^{s+1} \|\vec{u}\|_{\vec{H}^m(\Omega,D^*)} .$$

This result follows from Theorem 3.2 of section 3.2.[†] By assuming the regularity of the solution, we get the following estimates

---

[†]Since we can write $b(u-p_h u_h, v) = b(u-p_h u_h, v-p_h r_h v)$, we deduce that

$$\sup_{v\in U_0} \frac{|b(u-p_h u_h,v)|}{\|v\|_{U_0}} \le M\|u-p_h u_h\|\, e^X_{U_0}(p_h) \le M_2 e^X_{U_0}(p_h)$$

where $U_o \subset X$ and $e^X_U(p_h) = \sup_{u\in U_0} \|u-p_h r_h u\|_X \,/\, \|u\|_{U_0}$.

Here we take $X = \vec{H}^m(\Omega,D^*)$ and $U_o = \prod_{|q|\le m} H^{s+1}(\Omega,D^q)$.

Then $e^X_U(p_h) \le cM|h|^{s+1}$ by Theorem 3.2 of section 3.2 and

$$\sup_{v\in U_0} \frac{|b(u-p_h u_h,v)|}{\|v\|_{U_0}} = \|DD^*(\vec{u}-\vec{p}^{(s)}_{h,\Omega}\vec{u}_h) + \lambda A^{-1}(\vec{u}-\vec{p}^{(s)}_{h,\Omega}\vec{u}_h)\|_{U_0'} .$$

We obtain (4.13) by noticing that

$$\|\vec{f}\|_{U_0'} = \left(\sum_{|q|\le m} \|f^q\|^2_{H^{-s-1}(\Omega,D^q)} \right)^{1/2} .$$

Theorem 4.3. With the assumptions of Theorem 4.1 and the following result of regularity:

(4.14)

the solution $\vec{u}$ of (4.6) belongs to $\prod_{|q|\leqslant m} H^{s+1}(\Omega, D^q)$.

We obtain the following estimates: there exists a constant M independent of h such that

(4.15)

$$\text{i)} \quad \|\vec{u}-\vec{p}_{h,\Omega}^{(s)}\vec{u}_h\|_{\vec{H}^m(\Omega,D^*)} \leqslant M|h|^{s+1},$$

$$\text{ii)} \quad \|D^q D^*(u-\vec{p}_{h,\Omega}^{(s)}\vec{u}_h) + \lambda \sum_{|p|\leqslant m} \alpha_{pq}(x)(u^p - p_{h,\Omega}^{s_p} u_h^p\|_{H^{-s-1}(\Omega,D^q)} \leqslant M|h|^{2(s+1)}$$

This result follows from Theorem 3.2 of section 3.2, from Theorem 3.1 of [2] and from †.

Finally, let us use Theorem 2.2 of section 2.2 and the solution $\vec{u}_h$ of (4.6) in order to compute a posteriori esimates of the error $u-u^h$ between the solution u of the Dirichlet problem (4.5) and an approximant $u^h \in H_0^m(\Omega)$.

Theorem 4.4. We make the assumptions of Theorem 4.1. Let $u^h \in H^m(\Omega)$ be any approximant of the solution u of (4.5) satisfying $\gamma_j u^h = t_j$.

$$(4.16) \quad c\,\|u-u^h\|^2_{H^m(\Omega)} + \lambda\|u-u^h\|^2_{L^2(\Omega)} \leqslant$$

† See preceding footnote.

$$\leqslant c^{-1} \sum_{|q|\leqslant m} \| p^{s_q}_{h,\Omega} u^q_h - \sum_{|p|\leqslant m} a_{pq}(x) D^p u^h \|^2_{L^2(\Omega)}$$

$$+ \lambda^{-1} \| f - \lambda u^h - p^{(s)}_{h,\Omega} \mathbf{D}^* \vec{u}_h \|^2_{L^2(\Omega)} .$$

We end this section by looking back at the system (4.8). We introduce the matrices $I_h^{(s)}$ and $\mathcal{B}_h^{(s)p,q}$ defined by

(4.17)

$$\text{i)} \quad (I_h^{(s)} u_h, v_h)_h = \int_\Omega p_h^{(s)} u_h(x) p_h^{(s)} v_h(x) dx ,$$

$$\text{ii)} \quad (\mathcal{B}_h^{(s),p,q} u_h, v_h), = \int_h \alpha_{pq}(x) p_h^{s_p} u_h(x) p_h^{s_q} v_h(x) dx ,$$

whose entries are respectively given by

(4.18)

$$\text{i)} \quad (h_1 \cdots h_n)^{-1} \int_\Omega \pi_{(s),h} * \theta_{jh}(x) \cdot \pi_{(s),h} * \theta_{kh}(x) dx ,$$

$$\text{ii)} \quad (h_1 \cdots h_n)^{-1} \int_\Omega \alpha_{pq}(x) \pi_{s_p,h} * \theta_{jh}(x) \cdot \pi_{s_q,h} * \theta_{kh}(x) dx .$$

(See Theorem 4.2 of [1], page 112: these entries involve the integrals

$$\int \alpha_{pq}(x) x^k \theta_{jh}(x) dx \quad \text{for} \quad k \leqslant (2s) + p + q) .$$

Let us introduce the vector $\ell_h^{(s),q}$ whose components $(\ell_h^{(s),q})^k$ are defined by

$$(4.19) \quad (\ell_h^{(s),q})^k = \int_\Omega f(x) \nabla_h^q \pi_{(s),h} * \theta_{kh}(x) dx +$$

$$+ \lambda \sum_{0\leqslant j\leqslant m-1} \int_\Gamma t_j(x) \beta^q_{2m-1-j} [\pi_{s_q,h} * \theta_{kh}(x)] d\sigma(x)$$

(where we have written $\beta_j \cdot \vec{v} = \sum_{|q| \leq m} \beta_j^q v^q$) .

Let $\tilde{\nabla}_h^q = (-1)^{|q|}(\nabla_h^q)$. Then it is easy to verify the following result.

Theorem 4.5. The approximate problem (4.8) is equivalent to the following system of equations: find $\vec{u}_h = (u_h^q)_{|q| \leq m} \in \vec{H}_h^{m,s}(\Omega, D^*)$ satisfying for any q with $|q| \leq m$ the following equations:

(4.20)

$$\tilde{\nabla}_h^q I_h^{(s)}\left(\sum_{|p| \leq m} (-1)^{|p|} \nabla_h^p u_h^p\right) + \lambda \sum_{|p| \leq m} \mathcal{B}_h^{(s),p,q} u_h^p = \ell_h^{(s),q} .$$

In other words, $\vec{u}_h$ is solution of a linear system $\vec{A}_h \vec{u}_h = \vec{\ell}_h$. The matrix $\vec{A}_h$ can be viewed as a "block matrix" whose "block entries" are the matrices $(-1)^{|p|} \tilde{\nabla}_h^q I_h^{(s)} \nabla_h^p + \mathcal{B}_h^{(s),p,q}$.

Such a block entry has at most $(2(s+m)+1)^n$ non-zero entries in each row (or column).

4.2. External approximation of the conjugate Dirichlet problem

The bilinear form $b(\vec{u},\vec{v}) = (D^*\vec{u}, D^*\vec{v}) + \lambda\alpha(\vec{u},\vec{v})$ is the sum of $(D^*\vec{u}, D^*\vec{v})$, which is continuous on $\vec{H}^m(\Omega, D^*) \times \vec{H}^m(\Omega, D^*)$ and of $\lambda\alpha(\vec{u},\vec{v})$, which is continuous on $\vec{H} \times \vec{H}$ (where $\vec{H} = (L^2(\Omega))^M$).

Thus we can identify $\vec{H}^m(\Omega, D^*)$ with the diagonal of $\vec{H}^m(\Omega, D^*) \times \vec{H}$ and use approximations of this product space which are external convergent approximations of $\vec{H}^m(\Omega, D^*)$, (see [5]).

Let us describe the simplest example: we choose $s = 0$ and

(4.21) i) $V_h = \vec{H}_h^{m,0}(\Omega, D^*)$,

ii) $p_h u_h = ((p_{h,\Omega}^q u_h^q)_{|q| \leq m}$,

$(p_{h,\Omega}^0 u_h^q)_{|q| \leq m}) \in \vec{H}^m(\Omega, D^*) \times \vec{H}$

iii) $\hat{r}_h$ is the optimal restriction

associates with p_h in $\vec{H}^m(\Omega, D^*) \times H$.

It is routine to check that the approximations $(V_h, p_h, \hat{r}_h)$ of $\vec{H}^m(\Omega, D^*) \times \vec{H}$ defined by (4.21) are external convergent approximations of $\vec{H}^m(\Omega, D^*)$ (see the third section of [5] and the third section of this chapter).

For simplicity, let us consider the case of the homogeneous Dirichlet problem. Then the external approximation of the conjugate problem (4.8) is the following discrete variational system of equations:

(4.22)

i) $\vec{u}_h \in \vec{H}_h^{m,0}(\Omega, D^*)$,

ii) $(p_{h,\Omega}^0 D_h^* \vec{u}_h, \ p_{h,\Omega}^0 D_h^* \vec{v}_h)$

$$+ \lambda \sum_{|p|,|q| \leq m} \int_\Omega \alpha_{pq}(x) p_{h,\Omega}^0 u_h^p p_{h,\Omega}^0 v_h^q dx$$

$= (f, p_{h,\Omega}^0 D_h^* \vec{v}_h)$ for any $\vec{v}_h \in \vec{H}_h^{m,0}(\Omega, D^*)$.

Denoting by a^h the vector of components $(h_1 \cdot\cdot h_n)^{-1} \int_\Omega a(x)\theta_{kh}(x)dx$, it is not hard to check that (4.22) is equivalent to the system of finite-difference equations:

$$(4.23) \quad \text{i)} \quad \vec{u}_h \in \vec{H}_h^{m,0}(\Omega, D^*) ,$$

$$\text{ii)} \quad \sum_{|p| \leqslant m} \{(-1)^{|p|} \tilde{\nabla}_h^q \nabla_h^p u_h^p + \lambda \alpha_{pq}^h u_h^p\} = \tilde{\nabla}_h^q f^h$$

for any q such that $|q| \leqslant m$.

Thus, we deduce from Theorem 2.1, section 2.2 of [5], the following theorem of convergence.

Theorem 4.6. Let us assume (4.1) and that Ω is smooth. Let $f \in L^2(\Omega)$ be given, $t_j = 0$ for $0 \leqslant j \leqslant m-1$ and let u, $\vec{u}$, $\vec{u}_h$ be the solutions of (4.5), (4.6) and (4.23) respectively. Then

(4.24)

i) $p_{h,\Omega}^0 D_h^* \vec{u}_h$ converges to $D^* \vec{u}$ in $L^2(\Omega)$,

ii) $p_{h,\Omega}^0 u_h^q$ converges to u^q in $L^2(\Omega)$ for any $|q| \leqslant m$.

In other words, we have

$$(4.25) \quad \text{i)} \quad \lambda^{-1}(f - p_{h,\Omega}^0 D_h^* \vec{u}_h) \text{ converges to } u \text{ in } L^2(\Omega) ,$$

ii) $\sum_{|p|\leqslant m} \alpha_{pq}(x)p_h^0 u_h^p$ converges to $D^q u$ in $L^2(\Omega)$ for $|q| \leqslant m$.

4.3. Approximation of the conjugate Neumann problem

Let f be given in $L^2(\Omega)$ and u be the solution of the Neumann problem

$$(4.26) \qquad \Lambda u + \lambda u = f, \quad \delta_j u = 0 \quad \text{for} \quad m \leqslant j \leqslant 2m-1 .$$

Let $\vec{H}_0^m(\Omega,D^*) = \{\vec{v} \in \vec{H}^m(\Omega,D^*)$ such that $\beta_j \vec{v} = 0$ $(m \leqslant j \leqslant 2m-1)\}$. Then the variational formulation of the conjugate Neumann problem (4.26) is

$$(4.27) \quad \text{i)} \quad \vec{u} \in \vec{H}_0^m(\Omega,D^*) ,$$

$$\text{ii)} \quad (D^*\vec{u},D^*\vec{v}) + \lambda \sum_{|p|,|q|\leqslant m} \int_\Omega \alpha_{pq}(x)u^p v^q dx$$

$$= (f,D^*\vec{v}) \quad \text{for any} \quad \vec{v} \in \vec{H}_0^m(\Omega,D^*) .$$

The solutions u and $\vec{u}$ of (4.26) and (4.27) are related by formulas (4.7).

We consider the approximations $(\vec{H}_{0,h}^{m,s}(\Omega,D^*), p_h^{(s)}, \hat{r}_h)$ of $\vec{H}_0^m(\Omega,D^*)$ which are defined and studied in section 2.3. With this, the internal approximation of the conjugate problem (4.27) becomes the following system of discrete variational equations.

$$(4.28) \qquad \text{i)} \quad \vec{u}_h = (u_h^q)_{|q|\leqslant m} \in \vec{H}_{0,h}^{m,s}(\Omega,D^*) ,$$

$$\text{ii)}\quad (D^*\vec{p}_h^{(s)}\vec{u}_h,\ D^*\vec{p}_h^{(s)}\vec{v}_h)$$

$$+\ \lambda \sum_{|p|,|q|\leq m} \int_\Omega \alpha_{pq}(x)\, \overset{s}{p}_h{}^p u_h^p\, \overset{s}{p}_h{}^q v_h^q\, dx$$

$$= (f, D^*\vec{p}_h^{(s)}\vec{v}_h) \quad \text{for any} \quad \vec{v}_h \in \vec{H}^{m,s}_{0,h}(\Omega, D^*)$$

(this system can be written in a form analogous to (4.20)). We deduce the following result using on one hand Theorem 3.1 of [2] and, on the other hand, Theorems 3.3 and 3.4 of section 3.3.

<u>Theorem 4.7</u>. Let us assume (4.1) and that Ω is smooth and satisfies property (3.11) of [1], p. 96. Let f be given in $L^2(\Omega)$, u, $\vec{u}$ and $\vec{u}_h$ be the solutions of (4.26), (4.27), and (4.28) respectively. Then

$$(4.29)\qquad \vec{p}_h^{(s)}\vec{u}_h \ \text{converges to}\ \vec{u}\ \text{in}\ \vec{H}^m_0(\Omega, D^*)\ .$$

In other words, we obtain that

(4.30)

$$\text{i)}\quad \lambda^{-1}(f - p_h^{(s)} D_h^* \vec{u}_h)\ \text{converges to}\ u\ \text{in}\ L^2(\Omega)\ ,$$

$$\text{ii)}\quad \sum_{|p|\leq m} \alpha_{pq}(x)\ \overset{s}{p}_h{}^p u_h^p\ \text{converges to}\ D^q u\ \text{in}\ L^2(\Omega)\ .$$

Furthermore, the errors obey the following estimates:

$$(4.31)\qquad \| D^q D^*(\vec{u} - \vec{p}_h^{(s)}\vec{u}_h)$$

$$+ \lambda \sum_{|p| \leq m} \alpha_{pq}(x)(u^p - p_h^s q u_h^p) \| _{H^{-s-1}(\Omega, D^q)}$$

$$\leq M|h|^{s+1} \|\vec{u}\| _{\vec{H}_0^m(\Omega, D^*)} .$$

We can also prove the convergence of external approximations of the conjugate problem (4.27) of the Neumann problem, as in section 4.2.

REFERENCES

1. Aubin, J. P., "Approximation des espaces de distributions et des opérateurs différentiels." Bull. Soc. Math. Fr. Memoire 12 (1967), 1-139.
2. Aubin, J. P., "Behavior of the Error of the Approximate Solutions of Boundary Value Problems for Linear Elliptic Operators by Galerkin's and Finite Difference Methods." Ann. Scuo. Norm Pisa 21 (1967), 599-637.
3. Aubin, J. P., "Abstract Boundary Value Operators and Theor Adjoints," Rend. Sem. Mat. Padova (1969).
4. Aubin, J. P., "Approximation of Non-homogeneous Neumann Problems - Regularity of the Convergence and Estimates of Errors in Terms of n-width." MRC Report #924 (1968).
5. Aubin, J. P., "External and Partial Approximations," MRC Report #1004, (to appear).
6. Birkhoff, G., de Boor, C., Swartz, B., and Wendroff, B, "Rayleigh-Ritz Approximation by Piecewise Cubic Polynomials." SIAM J. Num. Anal. 3 (1966), 188-203.
7. de Boor, C., "The Method of Projections as Applied to the Numerical Solution of Two Point Boundary Value Problems Using Cubic Splines." Ph.D. thesis, University of Michigan, Ann Arbor, Michigan, 1966.
8. Burchard, H. G., "Theorie und Anwendung der Hyperkreismethode, Diplomarbeit," Universität Hamburg, 1963.
9. Céa, J., "Approximation variationelle des problèmes elliptiques," Ann. Inst. Fourier 14(1964), 343-444.
10. Ciarlet, P. G., Schultz, M. H., and Varga, R. S., "Numerical Methods of High Order Accuracy for Nonlinear Boundary Value Problems." IV. Periodic Boundary Conditions. Num. Math. 12, (1968), 120-133.

11. Diaz, J. B., and Weinstein, A., "Schwarz' Inequality and the Methods of Rayleigh-Ritz and Trefftz," J. Math. Phys. 26 (1947), 133-136.
12. Friedrichs, K. O., "Ein Verfahren der Variationsrechnung das Minimum eines Integrals als das Maximum eines anderen Ausdrucks darzustellen," Ges. Wiss. Göttingen, Nachrichten, Math. Phys. Kl. (1929), 13-20.
13. Golomb, M. and Weinberger, H. F., "Optimal Approximation and errorbounds." On Numerical Approximation. Proc. MRC Symp., R. E. Langer, ed., Univ. of Wisconsin Press, Madison, (1959), 117-190.
14. Kato, T., "On Some Approximate Methods Concerning the Operators T * T," Math. Annal. 126 (1953), 253-262.
15. Kato, T., "Estimation of Iterated Matrices, with Application to the von Neumann Condition," Num. Math. 2 (1960), 22-29.
16. Lions, J. L., and Magenes, E., "Problèmes aux limites non homogenes et applications" (two vol.). Dunod, Paris, 1968.
17. Noble, B., "Complementary Variational Principles for Boundary Value Problems I: Basic Principles, with an Application to Ordinary Differential Equations." MRC Report #473 (1964). II Nonlinear networks, MRC Report #643 (1966).
18. Payne, L. E. and Weinberger, H. F., "Bounds for Solutions of Second Order Elliptic Equations in Terms of Arbitrary Vector Fields," Archive Rat. Mech. Anal. 20 (1965), 95-106.
19. Prager W. and Synge, J. L., "Approximations in Elasticity Based on the Concept of Function Space," Quart. Appl. Math. 5 (1947), 241-269.
20. Schultz, M. H., "Multivariate Spline Functions and Elliptic Problems," Approximations with Special Emphasis on Spline Functions. Proc. MRC Symp., Academic Press, New York, I. J. Schoenberg, ed., (1969), 279-347.
21. Synge, J. L., The Hypercircle in Mathematical Physics. Cambridge Univ. Press, 1957.
22. Trefftz, E., "Ein Gegenstück zum Ritzschen Verfahren." Proc. 2nd Intern. Congress Appl. Mech. Zürich (1926), 131-137.
23. Weinberger, H. F., "A Variational Computation Method For Forced Vibration Problems," Proc. Symp. Appl. Math. Amer. Math. Soc. VIII, (1958).

THE FINITE ELEMENT METHOD FOR ELLIPTIC DIFFERENTIAL EQUATIONS*

Ivo Babuška**

1. Introduction

The finite element method is a special method for the numerical solution of partial differential equations. The name was originated by engineers who used the method in structural mechanics. The finite element method became a very widely used method in practice. See e.g. [1][2] and many others. The theoretical investigation of different aspects began a few years ago. Nevertheless today many fundamental results are known. There is a variety of methods called the finite element method. In this paper we shall analyze one special approach aimed especially at solving elliptical equations. To avoid incidental technical difficulties, we shall explain the results and present numerical experiments for simple model problems. The majority of the results mentioned in this paper is contained in [3]-[11]. We shall formulate some of the results here in a less general way for the sake of simplicity.

---

* This research was supported in part by the National Science Foundation under Grant No. NSF GU 2061 and in part by the Atomic Energy Commission under Contract No. AEC AT(40-1) 3443/3.

** Institute for Fluid Dynamics and Applied Mathematics, University of Maryland.

2. Approximation Theory in R_n

Approximation theory plays a very important role in the investigation of the finite element method.

We denoted by R_n the n-dimensional Euclidean space: $\underline{x} = (x_1,\ldots,x_n)$, $|\underline{x}|^2 = \sum_{i=1}^{n} x_i^2$. Let $S(R_n)$ be the set of all rapidly decreasing functions (at ∞) with the usual topology (see [14], p. 146). The space of generalized functions over $S(R_n)$ will be denoted by $S'(R_n)$. For any $\psi \in S(R_n)$ we define the Fourier transform $F(\psi)(\underline{\sigma})$.

$$(2.1) \qquad F(\psi)(\underline{\sigma}) = \tilde{\psi}(\underline{\sigma}) \doteq \int_{R_n} e^{i<\underline{x},\underline{\sigma}>} \psi(\underline{x})d\underline{x}$$

with

$$<\underline{x},\underline{\sigma}> = \sum_{j=1}^{n} x_j\sigma_j .$$

The Fourier transform $F(f)$ of $f \in S'$ will be defined by the equation

$$(2.2) \qquad (F(f),F(\psi)) = (2\pi)^n(f,\psi) .$$

For reference, see e.g. [15].

Let $f \in L_2 \subset S'$ where L_2 is the space of all square integrable functions on R_n. Then $F(f) \in L_2$ and

$$(2.3) \qquad \|F(f)\|_{L_2}^2 = (2\pi)^n \|f\|_{L_2}^2 .$$

Let $\chi(x)$, $x \in [0,\infty)$ be not decreasing function with

$\chi(0) \neq 0$. Then the space $W_2^{\chi}(R_n)$ will be the space of all functions $f \in S'$ that

$$F(f)(\underline{\sigma})\chi(\|\underline{\sigma}\|) \in L_2(R_n)$$

and

$$(2.4) \qquad (2\pi)^n \|f\|^2_{W_2^{\chi}(R_n)} = \|F(f)(\underline{\sigma})\chi(\|\underline{\sigma}\|)\|^2_{L_2}$$

The most important special case is for $0 \leqslant \alpha < \infty$ and $\chi(\sigma) = 1 + |\sigma|^{\alpha}$. For this space we shall write simply $W_2^{\alpha}(R_n)$ and it is the fractional Sobolev space. With proper function $\chi(x)$ we may characterize the spaces of analytic functions.

The fundamental problem of approximation, which plays an essential role in the finite element method, is the following.

Let A be a linear mapping of R_n on R_n . (We shall also denote by A the matrix of the mapping.) Let $\omega(\underline{x}) \in S'(R_n)$ be a function with compact support and $\xi(x)$, $x \in [0,1]$ a not increasing function. Let $0 \leqslant \alpha \leqslant \bar{\alpha} \leqslant \beta$. Then the question is whether there exists a mapping $B_{\xi}(h)$, $0 < h \leqslant 1$ of $W_2^{\beta}(R_n)$ into $W_2^{\bar{\alpha}}(R_n)$, such that

$$(2.5) \qquad B_{\xi}(h)(f) = \sum_{\underline{k}} c(h,f,\underline{k})\,\omega\left[\xi(h)\left\{\frac{\underline{x} - A\underline{k}h}{h}\right\}\right]$$

and

$$(2.6) \qquad \|f - B_{\xi}(h)f\|_{W_2^{\alpha}(R_n)} \leqslant K\,h^{\mu}\,\|f\|_{W_2^{\beta}(R_n)}$$

for all $0 \leqslant \alpha \leqslant \bar{\alpha}$, where $\underline{k}$ is a multiinteger and μ depends in general on $(\omega, \xi, \alpha, \beta)$ and K does not depend on f and h . The case $\xi(h) = 1$ was investigated recently by few authors. See [3][16][17][18][19]. When $\xi(h) = 1$ we shall simple write $B(h)$ instead of $B_\xi(h)$. In this case there exists a number t depending on ω such that $\mu = \min(t-\alpha, \beta-\alpha)$. More precisely

Theorem 2.1. Let $\Lambda(\underline{x}) = F(\omega)(\underline{x})$ and

$$\text{(2.7)} \qquad 1) \qquad \Lambda(\underline{o}) \neq 0$$

$$\text{(2.8)} \qquad 2) \qquad |\Lambda(\underline{x} - 2\pi(A^T)^{-1}\underline{k}| \leqslant Z(\underline{k}) \, \|\underline{x}\|^t \ , \quad \underline{k} \neq \underline{o}$$

for all $\underline{x}$ such that

$$\text{(2.9)} \qquad \|\underline{x}\| \leqslant \|(A^T)^{-1}\| \, \pi \, n^{\frac{1}{2}}$$

$$\text{(2.10)} \qquad 3) \qquad \sum_{\underline{k}} Z^2(\underline{k}) \, \|\underline{k}\|^{2\bar{\alpha}} = D < \infty$$

then in (2.6) $[\xi(h) = 1]$

$$\text{(2.11)} \qquad \mu = \min(t-\alpha, \beta-\alpha)$$

It is possible to show that (2.11) is optimal. More exactly, we have

Theorem 2.2.

1) μ can never be greater than $\beta - \alpha$ in (2.6).
2) Let $\beta - \alpha > \mu$ and $\Lambda(\underline{o}) \neq 0$. Then (2.8) holds.

It is also possible to show that for every ω with compact support $\mu < T < \infty$, i.e. for sufficiently large β the first term in (2.11) is the dominant term, see [3]. The situation can be essentially different in the general case. The situation is described by the next theorem.

Theorem 2.3. Let $\varepsilon > 0$. There exists a function ω with compact support and function $\xi(x)$ such that there exists the operator $B_\xi(h)$ of the form (2.5) which maps $W_2^0(R_n)$ into $W_2^\infty(R_n)$ so that for any $f \in W_2^\beta(R_n)$, $0 \leqslant \alpha \leqslant \beta < \infty$, we have

$$\|f - B_\xi(h)f\|_{W_2^\alpha(R_n)} \leqslant K(\alpha,\beta,\varepsilon)\ \|f\|_{W_2^\beta(R_n)} h^\mu$$

with

$$\mu = \beta - \alpha - \varepsilon .$$

Together with this theorem see also [12]. We see that the zero condition (2.8) is not necessary in this case.

Theorems 2.1-2.3 illustrate the situation for the Sobolev spaces $W_2^\alpha(R_n)$ (i.e., $\chi(\sigma) = 1 + |\sigma|^\alpha$). The problem may be solved more generally in the spaces W_2^χ . Then on the left hand side we will have the norm in $W_2^{\chi_1}(R_n)$ and on the right $W_2^{\chi_2}(R_n)$. A similar theory can be developed here. Under proper assumptions we shall have on the right hand side instead of Kh^μ , a more general function $\psi(h)$ which depends on χ_1 and χ_2 (and certainly on ω) .

We may construct the operator $B(h)$ in (2.5) so that theorem 2.1 holds and this operator has the additional property that if f has compact support, then the support of $B(h)f$ lies in an Lh neighborhood of the support of f, and L does not depend on f and h and similar is true for the operator $B_\xi(h)$ (with L being a function of h now). These results can be generalized in different directions. An important way is to study instead of (2.5) the approximations with functions of the form

$$\sum_{j=1}^{r} \sum_{\underline{k}} c_j(h,f,k)\omega_j\left[\xi_j(h)\left(\frac{\underline{x} - A\underline{k}h}{h}\right)\right]$$

For more about this see [3].

Consider now a special case, with great practical importance. For $x \in R_1$, let

$$(2.12)\qquad \begin{aligned} \phi_1(x) &= 1 \quad \text{for} \quad |x| \leq \frac{1}{2} \\ \phi_1(x) &= 0 \quad \text{for} \quad |x| > \frac{1}{2} \end{aligned}$$

By a recurrence formula define

$$(2.13)\qquad \phi_{k+1}(x) = \phi_k(x) * \phi_1(x) .$$

These functions are piecewise polynomial. For a numerical construction of $\phi_k(x)$ see [11].

For $\underline{x} = (x_1,\dots x_n) \in R_n$, let us define

$$(2.14)\qquad \phi_k(\underline{x}) = \prod_{j=1}^{n} \phi_k(x_j) .$$

The function $\phi_k(\underline{x})$ fulfills the assumption of theorem 2.1 with $t = k$ and $\bar{\alpha} = k + \frac{1}{2} - \varepsilon$ with $\varepsilon > 0$ arbitrary.

It is possible to prove a lot of interesting (and useful) theorems about function $\phi_k(x)$. For example

<u>Theorem 2.4</u>. $\lim_{k\to\infty} \sqrt{k}\, \phi_k(x\sqrt{k}) = \frac{\sqrt{6}}{\pi} e^{-6x^2}$, $x \in R_1$.

<u>Theorem 2.5</u>. Let Ω be a bounded Lipschitz domain, $h > 0$ fixed, $W_2^\alpha(\Omega)$ the Sobolev space on Ω ,† $f \in W_2^\alpha(\Omega)$, $\alpha \geq 0$, Then for every $\varepsilon > 0$, there is an $N(f,\varepsilon)$ such that for any $j > N(f,\varepsilon)$ there exists coefficients $C_j(k)$ such that

$$\left\| f - \sum_{\underline{k}} c_j(\underline{k}) \phi_j\left(\frac{\underline{x} - \underline{k}h}{h}\right)\right\|_{W_2^\alpha(\Omega)} < \varepsilon \quad .$$

For some special results of this type see [12]. This theorem has practical importance which we shall see later.

3. <u>Boundary Value Problems with Natural Boundary Conditions</u>

Let Ω be a bounded domain, $\Omega \in R_n$, $\dot{\Omega}$ the boundary of Ω . The domain will be called a Lipschitz domain if $\dot{\Omega}$ can be locally expressed as a Lipschitz function of $n-1$ variables. See e.g. [20] or [4]. Ω will be said to be $C^\infty(\Omega \in C^\infty)$ if $\dot{\Omega}$ can be locally expressed as a function of $n-1$ variables with all derivatives continuous.

Let us now define the Sobolev fractional spaces. The spaces $W_2^\alpha(\Omega)$ (resp $W_2^\alpha(R_m)$), $\alpha \geq 0$ will be the usual Sobolev space. For α an integer we have

† For $W_2^\alpha(\Omega)$ see (3.1), (3.2), (3.3).

$$(3.1) \qquad \|u\|^2_{W_2^\alpha(\Omega)} = \sum_{|\underline{k}|\leq\alpha} \|D^{\underline{k}}u\|^2_{L_2(\Omega)}$$

where the sum is over all derivatives of order $0 \leq |\underline{k}| \leq \alpha$. For $\alpha = [\alpha] + \sigma$, $0 < \sigma < 1$ the fractional space was introduced by Aronszajn [22] and Slobodetskii [23]. It has the following norm

$$(3.2) \qquad \|u\|^2_{W_2^\alpha(\Omega)} = \|u\|^2_{W_2^{[\alpha]}(\Omega)} + \sum_{|k|=[\alpha]} \|D^{\underline{k}}u\|^2_{W_2^\sigma(\Omega)}$$

where

$$(3.3) \qquad \|u\|^2_{W_2^\sigma(\Omega)} = \int_\Omega\int_\Omega \frac{|u(t)-u(\tau)|^2}{\|t-\tau\|^{n+2\sigma}}\, dtd\tau \quad .$$

The space $W_2^\alpha(\Omega)$ can also be considered as an interpolated space between $W_2^{[\alpha]}(\Omega)$ and $W_2^{[\alpha]+1}(\Omega)$. We introduce the spaces $W_2^\alpha(\Omega^\bullet)$ for $\Omega \in C^\infty$ in a similar manner. Locally we transform the boundary on the hyperplain and then we define the space for $n-1$ dimension. For more about it see e.g. [21]. Let $\alpha < 0$. Then we define $W_2^\alpha(\Omega) = [W_2^{-\alpha}(\Omega)]'$ resp $W_2^\alpha(\Omega^\bullet) = [W_2^{-\alpha}(\Omega^\bullet)]'$. We denote by $\mathcal{D}(\Omega)$ the set of all functions of the class C^∞ with compact support in Ω and by $\mathring{W}_2^\alpha(\Omega)$, $\alpha \geq 0$ the closure of $\mathcal{D}(\Omega)$ in $\mathring{W}_2^\alpha(\Omega)$.

As a model problem let us seek a weak solution for the Neumann problem for $\Omega \in C^\infty$.

$$(3.4) \qquad -\Delta u + u = f$$

with boundary condition

(3.5) $$\frac{\partial u}{\partial n} = g$$

where $f \in W_2^{\alpha}(\Omega)$ and $g \in W_2^{\beta}(\dot{\Omega})$. The finite element method will be now the following. We define a bilinear form over $W_2^1(\Omega) \times W_2^1(\Omega)$ by

(3.6) $$B(u,v) = \int_{\Omega} \left(\sum_{i=1}^{n} \frac{\partial u}{\partial x_i} \frac{\partial v}{\partial x_i} + uv \right) d\underline{x}$$

We chose $0 < h < 1$ and $j \geqslant 2$, integral. We take the manifold $M(h)$ of all functions of the form

(3.7) $$v_h(x) = \sum_{\underline{k}} C(h,k)\phi_j\left(\frac{\underline{x}-\underline{k}h}{h} \right)$$

where the sum is taken over all $\underline{k}$, such

(3.8) $$\operatorname{supp} \phi_j\left(\frac{\underline{x}-\underline{k}h}{h} \right) \cap \Omega \neq 0$$

and $\phi_j(\underline{x})$ is defined by (2.14). As an approximate solution we take the function $u_h(\underline{x}) \in M(h)$ such that

(3.9) $$B(u_h,v) = \int_{\Omega} fv\, d\underline{x} + \oint gv\, d\underline{s}$$

holds for all $v \in M(h)$. Such an element is uniquely determined. To find it we have to solve a system of linear algebraic equations.

Let us study now the error $\varepsilon(h) = u_h(\underline{x}) - u(\underline{x})$. There is a well known theorem

<u>Theorem 3.1</u>. Let u be solution of (3.4) and (3.5), for $\alpha > -1$, $\beta > -1/2$. Then

$$u(\underline{x}) \in W_2^{\mu}(\Omega), \quad \text{with} \quad \mu = \min(\alpha+2, \beta + \frac{3}{2})$$

and

$$(3.10) \quad \|u(\underline{x})\|_{W_2^{\mu}(\Omega)} \leq K\left[\|f\|_{W^{\alpha}(\Omega)} + \|g\|_{W^{\beta}(\Omega^{\bullet})} \right],$$

where K depends only on Ω.

Now two theorems (similar to the Lax-Milgrane theorem) are useful (see [4]).

Theorem 3.2. Let H_1 and H_2 be two Hilbert spaces with scalar product $(\cdot,\cdot)_{H_1}$ (resp $(\cdot,\cdot)_{H_2}$. Further let $B(u,v)$ be bilinear form on $H_1 \times H_2$, $u \in H_1$, $v \in H_2$ such that

$$(3.11) \quad |B(u,v)| \leq C_1 \|u\|_{H_1} \|v\|_{H_2},$$

$$(3.12) \quad \sup_{\substack{u \in H_1 \\ \|u\|_{H_1} \leq 1}} |B(u,v)| \geq C_2 \|v\|_{H_2},$$

$$(3.13) \quad \sup_{\substack{v \in H_2 \\ \|u\|_{H_2} \leq 1}} |B(u,v)| \geq C_3 \|u\|_{H_1}$$

with $C_2 > 0$, $C_3 > 0$, $C_1 < \infty$. Further let $f \in H_2'$, i.e., f is a linear functional on H_2. Then there exists exactly one element $u_0 \in H_1$ such that

$$B(u_0,v) = f(v)$$

for all $v \in H_2$ and

(3.14) $$\|u_0\|_{H_1} \leq \frac{1}{C_3} \|f\|_{H_2'(\Omega)} .$$

Theorem 3.3. Let the assumptions of theorem 3.2 be fulfilled. Further let there be given the closed subspaces $M_1 \in H_1$, $M_2 \in H_2$ and let for every $v \in M_2$

(3.15) $$\sup_{\substack{u \in M_1 \\ \|u\|_{H_1} \leq 1}} |B(u,v)| \geq C_2'(M_1,M_2) \|v\|_{H_2}$$

with $C_2'(M_1,M_2) > 0$, and for every $u \in M_1$

(3.16) $$\sup_{\substack{v \in M_2 \\ \|v\|_{H_2} \leq 1}} |B(u,v)| \geq C_3'(M_1,M_2) \|u\|_{H_1}$$

with

$$C_3'(M_1,M_2) > 0 .$$

Let $f \in H_2'$ be given, and let u_0 denote element of H_1 such that

(3.17) $$B(u_0,v) = f(v)$$

for all $v \in H_2$ (such an element exists and is unique by the theorem 2.1). Suppose there exists a $w \in M_1$ such that

$$(3.18) \qquad \|u_0 - w\|_{H_1} \leq \delta$$

Further let $\hat{u}_0 \in M_1$ be such that

$$(3.19) \qquad B(\hat{u}_0, v) = f(v)$$

for all $v \in M_2$. Then

$$(3.20) \qquad \|u_0 - \hat{u}_0\|_{H_1} \leq \left[1 + \frac{C_1}{C_3'(M_1, M_2)} \right] \delta$$

Using theorem 3.1, the Calderon continuation theorem and theorems 3.2, 3.3, 2.1 in together with (2.14) we get

Theorem 3.4. Let $\alpha > -1$, and $\beta > -1/2$. Then

$$(3.21) \qquad \|u_h(\underline{x}) - u(\underline{x})\|_{W_2^1(\Omega)} \leq C\left[h^{\mu_1} \|f\|_{W_2^{\alpha}(\Omega)} + h^{\mu_2} \|g\|_{W_2^{\beta}(\Omega^{\bullet})} \right]$$

where

$$(3.22) \qquad \mu_1 = \min[j-1, \alpha+1] ,$$

$$(3.23) \qquad \mu_2 = \min[j-1, \beta + \tfrac{1}{2}] .$$

Now the question is whether the rates μ_1 and μ_2 are optimal. To answer this question we shall study a concept called the n-widths. Let us introduce the notion of n-width. This notion which was introduced by Kolmogorov [24] has received considerable attention in recent years together with its applications in the theory of numerical methods. See e.g. [25][26][27][28]. Let H be a Hilbert space and $G \subset H$ a set. Then the m-width $d_m(H,G)$ of G

$$(3.24)\quad d_m(H,G) = \inf\{E(G,\mathcal{M});\ \mathcal{M} \subset H,\ \dim \mathcal{M} = m\}$$

where

$$(3.25)\qquad E(G,\mathcal{M}) = \sup_{f\in G}\ \inf_{g\in\mathcal{M}} \|f-g\|_H$$

and $\mathcal{M}$ is linear manifold. Putting $H = W_2^1(\Omega)$,

(3.26)
$$G = \{-\Delta u+u = f\ ,\ \frac{\partial u}{\partial n} = 0 \text{ on } \Omega^{\bullet}\ ,\ \|f\|_{W_2^{\alpha}(\Omega)} \leqslant 1\}$$

it is possible to show that in this case (see [5]).

$$(3.27)\qquad d_m(H,G) \geqslant Cm^{-\frac{\alpha+1}{n}}$$

We see from (3.27) that μ_1 in (3.22) is the optimal rate of convergence. To study the optimality of μ_2 we put $H = W_2^1(\Omega)$, and

$$(3.28)\qquad G = \left\{-\Delta u+u = 0,\ \left\|\frac{\partial u}{\partial n}\right\|_{W_2^{\beta}(\Omega^{\bullet})} \leqslant 1\right\}.$$

It is possible to show (see [5]) that now

$$(3.29)\qquad d_m(H,G) \geqslant Cm^{-\frac{\beta+\frac{1}{2}}{n-1}}\ .$$

By comparing (3.29) with (3.23) we see that we may, perhaps, get instead of μ_2 in (3.23) the rate $\mu_2' = \min\left[j-1, \left(\beta + \frac{1}{2}\right)\left(\frac{n}{n-1}\right)\right]$. Before we analyze this question let us make one remark. Why are we interested to get high order of convergence? It is because the parameter h is an indication of the amount of work that we have to do. Certainly, it is a very rough measure but it is of use. In the connection we are concerned with the number of unknowns, which are solution at the solution of a system of linear algebraic equations. The number of unknowns is obviously in order $|\Omega|h^{-n}$ where $|\Omega|$ is the volume of Ω. On the other hand in practice one very often uses the net (element) refinements. How to measure the amount of work then? The parameter h is now obviously an improper one. But we may introduce an effective h, that we shall denote by H. We mentioned that the number of unknowns $N = |\Omega|h^{-n}$. So neglecting a multiplication constant we have

$$(3.30) \qquad h = \left(\frac{|\Omega|}{N}\right)^{\frac{1}{n}}$$

Now let us assume that by using refinement techniques we have to find N unknown by solving a linear algebraic system. So we may introduce an effective h by the formula

$$(3.31) \qquad H = \left(\frac{|\Omega|}{N}\right)^{\frac{1}{n}}$$

Let us go back to theorem 3.3 and to the question of the optimality of the rate of convergence.

It is possible to show that with a regular net it is impossible to improve the rate. But another situation will happen if we investigate a proper refinement. For this reason instead of (3.7) we take

$$v_h(\underline{x}) = \sum_{\ell=0}^{N(h)} \sum_{k} C\,(h,\underline{k})\phi_j\left(\frac{x-h\lambda^{\ell}\underline{k}}{\lambda^{\ell}h}\right) \tag{3.32}$$

where $0 < \lambda < 1$,

$$N(h) = \left[\frac{1}{n-1}\left|\frac{\ell gh}{\ell g\lambda}\right|\right]. \tag{3.33}$$

The second sum for $\ell = 0$ is taken for all $\underline{k}$ such that

$$\operatorname{supp}\phi_j\left(\frac{\underline{x}-h\underline{k}}{h}\right) \cap \Omega \neq \emptyset \tag{3.34}$$

and sum for $0 < \ell \leqslant N(h)$ is taken for all $\underline{k}$ such that

$$\operatorname{supp}\phi_j\left(\frac{\underline{x}-h\underline{k}\lambda^{\ell}}{h\lambda^{\ell}}\right) \cap (\Omega-\Omega_{\lambda}n\ell) \neq \emptyset \tag{3.35}$$

where $\Omega_s = E[\underline{x}\ \ \Omega,\ \rho(\underline{x}_1\Omega^{\bullet}) > s]$ and $\rho(x_1\Omega^{\bullet})$ is the distance of x to $\Omega^{\bullet}$.

Now using a procedure analogous to the previous case we get from certain basic properties (see [30]) of the solutions of elliptic equations the following theorem.

Theorem 3.5. Let $\alpha > -1$, $\beta > 1/2$. Then

$$\|u_h-u\|_{W_2^{1}(\Omega)} \leqslant C(\varepsilon)\left[H^{\mu_1-\varepsilon}\|f\|_{W_2^{\alpha}(\Omega)} + H^{\mu_2-\varepsilon}\|g\|_{W_2^{\beta}(\Omega^{\bullet})}\right] \tag{3.36}$$

where

$$\mu_1 = \min[j-1,\ \alpha+1] \tag{3.37}$$

$$\mu_2 = \min\left[j-1,\ \left(\beta+\frac{1}{2}\right)\frac{n}{n-1}\right] \tag{3.38}$$

for arbitrary $\varepsilon > 0$.

We see now that we get the optimal rate of convergence (provided that we neglect ε). Let us make several remarks. It is very important that the ratio between the greatest and smallest meshsize converge to infinite. This is a necessary condition to get the optimal rate of convergence. All that we have said may be generalized for higher order equations and systems of elliptic type. The norm on the left hand side of (3.36) and (3.21) will be the natural "energy" norm. It is important that the boundary conditions are natural.

In all that we have said the natural error bound is in the "energy" norm. In our problem however we have the norm in the space $W_2^1(\Omega)$. Now there is the problem to determine error bounds in other spaces. Let us only discuss here error-bounds in the spaces $W_2^\alpha(\Omega)$. Let us assume that the "energy" norm is equivalent with the norm in $W_2^{\alpha_0}(\Omega)$. Then we have to distinguish two different cases. The norm in $W_2^\alpha(\Omega)$ for $0 \leqslant \alpha \leqslant \alpha_0$ and for $\alpha \geqslant \alpha_0$. If we go "down" i.e., consider $0 \leqslant \alpha \leqslant \alpha_0$ then in general we get better error bounds of the type

$$\|u_h-u\|_{W_2^\alpha(\Omega)} \leqslant Ch^{\mu+\alpha_0-\alpha}\,\|f\|_{W_2^\beta(\Omega)} \tag{3.39}$$

where μ is the rate of convergence for the "energy" norm of the error. So e.g. in our case the estimate (3.21) in theorem 3.4 in the space $L_2 = W_2^0$ will be

$$(3.40)\qquad \|u_h-u\|_{L_2(\Omega)} \leq C\left[h^{\mu_1+1}\|f\|_{W_2^\alpha(\Omega)} + h^{\mu_2+1}\|g\|_{W_2^\beta(\Omega^\bullet)}\right].$$

We will not discuss the matter in detail. We only emphasize that for the estimate (3.40) we need a proper j in (3.7). If the j is not high enough (with resp of the problem) then (3.39) would not be true. In our model problem, $j \geq 2$ is enough. Similar results can be obtained in the case of refined mesh size. The case for $\alpha \geq \alpha_0$ is more complicated. In general it is possible to prove that the following theorem

<u>Theorem 3.6</u>. Let $\alpha > -1$, $\beta > -1/2$, $0 \leq s \leq \alpha+1$, $0 \leq s \leq \beta+1/2$, $0 \leq s \leq j-1$. Then

$$(3.41)\qquad \|u_h-u\|_{W_2^{s+1}(\Omega^*)} \leq C(\Omega^*)\left[h^{\mu_1-s}\|f\|_{W_2^\alpha(\Omega)} + h^{\mu_2-s}\|g\|_{W_2^\beta(\Omega^\bullet)}\right],$$

with μ_1 and μ_2 given by (3.22) and (3.23), and Ω^* any domain such that $\bar{\Omega}^* \subset \Omega$. A different problem will occur if we are interested in the norm of $W_2^{s+1}(\Omega)$. It is not known whether (3.41) holds for $\Omega^* = \Omega$ for general Ω.

Let us follow another way. Let $\psi(\underline{x})$ $j = 1, 2, \ldots, p$ be p functions with continuous derivatives of all orders, compact support and $\sum_{j=1}^{p} \psi_j(\underline{x}) = 1$ on Ω and supp $\psi_j(\underline{x}) \cap \dot{\Omega}$ can be expressed as a function of $n-1$ variables. Further let $u_h(x) = \sum C(h,k)\phi_j\left(\frac{\underline{x}-\underline{k}h}{h}\right)$. Let $0 \leqslant s = \min(\alpha+1, \beta+1/2)$ and $q \geqslant s$ be an integer. Then we may create p functions

$$(3.42) \qquad u_{h,q,\ell}(\underline{x}) = \sum_{\underline{k}} d_\ell(h,\underline{k})\phi_{j+q}\left(\frac{\underline{x}-\underline{k}h}{h}\right)$$

with

$$(3.43) \qquad d_\ell(h,\underline{k}) = \sum_{\underline{r}} a_\ell(\underline{r})C(h,\underline{k}-\underline{r})$$

(with only a few terms in (3.43)). Then putting

$$(3.44) \qquad \hat{u}_h(\underline{x}) = \sum_{\ell=1}^{p} \psi_\ell(\underline{x})u_{h,q,\ell}(\underline{x})$$

we get theorem 3.7

Theorem 3.7. Let $\alpha > -1$, $\beta > -1/2$, $s_0=\min(\alpha+1, \beta+1/2)$, $q \geqslant s_0$ an integer. Then for $s < s_0$

$$\|\hat{u}_h(\underline{x})-u(\underline{x})\|_{W_2^{s+1}(\Omega)}$$

$$\leqslant C\left[h^{\mu_1-s}\|f\|_{W_2^{\alpha}(\Omega)} + h^{\mu_2-s}\|g\|_{W_2^{\beta}(\dot{\Omega})}\right]$$

with μ_1 and μ_2 given by (3.22) and (3.23). To create the function $u_h(\underline{x})$ is a very cheap operation from a practical standpoint. We will not go into details concerning the coefficients $a_\ell(\underline{r})$, see [9].

We have been interested in error bounds in terms of rates of convergence as a function of h i.e. for $h \to 0$ with fixed j in (3.7). From a practical point of view there is a very important question of how to choose an optimal (or good) j . This question is very important and very delicate. We shall discuss it together with some numerical experiment later. Here we shall only formulate one theorem.

Theorem 3.8. Let $\alpha > -1$, $\beta > -1/2$, $h_0 > 0$ fixed, $\Omega \in C^\infty$. Let $u_j(\underline{x})$ be the solution $u_{h_0}(\underline{x})$ in the form (3.7) with given h_0 and $j = 2,3,\ldots$. Then

$$\lim_{j\to\infty} \|u_j(\underline{x})-u(\underline{x})\|_{W_2^1(\Omega)} = 0 .$$

4. Boundary Value Problem with Natural Boundary Conditions and Singularities on the Boundary

A case which is ve ' important in practice is one where the domain has corners and edges, i.e. the case when the boundary is piecewise very smooth. Another case which is essentially similar is the case of mixed boundary conditions, different type on different parts of the boundary.

Many of these cases especially in two dimensions may be processed by the approach of weighted spaces. So let Ω be a Lipschitz domain. Let $\underline{\underline{y}} \equiv (\underline{y}_1, \ldots, y_p)$,

$\underline{y}_i \in \dot{\Omega}$, $\underline{y}_i \neq \underline{y}_j$ for $i \neq j$, $\underline{\alpha} \equiv (\alpha_1, \ldots, \alpha_p)$, $\alpha_i \geq 0$, k an integer. Let us introduce the space $W^{k,s}_{2,\underline{\underline{y}},\underline{\alpha}}(\Omega)$ $s \geq 0$ as the space of all functions with finite norm $\|u\|_{W^{k,s}_{2,\underline{\underline{y}},\underline{\alpha}}(\Omega)}$ where

$$(4.1)\qquad \|u\|^2_{W^{k,s}_{2,\underline{\underline{y}},\underline{\alpha}}(\Omega)} = \|u\|^2_{W^s_2(\Omega)} + \sum_{|q|=k} \int_\Omega \left[\min\left(1, \prod_{j=1}^{p} \rho^{\alpha_j}(x,y_j)\right) \right]^2 [D^q u]^2 \, dx$$

and where $\rho(x,y_j)$ is the distance of the points x and y_j. As a model problem let us now investigate the equation

$$(4.2)\qquad -\Delta u + u = f$$

with the homogeneous boundary condition

$$(4.3)\qquad \frac{\partial u}{\partial n} = 0 \quad \text{on} \quad \dot{\Omega}\ .$$

Theorem 3.1 does not hold for domains with corners. Many times an analogous theorem will hold using the spaces $W^{\ell,s}_{2,\underline{\underline{y}},\underline{\alpha}}$.

For our purposes let us introduce a (k,s,y,β) regular Neumann problem. Let Ω be a Lipschitz domain, $\underline{\underline{y}} = (\underline{y}_1, \ldots, \underline{y}_p)$, $\underline{y}_i \in \dot{\Omega}$, $y_i \neq y_j$ for $i \neq j$, $\underline{\beta} = (\beta_1, \ldots, \beta_p)$, $\beta_i \geq 0$.

Then the Neumann problem (4.2), (4.3) will be said to be $(k,s,y,\underline{\beta})$ regular if its weak solution $u \in W_2^1(\Omega)$ is such that

(4.4) $$\max_{j=2,\ldots,k}\left[\,\|u\|_{W^{j,s}_{2,\underline{\underline{y}},\underline{\beta}+j-2}}\right] \leq C\,\|f\|_{W_2^{k-2}(\Omega)}$$

where

$$\beta+j-2 = (\beta_1+j-2,\ \ldots,\ \beta_p+j-2)\ .$$

The Neumann problem is regular in the mentioned sense, for example in two dimensions when the boundary is piecewise smooth and is composed of straight lines at corners. See [32].

In this case we may utilize the idea of refining the meshsize in the neighborhood of the points y_i. For details see [6].

It is possible to prove the following theorem.

Theorem 4.1. Let $u_h(\underline{x})$ be the approximate solution of the $(k,s,\underline{y},\underline{\beta})$ regular Neumann (4.2), (4.3) problem with $k \geq 2$, $\beta = (\beta_1,\ \ldots,\ \beta_p)$, $\beta_1 \leq \beta_0 < 1$, $s > 1$ by the finite element method with the right refinement around the points y_i. Then

(4.5) $$\|u_h-u\|_{W_2^1(\Omega)} \leq CH^{k+1}\,\|f\|_{W_2^k(\Omega)}$$

for $k \geq 0$.

We see that in many cases (e.g. domains with corners on the boundary) using a proper refinement we are able to get the same rate of convergence as in the case without singularities, (e.g. smooth domains).

A similar approach can be used in many cases and this creates a background for dealing with singularities of very different kind. Using this approach it is possible to solve the problem with unbounded domain and many other cases.

5. Boundary Value Problem with Dirichlet Boundary Conditions -- The Case of the Lipschitz Domain

Let Ω be a Lipschitz domain and let us be interested in the approximate solution of the Dirichlet problem for the Laplace's equation as a model problem

$$(5.1) \qquad -\Delta u = f$$

$$(5.2) \qquad u = 0 \quad \text{on} \quad \dot{\Omega} .$$

The following theorem is valid (see [31]).

Theorem 5.1. Let $-1/2 < \alpha < 1/2$. Then there exists a $\mathring{W}_2^{1+\alpha}(\Omega)$ for every $f \in W_2^{-1+\alpha}(\Omega)$ and

$$(5.3) \qquad \|u\|_{W_2^{1+\alpha}(\Omega)} \leq C \, \|f\|_{W_2^{-1+\alpha}(\Omega)}$$

where C does not depend on f.

Let us describe the finite element method for this case. In a manner similar to (3.7) we take for $j \geq 3$

$$v_h(\underline{x}) = \sum_{\underline{k}} C(h,\underline{k})\phi_j\left(\frac{\underline{x}-\underline{k}h}{h}\right) \tag{5.4}$$

and the sum in (5.4) is over all $\underline{k}$ such that

$$\text{supp } \phi_j\left(\frac{\underline{x}-\underline{k}h}{h}\right) \subset \bar{\Omega} . \tag{5.5}$$

The solution $u_h(\underline{x})$ will be the function of the form (5.4) such that

$$\int_\Omega \sum_{i=1}^{n} \frac{\partial u_h}{\partial x_i}\frac{\partial v_h}{\partial x_i}\, d\underline{x} = \int f\, v_h\, d\underline{x} \tag{5.6}$$

holds for all functions of the form (5.4). As in Chapter 3 to determine the coefficients $C(h,\underline{k})$ we have to solve a system of linear algebraic equations. On the right hand side of the system we have the terms

$$\int f\, \phi_j\left(\frac{\underline{x}-h\underline{k}}{h}\right) d\underline{x} .$$

Because $\phi_j \in W_2^2(\Omega)$ the integral makes sense for $f \in W_2^{-2}(\Omega)$. So in R_2, f could be a Dirac's function and the solution would be the Green function. But let us state the theorem about the error of the approximate solution.

Theorem 5.2. We have

$$\|u-u_h\|_{W_2^\beta(\Omega)} \leq Ch^\mu \|f\|_{W_2^\alpha(\Omega)} \tag{5.7}$$

where

(5.8) $\quad$ for $-1/2 > \alpha \geqslant -1, \quad 3/2 > \beta \geqslant 1,$

$$\alpha - 2\beta + 3 \geqslant 0, \quad \mu = \alpha + 1 - 2(\beta - 1)$$

(5.9) $\quad$ for $-1/2 > \alpha \geqslant -1, \quad 0 \leqslant \beta \leqslant 1,$

$$\mu = \alpha + 1 + (1-\beta)/2 - \varepsilon \quad \text{for} \quad \varepsilon > 0$$

and arbitrary,

(5.10)

for $\alpha \leqslant -1, \quad 0 \leqslant \beta \leqslant \alpha + 2, \quad -3/2 < \alpha, \quad 5/2 + 2\alpha > \dfrac{(\frac{3}{2} + \alpha)\beta}{2+\alpha}$

$$\mu = 5/2 + 2\alpha - \frac{(\frac{3}{2} + \alpha)\beta}{2+\alpha} - \varepsilon \quad \text{for} \quad \varepsilon > 0$$

and arbitrary, see [4].

Let us mention some special cases of theorem 5.2, using some well known embedding theorems.

For $n = 2$ (i.e. in the plane) we get

(5.11) $$\|u-u_h\|_{L_\infty} \leqslant Ch^{\mu'} \|f\|_{W_2^{\alpha'}(\Omega)}$$

for $-1/2 > \alpha' > -1, \quad \mu' = \alpha' + 1 - \varepsilon, \quad \varepsilon > 0$ and arbitrary. It is interesting to compare this result with Laasonen's [33], [34] upper bound and Volkov's [35] lower bound for the rate of convergence of the finite difference method commonly called the five point formula. Using arguments similar to those of Laasonen and Volkov we see that our result

$$\|u-u_h\|_{L_\infty} \leq Ch^{\frac{1}{2}-\varepsilon} \|f\|_{W_2^{-\frac{1}{2}}(\Omega)} \tag{5.12}$$

is an almost optimal one (see also [36]). Another interesting case for $n = 2$ is the case when $f(\underline{x}) = \Theta(x)$ and $\Theta(x)$ in a Dirac function. Then the exact solution is the Green function and we see that the convergence in L_2 is $h^{\frac{1}{2}-\varepsilon}$.

Let us mention, that we have considered only a model problem. But these arguments may be used to prove more general theorems.

6. The Solution of the Dirichlet Problem by the Perturbed Variational Principle

The Dirichlet boundary conditions are essential boundary conditions; using classical approach we must fulfill them exactly. This causes many troubles in practice. We shall show an approach that avoids these difficulties. For simplicity we shall show it on the model problem

$$-\Delta u = f \tag{6.1}$$

$$u = g \quad \text{on} \quad \dot{\Omega} \tag{6.2}$$

where $\dot{\Omega} \in C^\infty$; for more about this see [7].

Let us describe the method. Take as in Chapter 3

$$(6.3) \qquad v_h(\underline{x}) = \sum_k C(h,\underline{k})\phi_j\left(\frac{\underline{x}-\underline{k}h}{h}\right)$$

where the sum will be over all h such that

$$\operatorname{supp} \phi_j \left(\frac{\underline{x}-\underline{k}h}{h}\right) \cap \Omega \neq 0$$

The solution $u_h(\underline{x})$ will be the solution of the form 6.3 such that

$$(6.4) \qquad \int_\Omega \sum_{i=1}^n \frac{\partial u_h}{\partial x_i}\frac{\partial v_h}{\partial x_i}\, d\underline{x} = \int_\Omega f\, v_h\, d\underline{x} - \psi(h)\oint_{\dot\Omega} (u_h - g)v_h\, d\underline{s}\ .$$

holds for all $v_h(x)$ of the form (6.3) where $\psi(h) = Ch^{-\sigma}$, $C > 0$, $\sigma > 0$. Now the following theorem holds (see [7]).

Theorem 6.1. Let $k > -1/2$, $s > 1$, $j \geq 2$, $f \in W_2^k(\Omega)$, $g \in W_2^s(\dot\Omega)$ in (6.1) and (6.2). Then

$$(6.5) \qquad \|u-u_h\|_{W_2^1(\Omega)} \leq Ch^\mu \left[\|f\|_{W_2^k(\Omega)} + \|g\|_{W_2^s(\Omega)} \right]$$

with

$$\mu = \min\left[k+1,\ s-\frac{1}{2},\ k+\frac{3}{2}-\frac{\sigma}{2}-\varepsilon,\ s-\frac{\sigma}{2}-\varepsilon,\ \frac{\sigma}{2},\ k-1,\ k-\frac{1}{2}-\frac{\sigma}{2}-\varepsilon\right]$$

for $\varepsilon > 0$ and arbitrary.

This principle which may also be called the penalty approach may be used very generally. As a special case let us mention the approach for solving elliptical equations with discontinuous coefficients, higher order equations and other cases; see [8].

We have shown on a few model problems the different ideas and results connected with the finite element method. It is possible to combine these ideas in different ways.

7. Computational Aspects and Numerical Experiments

Computation with the finite elements method has four stages.

1. Computing of the function $\phi_j(x)$.
2. Creating the system of linear equations.
3. Solution of the system.
4. Computing the approximate solution.

Part 1 is very well covered by procedure described in [11]. Part 2 has difficulties especially in programming. Part 3 is the problem of solving general space matrices. In the case refinement of the elements is used, the system is generally very complicated. It seems to be, that it is very difficult to find effective iterative methods in general cases. So it appears to be that special elimination procedures for the sparce matrices have promise for the future. See e.g. [13], [37] and others. Part 4 has in general no difficulties.

All that we have said has been concerned with exact computations, i.e. computations without round-off errors. Round-off errors change the situation. These questions have not been studied enough yet. But it is possible to apply to them the general ideas explained in [38].

Let us now show some results of numerical experiments; for more about this see [10].

We shall be interested in the problem

$$(7.1) \qquad -y'' + Cy = -\sin D\left(x - \frac{\pi}{2} \right) ,$$

$$C > 0, \quad D > 0,$$

with boundary conditions

$$(7.2) \qquad y'(0) = y'(\pi) = 0 .$$

The method for solving this problem was described in Chapter 3. The method is without refinement. The system of linear equations was solved by elimination method (band-matrix procedure).

First let us study the case of exact computation (without round-off errors).

On the basis of results in Chapter 3 and because of the analyticity of the right hand side we have

$$(7.3) \qquad \|u_h - u\|_{W_2^\alpha(\Omega)} \leqslant Ch^{j-\alpha} \text{ for } \alpha \leqslant j, \quad j \geqslant 2$$

and C does not depend on h.

From experiments we have that

$$(7.4) \qquad \frac{\|u_h - u\|_{W_2^\alpha(\Omega)}}{h^{j-\alpha}} \approx C$$

By Banach space interpolation theory theorems we have

$$(7.5) \qquad \|u_h - u\|_{L_\infty} \leqslant Ch^{j-\frac{1}{2}}$$

The experiments show that

$$(7.6) \qquad \|u_h - u\|_{L_\infty} \approx Ch^{j} .$$

Typical behavior may be seen in Figures 1a and 1b.

For the error at one fixed point x (7.6) does not hold. See Figure 2 in this case.

It is advantageous to use high order method (big j) and in Figure 3 we see the error depending on j.

The character of the error is oscillatory. In Figures 4a,b,c we see the error (scale on the left hand side, and the exact solution with scale on the right hand side. The grid points are the asterisks on the axes.

For k odd the grid points coincide (inside the interval) with the zeros of the error. The rate of convergence in L_2 and L_∞ in grid points are of order h^{j+1} (instead of h^j), provided that the grid point is not close to the boundary.

There is a boundary layer effect with the width of order h, which changes the behavior in comparison with the behavior inside. The boundary layer can best be seen on derivatives. A typical case is presented in Figure 5, which shows these details for the second derivative.

Let us study the influence of the round-off errors.

There exists a critical h (depending on the length of the computer word) for which we get the minimal error. In Figure 6 we have the error for the computation in simple and double precision.

The error has a similar behavior in its dependence on j (order of method).

Because of the positive definiteness of the matrix no great troubles arise with very small pivots through the elimination method (with computation in normalized floating points). For more about these questions see [10].

REFERENCES

1. Zienkiewicz, O. C., The Finite Element Method in Structural and Continuum Mechanics, London, McGraw-Hill, 1967.
2. Rashid, Y. R., "On Computational Methods in Solid Mechanics and Stress Analysis," Conf. on Effective Use of Comp. in the Nuclear Industry, April 21-23, 1969, Knoxville.
3. Babuška, I., "Approximation by Hill Functions," Tech. Note BN-648, 1970, University of Maryland, Institute for Fluid Dynamics and Applied Mathematics.
4. Babuška, I., "Error Bounds for Finite Element Method," Technical Note BN-630, 1969, University of Maryland, Institute for Fluid Dynamics and Applied Mathematics.
5. Babuška, I., "The Rate of Convergence for the Finite Element Method," Technical Note BN 646, 1970, University of Maryland, Institute for Fluid Dynamics and Applied Mathematics.
6. Babuška, I., "Finite Element Method for Domains with Corners," Tech. Note BN-636, 1970, University of Maryland, Institute for Fluid Dynamics and Applied Mathematics.
7. Babuška, I., "Numerical Solution of Boundary Value Problems by the Perturbed Variational Principle, Tech. Note BN-624, 1969, University of Maryland, Institute for Fluid Dynamics and Applied Mathematics.

8. Babuška, I., "The Finite Element Method for Elliptic Equations with Discontinuous Coefficients," Tech. Note BN-631, 1969, University of Maryland, Institute for Fluid Dynamics and Applied Mathematics.
9. Babuška, I., "Computation of Derivatives in the Finite Element Method," Tech. Note BN-650, 1970, University of Maryland, Institute for Fluid Dynamics and Applied Mathematics.
10. Babuška, I., K. Segeth and J. Segethová, "Numerical Experiments and Problems Connected with Finite Element Method," to appear.
11. Segethová, J., "Numerical Construction of the Hill Functions," Tech. Rep. 70-110-NGL-21-002-008, 1970, University of Maryland, Computer Science Center.
12. Segeth, K., "Problems of Universal Approximation by Hill Functions," Communication, SYNSPADE, 1970.
13. Segethová, J., "Elimination for Sparce Symmetric Systems of Special Structure," Communication, SYNSPADE, 1970.
14. Yosida, K., Functional Analysis, New York, Academic Press, 1965.
15. Gel'fand, I. M. and G. M. Shilov, Generalized Functions (Translated from Russian), Vol. 1, Vol. 2, Academic Press, New York.
16. Fix, G and G. Strang, "Fourier Analysis of the Finite Element Method in Ritz-Galerkin Theory," Studies in Applied Mathematics, 48 (1969), 265-273.
17. Strang, G. and G. Fix, "A Fourier Analysis of the Finite Element Variational Method." To appear.
18. Strang, G., "The Finite Element Method and Approximation Theory," Proc. of SYNSPADE, 1970. To appear.
19. Di Guglielmo, F., "Construction d'approximations des espaces de Sobolev sur des réseaux en simplexes." Calcolo, Vol. 6 (1969), 279-331.
20. Nečas, J., Les méthods directes en théorie des équations elliptiques, Academia, Prague, 1967.
21. Lions, J. L, and E. Magenès, "Problèmes aux limites non homogènes, Dunod, 1968.
22. Aronszajn, N., "Boundary Value of Functions with Finite Dirichlet Integral," Conf. on Partial Differential Equations, No. 14, University of Kansas, 1955.
23. Slobodeckiĭ, M. I., "Generalized Sobolev space and Their Application to Boundary Problems for Partial Differential Equations," Leningr. goš. Univ. 197 (1958), pp. 54-112.

24. Kolmogorov, A. N., "Über die beste Annäherung von Funktionen einer gegebenen Funktionklasse," Ann. Math. (2), 37, 1936, 107-111.
25. Golomb, M., "Splines, n-widths and Optimal Approximations MRC," Tech. Summary Rep. 784, Sept. 1967.
26. Babuška, I., S. L. Sobolev, "The Optimization of Numerical Processes," (Russian), Aplikace Mat. 96-129 (1965).
27. Jerome, J., "On the L_2 n-width of Certain Classes of Functions of Several Variables," J. Math. Anal. Appl. 20 (1967), 110-123.
28. Jerome, J., "On n-width in Sobolev Spaces and Applications to Elliptic Boundary Value Problems MRC," Tech. Sum. Rep. 917, August, 1968.
29. Aubin, J. P., "Approximations of Non-homogeneous Neumann Problems - Regularity of the Convergence and Estimates of Errors in Terms of n-width," MRC, Tech. Summary Rep. 924, August 1968.
30. Focht, A. S., "Some Inequalities for the Solution and Its Derivatives of an Equation of Elliptic Type in L_2 Norm," (Russian), Proc. of the Steklov, Inst. of Math., 1965, 77, 160-191.
31. Nečas, J., "Sur la coercivité des formes sesquiliveaires elliptiques," Rev. Roumaine de Math. Pure et App. Vol. TX, No. 1, 1964, pp. 47-69.
32. Kondrat'ev, V. A., "Boundary Values Problems for Elliptical Equations on the Domains with Cones" (in Russian), Proc. of the Moscow Math. Soc. v. 16 (1967), 209-292.
33. Laasonen, P., "On the Degree of Convergence of Discrete Approximation for the Solution of the Dirichlet Problem," Ann. Acad. Sci. Finn. Ser. A, I, No. 246 (1957).
34. Laasonen, P., "On the Truncation Error of Discrete Approximations into the Solution of Dirichlet Problems in a Domain with Corners," J. Assoc. Comp. Math. 5 (1958), pp. 32-38.
35. Volkov, E. A., "Method of Composit Meshes for Bounded and Unbounded Domain with Piecewise Smooth Boundary," (in Russian) Proc. of the Steklov, Institute of Math., Moscow, No. 96 (1968), pp. 117-148.
36. Veidinger, L., "On the Order of Convergence of Finite Difference Approximation to the Solution of Dirichlet Problem in a Domain with Corners," Studia Sc. Mat. Hung. 1968, 3, No. 1-3, 337-343.

37. Brayton, R. K., F. G. Gustavson, and R. A. Willoughby, "Some Results on Sparce Matrices," IBM Watson Res. Center, to appear.
38. Babuška, I., M. Práger, E. Vitásek, Numerical Processes in Differential Equations, J. Wiley, London, New York, Sydney, 1966.

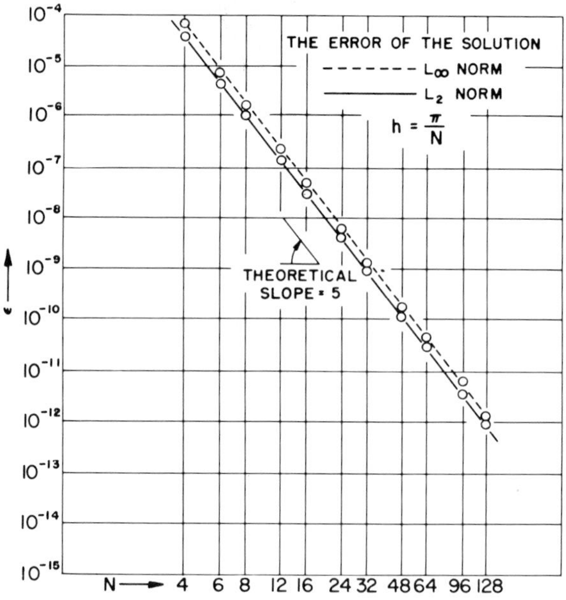

Fig. 1a Approximate Solution for j=5, D=1, $C=\frac{1}{4}$

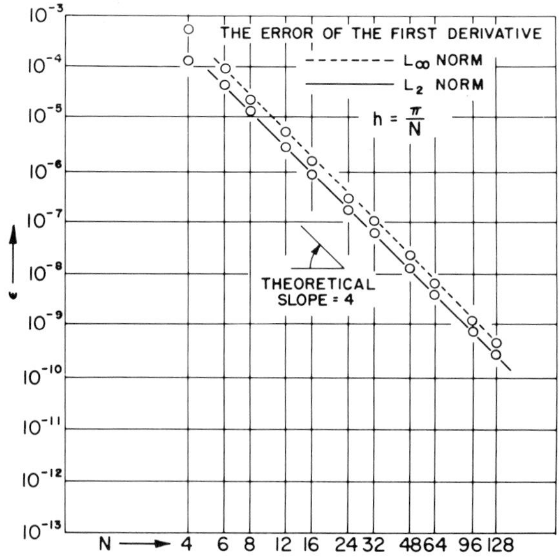

Fig. 1b Approximate Solution for j=5, D=1, $C=\frac{1}{4}$

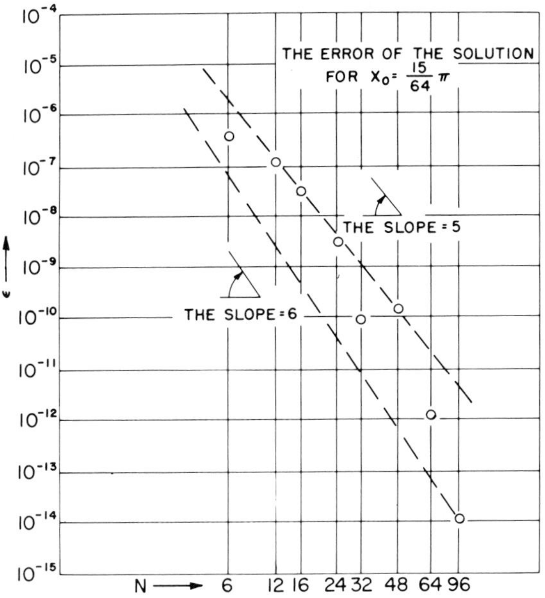

Fig. 2 Approximate Solution for $j=5$, $D=1$, $C=\frac{1}{4}$

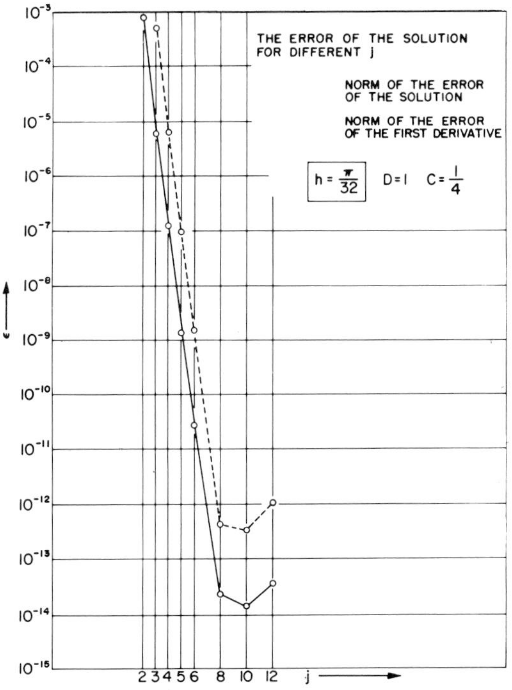

Fig. 3a The Error of the Solution for Different j

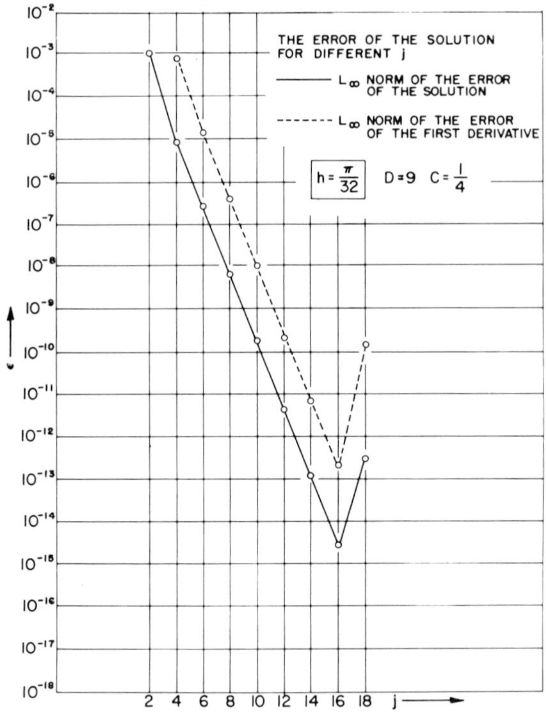

Fig. 3b   The Error of the Solution for Different j

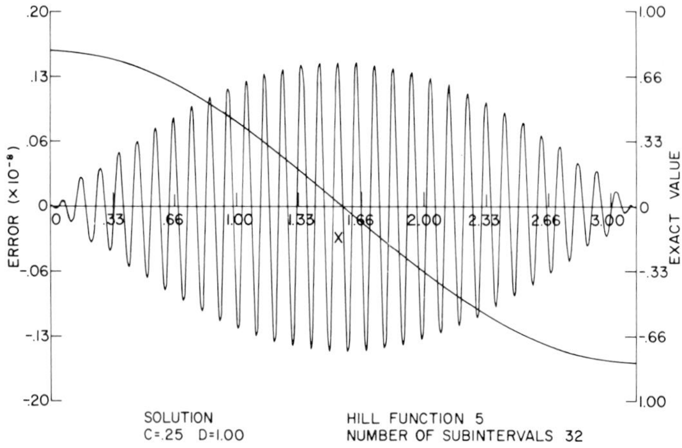

Fig. 4a   The Error of the Solution for j=5, D=1, C=$\frac{1}{4}$

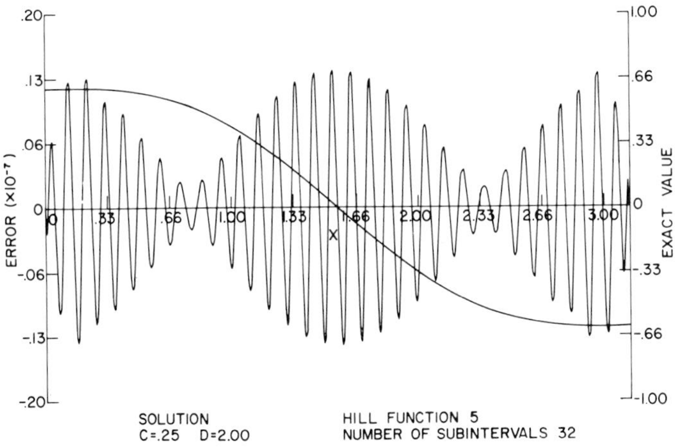

Fig. 4b The Error of the Solution for $j=5$, $D=2$, $C=\frac{1}{4}$

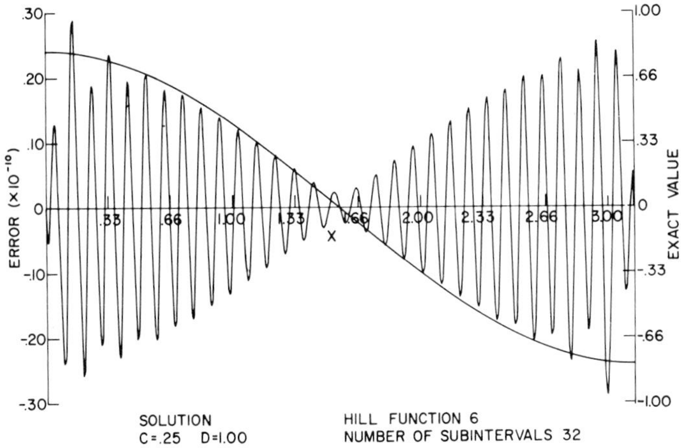

Fig. 4c The Error of the Solution for $j=6$, $D=1$, $C=\frac{1}{4}$

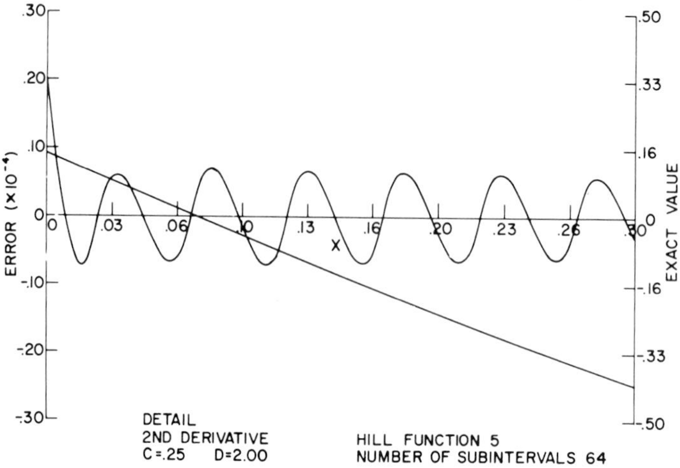

Fig. 5 The Error of the Second Derivative for $j=5$, $D=2$, $C=\frac{1}{4}$

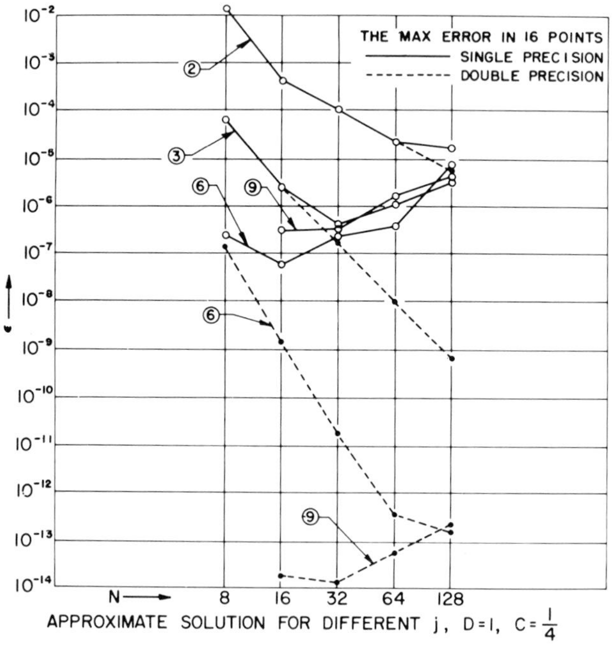

Fig. 6 Approximate Solution for Different j, $D=1$, $C=\frac{1}{4}$

ON THE NUMERICAL SOLUTION OF ELLIPTIC BOUNDARY VALUE PROBLEMS BY LEAST SQUARES APPROXIMATION OF THE DATA*

J. H. Bramble and A. H. Schatz**

1. Introduction

In the approximate solution of boundary value problems arising in the theory of elliptic partial differential equations several rather general approaches have been taken. The method perhaps most extensively used has been the method of finite differences. One theoretical and practical difficulty with this method has been the treatment of the boundary, especially in problems of higher order. A second method which has been considered quite widely is the so-called variational approach. This usually consists in reformulating the boundary value problem as a problem in the minimization of a certain functional (usually an "energy" integral) over an appropriate space of function. The corresponding approximate problem would then be to look for the minimum only over a finite dimensional subspace. Again, except in

---

* This research was supported in part by the National Science Foundation under Grant Number NSF-GP-22936.

** Cornell University.

the case of natural boundary conditions, one difficulty has been the selection of an appropriate finite dimensional subspace. Although these methods were first considered many years ago their theoretical investigation has been undertaken only in the last few years, particularly from the point of view of error estimation. The approximation theoretic study of special sets of approximating functions (e.g. piecewise polynomials) has greatly encouraged researchers to investigate more thoroughly variational procedures for the solution of boundary problems since with such choices of trial functions the resulting linear systems retain many of the desirable features of those arising from difference equations. Rather than give extensive references on these subjects we refer the reader to related papers in this volume and references cited therein.

Another approach, though closely related to the second, is the so-called "least squares" method. Roughly speaking, it consists of finding that element of a finite dimensional subspace which best fits the data of a given problem in the least squares sense. The purpose of this paper is to survey some results obtained recently by the authors concerning the behavior of the error resulting from the use of such methods. A great advantage in the use of these methods lies in the fact that the subspaces chosen do not have to satisfy any boundary condition, so that the choice of approximating functions is easy.

Before describing in more detail the outline of this paper we would like to mention some other closely related work. Babuška [4] has introduced a boundary perturbation of the usual energy principle in such a way that his trial functions need not satisfy boundary conditions. His method is a "penalty method" and the penalty for unrestricted trial functions is loss of accuracy. In [6] he uses an analogous approach to treat certain cases in which the coefficients are discontinuous.

A brief outline of the present paper is as follows. Section 2 contains definitions and notations. In the third section we introduce certain classes of finite dimensional subspaces of Sobolev spaces and state a theorem concerning their approximation theoretic properties relative to general boundary problems. Next we state an approximation scheme of least squares type for $2m^{th}$ order elliptic equations and general boundary conditions. Estimates for the error are given. In the fifth section we discuss a second order equation with coefficients discontinuous across an interface. Again an approximation scheme is given together with corresponding error estimates. None of the schemes considered here requires trial functions satisfying boundary conditions and the problems are not required to be self-adjoint. The last section contains examples of specific problems and specific choices of subspaces. Since the purpose of this paper is to outline some recent work of the authors we refer the reader to [12, 13, 14, 15] for detailed proofs.

2. Preliminaries

Let R be a bounded domain in $\mathbb{R}^N$ with boundary ∂R. We shall assume (for convenience) that ∂R is of class C^∞ and shall consider in R the operator A of order $2m$ with infinitely differentiable real coefficients:

$$(2.1) \qquad Au = A(x,D)u = \sum_{|\alpha||\beta|\leq m} (-1)^{|\alpha|} D^\alpha(a_{\alpha\beta}(x)\, D^\beta u)$$

where $\alpha = (\alpha_1,\ldots,\alpha_N)$ and $\beta = (\beta_1,\ldots,\beta_N)$ are multi-indices, $D^\alpha = \left(\frac{\partial}{\partial x_1}\right)^{\alpha_1} \cdots \left(\frac{\partial}{\partial x_N}\right)^{\alpha_N}$. We note that A is not necessarily formally self-adjoint. Set

$$A_0(x,\xi) = \sum_{|\alpha|,|\beta|=m} (-1)^m a_{\alpha\beta}(x)\zeta^{\alpha+\beta}$$

where

$$\zeta^{\alpha+\beta} = \zeta_1^{\alpha_1+\beta_1} \cdots \zeta_N^{\alpha_N+\beta_N} .$$

We shall assume that A is uniformly elliptic; i.e. there is a constant $a > 0$ independent of x such that

$$a^{-1}|\xi|^{2m} \leqslant |A_0(x,\xi)| \leqslant a|\xi|^{2m}$$

for all $x \in \overline{R}$ and all $\xi \in R^N$.

We shall consider the boundary value problem

$$\text{(2.2)} \qquad \begin{aligned} &Au = f \quad \text{in} \quad R , \\ &B_j u = g_j , \quad j = 0,\ldots,m-1 \quad \text{on} \quad \partial R \end{aligned}$$

where the B_j are boundary differential operators of order m_j, $0 \leqslant m_j \leqslant 2m-1$, $j = 0,\ldots,m-1$, and f and g_j are given. The operators B_j are defined by

$$\text{(2.3)} \qquad B_j\phi = \sum_{|\alpha|\leqslant m_j} b_{j\alpha}(x)D^\alpha\phi$$

for $x \in \partial R$ and $D^\alpha\phi$ is defined on ∂R by continuity or if necessary as a trace.

All functions considered in this paper will be real valued. The conditions which we shall place on our problem are as follows.

Condition A:

i) A is uniformly elliptic with coefficients in $C^\infty(\overline{R})$.

ii) The boundary system $\{B_j\}$ is normal and covers the operator A (see e.g. [23]) and has coefficients in $C^\infty(\partial R)$. The order of B_j is m_j , $0 \leqslant m_j \leqslant 2m-1$ where for simplicity we may assume that $m_0 < m_j$ for $j \neq 0$.

iii) The only solution of (2.2) in $C^\infty(\overline{R})$ with zero data is the zero solution.

We shall make use of the following function spaces.

1) $H^s(R)$, for s a non-negative integer, is the Sobolev space of order s on R with the norm $\|\cdot\|_s$ derived from the inner product

$$(u,v)_s = \sum_{|\alpha|\leqslant s} (D^\alpha u, D^\alpha v)_0 ,$$

where

$$(u,v)_0 = \int_R uv\,dx .$$

For $s > 0$ and not an integer $H^s(R)$ is defined by real interpolation between successive integers (cf. [23]). For $s < 0$, $H^s(R)$ is the completion of $C^\infty(\overline{R})$ with respect to the norm

$$\|u\|_s = \sup_{v \in C^\infty(\overline{R})} \frac{(u,v)_0}{\|v\|_{-s}} .$$

2) $H^s(\partial R)$ for all real s is the Sobolev space of order s on ∂R. For a precise definition we refer the reader to [23]. We shall denote by $|\cdot|_s$ the norm on $H^s(\partial R)$. Note that $H^0(\partial R) = L_2(\partial R)$ and in this case we denote the inner product by

$$\langle u,v \rangle_0 = \int_{\partial R} uv\,d\sigma .$$

In this discussion we shall restrict ourselves to the approximation of solutions u of the boundary value problem (2.2) which have a certain degree of regularity. The conditions that we impose on the data here can be relaxed considerably (see [12], [14]). For simplicity of presentation we shall assume that $u \in H^\beta(R)$ for some $\beta \geq 2m$. The following result, due to a number of authors, gives precise conditions on the data under which such solutions can be found and is essential in what follows.

Theorem (cf. [23],[25]). Suppose that Condition A is satisfied. The mapping $\mathcal{P}u = (Au, B_0u, \ldots, B_{m-1}u)$ of $C^\infty(\overline{R})$ into $C^\infty(R) \times \overbrace{C^\infty(\partial R) \times \cdots \times C^\infty(\partial R)}^{m\text{-times}}$, completed by continuity, is a homeomorphism of $H^\beta(R)$ onto $H^{\beta-2m}(R) \times \prod_{j=0}^{m-1} H^{\beta-m_j-\frac{1}{2}}(\partial R)$ for all real numbers $\beta \geq 2m$.

The norms $\|u\|_\beta$ and $\|Au\|_{\beta-2m} + \sum_{j=0}^{m-1} |B_j u|_{\beta-m_j-\frac{1}{2}}$ are equivalent.

3. Finite Dimensional Subspaces of $H^k(R)$

We shall now discuss the subspaces of functions which will be used to approximate the solution of (2.2). We shall first state our basic assumption concerning them and then show that as a consequence they have certain other approximation theoretic properties relative to the "data spaces" of the differential operators which we are considering.

Let h, $0 < h < 1$, be a parameter. For any two non-negative integers k and r with $k < r$, let $S^h_{k,r}(R) = S^h_{k,r}$ be any finite dimensional subspace of $H^k(R)$ which satisfies the following condition:

(*) For each $u \in H^r(R)$ there exists a $\bar{u} \in S^h_{k,r}$ and a constant C, independent of h and u, such that

$$\|u-\bar{u}\|_k \leq Ch^{r-k} \|u\|_r . \tag{3.1}$$

This is obviously equivalent to the condition that

$$\inf_{\chi \in S^h_{k,r}} \|u-\chi\|_k \leq Ch^{r-k} \|u\|_r . \tag{3.2}$$

Subspaces having the property (*) have been constructed by many authors, see for example Hilbert [22], Schultz [27], Aubin [3], DiGuglielmo [18], Babuška [5], Fix and Strang [20] and Bramble and Zlámal [16]. One possible example of such subspaces is the restriction to $\overline{R}$ of "spline functions" defined with respect to a uniform mesh of width h on $\mathbb{R}^N$.

The following approximation theoretic result is crucial to our analysis of approximation schemes for $2m^{th}$ order elliptic boundary value problems which will be discussed in the next section. This result states that the subspaces $S^h_{k,r}$ satisfying (*) have further approximation theoretic properties relative to the "data spaces" occurring in the boundary value problems. It is a special case of a general theorem found in [13].

Theorem 3.1. Let $S^h_{k,r}$ satisfy (*) with $2m \leqslant k < r$. Then for all $F = (f, g_0, \ldots, g_{m-1}) \in H^{\lambda}(R) \times H^{\lambda_0}(\partial R) \times \cdots \times H^{\lambda_{m-1}}(\partial R)$, $0 \leqslant \lambda \leqslant r-2m$, $0 \leqslant \lambda_j \leqslant r - m_j - \frac{1}{2}$, $j = 0, \ldots, m-1$, there exists a constant C independent of F and h such that

$$\inf_{\chi \in S^h_{k,r}} \left(\|f - A\chi\|_0 + \sum_{j=0}^{m-1} h^{-(2m-m_j-\frac{1}{2})} |g_j - B_j \chi|_0 \right)$$

$$\leqslant C \left(h^{\lambda} \|f\|_{\lambda} + \sum_{j=0}^{m-1} h^{-(2m-m_j-\frac{1}{2})+\lambda_0} |g|_{\lambda_0} \right).$$

4. Least Squares Methods for $2m^{th}$ Order Boundary Value Problems

Let u be the solution of (2.2). The approximation scheme we shall consider is as follows:

Let $S^h_{k,r}$ be given satisfying (*) with $2m \leq k < r$. Find $w \, S^h_{k,r}$ such that

$$(4.1) \qquad (f-Aw, A\phi)_0 + \sum_{j=0}^{m-1} h^{-2(2m-m_j-\frac{1}{2})} \langle g_j - B_j w, B_j \phi \rangle_0 = 0$$

or

$$\int_R (f-Aw)A\phi dx + \sum_{j=0}^{m-1} h^{-2(2m-m_j-\frac{1}{2})} \int_{\partial R} (g_j - B_j w) B_j \phi \, d\sigma = 0$$

for all $\phi \in S^h_{k,r}$.

Since $\mathcal{P}(S^h_{k,r})$ (the image of $S^h_{k,r}$ under the mapping $\mathcal{P}\phi = (A\phi, B_0\phi, \ldots, B_{m-1}\phi)$) is a finite dimensional subspace of $H^0(R) \times \overbrace{H^0(\partial R) \times \cdots \times H^0(\partial R)}^{m\text{-times}}$, by iii) of Condition A (uniqueness) w exists and is unique. It is determined by solving a linear system of algebraic equations whose coefficients depend only on f, g and $S^h_{k,r}$. An alternative way of stating (4.1) is the following: Among all $\chi \in S^h_{k,r}$ find the one which minimizes the functional

$$(4.2) \qquad \|f - A\chi\|_0^2 + \sum_{j=0}^{m-1} h^{-2(2m-m_j-\frac{1}{2})} |g - B_j \chi|_0^2 .$$

Note that the above scheme is a least squares method involving only L_2 norms of the data. For a given $S^h_{k,r}$ (with h fixed) the weighting factors $h^{-2(2m-m_j-\frac{1}{2})}$ are constants, but the manner in which they depend on h is critical for our estimates.

We shall now consider some results concering the rate of convergence of the approximate solution w determined by (4.1) to the solution u of (2.2). We shall first consider error estimates on R.

<u>Theorem 4.1</u>. Suppose that Condition A is satisfied and u is the solution of (2.2) with $F = (f, g_0, \ldots, g_{m-1})$ $\in H^{\beta-2m}(R) \times \prod_{j=0}^{m-1} H^{\beta-m_j-\frac{1}{2}}(\partial R)$. For given $S_{k,r}^h$ satisfying (*) with $2m \leqslant k < r$, let w be the solution of the approximate problem (4.1) and set $e = u - w$.

<u>Case 1</u>. Suppose that $4m \leqslant r$. Then there exists a constant C independent of F and h such that

$$\|e\|_\rho \leqslant Ch^{\beta-\rho} \|u\|_\beta \tag{4.3}$$

for each ρ and β satisfying $4m - r \leqslant \rho \leqslant m_0 + \frac{1}{2}$ and $2m \leqslant \beta \leqslant r$.

<u>Case 2</u>. Suppose that $2m < r \leqslant 4m$. Then if $4m - r \leqslant m_0 + \frac{1}{2}$

$$\|e\|_\rho \leqslant Ch^{\beta-\rho} \|u\|_\beta \tag{4.4}$$

for each ρ and β satisfying $4m - r \leqslant \rho \leqslant m_0 + \frac{1}{2}$ and $2m \leqslant \beta \leqslant r$. If $m_0 + \frac{1}{2} < 4m - r$ then

$$\|e\|_{m_0+\frac{1}{2}} \leqslant Ch^{(r-2m)+(\beta-2m)} \|u\|_\beta \tag{4.5}$$

for each β satisfying $2m \leqslant \beta \leqslant r$. In (4.4) and (4.5)

C is a constant which is independent of h and F.

Remark. The norm $\|u\|_\beta$ may be replaced by the data norm $\|f\|_{\beta-2m} + \sum_{j=0}^{m-1} |g_j|_{\beta-m_j-\frac{1}{2}}$ in the right hand sides of (4.3), (4.4) and (4.5).

For the proof of Theorem 4.1 the reader is referred to [14]. We just wish to draw attention to the connection between the definition of the approximate problem (4.1) in the form (4.2) and the statement of Theorem 3.1. The relevance of Theorem 3.1 to the proof of Theorem (4.1) is then apparent.

Let us now briefly discuss the results contained in Theorem 4.1. In the case that $r \geq 4m$ the inequality (4.3) says that asymptotically the approximation scheme (4.1) gives the best possible results relative to the assumed approximation theoretic properties of the spaces $S^h_{k,r}$.

In order that we may illustrate this simply, let us consider the case of estimating the error in the $H^0(R) = L_2(R)$ norm. According to our assumptions concerning $S^h_{k,r}$ one can show that the best approximation $\bar{u}$ in $S^h_{k,r}$ to $u \in H^\beta(R)$ in the $L^2(R)$ norm, in general only satisfies the inequality

$$(4.6) \qquad \|u-\bar{u}\|_0 \leq C_1 h^\beta \|u\|_\beta$$

where for simplicity we have taken $2m \leq \beta \leq r-2m$ and where C_1 is independent of u and h. The estimate (4.3) with $\rho = 0$ shows that

(4.7) $$\|u-w\|_0 \leq Ch^{\beta} \|u\|_{\beta}$$

where w is the solution of the approximate problem (4.1). Hence the property (4.6) is essentially reproduced.

The case when $2m < r < 4m$ is more difficult to discuss. Suppose first that $4m-r \leq m_0+\frac{1}{2}$ (where m_0 is order of the boundary differential operator of lowest order). In this case the inequality (4.4) says if we measure the error in an appropriate norm $\|\cdot\|_{\rho}$, $4m-r \leq \rho \leq m_0+\frac{1}{2}$ then w essentially has the same approximation properties as the best approximation to u in $S^h_{k,r}$ in that norm. In the case when $m_0+\frac{1}{2} < 4m-r$, the inequality (4.5) indicates that the properties of w are not as good as those of the best approximation when measured in any norm up to $m_0+\frac{1}{2}$. Specific examples will be given in Section 6.

Interior estimates. Suppose that R_1 is any compact subset of R. We shall denote by $\|\cdot\|^{R_1}_s$ the norm on the Sobolev space $H^s(R_1)$ which is defined in the same manner as $H^s(R)$ except now with respect to R_1. We shall now state some estimates for the error $e = u-w$ on R_1. The proof may be found in [14].

Theorem 4.2. Suppose that the conditions of Theorem 4.1 are satisfied. Let R_1 be any compact subset of R. Then there exists a constant C independent of h and F such that

(4.8) $$\|e\|^{R_1}_{\rho} \leq Ch^{\beta-\rho} \|u\|_{\beta}$$

for each ρ and β satisfying $4m-r \leqslant \rho \leqslant 2m$, $\rho \leqslant \beta$ and $2m \leqslant \beta \leqslant r$.

We first note that Theorem 4.2 gives estimates for the derivatives of the error up to order $2m$ over R. This is in contrast to Theorem 4.1 where we are restricted to estimates for the derivatives of the error up to order $m_0+\frac{1}{2}$ over R.

In the case when $4m \leqslant r$ the estimate (4.8) says that we are able to estimate the error in all norms $\|\cdot\|_\rho^{R_1}$ for $0 \leqslant \rho \leqslant 2m$ with the best possible power of h. If $2m < r < 4m$, we are also able to do this provided we restrict ourselves to an appropriately high norm $\|\cdot\|_\rho$ where $4m-r \leqslant \rho \leqslant 2m$. In contrast to Theorem 4.1 the analogous case of the estimate (4.5) does not occur here since, as mentioned above, we are not restricted to estimates only involving Sobolev norms up to order $m_0+\frac{1}{2}$. A phenomenon similar to that given in Theorem 4.2 has been observed by Fix and Strang [20] in their study of Rayleigh-Ritz-Galerkin methods over R^N. We further note that maximum norm estimates over interior subdomains can be easily obtained from Theorem 4.2 using Sobolev inequalities.

5. Interface Problems - Equations with Discontinuous Coefficients

The approximation scheme which was used to approximate the solution of the boundary value problem can be generalized to yield approximate solutions of so-called interface problems (or equations with discontinuous coefficients). For simplicity of presentation we shall not treat such problems with $2m^{th}$ order operators and general boundary conditions but rather restrict ourselves here to a discussion of a specific example.

Let R_1 be an open domain with C^∞ boundary ∂R_1 (the interface) and suppose that $\overline{R}_1 \subset R$. Let $R_2 = R - \overline{R}_1$. As an example of an interface problem we shall consider the problem

$$
\begin{aligned}
&\Delta u_i = f_i \quad \text{in } R_i, \quad i = 1, 2, \\
&a_1 u_1 - u_2 = \phi_1 \quad \text{on } \partial R_1, \\
(5.1) \qquad &a_2 \frac{\partial u_1}{\partial n} - \frac{\partial u_2}{\partial n} = \phi_2 \quad \text{on } \partial R_1, \\
&u_2 = g \quad \text{on } \partial R,
\end{aligned}
$$

where $a_1, a_2 > 0$ are constants and f_1, f_2, ϕ_1, ϕ_2 and g are given functions. Let us denote by $|||u|||_s$ the norm on the product space $H^s(R_1) \times H^s(R_2)$. The following result concerning existence uniqueness and regularity of solutions of (5.1), is basic to our investigation (cf. [25]).

<u>Theorem 5.1</u>. The mapping $\mathcal{P}u = (\Delta u_1, \Delta u_2, a_1 u_1 - u_2, a_2 \frac{\partial u_1}{\partial n} - \frac{\partial u_2}{\partial n}, u_2)$ of $C^\infty(\overline{R}_1) \times C^\infty(\overline{R}_2)$ into $C^\infty(\overline{R}_1) \times C^\infty(\overline{R}_2) \times C^\infty(\partial R_1) \times C^\infty(\partial R_1) \times C^\infty(\partial R)$, completed by continuity, is a homeomorphism of $H^\beta(R_1) \times H^\beta(R_2)$ onto

$$
H^{\beta-2}(R_1) \times H^{\beta-2}(R_2) \times H^{\beta-\frac{1}{2}}(\partial R_1) \times H^{\beta-\frac{3}{2}}(\partial R_1) \times H^{\beta-\frac{1}{2}}(\partial R) \quad \text{for all } \beta \geq 2.
$$

Then norms $|||u|||_\beta$ and

$$\|\Delta u_1\|_{\beta-2}^{R_1} + \|\Delta u_2\|_{\beta-2}^{R_2} + |u_1-u_2|_{\beta-\frac{1}{2}}^{\partial R_1}$$

$$+ \left|a\frac{\partial u_1}{\partial n} - \frac{\partial u_2}{\partial n}\right|_{\beta-\frac{3}{2}}^{\partial R_1} + |u_1-u_2|_{\beta-\frac{1}{2}}^{\partial R}$$

are equivalent.

We shall consider an approximation scheme which is a generalization of the scheme (4.1). The approximating functions we shall use will be any finite dimensional subspace of $H^k(R_1) \times H^k(R_2)$ of the type $S_{k,r}^h(R_1) \times S_{k,r}^h(R_2) = \tilde{S}_{k,r}^h$ where each $S_{k,r}^h(R_i)$, $i = 1,2$, has the property (*) in their respective domains. Hence we shall assume that for all $u \in H^r(R_1) \times H^r(R_2)$ there exists a constant C independent of u such that

$$(**) \qquad \inf_{\chi \in \tilde{S}_{k,r}^h} |||u-\chi|||_k \leq Ch^{r-k} |||u|||_r$$

where $\chi = (\chi_1, \chi_2)$.

One can show, in analogy with Theorem 3.1 that the subspaces $\tilde{S}_{k,r}^h$ have the following approximation theoretic property relative to the "data spaces" of the problem considered here.

<u>Theorem 5.2</u>. Let $\tilde{S}_{k,r}^h$ satisfy (**) with $2 \leq k < r$. Then for all

$$F = (f_1, f_2, \phi_1, \phi_2, g) \in H^{\lambda}(r_1) \times H^{\lambda}(R_2) \times H^{\lambda_1}(\partial R_1) \times H^{\lambda_2}(\partial R_1) \times H^{\lambda_3}(\partial R)$$

satisfying

$$0 \leq \lambda \leq r-2, \quad 0 \leq \lambda_1, \quad \lambda_3 \leq r-\tfrac{1}{2}, \quad 0 \leq \lambda_2 \leq r-\tfrac{3}{2},$$

there exists a constant C independent of h and F such that

$$\|f_1-\Delta\chi_1\|_0^{R_1} + \|f_2-\Delta\chi_2\|_0^{R_2} + h^{-\frac{3}{2}}|\phi_1-(a_1\chi_1-\chi_2)|_0^{\partial R_1} + h^{-\frac{1}{2}}\left|\phi_2-\frac{\partial(a_2\chi_1-\chi_2)}{\partial n}\right|_0^{\partial R_1} + h^{-\frac{3}{2}}|g-\chi_2|_0^{\partial R}$$
$$\leq C(\|f_1\|_\lambda^{R_1} + \|f_2\|_\lambda^{R_2} + h^{-\frac{3}{2}+\lambda_1}|\phi_1|_{\lambda_1}^{\partial R_1} + h^{-\frac{1}{2}+\lambda_2}|\phi_2|_{\lambda_2}^{\partial R_1} + h^{-\frac{3}{2}+\lambda_3}|g|_{\lambda_3}^{\partial R}) \tag{5.2}$$

Let u be the solution of (5.1) for given data $F = (f_1, f_2, \phi_1 \phi_2, g)$. The approximation scheme which we shall consider is the following:

Let $\tilde{S}_{k,r}^h = S_{k,r}^h(R_1) \times S_{k,r}^h(R_2)$ be given with $2 \leq k < r$. Find $w \in \tilde{S}_{k,r}^h$ such that

$$\int_{R_1}(f_1-\Delta w_1)\Delta\chi dx + \int_{R_2}(f_2-\Delta w_2)\Delta\chi dx \tag{5.3}$$

$$+ h^{-3}\int_{\partial R_1} (\phi_1(a_1w_1-w_2))(a_1\chi_1-\chi_2)\,d\sigma_1$$

$$+ h^{-1}\int_{\partial R_1} \left(\phi_2 - \frac{\partial(a_2w_1-w_2)}{\partial n}\right) \frac{\partial(a_2\chi_1-\chi_2)}{\partial n}\,d\sigma_1$$

$$+ h^{-3}\int_{\partial R} (g-w_2)\,\chi_2\,d\sigma_2 \quad = \quad 0$$

for all $\chi = (\chi_1,\chi_2) \in \tilde{S}^h_{k,r}$.

It is easily seen that the solution of (5.3) exists and is unique and, as in the scheme (4.1), is determined by solving a linear system of algebraic equations whose coefficients depend only on the data and $\tilde{S}^h_{k,r}$.

Let w be the solution of (5.3). Then using methods analogous to those used in the proof of Theorem 4.1 we can (restricting ourselves for simplicity to estimates for $|||u-w|||_0$) obtain the following error estimates.

<u>Theorem 5.3</u>. Suppose that u is the solution of (5.1) for

$$F = (f_1,f_2,\phi_1,\phi_2,g) \in H^{\beta-2}(R_1) \times H^{\beta-2}(R_2) \times H^{\beta-\frac{1}{2}}(\partial R_1)$$

$$\times H^{\beta-\frac{3}{2}}(\partial R_1) \times H^{\beta-\frac{1}{2}}(\partial R) .$$

For given $\tilde{S}^h_{k,r}$, with $2 \leqslant k < r$, $r \geqslant 4$, satisfying (**) let w be the solution of the approximate problem (5.3). Then there exists a constant C independent of h and F such that

$$(5.4)\qquad |||u-w|||_0 \leq Ch^\beta |||u|||_\beta$$
$$\leq Ch^\beta(\|f_1\|^{R_1}_{\beta-2} + \|f_2\|^{R_2}_{\beta-2} + |\phi_1|^{\partial R_1}_{\beta-\frac{3}{2}}$$
$$+ |\phi_2|^{\partial R_2}_{\beta-\frac{3}{2}} + |g|^{\partial R}_{\beta-\frac{1}{2}})$$

for each $2 \leq \beta \leq r$.

Analogous schemes for more general problems with discontinuous coefficients can be treated in much the same manner as in the present example. A full account is to be found in [15]. A specific application of the above scheme is given in example 3 in the next section.

6. Examples

In this section we shall illustrate the main results of the previous sections with specific examples in special cases.

Example 1. Consider Dirichlet's problem

$$(6.1)\qquad \begin{aligned} \Delta u + qu &= f \quad \text{in} \quad R \\ u &= g \quad \text{on} \quad \partial R \end{aligned}$$

where $q = q(x)$ is any smooth function for which the solution of (5.1) is unique.

For given $S^h_{k,r}$ the approximation scheme (4.1) consists in finding $w \in S^h_{k,r}$ such that

$$\int_R (f-(\Delta w+qw))(\Delta\phi+q\phi)dx + h^{-3}\int_{\partial R}(g-w)\phi d\sigma = 0$$

for all $\phi \varepsilon S^h_{k,r}$.

Let us first consider error estimates over R .

Case 1. Suppose that we take $k = 2$, $r = 4$ and $S^h_{k,r}$ to be say subic splines or cubic Hermite polynomials (cf. [10],[11]). If w is the solution of the approximate problem (4.1), then (4.3) of Theorem 4.1 yields

$$\|u-w\|_0 \leqslant Ch^\beta \|u\|_\beta$$

for any $2 \leqslant \beta \leqslant 4$. In particular the maximum rate of convergence is given by

$$\|u-w\|_0 \leqslant Ch^4 \|u\|_4 .$$

If the data are smoother and we take $k = 2$, $r = 6$ and $S^h_{k,r}$ to be say quintic splines then we obtain

$$\|u-w\|_0 \leqslant Ch^6 \|u\|_6 .$$

In general if $u \varepsilon H^r(R)$ then we obtain using any $S^h_{2,r}$ for $r \geqslant 4$

$$\|u-w\|_0 \leqslant Ch^r \|u\|_r .$$

If we take $k = 2$, $r = 3$ and $S_{k,r}$ to be say quadratic splines, then (4.5) of Theorem 4.1 yields

$$\|u-w\|_{\frac{1}{2}} \leqslant Ch^2 \|u\|_3 .$$

Interior estimates for the error may be obtained from Theorem 4.2. We again take w to be the solution of the approximate problem (4.1).

If we take $k = 2$ and $r = 4$ we then have that

$$\|u-w\|_{\rho}^{R_1} \leq Ch^{4-\rho} \|u\|_4$$

for any $0 \leq \rho \leq 2$. If we were to take $k = 2$ and $r = 3$ then we obtain

$$\|u-w\|_{\rho}^{R_1} \leq Ch^{3-\rho} \|u\|_3$$

where in this case ρ is restricted to $1 \leq \rho \leq 2$.

We remark that in this example the function q is not required to satisfy a sign condition and hence, for example q could be a positive constant which lies between two fixed membrane eigenvalues. In this case the usual bilinear form associated with the problem (6.1) with $g = 0$ would be

$$\sum_{j=0}^{N} \int_R \frac{\partial\phi}{\partial x_i} \frac{\partial\psi}{\partial x_i} dx - q\int_R \phi\psi dx = 0$$

which is not positive definite. Hence in the classical Rayleigh-Ritz method (for a given h) even the existence is unclear. This difficulty is not present in our method, since, in view of the uniqueness of solutions, the forms associated with (4.1) are always positive definite. The results of this example do not change if instead of $\Delta u + qu$ we take any second order uniformly elliptic operator with smooth coefficients for which the solution of Dirichlet's problem is unique.

Example 2. Let us briefly consider Dirichlet's problem for the biharmonic operator

$$\Delta^2 u = f \quad \text{in} \quad R$$

(6.2)
$$\left.\begin{aligned} u &= g_0 \\ \frac{\partial u}{\partial n} &= g_1 \end{aligned}\right\} \quad \text{on} \quad R .$$

In this case the approximation scheme, for given $S^h_{k,r}$, consists in finding a $w \in S^h_{k,r}$ such that

$$\int_R (f-\Delta^2 w)\Delta^2\phi dx + h^{-7}\int_{\partial R}(g_0-w)\phi d\sigma + h^{-5}\int_{\partial R}\left(g_1 - \frac{\partial w}{\partial n}\right)\frac{\partial\phi}{\partial n}\, d\sigma = 0$$

for all $\phi \in S^h_{k,r}$.

If we take $k = 4$, $r = 6$ and $S^h_{k,r}$ to be say quintic splines (cf. [10]) we obtain from (4.5) of Theorem 4.1 that

$$\|u-w\|_{\frac{1}{2}} \leq Ch^4 \|u\|_6 .$$

If $k = 4$ and $r = 8$ we obtain from (4.3) that

$$\|u-w\|_0 \leq Ch^8 \|u\|_8 .$$

Interior estimates may be obtained from Theorem 4.1. Let R_1 be any compact subset of R. In the case $k = 4$ and $r = 6$ we obtain

$$\|u-w\|_\rho^{R_1} \leqslant Ch^{6-\rho} \|u\|_6$$

for any $2 \leqslant \rho \leqslant 4$. If we take $k = 4$ and $r = 8$ then

$$\|u-w\|_\rho^{R_1} \leqslant Ch^{8-\rho} \|u\|_8$$

for any $0 \leqslant \rho \leqslant 4$.

Example 3. In the approximation scheme (5.3) for interface problems, we have used two independent sets of trial function, namely $S_{k,r}^h(R_1)$ and $S_{k,r}^h(R_2)$. There are situations which arise in approximating the solution of (2.2) in which the scheme (5.3) will lead to better approximations than those obtained from (4.1). For example consider the Dirichlet problem

$$(6.3)\qquad \begin{aligned} \Delta u &= f \quad \text{in} \quad R \\ u &= 0 \quad \text{on} \quad \partial R \end{aligned}$$

for which f is a piecewise smooth function. Say for simplicity that

$$f = \begin{cases} 1 & \text{on} \quad R_1 \\ 0 & \text{on} \quad R_2 \end{cases},$$

where R_1 and R_2 are as described in the interface problem.

Now certainly $f \in H^0(R) = L_2(R)$ and $u \in H^2(R)$. If $w \in S^h_{k,r}(R)$ with $k \geq 2$, $r \geq 4$ is the solution of the approximation scheme (4.1) we obtain the estimate

$$\|u-w\|_0 \leq Ch^2 \|u\|_2 . \tag{6.4}$$

However we have not made full use of the smoothness properties of f. If we set $u = u_i$ in R_i, $i = 1,2$. Then u is the solution of the interface problem

$$\begin{aligned} &\Delta u_1 = 1 \text{ in } R_1 , \quad \Delta u = 0 \text{ in } R_2 , \\ &u_1 - u_2 = 0 , \frac{\partial u_1}{\partial n} - \frac{\partial u_2}{\partial n} = 0 \text{ on } \partial R_1 , \\ &u_2 = 0 \text{ on } \partial R_2 , \end{aligned} \tag{6.5}$$

where in this case $u_i \in C^\infty(\overline{R}_i)$, $i = 1,2$. If we now use the approximation scheme (5.3) for given $\tilde{S}^h_{k,r}$ with $k \geq 2$ and $r \geq 4$ we obtain

$$|||u-w|||_0 \leq Ch^r |||u|||_r \tag{6.6}$$

where $w = (w_1, w_2)$ is the solution of (5.3). Thus in contrast to (6.4) the order of convergence in (6.6) is limited only by the "goodness" of the subspaces used.

REFERENCES

1. Agmon, S., Lectures on Elliptic Boundary Value Problems, Van Nostrand, Princeton, 1965.
2. Aubin, J.-P., "Behavior of the Error of the Approximate Solutions of Boundary Value Problems for Linear Elliptic

Operators by Galerkin's and Finite Difference Methods," Annali della Scuola Normale di Pisa, 21 (1967), 599-637.
3. Aubin, J-P., "Interpolation et approximation optimales et 'Spline Functions'," J. Math. Anal. Appl. 24 (1968), 1-24.
4. Babuška, I., "Numerical Solution of Boundary Value Problems by the Perturbed Variational Principle," Univ. of Maryland Tech. Note BN-624, Oct. 1969.
5. Babuška, I., "Approximation by Hill Functions," Univ. of Maryland Tech. Note BN-648.
6. Babuška, I., "The Finite Element Method for Elliptic Equations with Discontinuous coefficients," Univ. of Maryland Tech. Note BN-631.
7. Babuška, I., "The Rate of Convergence for the Finite Element Method," Univ. of Maryland Tech. Note BN-646.
8. Babuška, I., "Computation of Derivatives in the Finite Element Method," Univ. of Maryland Tech. Note BN-650.
9. Babuška, I., "Finite Element Method for Domains with Corners," Univ. of Maryland Tech. Note BN-636.
10. Bramble, J. H., and Hilbert, S., "Estimation of Linear Functionals on Sobolev Spaces with Application to Fourier Transforms and Spline Interpolation," SIAM Num. Anal. SIAM Num. Anal. Vol. 7 (1970), 112-124.
11. Bramble, J. H., and Hilbert, S., "Bounds for a Class of Linear Functionals with Application to Hermite Interpolation," Num. Math. (in print).
12. Bramble, J. H., and Schatz, A. H., "Rayleigh-Ritz-Galerkin Methods for Dirichlet's Problem Using Subspaces without Boundary Conditions," Comm. Pure Appl. Math. (to appear).
13. Bramble, J. H., and Schatz, A. H., "On Some Finite Dimensional Subspaces of Certain Banach Spaces and their Properties Relative to Best Approximation," (to appear).
14. Bramble, J. H., and Schatz, A. H., "Least Squares Method for $2m^{th}$ Order Elliptic Boundary Value Problems," (to appear).
15. Bramble, J. H., and Schatz, A. H., "Least Squares Methods for Elliptic Equations with Discontinuous Coefficients," (to appear).
16. Bramble, J. H., and Zlámal, "Triangular Elements in the Finite Element Method," Math. Comp. (to appear).
17. Ciarlet, P., Schultz, M., and Varga, R., "Numerical Methods of High-Order Accuracy for Nonlinear Boundary Value Problems I. One Dimensional Problems," Num. Math. 9 (1967), 394-430.

18. DiGuglielmo, F., "Construction d'approximations des espaces de Sobolev sur des réseaux en simplexes," Calcolo, Vol. 6 (1969), 279-331.
19. Fix, G., and Strang, G., "Fourier Analysis of the Finite Element Method in Ritz-Galerkin Theory," Studies in Appl. Math. Vol. 48, No. 3 (1969), 265-273.
20. Strang, G., "The Finite Element Method and Approximation Theory," these proceedings.
21. Friedrichs, K., and Keller, H.,"A Finite Difference Scheme for Generalized Neumann Problems," Numerical Solution of Partial Differential Equations (J. H. Bramble, editor), Academic Press, New York, 1966.
22. Hilbert, S., "Numerical Methods for Elliptic Boundary Problems," Ph.D. thesis, Univ. of Maryland, 1969.
23. Lions, J. L., and Magenes, E., Problemes aus limites non homogenes et applications, Vol. 1, Dunod, Paris, 1968.
24. Nitsche, J., "Lineare Spline-Funktionen und die Methoden von Ritz für elliptische Randwert probleme (preprint).
25. Roitberg, J. A., and Seftel, Z. G., "A Theorem on Homeomorphisms for Elliptic Systems and its Applications," Math. USSR Sbornik, Vol. 7 (1969), No. 3.
26. Schechter, M., "On L_p Estimates and Regularity II," Math. Scandinavica 13 (1963), 47-69.
27. Schultz, M., "Multivariate Spline Functions and Elliptic Problems," SIAM Num. Anal. Vol. 6 (1969), 523-538.
28. Schultz, M., "Rayleigh-Ritz-Galerkin Methods for Multidimensional Problems," SIAM Num. Anal. Vol. 6 (1969), 570-582.
29. Varga, R., "Hermite Interpolation-type Ritz Methods for Two-Point Boundary Value Problems," Numerical Solution of Partial Differential Equations (J. H. Bramble, editor), Academic Press, New York, 1966.
30. Zlámal, M., "On the Finite Element Method," Num. Math. 12 (1968), 394-409.

ALTERNATING-DIRECTION GALERKIN METHODS ON RECTANGLES

Jim Douglas, Jr., and Todd Dupont*

1. Introduction

Alternating-direction methods in several forms have proved to be very valuable in the approximate solution of partial differential equations problems involving several space variables by finite differences. The methods have been applied to transient problems directly and to stationary problems as iterative procedures. Their efficiency within the class of finite difference methods is well established practically, although many gaps remain in the rigorous demonstration of the observed behavior, particularly with respect to the analysis of alternating-direction iterative procedures.

Galerkin methods, along with other variational methods, combined with some recent advances in practical approximation theory offer a great theoretical increase in accuracy over standard finite-difference methods. It is clear that this improvement is relatively easily implemented in single

* University of Chicago

space variable problems, but it has not been quite as clear how to use the methods competively for several space variables.

The object of this paper is to combine alternating-direction methods with Galerkin methods to obtain highly efficient procedures for the numerical solution of second order parabolic and hyperbolic problems in two or more space variables and for the iterative soluton of the algebraic equations arising from the Galerkin treatment of elliptic problems. The results to be presented here will be limited to rectangular domains. There are several reasons for this. First, the methods are complicated enough to define on rectangles that the authors feel that is is of interest to present the special case. Second, we are not yet certain that we have the best possible formulation for non-rectangular domains. Our present formulation does not follow as a direct extension of the formulation for a rectangle; however, it does reduce to the same algebraic problem in that case. Third, and most important, we have not yet obtained proofs in the non-rectangular case.

The heat equation on a rectangle will be treated first and then extensions to variable coefficients and non-linear parabolic equations and systems will be given. Then a formulation of an alternating-direction Galerkin method for a wave equation will be presented. Finally, an iterative procedure for elliptic equations will be described. The emphasis here will be on the parabolic case.

There are several versions of alternating-direction methods [3, 6-9, 24, 28, 29]; we shall employ the fractional step procedure [28, 29] as a basis for our development.

Consider the heat equation on the unit square:

(1.1)

$$\frac{\partial u}{\partial t} = \frac{\partial^2 u}{\partial x^2} + \frac{\partial^2 u}{\partial y^2}\ , \quad (x,y) \in D = [0,1]^2\ , \quad 0 < t \leq T\ ,$$

$$u(x,y,t) = 0\ , \qquad (x,y) \in \partial D\ , \quad 0 < t \leq T\ ,$$

$$u(x,y,0) = u_o(x,y)\ , \quad (x,y) \in D\ .$$

Let

(1.2) $$0 = t_o < t_1 < t_2 < \cdots < t_m = T\ .$$

The continuous time functional step procedure is the following (with the space coordinates supressed in writing the arguments):

(1.3) $$v(0) = u_o\ ,$$

and

$$\frac{\partial v}{\partial t} = \frac{\partial^2 v}{\partial x^2}\ , \quad (x,y) \in D\ , \quad t_n < t \leq t_{n+1}\ ,$$

(1.4) $$v = 0\ , \quad (x,y) \in \partial D\ ,$$

$$v(t_n) = \lim_{t \uparrow t_n} w(t)\ , \quad (x,y) \in D\ , \quad n \geq 1\ .$$

and

$$\frac{\partial w}{\partial t} = \frac{\partial^2 w}{\partial y^2}\ , \quad (x,y) \in D\ , \quad t_n < t \leqslant t_{n+1}\ ,$$

(1.5) $$w = 0\ , \quad (x,y) \in \partial D\ ,$$

$$w(t_n) = \lim_{t \uparrow t_{n+1}} v(t)\ , \quad n \geqslant 0\ .$$

Note that on each time interval the solution of (1.4) is obtained by solving a family of single space variable problems parameterized by y, $0 < y < 1$. Similarly, the evaluation of w is given by solving single space variable problems parameterized by x, $0 < x < 1$. Obviously, this implies a very strong simplification in the evaluation of the solution. It is also important to note that the solution $v(t_n)$, $n \geqslant 0$ is exact for smooth data. This is easily seen by separation of variables. Let

(1.6) $$u_0(x,y) = \sum_{p,q=1}^{\infty} c_{pq} \sin \pi px \sin \pi qy\ .$$

Then,

(1.7) $$v(x,y,t_1) = \sum_{p,q} c_{pq} e^{-\pi^2 p^2 t_1} \sin \pi px \sin \pi qy$$

and

(1.8) $$w(x,y,t_1) = u(x,y,t_1)$$

$$= \sum_{p,q} c_{pq}{}^{-\pi^2(p^2+q^2)t_1} \sin \pi px \sin \pi qy\ .$$

Obviously, the equality repeats at each t_k.

The Galerkin method is a variational method for approximating the solution of differential equations based on the weak form of the differential equation. Let

$$\langle f,g \rangle = \int_D fg \, dx \, dy \,. \tag{1.9}$$

Let $H_0^1(D)$ indicate the Sobolev space of functions having gradients in $L^2(D)$ and vanishing on ∂D [22]. A weak form of (1.1) is given by seeking $u(t) \in H_0^1(D)$ such that $\partial u/\partial t \in L^2(D)$ exists as a distribution and such that

$$\begin{gathered} \langle \frac{\partial u}{\partial t}(t),v \rangle + \langle \nabla u(t),\nabla v \rangle = 0 \,, \\ v \in H_0^1(D) \,, \quad 0 < t \leq T \,, \quad u(0) = u_0 \,. \end{gathered} \tag{1.10}$$

The gradient operator refers to the space coordinates only. Let $\mathcal{M}$ be a finite-dimensional subspace of $H_0^1(D)$. The Galerkin method consists of restricting (1.10) to the subspace $\mathcal{M}$; i.e., find a differentiable function $U(t) \in \mathcal{M}$ such that

(1.11)

$$\langle \frac{\partial U}{\partial t}(t),V \rangle + \langle \nabla U(t),\nabla V \rangle = 0 \,, \quad V \in \mathcal{M} \,, \quad 0 < t \leq T \,,$$

$$\langle U(0),V \rangle = \langle u_0,V \rangle \,, \quad V \in \mathcal{M} \,.$$

When a specific basis, say $\{V_1,V_2,\cdots,V_N\}$, is chosen for $\mathcal{M}$ (1.11) reduces to a system of ordinary differential equations. Let

$$(1.12) \quad U(t) = \sum_{i=1}^{N} \zeta_i(t) V_i \ , \quad \zeta = (\zeta_1, \cdots, \zeta_N)^T \ .$$

If

$$(1.13) \qquad C = (c_{ij}) \ , \quad c_{ij} = \langle V_i, V_j \rangle \ ,$$

and

$$(1.14) \qquad A = (a_{ij}) \ , \quad a_{ij} = \langle \nabla V_i, \nabla V_j \rangle \ ,$$

then (1.11) can be written in the form

$$(1.15) \qquad C\zeta'(t) + A\zeta(t) = 0 \ , \quad \zeta(0) \text{ specified } .$$

There are extensive treatments of the accuracy of the approximation U, both in the continuous time version (1.10)-(1.15) and in the discrete time case [5, 10, 12, 25].

2. An Alternating-Direction Galerkin Method for the Heat Equation on a Rectangle

Since the object of any alternating-direction approach is to reduce a multi-dimensional problem to a collection of one-dimensional ones, it is appropriate to choose a subspace $\mathcal{M}$ having a representation by a tensor product basis in the Galerkin process. Let $\{\alpha_i(x) \mid i=1, \cdots, N_x\}$, $\{\beta_j(y) \mid j=1, \cdots, N_y\} \subset H_0^1([0,1])$, and let

$$M_x = \text{Span}(\alpha_1, \cdots, \alpha_{N_x}) ,$$

(2.1) $$M_y = \text{Span}(\beta_1, \cdots, \beta_{N_y}) ,$$

$$M = M_x \otimes M_y = \text{Span}(\alpha_1\beta_1, \cdots, \alpha_{N_x}\beta_{N_y}) .$$

Also, let

$$\langle f,g\rangle_x = \int_0^1 fg\ dx , \quad \langle f,g\rangle_y = \int_0^1 fg\ dy .$$

First, project u_o into $\mathcal{M}$:

(2.3) $$U_o = \sum_{p,q} c_{pq}\alpha_p(x)\beta_q(y) ,$$

where

(2.4) $$\langle U_o,\alpha_k\beta_\ell\rangle = \langle u_o,\alpha_k\beta_\ell\rangle ;$$

$$k = 1, \cdots, N_x , \quad \ell = 1, \cdots, N_y .$$

Now, the continuous time Galerkin fractional steps procedure corresponding to (1.3)-(1.5) can be formulated as follows $(V(t),W(t) \in \mathcal{M}$ for all $t)$:

(2.5) $$V(0) = U_o ,$$

and

$$\langle \frac{\partial V}{\partial t}(t),Z\rangle + \langle \frac{\partial V}{\partial x}(t), \frac{\partial Z}{\partial x}\rangle = 0 ,$$

$$Z \in M, \quad t_n < t \leqslant t_{n+1}, \tag{2.6}$$

$$V(t_n) = \lim_{t \uparrow t_n} W(t), \quad n \geqslant 1,$$

and

$$< \frac{\partial W}{\partial t}(t), Z> + < \frac{\partial W}{\partial y}(t), \frac{\partial Z}{\partial y}> = 0,$$

$$Z \in M, \quad t_n < t \leqslant t_{n+1}, \tag{2.7}$$

$$W(t_n) = \lim_{t \uparrow t_{n+1}} V(t), \quad n \geqslant 0.$$

For there to be any advantage to (2.5)-(2.7), it is necessary that the system reduce to a collection of single-space variable Galerkin problems. That this occurs can be seen as follows. Let

$$\begin{aligned} C_x &= (<\alpha_i, \alpha_j>_x), \quad C_y = (<\beta_i, \beta_j>_y), \\ A_x &= (<\alpha_i', \alpha_j'>_x), \quad A_y = (<\beta_i', \beta_j'>_y). \end{aligned} \tag{2.8}$$

If

$$\begin{aligned} V(t) &= \sum_{p,q} \xi_{pq}(t)\alpha_p(x)\beta_q(y), \\ W(t) &= \sum_{p,q} \eta_{pq}(t)\alpha_p(x)\beta_q(y), \end{aligned} \tag{2.9}$$

then (2.6) can be written in the form

$$C_x \otimes C_y \frac{d\xi}{dt}(t) + A_x \otimes C_y \xi(t) = 0 \ , \quad t_n < t \leqslant t_{n+1} \ ,$$

(2.10)

$$\xi(t_n) = \lim_{t \uparrow t_n} \eta(t) \ ,$$

since

$$\langle\alpha_p\beta_q, \alpha_k\beta_\ell\rangle = \langle\alpha_p, \alpha_k\rangle_x \ \langle\beta_q, \beta_\ell\rangle_y \ ,$$

(2.11)

$$\langle\alpha'_p\beta_q, \alpha'_k\beta_\ell\rangle = \langle\alpha'_p, \alpha'_k\rangle_x \ \langle\beta_q, \beta_\ell\rangle_y \ .$$

Note the C_x, C_y, A_x, and A_y are positive-definite:

$$(C_x\zeta, \zeta) = \|f\|^2_{L^2([0,1])} \ ,$$

$$\text{where} \quad f(x) = \sum_p \zeta_p \alpha_p(x) \ ,$$

(2.12)

$$(A_x\zeta, \zeta) = \|f\|^2_{H^1_0([0,1])} \ , \qquad \text{etc.}$$

Thus, C_y^{-1} exists and the differential equation in (2.10) can be premultiplied by $I \otimes C_y^{-1}$ to give

(2.13) $$C_x \otimes I \frac{d\xi}{dt} + A_x \otimes I\xi = 0 \ , \quad t_n < t \leqslant t_{n+1} \ ;$$

i.e., for each q, $q=1, \cdots, N_y$, (2.13) provides a system of ordinary differential equations for the coefficients ξ_{pq}, $p=1, \cdots, N_x$; and (2.6) is equivalent to a collection of N_y one-dimensional Galerkin problems. Similarly, (2.7) can be reduced to the system

$$I \otimes C_y \frac{d\eta}{dt}(t) + I \otimes A_y \eta(t) = 0 ,$$

(2.14) $$t_n < t \leqslant t_{n+1} ,$$

$$\eta(t_n) = \lim_{t \uparrow t_{n+1}} \xi(t) .$$

In the Introduction it was pointed out that the solution of the continuous time fractional step procedure was exact for (1.1) at the time nodes; this property carries over to the Galerkin approximations in the following sense. For any tensor product basis $\mathcal{M} = \mathcal{M}_x \otimes \mathcal{M}_y$ the solution of (2.6)-(2.7) coincides with the solution of (1.11) at the times $t_1, t_2, \cdots,$ provided that the same initial condition is employed for the two methods. This can be seen as follows. First, it is obviously sufficient to show that they coincide at time t_1. Now, if

(2.15) $$\xi(0) = c ,$$

it follows that

(2.16) $$\xi(t) = \exp[-tC_x^{-1}A_x \otimes I]c , \quad 0 < t \leqslant t_1 .$$

Then

$$\eta(t) = \exp[-tI \otimes C_y^{-1}A_y]\xi(t_1) , \quad 0 < t \leqslant t_1 .$$

Since $P \otimes I$ and $I \otimes Q$ always commute,

$$\lim_{t \uparrow t_1} \eta(t) = \exp[-t_1(C_x^{-1}A \otimes I + I \otimes C_y^{-1}A_y)]c \ . \tag{2.18}$$

For the choice of M being used, (1.15) takes the form

$$\begin{aligned} & C_x \otimes C_y \frac{d\zeta}{dt} + (A_x \otimes C_y + C_x \otimes A_y)\zeta = 0 \ , \\ & \zeta(0) = c \ . \end{aligned} \tag{2.19}$$

Hence,

$$\frac{d\zeta}{dt} + (C_x^{-1}A_x \otimes I + I \otimes C_y^{-1}A_y)\zeta = 0 \ , \tag{2.20}$$

and

$$\zeta(t_1) = \lim_{t \uparrow t_1} \eta(t) \ ; \tag{2.21}$$

thus, the proposition is proved. This result has the immediate consequence that most of the error estimates [12, 25] that have been established for the standard continuous time Galerkin methods can be interpreted to give estimates valid at time $\{t_k\}$ for the system (2.6)-(2.7); these interpretations will be left to the reader.

For practical computation it is necessary to discretize (2.6)-(2.7) in time. One simple way to do this is to employ the Crank-Nicolson difference system. Let the superscript n refer to time $t_n = n\Delta t$, where for convenience the time steps will be taken uniform. Then, (2.6)-(2.7) can be replaced by the system

(2.22)

$$\eta^0 = \xi^0 = c \ ,$$

$$C_x \otimes I \, \frac{\xi^{n+1} - \eta^n}{\Delta t} + A_x \otimes I \, \frac{\xi^{n+1} + \eta^n}{2} = 0 \ , \quad n \geq 0 \ ,$$

$$I \otimes C_y \, \frac{\eta^{n+1} - \xi^{n+1}}{\Delta t} + I \otimes A_y \, \frac{\eta^{n+1} + \xi^{n+1}}{2} = 0 \ , \quad n \geq 0 \ .$$

Some practical remarks regarding computation can be made here before turning to the analysis of the error of approximation of the solution of (1.1) given by that of (2.22). No mention has been made of the choice of the bases $\{\alpha_k(x)\}$ and $\{\beta_\ell(y)\}$. The usual choices are smooth splines, Hermite splines, or other "localized" bases. With any of these choices the solution of the algebraic equations in (2.22) becomes the solution of a collection of (block) band matrix equations. Not only are these algebraic systems trivial by Gaussian elimination, but the left-hand side manipulations need be done only once (for fixed Δt) for $\{\alpha_k\}$ and $\{\beta_\ell\}$, respectively. This fact will strongly motivate the extensions to nonlinear parabolic and elliptic equations to be discussed later.

In order to discuss the error caused by the use of (2.22), it is helpful to eliminate the ξ-variable. It is easy to see that

(2.23)

$$(C_x \otimes I + \tfrac{1}{2} \Delta t A_x \otimes I)(I \otimes C_y + \tfrac{1}{2} \Delta t I \otimes A_y)\eta^{n+1}$$

$$= (C_x \otimes I - \frac{1}{2}\Delta t A_x \otimes I)(I \otimes C_y - \frac{1}{2}\Delta t\ I \otimes A_y)\eta^n ,$$

or

(2.24)

$$[C_x \otimes C_y + \frac{1}{2}\Delta t(C_x \otimes A_y + A_x \otimes C_y) + \frac{1}{4}(\Delta t)^2 A_x \otimes A_y]\eta^{n+1}$$

$$= [C_x \otimes C_y - \frac{1}{2}\Delta t(C_x \otimes A_y + A_x \otimes C_y) + \frac{1}{4}(\Delta t)^2 A_x \otimes A_y]\eta^n .$$

The $A_x \otimes A_y$ terms represent the perturbation of the standard two-dimensional Crank-Nicolson Galerkin process.

At this point it is easy to see how to generalize the method in a small but important way. Consider the inhomogeneous equation

$$\text{(2.25)} \qquad \frac{\partial u}{\partial t} = \Delta u + f(x,y,t) , \quad (x,y) \in D , \quad 0 < t \leq T .$$

Let

$$\text{(2.26)} \qquad f^{n+\frac{1}{2}} = f(x,y,t^{n+\frac{1}{2}}) \quad \text{or} \quad \frac{1}{2}[f(x,y,t^n)+f(x,y,t^{n+1})] .$$

Project $f^{n+\frac{1}{2}}$ into the space $\mathcal{M}$:

$$\text{(2.27)} \qquad (\phi^{n+\frac{1}{2}})_{pq} = \langle f,\alpha_p\beta_q\rangle , \quad \forall\ p,q .$$

Then, (2.24) can be modified as below to include (2.25):

(2.28)

$$[C_x \otimes C_y + \frac{1}{2}\Delta t(C_x \otimes A_y + A_x \otimes C_y) + \frac{1}{4}(\Delta t)^2 A_x \otimes A_y]\eta^{n+1}$$

$$= [C_x \otimes C_y - \frac{1}{2}\Delta t(C_x \otimes A_y + A_x \otimes C_y) + \frac{1}{4}(\Delta t)^2 A_x \otimes A_y]\eta^n + \Delta t\phi^{n+\frac{1}{2}} .$$

Note that no essential complication arises in the evaluation of η^{n+1} ; it can still be otained as a two step one-dimensional calculation. For the purpose of analysis, it is valuable to convert (2.28) back to inner product form:

$$(2.29)\quad \begin{aligned} &< \frac{W^{n+1} - W^n}{\Delta t}, Z > + < \frac{1}{2}\nabla(W^{n+1} + W^n), \nabla Z > \\ &\qquad + \frac{\Delta t}{4} < \frac{\partial^2}{\partial x \partial y}(W^{n+1} - W^n), \frac{\partial^2}{\partial x \partial y} Z > \\ &= < f^{n+\frac{1}{2}}, Z > , \quad Z \in \mathcal{M} . \end{aligned}$$

Now,

$$(2.30)\quad \begin{aligned} &< \frac{u^{n+1} - u^n}{\Delta t} + \delta_1^n, z > + < \frac{1}{2}\nabla(u^{n+1} + u^n + \delta_2^n), \nabla z > \\ &\qquad + \frac{\Delta t}{4} < \frac{\partial^2}{\partial x \partial y}(u^{n+1} - u^n), \frac{\partial^2 z}{\partial x \partial y} \\ &= < f^{n+\frac{1}{2}} + \delta_3^n, z > + \frac{\Delta t}{4} < \frac{\partial^2}{\partial x \partial y}(u^{n+1} - u^n), \frac{\partial^2 z}{\partial x \partial y} , \\ &\qquad z \in H_0^1(D) , \end{aligned}$$

where

$$(2.31)\quad \|\delta_1^n\|_{L^2(D)} , \quad \|\nabla \delta_2^n\|_{L^2(D)} , \quad \|\delta_3^n\|_{L^2(D)} = O((\Delta t)^2)$$

if $u \in C^4(\bar{D} \times [0,T])$ and $f \in C^2(\bar{D} \times [0,T])$. In fact, somewhat weaker assumptions would suffice. Let

$$(2.32)\quad \delta u = \tilde{u} - u , \quad \tilde{u}(t) \in \mathcal{M} ,$$

where $\tilde{u}$ is an arbitrary map of $[0,T]$ into $\mathcal{M}$. Let

(2.33) $$e^n = u^n - W^n, \quad e^{n+\frac{1}{2}} = \frac{1}{2}(e^n + e^{n+1}).$$

Take $z = Z = e^{n+\frac{1}{2}} + \delta u^{n+\frac{1}{2}} = e^{n+\frac{1}{2}} + \frac{1}{2}(\delta u^n + \delta u^{n+1})$ and subtract (2.29) from (2.30):

$$< \frac{e^{n+1} - e^n}{\Delta t} + \delta_1^n, e^{n+\frac{1}{2}} + \delta u^{n+\frac{1}{2}} >$$

$$+ < \nabla e^{n+\frac{1}{2}} + \nabla\delta_2^n, \nabla e^{n+\frac{1}{2}} + \nabla\delta u^{n+\frac{1}{2}} >$$

$$+ \frac{\Delta t}{4} < \frac{\partial^2}{\partial x \partial y}(e^{n+1} - e^n), \frac{\partial^2}{\partial x \partial y} e^{n+\frac{1}{2}} + \frac{\partial^2}{\partial x \partial y} \delta u^{n+\frac{1}{2}} >$$

$$= < \delta_3^n, e^{n+\frac{1}{2}} + \delta u^{n+\frac{1}{2}} >$$

$$+ \frac{\Delta t}{4} < \frac{\partial^2}{\partial x \partial y}(u^{n+1} - u^n), \frac{\partial^2}{\partial x \partial y} e^{n+\frac{1}{2}} + \frac{\partial^2}{\partial x \partial y} \delta u^{n+\frac{1}{2}} >$$

Now, with ε as a generic small constant and C a generic (sometimes large) one,

$$|<\delta_1^n, e^{n+\frac{1}{2}} + \delta u^{n+\frac{1}{2}} >|$$

(2.35) $$\leq \varepsilon \|e^{n+\frac{1}{2}}\|^2_{L^2(D)} + C\|\delta u^{n+\frac{1}{2}}\|^2_{L^2(D)} + C\|\delta_1^n\|^2_{L^2(D)}$$

$$\leq \varepsilon \|e^{n+\frac{1}{2}}\|^2_{H_0^1(D)} + C\|\delta u^{n+\frac{1}{2}}\|^2_{L^2(D)} + C\|\delta_1^n\|^2_{L^2(D)},$$

$$|<\nabla e^{n+\frac{1}{2}}, \nabla\delta u^{n+\frac{1}{2}}>| \leqslant \varepsilon\|e^{n+\frac{1}{2}}\|^2_{H^1_o(D)} + C\|\delta u^{n+\frac{1}{2}}\|^2_{H^1_o(D)} ,$$

$$|<\nabla\delta^n_2, \nabla e^{n+\frac{1}{2}} + \nabla\delta u^{n+\frac{1}{2}}>|$$

$$\leqslant \varepsilon\|e^{n+\frac{1}{2}}\|^2_{H^1_o(D)} + C\|\delta^n_2\|^2_{H^1_o(D)} + C\|\delta u^{n+\frac{1}{2}}\|^2_{H^1_o(D)} ,$$

$$|<\delta^n_3, e^{n+\frac{1}{2}} + \delta u^{n+\frac{1}{2}}>|$$

$$\leqslant \varepsilon\|e^{n+\frac{1}{2}}\|^2_{H^1_o(D)} + C\|\delta^n_3\|^2_{L^2(D)} + C\|\delta u^{n+\frac{1}{2}}\|^2_{L^2(D)} ,$$

where $\|e\|_{H^1_o(D)} = \|\nabla e\|_{L^2(D)} \geqslant C(D)\|e\|_{L^2(D)}$. Also,

$$\frac{\Delta t}{4} < \frac{\partial^2}{\partial x\partial y} (u^{n+1} - u^n), \frac{\partial^2}{\partial x\partial y} e^{n+\frac{1}{2}} >$$

$$= - \frac{\Delta t}{4} < \frac{\partial^3}{\partial x^2\partial y} (u^{n+1} - u^n), \frac{\partial}{\partial y} e^{n+\frac{1}{2}} >$$

(2.36)

$$+ \frac{\Delta t}{4} \int_0^1 \frac{\partial^2}{\partial x\partial y} (u^{n+1} - u^n) \frac{\partial}{\partial y} e^{n+\frac{1}{2}} \Big|_{x=0}^{x=1} dy$$

$$= - \frac{\Delta t}{4} < \frac{\partial^3}{\partial x^2\partial y} (u^{n+1} - u^n), \frac{\partial}{\partial y} e^{n+\frac{1}{2}} > ,$$

since $\partial e^{n+\frac{1}{2}}/\partial y = 0$ on the boundaries $\{x = 0\}$ and $\{x = 1\}$. Thus,

$$|\frac{\Delta t}{4} <\frac{\partial^2}{\partial x \partial y} (u^{n+1}-u^n), \frac{\partial^2}{\partial x \partial y} e^{n+\frac{1}{2}}>| \tag{2.37}$$
$$\leqslant \varepsilon \|e^{n+\frac{1}{2}}\|^2_{H_0^1(D)} + C(\Delta t)^4 .$$

Similarly, $\partial\delta u^{n+\frac{1}{2}}/\partial y = 0$ on $\{x = 0\}$ and $\{x = 1\}$

$$|\frac{\Delta t}{4} <\frac{\partial^2}{\partial x \partial y} (u^{n+1}-u^n), \frac{\partial^2}{\partial x \partial y} \delta u^{n+\frac{1}{2}}>| \tag{2.38}$$
$$\leqslant C(\Delta t)^4 + C \|\delta u^{n+\frac{1}{2}}\|^2_{H_0^1(D)} .$$

It follows from (2.31), (2.34)-(2.38) that

$$\frac{\|e^{n+1}\|^2_{L^2} - \|e^n\|^2_{L^2}}{2\Delta t} + (1-\varepsilon) \|e^{n+\frac{1}{2}}\|^2_{H_0^1}$$
$$+ \frac{\Delta t}{8} \left(\left\|\frac{\partial^2 e^{n+1}}{\partial x \partial y}\right\|^2_{L^2} - \left\|\frac{\partial^2 e^n}{\partial x \partial y}\right\|^2_{L^2}\right) \tag{2.39}$$
$$\leqslant <\frac{e^n-e^{n+1}}{\Delta t}, \delta u^{n+\frac{1}{2}}>$$
$$+ \frac{\Delta t}{4} <\frac{\partial^2}{\partial x \partial y} (e^n-e^{n+1}), \frac{\partial^2}{\partial x \partial y} \delta u^{n+\frac{1}{2}}>$$
$$+ C[\|\delta u^{n+\frac{1}{2}}\|^2_{H_0^1} + (\Delta t)^4] .$$

Multiply by Δt and add for $n = 0,\cdots,k-1$:

$$\frac{1}{2} \|e^k\|^2_{L^2} + (1-\varepsilon) \sum_{n=0}^{k-1} \|e^{n+\frac{1}{2}}\|^2_{H_0^1} \Delta t + \frac{(\Delta t)^2}{8} \left\|\frac{\partial^2 e^k}{\partial x \partial y}\right\|^2_{L^2}$$

(2.40)

$$\leq Ct_k(\Delta t)^4 + C \sum_{n=0}^{k-1} \|\delta u^{n+\frac{1}{2}}\|^2_{H^1_0} \Delta t$$

$$+ \frac{1}{2}\|e^o\|^2_{L^2} + \frac{(\Delta t)^2}{8} \left\|\frac{\partial^2 e^o}{\partial x \partial y}\right\|^2_{L^2}$$

$$+ \sum_{n=0}^{k-1} \left\{ \left\langle \frac{e^n - e^{n+1}}{\Delta t}, \delta u^{n+\frac{1}{2}} \right\rangle \right.$$

$$\left. + \frac{(\Delta t)^2}{4} \left\langle \frac{\partial^2}{\partial x \partial y}\left(\frac{e^n - e^{n+1}}{\Delta t}\right), \frac{\partial^2}{\partial x \partial y} \delta u^{n+\frac{1}{2}} \right\rangle \right\} \Delta t \ .$$

Now, for $\tilde{D} = 1$ or $\partial^2/\partial x \partial y$,

$$\left| \sum_{n=0}^{k-1} \left\langle \tilde{D} \frac{e^n - e^{n+1}}{\Delta t}, \tilde{D} \delta u^{n+\frac{1}{2}} \right\rangle \Delta t \right|$$

$$= \left| \sum_{n=1}^{k-1} \left\langle \tilde{D} e^n, \tilde{D}\delta \frac{u^{n+\frac{1}{2}} - u^{n-\frac{1}{2}}}{\Delta t} \right\rangle \Delta t \right.$$

$$\left. + \langle \tilde{D} e^o, \tilde{D}\delta u \rangle - \langle \tilde{D} e^k, \tilde{D}\delta u^{k-\frac{1}{2}} \rangle \right|$$

(2.41)

$$\leq \frac{1}{2} \sum_{n=1}^{k-1} \left\{ \|\tilde{D} e^n\|^2_{L^2} + \left\|\tilde{D} \frac{u^{n+\frac{1}{2}} - u^{n-\frac{1}{2}}}{\Delta t}\right\|^2_{L^2} \right\} \Delta t$$

$$+ \frac{1}{2}\|\tilde{D} e^o\|^2_{L^2} + \frac{1}{2}\|\tilde{D}\delta u\|^2_{L^2} + \frac{1}{4}\|\tilde{D} e^k\|^2_{L^2}$$

$$+ \|\tilde{D}\delta u^{k-\frac{1}{2}}\|^2_{L^2} \ .$$

Thus,

$$\|e^k\|^2_{L^2} + \sum_{n=1}^{k-1} \|e^{n+\frac{1}{2}}\|^2_{H^1_0} \Delta t + (\Delta t)^2 \left\|\frac{\partial^2}{\partial x \partial y} e^k\right\|^2_{L^2}$$

$$
\begin{aligned}
&\leq C \sum_{n=1}^{k-1} \left\{ \|e^n\|^2_{L^2} + (\Delta t)^2 \left\|\frac{\partial^2}{\partial x \partial y} e^n\right\|^2_{L^2}\right\} \Delta t \\
&+ C\Bigg[(\Delta t)^4 + \sum_{n=1}^{k-1} \|\delta u^{n+\frac{1}{2}}\|^2_{H_0^1} \Delta t \\
&+ \sum_{n=1}^{k-1} \left\{ \left\|\delta \frac{u^{n+\frac{1}{2}} - u^{n-\frac{1}{2}}}{\Delta t}\right\|^2_{L^2} \right. \\
&\left. + (\Delta t)^2 \left\|\frac{\partial^2}{\partial x \partial y} \delta \frac{u^{n+\frac{1}{2}} - u^{n-\frac{1}{2}}}{\Delta t}\right\|^2_{L^2}\right\} \Delta t \\
&+ \|e^0\|^2_{L^2} + (\Delta t)^2 \left\|\frac{\partial^2}{\partial x \partial y} e^0\right\|^2_{L^2} + \|\delta u^{\frac{1}{2}}\|^2_{L^2} \\
&+ (\Delta t)^2 \left\|\frac{\partial^2}{\partial x \partial y} \delta u^{\frac{1}{2}}\right\|^2_{L^2} + \|\delta u^{k-\frac{1}{2}}\|^2_{L^2} \\
&+ (\Delta t)^2 \left\|\frac{\partial^2}{\partial x \partial y} \delta u^{k-\frac{1}{2}}\right\|^2_{L^2}\Bigg],
\end{aligned}
\tag{2.42}
$$

where C is independent of k for $t_k \leq T$. Let

$$
\begin{aligned}
\|\tilde{v}\|^2_{X \times L^2} &= \sum_{n=0}^{\frac{T}{\Delta t}-1} \|v^{n+\frac{1}{2}}\|^2_X \Delta t \\
\|v\|^2_{X \times L^2} &= \sum_{n=0}^{T/\Delta t} \|v^n\|^2_X \Delta t, \\
\|v\|_{X \times L^\infty} &= \max_{0 \leq t_n \leq T} \|v^n\|_X .
\end{aligned}
\tag{2.43}
$$

The discrete Gronwall lemma and (2.42) imply that

$$
\begin{aligned}
&\|e\|^2_{L^2 \times L^\infty} + \|\tilde{e}\|^2_{H^1_0 \times L^2} + (\Delta t)^2 \left\|\frac{\partial^2 e}{\partial x \partial y}\right\|^2_{L^2 \times L^\infty} \\
&\quad \leq C\Big[(\Delta t)^4 + \|\delta u\|^2_{H^1_0 \times L^2} \\
&\quad + \|\delta \partial_t u\|^2_{L^2 \times L^2} + \|\delta u\|^2_{L^2 \times L^\infty} \\
&\quad + (\Delta t)^2 \left\|\frac{\partial^2}{\partial x \partial y} \delta u\right\|^2_{L^2 \times L^\infty} \qquad (2.44) \\
&\quad + (\Delta t)^2 \left\|\frac{\partial^2}{\partial x \partial y} \delta \partial_t u\right\|^2_{L^2 \times L^2} \\
&\quad + \|e^0\|^2_{L^2} + (\Delta t)^2 \left\|\frac{\partial^2}{\partial x \partial y} e^0\right\|^2_{L^2}\Big],
\end{aligned}
$$

where $\partial_t u^n = (u^{n+1} - u^{n-1})/2\Delta t = (u^{n+\frac{1}{2}} - u^{n-\frac{1}{2}})/\Delta t$. In the $L^2 \times L^2$ norms of terms involving ∂_t the sum starts at $n = 1$.

The above argument can be summarized in the following theorem.

<u>Theorem 1</u>. Let $f \in C^2(\bar{D} \in [0,T])$ and assume that the solution u of

$$\frac{\partial u}{\partial t} = \Delta u + f(x,y,t) , \qquad (x,y) \in D = [0,1]^2, \quad 0 < t \leq T ,$$

$$u(x,y,t) = 0 , \qquad (x,y) \in \partial D, \quad 0 < t \leq T ,$$

$$u(x,y,0) = u_0(x,y) , \qquad (x,y) \in D ,$$

is four times continuously differentiable. Let the alternating-direction Galerkin approximate solution be defined by (2.28). Then, the approximation error can be estimated by (2.44).

A few remarks are in order. First, homogeneous Neumann conditions could have been specified instead of homogeneous Dirichlet conditions, or a mixture of the two. With Neumann conditions the space $H^1(D)$ would have been employed rather than $H_0^1(D)$. The proof is essentially unaltered; the only significant change occurs in the integration by parts in (2.36), but the other term vanishes in the boundary integral and the result is unaltered.

Inhomogeneous boundary conditions can be treated several ways. For the rectangular domain of this paper, it is particularly easy to modify the inhomogeneous term f in order to subtract out either values or flows specified on ∂D. It is equally easy with the usual choices for basis functions for $\mathcal{M}$ to include the basis functions affecting the boundary nodes that had been supposed in setting up $\mathcal{M}$ in order to satisfy the boundary conditons (i.e., for a Hermite basis, put the "value" functions back into $\mathcal{M}$). These two methods are frequently computationally equivalent.

3. General Parabolic Equations

In this section alternating-direction methods for the nonlinear (or linear with variable coefficients) parabolic equation

$$(3.1) \quad \frac{\partial u}{\partial t} = \nabla \cdot (a(x,y,t,u)\nabla u) + f(x,y,t,u,\nabla u) \, , \quad (x,y) \in D \, ,$$

will be devised using very strongly the results of the last section. The region D will remain a rectangle for the alternating-direction results. Two procedures will be studied. The first will be based on a Laplace-modified Galerkin analogue of the forward difference equation and the other on a Laplace-modified unconditionally unstable, centered-in-time, explicit difference equation. Both methods lead to simple algebraic problems essentially identical with the one arising in (2.28); the second is more accurate or more efficient, but it requires stronger hypotheses.

The Laplace-modified forward Galerkin equation can be expressed as follows, with no restriction to either a tensor product basis or to a rectangle:

$$(3.2) \quad \begin{aligned} &\left\langle \frac{U^{n+1}-U^n}{\Delta t}, V \right\rangle + \langle a^n(U)\nabla U^n, \nabla V\rangle \\ &\qquad + \lambda \langle \nabla(U^{n+1}-U^n), \nabla V\rangle \\ &\qquad = \langle f^n(U), V\rangle \, , \quad V \in M \, , \end{aligned}$$

where $M \subset H_0^1(D)$ and

$$(3.3) \quad a^n(U) = a(t_n, U^n) \, , \quad f^n(U) = f(t_n, U^n, \nabla U^n) \, .$$

The Laplace-modification of the centered equation is

$$(3.4)\quad \begin{aligned} &< \frac{U^{n+1}-U^{n-1}}{2\Delta t}, V> + <a^n(U)\nabla U^n, \nabla V> \\ &+ \lambda <\nabla(U^{n+1}-2U^n+U^{n-1}), \nabla V> \\ &= <f^n(U), V>, \quad V \in M . \end{aligned}$$

The parameter λ is subject to a lower bound in both cases in order that stability can occur. The time-discretization is locally first order for (3.2) and second order for (3.4); the analysis of (3.4) will be limited to the case $a(x,y,t,u) = a(x,y,t)$.

Before perturbing (3.2) and (3.4) to obtain alternating-direction versions, let us consider the accuracy of the solution of (3.2), assuming the same boundary and initial conditons as before. Then, for $v \in H_0^1(D)$,

$$(3.5)\quad \begin{aligned} &< \frac{u^{n+1}-u^n}{\Delta t} + \delta_1^n, v> + <a^n(u)\nabla u^n, \nabla v> \\ &+ \lambda <\nabla(u^{n+1}-u^n), \nabla v> \\ &= <f^n(u), v> \\ &+ \lambda <\nabla(u^{n+1}-u^n), \nabla v> , \end{aligned}$$

and

$$(3.6)\quad \|\delta_1^n\|_{L^2} = O(\Delta t) ,$$

$$\|\nabla(u^{n+1}-u^n)\|_{L^2} = O(\Delta t) ,$$

if $u \in C^2(\bar{D} \times [0,T])$. Let $v = V = e^{n+1} + \delta u^{n+1}$ and subtract (3.2) from (3.5):

$$< \frac{e^{n+1}-e^n}{\Delta t} ,e^{n+1}+\delta u^{n+1}> + <a^n(U)\nabla e^n,\nabla e^{n+1}+\nabla\delta u^{n+1}>$$

$$+ <\{a^n(u)-a^n(U)\}\nabla u^n,\nabla e^{n+1}+\nabla\delta u^{n+1}>$$

$$+ \lambda <\nabla(e^{n+1}-e^n),\nabla e^{n+1}+\nabla\delta u^{n+1}>$$

$$= <f^n(u)-f^n(U),e^{n+1}+\delta u^{n+1}>$$

$$+ \lambda <\nabla(u^{n+1}-u^n),\nabla e^{n+1}+\nabla\delta u^{n+1}>$$

$$- <\delta_1^n,e^{n+1}+\delta u^{n+1}> .$$

Assume that the functions a and f are Lipschitz continuous and bounded with respect to u and ∇u. Then,

$$<e^{n+1}-e^n,e^{n+1}> \geq \frac{1}{2}\|e^{n+1}\|^2_{L^2} - \frac{1}{2}\|e^n\|^2_{L^2} ,$$

$$|<\{a^n(u)-a^n(U)\}\nabla u^n,\nabla e^{n+1}+\nabla\delta u^{n+1}>|$$

$$\leq \|\nabla u^n\|_{L^\infty}\|a^n(u)-a^n(U)\|_{L^2}\left[\|e^{n+1}\|_{H_0^1} + \|\delta u^{n+1}\|_{H_0^1}\right]$$

$$\leq \varepsilon\|e^{n+1}\|^2_{H_0^1} + C\|e^n\|^2_{L^2} + C\|\delta u^{n+1}\|^2_{H_0^1} ,$$

$$|\langle a^n(U)\nabla e^n, \nabla\delta u^{n+1}\rangle| \leqslant \varepsilon \, \|e^n\|^2_{H^1_0} + C\, \|\delta u^{n+1}\|^2_{H^1_0} \quad ,$$

$$|\langle\nabla(e^{n+1}-e^n), \nabla\delta u^{n+1}\rangle| \leqslant \varepsilon \, \|e^{n+1}\|^2_{H^1_0} + \varepsilon\, \|e^n\|^2_{H^1_0} \tag{3.8}$$

$$+ C\, \|\delta u^{n+1}\|^2_{H^1_0} \quad ,$$

$$|\langle f^n(u) - f^n(U), e^{n+1}+\delta u^{n+1}\rangle|$$

$$\leqslant \varepsilon\, \|e^n\|^2_{H^1_0} + C\, \|e^n\|^2_{L^2} + C\, \|e^{n+1}\|^2_{L^2} + C\, \|\delta u^{n+1}\|^2_{L^2} ,$$

$$|\langle\nabla(u^{n+1}-u^n), \nabla e^{n+1}+\nabla\delta u^{n+1}\rangle|$$

$$\leqslant \varepsilon\, \|e^{n+1}\|^2_{H^1_0} + C\, \|\delta u^{n+1}\|^2_{H^1_0} + C(\Delta t)^2 \quad ,$$

$$|\langle\delta_1^n, e^{n+1}+\delta u^{n+1}\rangle| \leqslant C\, \|e^{n+1}\|^2_{L^2} + C\, \|\delta u^{n+1}\|^2_{L^2} + C(\Delta t)^2 \; .$$

The parameter λ can be chosen as follows. Assume

$$0 < a_{min} \leqslant a(x,y,t,u) \leqslant a_{max} < \infty \; . \tag{3.9}$$

If $\lambda > 0$,

$$\|a^n(U)-\lambda\|_{L^\infty} \leqslant \max(\lambda - a_{min}, \, a_{max} - \lambda) \; . \tag{3.10}$$

Then

$$\lambda - \frac{1}{2}\, \|a^n(U)-\lambda\|_{L^\infty} - \frac{1}{2}\, \|a^{n+1}(U)-\lambda\|_{L^\infty} - \varepsilon = \tag{3.11}$$

$$= \min(a_{min}, 2\lambda - a_{max}) - \varepsilon = q > 0 ,$$

provided that

$$\lambda > \frac{1}{2} a_{max} . \tag{3.12}$$

Since

$$<\lambda \nabla e^{n+1} + \{a^n(U) - \lambda\} \nabla e^n, \nabla e^{n+1}>$$

$$\geqslant \{\lambda - \frac{1}{2} \|a^n(U)-\lambda\|_{L^\infty}\} \|e^{n+1}\|^2_{H^1_0} - \frac{1}{2} \|a^n(U)-\lambda\|_{L^\infty} \|e^n\|^2_{H^1_0} , \tag{3.13}$$

it follows that

$$\frac{\|e^{n+1}\|^2_{L^2} - \|e^n\|^2_{L^2}}{2\Delta t} + \{\lambda - \frac{1}{2} \|a^n(U)-\lambda\|_{L^\infty} - \varepsilon\} \|e^{n+1}\|^2_{H^1_0} - \{\frac{1}{2} \|a^n(U)-\lambda\|_{L^\infty} + \varepsilon\} \|e^n\|^2_{H^1_0} + <\frac{e^{n+1}-e^n}{\Delta t}, \delta u^{n+1}> \tag{3.14}$$

$$\leqslant C \|e^{n+1}\|^2_{L^2} + C \|e^n\|^2_{L^2} + C \|\delta u^{n+1}\|^2_{H^1_0} + C(\Delta t)^2 .$$

Multiply by $2\Delta t$ and add for $n = 0,\cdots,N-1$:

$$\|e^N\|^2_{L^2} + 2q \sum_1^N \|e^n\|^2_{H^1_0} \Delta t + 2 \sum_0^{N-1} <\frac{e^{n+1}-e^n}{\Delta t}, \delta u^{n+1}> \delta t$$

(3.15)

$$\leqslant \|e^o\|^2_{L^2} + C\Delta t \, \|e^o\|^2_{H^1_0} + C \sum_0^N \|e^n\|^2_{L^2} \Delta t + C \sum_1^N \|\delta u^n\|^2_{H^1_0} \Delta t + C(\Delta t)^2 .$$

Since

$$\left| \sum_0^{N-1} < \frac{e^{n+1}-e^n}{\Delta t}, \delta u^{n+1} > \Delta t \right|$$

$$(3.16) \quad = \left| <e^N, \delta u^N> - <e^o, \delta u^1> - \sum_1^{N-1} <e^n, \frac{\delta u^{n+1}-\delta u^n}{\Delta t} > \Delta t \right|$$

$$\leqslant \frac{1}{2} \left[\|e^N\|^2_{L^2} + \|e^o\|^2_{L^2} + \|\delta u^N\|^2_{L^2} + \|\delta u^1\|^2_{L^2} \right] + \frac{1}{2} \sum_1^{N-1} \left\| \frac{\delta u^{n+1}-\delta u^n}{\Delta t} \right\|^2_{L^2} \Delta t + \frac{1}{2} \sum_1^{N-1} \|e^n\|^2_{L^2} \Delta t ,$$

the Gronwall lemma implies that

$$\|e\|^2_{L^2 \times L^\infty} + \|e\|^2_{H^1_0 \times L^2}$$

$$(3.17) \quad \leqslant C \Big[\|e^o\|^2_{L^2} + \Delta t \, \|e^o\|^2_{H^1_0} + \|\delta u\|^2_{H^1_0 \times L^2} + \|\Delta_t \delta u\|^2_{L^2 \times L^2} + \|\delta u\|^2_{L^2 \times L^2} + (\Delta t)^2 \Big] ,$$

where $\Delta_t \delta u^n = (\delta u^{n+1} - \delta u^n)/\Delta t$. Thus the Laplace-modified procedure (3.2) leads to the satisfactory error estimate (3.17) under very reasonable restrictions on the equation (3.1) and its solution, provided only that the restriction

(3.12) is observed. Recall that no constraint on the domain D was imposed nor was there any on the subspace M of $H_0^1(D)$. It is also obvious that the estimate (3.17) was independent of the dimension of D.

In two or more space variables the obtaining of the solution of (3.2) at each time level would require the solution of the same type of algebraic system as arises in elliptic problems. The alternating-direction modification avoids this. Let D again be the unit squire and let M have the tensor product basis $\{\alpha_k\} \otimes \{\beta_\ell\}$ as before. Then the algebraic system for (3.2) can be written in the form

$$
\begin{aligned}
&[C_x \otimes C_y + \lambda\Delta t(A_x \otimes C_y + C_x \otimes A_y)]\eta^{n+1} \\
&\qquad = [C_x \otimes C_y - \lambda\Delta t(A_x \otimes C_y + C_x \otimes A_y)]\eta^n + \Delta t\psi_2^n
\end{aligned}
\tag{3.18}
$$

where

$$
\begin{aligned}
&(\psi^n)_{pq} = - \langle a^n(U)\nabla U^n, \nabla(\alpha_p\beta_q)\rangle + \langle f^n(U), \alpha_p\beta_q\rangle , \\
&\qquad \alpha_p \in M_x , \quad \beta_q \in M_y ,
\end{aligned}
\tag{3.19}
$$

and C_x, A_x, etc., are still given by (2.8). This relation needs to be perturbed by the term

$$\lambda^2(\Delta t)^2 A_x \otimes A_y(\eta^{n+1} - \eta^n) \tag{3.20}$$

so that the left-hand side factors and so that the

perturbation term is sufficiently high order in time. Thus, an alternating-direction modification of (3.2) can be defined as below:

(3.21)

$$(C_x \otimes I + \lambda\Delta t A_x \otimes I)(I \otimes C_y + \lambda\Delta t I \otimes A_y)(\eta^{n+1}-\eta^n) = \Delta t\psi^n .$$

The left-hand side in (3.21) was written in terms of $\eta^{n+1}-\eta^n$ deliberately; solving for the change in η both reduces the number of arithmetic operations per time step and the relative size of the rounding error in η^{n+1} incurred. It also emphasizes a practical aspect of (3.21). Note that solving (3.21) is accomplished by solving one collection of identical one-dimensional problems followed by a second collection. Thus, the forward solution in Gaussian elimination can be done just once for all the time steps of the entire problem; only the right-hand side must be calculated and the elimination solution consists only of right-hand side manipulations on small problems and the back solutions, even though the differential equation has time-dependent or nonlinear coefficients. Consequently, (3.21) provides a better numerical method than a first-order correct in time direct version of alternating-directions. Another practical comment is that the quadratures, if done dumbly, can destroy any computational advantages inherent in Galerkin procedures. A very significant reduction [11, 13] in computing can be obtained by projecting $a^n(U)$ and $f^n(U)$ into the space $\mathcal{M}$ and then evaluating the integrals by formula; however, the analysis to follow will not consider this.

The analysis of (3.21) is again best carried out using the equivalent inner product form

$$
\begin{aligned}
&< \frac{U^{n+1}-U^{n}}{\Delta t}, Z> + <a^{n}(U)\nabla U^{n}, Z> + \lambda <\nabla(U^{n+1}-U^{n}), \nabla Z> \\
(3.22)\quad & + \lambda^{2}\Delta t < \frac{\partial^{2}}{\partial x \partial y}(U^{n+1}-U^{n}), \frac{\partial^{2}}{\partial x \partial y} Z> = <f^{n}(U), Z>, \\
& \qquad Z \in M .
\end{aligned}
$$

If u $C^4(\bar{D} \times [0,T])$ and a and f are Lipschitz continuous with respect to u and ∇u, for $z \in H_0^1(D)$

$$
\begin{aligned}
&< \frac{u^{n+1}-u^{n}}{\Delta t} + \delta_1^n, z> + <a^{n}(u)\nabla u^{n}, \nabla z> + \lambda <\nabla(u^{n+1}-u^{n}), \nabla z> \\
(3.23)\quad & + \lambda^{2}\Delta t < \frac{\partial^{2}}{\partial x \partial y}(u^{n+1}-u^{n}), \frac{\partial^{2} z}{\partial x \partial y} > \\
& = <f^{n}(u), z> + \lambda <\nabla(u^{n+1}-u^{n}), \nabla z> \\
& + \lambda^{2}\Delta t < \frac{\partial^{2}}{\partial x \partial y}(u^{n+1}-u^{n}), \frac{\partial^{2} z}{\partial x \partial y} > .
\end{aligned}
$$

Again take $z = Z = e^{n+1} + \delta u^{n+1}$. All of the terms treated in (2.7)-(3.13) can be handled in exactly the same fashion. The last term can be integrated by parts analogously to (2.37) to give the bound

$$|\langle \frac{\partial^2}{\partial x\partial y}(u^{n+1}-u^n), \frac{\partial^2 e^{n+1}}{\partial x\partial y} + \frac{\partial^2}{\partial x\partial y}\delta u^{n+1}\rangle|$$

(3.24)

$$\leqslant \varepsilon \|e^{n+1}\|^2_{H^1_0} + C\|\delta u^{n+1}\|^2_{H^1_0} + C(\Delta t)^2 .$$

Thus

$$\frac{\|e^{n+1}\|^2_{L^2} - \|e^n\|^2_{L^2}}{2\Delta t} + \left\{\lambda - \frac{1}{2}\|a^n(U)-\lambda\|_{L^\infty} - \varepsilon\right\}\|e^{n+1}\|^2_{H^1_0}$$

$$- \left\{\frac{1}{2}\|a^n(U)-\lambda\|_{L^\infty} + \varepsilon\right\}\|e^n\|^2_{H^1_0}$$

(3.25)

$$+ \frac{1}{2}\lambda^2\Delta t\left[\|\frac{\partial^2 e^{n+1}}{\partial x\partial y}\|^2_{L^2} - \|\frac{\partial^2 e^n}{\partial x\partial y}\|^2_{L^2}\right] + \langle \frac{e^{n+1}-e^n}{\Delta t}, \delta u^{n+1}\rangle$$

$$+ \lambda^2(\Delta t)^2 \langle \frac{\partial^2}{\partial x\partial y}\left(\frac{e^{n+1}-e^n}{\Delta t}\right), \frac{\partial^2}{\partial x\partial y}\delta u^{n+1}\rangle$$

$$\leqslant C\|e^n\|^2_{L^2} + C\|e^{n+1}\|^2_{L^2} + C\|\delta u^{n+1}\|^2_{H^1_0} + C(\Delta t)^2 .$$

The arguments given in (2.40)-(2.44) and (3.14)-(3.17) can be combined to yield the estimate (again assuming that (3.12) holds)

$$\|e\|^2_{L^2\times L^\infty} + \|e\|^2_{H^1_0\times L^2} + (\Delta t)^2\|\frac{\partial^2 e}{\partial x\partial y}\|^2_{L^2\times L^\infty}$$

$$\leqslant C\left[(\Delta t)^2 + \|e^0\|^2_{L^2} + \Delta t\|e^0\|^2_{H^1_0} + (\Delta t)^2\|\frac{\partial^2 e^0}{\partial x\partial y}\|^2_{L^2}\right.$$

(3.26)

$$+ \|\delta u\|^2_{L^2\times L^\infty} + \|\delta u\|^2_{H^1_0\times L^2} + \|\Delta_t \delta u\|^2_{L^2\times L^2}$$

$$+ (\Delta t)^2 \left\|\frac{\partial^2}{\partial x\partial y}\delta u\right\|^2_{L^2\times L^\infty} + (\Delta t)^2 \left\|\Delta_t \frac{\partial^2}{\partial x\partial y}\delta u\right\|^2_{L^2\times L^2}\Bigg] .$$

Note that the three new terms introduced by the alternating-directions are multiplied by $(\Delta t)^2$; thus, the accuracy of approximation holds if it is possible to approximate u as a function of time in M boundedly for the three norms multiplied by $(\Delta t)^2$. The above derivation leads to the following theorem.

Theorem 2. Let $u \in C^4(\overline{D} \times [0,T])$, $D = [0,1]^2$, be the solution of

$$\frac{\partial u}{\partial t} = \nabla\cdot(a(x,y,t,u)\nabla u) + f(x,y,t,u,\nabla u) ,$$

$$(x,y) \in D , \quad 0 < t \leq T ,$$

$$u = 0 , \qquad (x,y) \in \partial D , \quad 0 < t \leq T ,$$

$$u = u_0(x,y) , \qquad (x,y) \in D , \quad t = 0 ,$$

where f and a are bounded and Lipschitz continuous with respect to u and ∇u and $f \in C^2$ as a function of t. Let u be approximated by the solution U of (3.21)-(3.22) and let $e = u - U$. Then e can be estimated by (3.26).

Consider now equation (3.4) when the coefficient a is linear:

$$\left< \frac{U^{n+1}-U^{n-1}}{2\Delta t}, V\right> + \left<a^n \nabla U^n, \nabla V\right>$$

$$+ \lambda \left<\nabla(U^{n+1}-2U^n+U^{n-1}), \nabla V\right> \tag{3.27}$$

$$= \left<f^n(U), V\right>, \quad V \in M,$$

where $f^n(U)$ is as in (3.3),

$$a^n = a(x,y,t_n), \tag{3.28}$$

D is arbitrary, and $M \subset H_0^1(D)$. Then, if $u \in C^4(\bar{D} \times [0,T])$ and a and f satisfy the previous conditions on them,

$$\left< \frac{u^{n+1}-u^{n-1}}{2\Delta t} + \delta_1^n, v\right> + \left<a^n \nabla u^n, \nabla z\right>$$

$$+ \lambda \left<\nabla(u^{n+1}-2u^n+u^{n-1}) + \nabla\delta_2^n, \nabla v\right> \tag{3.29}$$

$$= \left<f^n(u), v\right>, \quad v \in H_0^1(D),$$

where

$$\|\delta_1^n\|_{L^2}, \quad \|\nabla\delta_2^n\|_{L^2} = O((\Delta t)^2). \tag{3.30}$$

The test function that would correspond to the choices made earlier would be $\lambda e^{n+1} + (a^n-2\lambda)e^n + \lambda e^{n-1}$ plus the terms to put it back into M, but the multiplier $a^n - 2\lambda$ in general removes the function from M. A new choice must be made, and it is $v = V = e^{n+1} + \delta u^{n+1} - e^{n-1} - \delta u^{n-1}$. Then,

$$\|\partial e^n\|^2_{L^2} + \frac{\lambda}{2\Delta t}\left[\|e^{n+1}\|^2_{H^1_0} - \|e^{n-1}\|^2_{H^1_0} - 2\Delta t <(2-\frac{a^n}{\lambda})\nabla e^n, \nabla\partial e^n>\right]$$

$$(3.31) \qquad \leqslant \varepsilon\|\partial e^n\|^2_{L^2} + C\left[\|e^{n+1}\|^2_{H^1_0} + \|e^n\|^2_{H^1_0} + \|e^{n-1}\|^2_{H^1_0} + \|\partial\delta u^n\|^2_{H^1_0} + (\Delta t)^4\right],$$

since $u \in C^4$ implies that $-\Delta\delta^n_2 = \Delta(u^{n+1} - 2u^n + u^{n-1}) = O((\Delta t)^2)$ and

$$|<\nabla\delta^n_2, \nabla\partial e^n>| = |<\Delta\delta^n_2, \partial e^n>|$$

$$(3.32) \qquad \leqslant \varepsilon\|\partial e^n\|^2_{L^2} + C(\Delta t)^4 .$$

Multiply by $2\Delta t$ and add for $n = 1,\cdots,N-1$:

$$(2-\varepsilon)\sum_{n=1}^{N-1}\|\partial e^n\|^2_{L^2}\Delta t + \lambda\left[\|e^N\|^2_{H^1_0} + \|e^{N-1}\|^2_{H^1_0}\right.$$

$$- 2 \sum_{n=1}^{N-1} \langle (2- \frac{a^n}{\lambda})\nabla e^n, \nabla\partial e^n\rangle \Delta t \Big]$$

(3.33)

$$\leq \lambda\Big[\|e^1\|^2_{H^1_0} + \|e^0\|^2_{H^1_0}\Big] + C\Big[(\Delta t)^4 + \sum_{n=1}^{N-1} \|\partial\delta u^n\|^2_{H^1_0} \Delta t$$

$$+ \sum_{n=0}^{N} \|e^n\|^2_{H^1_0} \Delta t\Big] .$$

Now, since a is differentiable in t,

$$\Big| 2 \sum_{n=1}^{N-1} \langle (2- \frac{a^n}{\lambda})\nabla e^n, \nabla\partial e^n\rangle \Delta t \Big|$$

$$= \Big| \langle (2- \frac{a^{N-1}}{\lambda})\nabla e^{N-1}, \nabla e^N\rangle - \langle (2- \frac{a^1}{\lambda})\nabla e^1, \nabla e^0\rangle$$

$$+ \sum_{n=1}^{N-1} \langle \frac{a^n - a^{n+1}}{\lambda\Delta t} \nabla e^n, \nabla e^{n+1}\rangle \Delta t \Big|$$

(3.34)

$$\leq \frac{1}{2} \|2- \frac{a^{N-1}}{\lambda}\|_{L^\infty} \Big\{ \|e^N\|^2_{H^1_0} + \|e^{N-1}\|^2_{H^1_0}\Big\}$$

$$+ \frac{1}{2} \|2- \frac{a^1}{\lambda}\|_{L^\infty} \Big\{ \|e^1\|^2_{H^1_0} + \|e^0\|^2_{H^1_0}\Big\}$$

$$+ C \sum_{n=1}^{N} \|e^n\|^2_{H^1_0} \Delta t .$$

If

(3.35) $$\lambda > \frac{1}{4} a_{max} ,$$

then $\| 2 - \frac{a^n}{\lambda} \|_{L^\infty} \leqslant 2-q$ for some $q > 0$ and Gronwall implies that

$$(3.36)\qquad \begin{aligned} \|\partial e\|^2_{L^2\times L^2} + \|e\|^2_{H^1_0\times L^\infty} \leqslant C\Big[\|e^0\|^2_{H^1_0} + \|e^1\|^2_{H^1_0} \\ + \|\partial\delta u\|^2_{H^1_0\times L^2} + (\Delta t)^4 \Big] . \end{aligned}$$

The appearance of e^1 in the error estimate is quite natural, since (3.27) requires U^0 and U^1 as initial values. It is clear that U^1 must be produced by some other procedure; discussion of means of obtaining U^1 will be postponed until after the introduction of an alternating-direction version of (3.27).

The algebraic form of (3.27) is the following, with C_x, A_x, etc., as before:

$$(3.37)\qquad \begin{aligned} (C_x \otimes C_y)(\eta^{n+1} - \eta^{n-1}) + 2\lambda\Delta t(A_x \otimes C_y \\ + C_x \otimes A_y)(\eta^{n+1} - 2\eta^n + \eta^{n-1}) = 2\Delta t\Phi^n , \end{aligned}$$

where

$$(3.38)\qquad (\Phi^n)_{pq} = -\langle a^n\nabla U^n, \nabla\alpha_p\beta_q\rangle + \langle f^n(U), \alpha_p\beta_q\rangle .$$

In order that the operator on η^{n+1} factor, it is necessary to perturb the equation by adding the term $4\lambda^2(\Delta t)^2 A_x \otimes A_y \eta^{n+1}$. To preserve the accuracy it is necessary to subtract off similar terms involving η^n and/or η^{n-1}. The following perturbation accomplishes those objectives:

$$\begin{aligned}(3.39)\qquad &(C_x \otimes C_y + 4\lambda^2(\Delta t)^2 A_x \otimes A_y)(\eta^{n+1} - \eta^{n-1}) \\ &+ 2\lambda\Delta t(A_x \otimes C_y + C_x \otimes A_y)(\eta^{n+1} - 2\eta^n + \eta^{n-1}) \\ &= 2\Delta t\Phi^n .\end{aligned}$$

It follows in the usual manner with the test function $\partial e^n + \partial\delta u^n$ that, with $\|\delta_1^n\|_{L^2} = O((\Delta t)^2)$,

$$\begin{aligned}&<\partial e^n+\delta_1^n, \partial e^n+\partial\delta u^n> + <a^n\nabla e^n, \nabla\partial e^n+\nabla\partial\delta u^n> \\ &+ \lambda<\nabla(e^{n+1}-2e^n+e^{n-1}), \nabla\partial e^n+\nabla\partial\delta u^n> \\ &+ 4\lambda^2(\Delta t)^2 < \frac{\partial^2}{\partial x\partial y}\partial e^n, \frac{\partial^2}{\partial x\partial y}(\partial e^n+\partial\delta u^n)> \\ (3.40)\qquad &= <f^n(u)-f^n(U), \partial e^n+\partial\delta u^n> \\ &+ \lambda<\nabla(u^{n+1}-2u^n+u^{n-1}), \nabla\partial e^n+\nabla\partial\delta u^n> \\ &+ 4\lambda^2(\Delta t) < \frac{\partial^2}{\partial x\partial y}\partial u^n, \frac{\partial^2}{\partial x\partial y}(\partial e^n+\partial\delta u^n)>\end{aligned}$$

for $u \in C^5(\bar{D} \times [0,T])$. The last terms on the left-hand and right-hand sides are new and can be treated as below. First,

$$(3.41)\qquad \begin{aligned} &< \frac{\partial^2}{\partial x\partial y}\,\partial e^n,\ \frac{\partial^2}{\partial x\partial y}\,(\partial e^n+\partial\delta u^n)> \\ &\geq \frac{1}{2}\,\Big\|\frac{\partial^2}{\partial x\partial y}\,\partial e^n\Big\|^2_{L^2} - \frac{1}{2}\,\Big\|\frac{\partial^2}{\partial x\partial y}\,\partial\delta u^n\Big\|^2_{L^2}\,. \end{aligned}$$

The right-hand side term can be integrated by parts twice without introducing a boundary term; hence,

$$(3.42)\qquad \begin{aligned} &\Big|(\Delta t)^2 < \frac{\partial^2}{\partial x\partial y}\,\partial u^n,\ \frac{\partial^2}{\partial x\partial y}\,(\partial e^n+\partial\delta u^n)>\Big| \\ &= \Big|(\Delta t)^2 < \frac{\partial^4}{\partial x^2\partial y^2}\,\partial u^n, \partial e^n+\partial\delta u^n>\Big| \\ &\leq C(\Delta t)^4 + \varepsilon\,\|\partial e^n\|^2_{L^2} + C\,\|\partial\delta u^n\|^2_{L^2}\,. \end{aligned}$$

Thus,

$$(3.43)\qquad \begin{aligned} &\|\partial e^n\|^2_{L^2}\,\Delta t + \lambda\,\Big\{\|e^{n+1}\|^2_{H^1_0} - \|e^{n-1}\|^2_{H^1_0} \\ &\quad + 2\Delta t < \Big(\frac{a^n}{\lambda} - 2\Big)\nabla e^n, \nabla\partial e^n>\Big\} \\ &\quad + \lambda^2(\Delta t)^2\,\Big\|\frac{\partial^2}{\partial x\partial y}\,\partial e^n\Big\|^2_{L^2}\,\Delta t \\ &\leq C\Big[(\Delta t)^4 + \|\partial\delta u^n\|^2_{H^1_0} + \|e^{n+1}\|^2_{H^1_0} \\ &\quad + \|e^n\|^2_{H^1_0} + \|e^{n-1}\|^2_{H^1_0} \end{aligned}$$

$$+ (\Delta t)^2 \|\frac{\partial^2}{\partial x \partial y} \partial\delta u^n\|^2_{L^2}\Big]\Delta t \;.$$

If again $\lambda > \frac{1}{4} a_{max}$,

$$\|\partial e\|^2_{L^2 \times L^2} + \|e\|^2_{H_0^1 \times L^\infty} + (\Delta t)^2 \|\frac{\partial^2}{\partial x \partial y} \partial e\|^2_{L^2 \times L^2}$$

$$\leq C\Big[(\Delta t)^4 + \|e^0\|^2_{H_0^1} + \|e^1\|^2_{H_0^1} + \|\partial\delta u\|^2_{H_0^1 \times L^2} \tag{3.44}$$

$$+ (\Delta t)^2 \|\frac{\partial^2}{\partial x \partial y} \partial\delta u\|^2_{L^2 \times L^2}\Big].$$

<u>Theorem 3</u>. Let u $C^5(\overline{D} \times [0,T])$ be the solution of

$$\frac{\partial u}{\partial t} = \nabla\cdot(a(x,y,t)\nabla u) + f(x,y,t,u,\nabla u) \;,$$

$$(x,y) \in D \;, \quad 0 < t \leq T \;,$$

$$u = 0 \;, \qquad (x,y) \in \partial D \;, \quad 0 < t \leq T \;,$$

$$u = u_0(x,y) \;, \qquad (x,y) \in D \;, \quad t = 0 \;,$$

where f is Lipschitz continuous with respect to u and ∇u and is twice continuously differentiable with respect to t. Let u be approximated by the function

$$U^n = \sum_{p,q} \eta^n_{pq} \alpha_p(x) \beta_q(y) \;,$$

where η^n is determined by (3.39) for $n \geq 2$. Then the error $e = u - U$ can be estimated by (3.44).

It is necessary to specify U^1 as well as U^0 to start the process (3.39). For finite difference methods employing three or more levels the same initialization problem occurs; however, most estimates for these methods bound only the L_2-norm of the error. Consequently, almost any difference method that is globally first-order correct in Δt can be used to supply initial values for a globally second-order, multi-level method. The ability to bound the spatial derivatives of the error is highly desirable, but, of course, it imposes the requirement that U^1 be a better approximation of u^1 than would be needed to maintain second-order accuracy were only the L_2-norm being estimated. A glance at the error propagation relation (3.31) quickly convinces one that $\|e^1\|_{H_0^1}$ is not in general $O((\Delta t)^2)$; thus, it is not adequate to start with one step utilizing (3.21) and then shift to (3.39). The point then is that the first step should also be second-order correct; the problem is to find such a procedure subject to the restriction that no new large chunk of program would be required to evaluate U^1. This can be accomplished in the following way. Use standard Crank-Nicolson for one step:

$$\left\langle \frac{U^1-U^0}{\Delta t} \right\rangle + \frac{1}{2} \langle a^1 \nabla U^1 + a^0 \nabla U^0, \nabla V \rangle$$

(3.45)

$$= \frac{1}{2} \langle f^1(U) + f^0(U), V \rangle , \quad V \in M .$$

The evaluation of η^1 can be obtained iteratively by an alternating-direction iteration using exactly the same major pieces of code that are needed to evaluate the solution of (3.39) for U^2, U^3, etc. Obviously, the iterative first step will take longer on the machine than each succeeding step. The iteration will be discussed in the last section. That (3.45) does give $\|e^1\|_{H_0^1}$ to the required accuracy is a trivial consequence of [12].

A number of generalizations of both techniques treated above are easily seen. First, the operator $\nabla\cdot(a\nabla)$ is a special case of the general symmetric second-order uniformly elliptic operator (summation convention assumed) $(a_{ij}, u_{x_i})_{x_j}$, where $a_{ij} = a_{ji}$ and

$$(3.46)\qquad \begin{aligned} &a_{min}|\xi|^2 \leqslant a_{ij}\xi_i\xi_j \leqslant a_{max}|\xi|^2 , \\ &0 < a_{min} \leqslant a_{max} < \infty , \quad \xi \in R^2 , \end{aligned}$$

(3.46) holding for all (x,y,t,u). Thus, if equation (3.1) is extended to

$$(3.47)\qquad \begin{aligned} &\frac{\partial u}{\partial t} = \frac{\partial}{\partial x_j}\left(a_{ij}\frac{\partial u}{\partial x_i}\right) + f(x_1,x_2,t,u,\nabla u) , \\ &\qquad\qquad\qquad (x_1,x_2) \in D , \end{aligned}$$

then (3.21) is modified only in changing the definition of the right-hand side term ψ^n to

$$(3.48)\qquad (\psi^n)_{pq} = -\langle a_{ij}^n(U)\frac{\partial U^n}{\partial x_i}, \frac{\partial}{\partial x_j}\alpha_p\beta_q\rangle + \langle f^n(U),\alpha_p\beta_q\rangle .$$

A similar modification in the definition of Φ^n provides an extension of (3.39) to handle (3.47); the coefficients a_{ij} should be assumed independent of u for (3.39). The proofs are modified only trivially, since uniform ellipticity was all that was used about the special form of the operator. One small additional piece of argument will be explained in the treatment of parabolic systems.

The restriction to two space variables clearly was of no great importance. Consider the domain $D = [0,1]^J$ and the equation

$$(3.49)\quad \begin{aligned} &\frac{\partial u}{\partial t} = \frac{\partial}{\partial x_j}\left(a_{ij}(x,t,u)\frac{\partial u}{\partial x_i}\right) + f(x,t,u,\nabla u), && x \in D,\quad 0 < t \leq T,\\ &u = 0 && x \in \partial D,\quad 0 < t \leq T,\\ &u = u_0(x) && x \in D,\quad t = 0. \end{aligned}$$

Let $\mathcal{M}_i = \mathrm{Span}\left(\left\{\alpha_k^i(x_i) \mid k = 1,\cdots,M_i\right\}\right)$ and set

$$(3.50)\qquad \mathcal{M} = \mathcal{M}_1 \otimes \cdots \otimes \mathcal{M}_J .$$

Let

$$(3.51)\qquad \begin{aligned} C_i &= (\langle\alpha_p^i, \alpha_q^i\rangle_{x_i}),\\ A_i &= \left(\left\langle \frac{d\alpha_p^i}{dx_i}, \frac{d\alpha_q^i}{dx_i}\right\rangle_{x_i}\right), \end{aligned}$$

and set

$$(3.52)\qquad \begin{aligned} P_i &= I \otimes \cdots \otimes I \otimes C_i \otimes I \otimes \cdots \otimes I \ , \\ Q_i &= I \otimes \cdots \otimes I \otimes A_i \otimes I \otimes \cdots \otimes I \ . \end{aligned}$$

Set $k = (k_1, \cdots, K_J)$, $k_i \in \{1, \cdots, M_i\}$ and let

$$(3.53)\qquad \alpha_k(x) = \prod_{i=1}^{J} \alpha_{k_i}^{i}(x_i) \ , \qquad x \in D \ .$$

Let Φ^n be the vector with components

$$(3.54)\qquad (\Phi^n)_k = -\langle a_{ij}^n \frac{\partial U^n}{\partial x_i}, \frac{\partial \alpha_k}{\partial x_j} \rangle + \langle f^n, \alpha_k \rangle \ .$$

Then, the generalization of (3.21) to J space variables is given by

$$(3.55)\qquad \prod_{i=1}^{J} (P_i + \lambda \Delta t Q_i)\,(\eta^{n+1} - \eta^n) = \Delta t \Phi^n \ .$$

The extension of (3.39) can be written in the form

$$(3.56)\qquad \prod_{i=1}^{J} (P_i + 2\lambda \Delta t Q_i)(\eta^{n+1} - \eta^{n-1}) = 2\Delta t \Psi^n \ ,$$

where

$$(3.57)\qquad (\Psi^n)_k = (\Phi^n)_k + 2\lambda \langle \nabla(U^n - U^{n-1}), \nabla \alpha_k \rangle \ .$$

The proofs of Theorems 2 and 3 can be modified to cover these equations; obviously, the perturbation terms arising on both sides of the estimates become more complicated.

Some parabolic systems can be handled by the alternating-direction procedures above with a very strong efficiency. In fact, the methods will decouple the algebraic system into a collection of algebraic problems each exactly equivalent to a scalar problem. More remarks will be made on the practical aspects later. The general symmetric parabolic system can be attacked by the techniques to be given, but the morass of notation that would be generated would be more than either the readers or the authors are willing to wade through. The same remark applied to J space variables. So, consider the special case

$$(3.58)\quad \begin{aligned} &\frac{\partial u_i}{\partial t} = \nabla\cdot(a_{ij}(x,y,t,u)\nabla u_j) + f_i(x,y,t,u,\nabla u) , \\ &\qquad\qquad i=1,\cdots,K , \qquad (x,y)\in D , \quad 0<t\leq T , \\ &u = 0 , \qquad\qquad\qquad\qquad (x,y)\in \partial D , \quad 0<t\leq T , \\ &u = u(x,y) , \qquad\qquad\qquad (x,y)\in D , \quad t=0 . \end{aligned}$$

Assume, in addition to the usual constraints on a_{ij} and f_i, that $a_{ij} = a_{ji}$. Let $M = \{\alpha_p(x)\}\otimes\{\beta_q(y)\}$ as usual and seek an approximate solution $U = (U_1,\cdots,U_K) \in N = M^K$. One possible generalization of (3.21) is the following:

(3.59)

$$(C_x + \lambda\Delta t A_x) \otimes (C_y + \lambda\Delta t C_y)(\eta_i^{n+1} - \eta_i^n) = \Delta t \Phi_i^n \ ,$$

$$i = 1,\cdots,K \ ,$$

where

(3.60)
$$(\Phi_i^n)_{pq} = -\langle a_{ij}^n(U)\nabla U_j^n, \nabla\alpha_p\beta_q\rangle + \langle f_i^n(U), \alpha_p\beta_q\rangle \ , \quad i = 1,\cdots,K \ .$$

Note that only the right-hand side of (3.59) involves more than one of the dependent variables U_j; thus, arithmetically, (3.59) is solved as a sequence of K independent problems exactly of the same form as (3.21) apply here. The inner product form equivalent to (3.59)-(3.60) is

(3.61)
$$\left\langle \frac{U_i^{n+1} - U_i^n}{\Delta t}, V\right\rangle + \lambda \langle \nabla(U_i^{n+1} - U_i^n), \nabla V\rangle$$
$$+ \lambda^2 \Delta t \left\langle \frac{\partial^2}{\partial x \partial y}(U_i^{n+1} - U_i^n), \frac{\partial^2}{\partial x \partial y} V\right\rangle$$
$$+ \langle a_{ij}^n(U)\nabla U_j^n, \nabla V\rangle = \langle f_i^n(U), V\rangle \ , \quad V \in M \ .$$

Choose $v = V = e_i^{n+1} + \delta u_i^{n+1}$ in the i-th equation and subtract (3.61) from the corresponding relation for u_i. After a bit of manipulation using in particular (3.24), it follows that

$$\frac{1}{2\Delta t}\left[\|e_i^{n+1}\|_{L^2}^2 - \|e_i^n\|_{L^2}^2\right] + \lambda\|e_i^{n+1}\|_{H_0^1}^2$$

$$+ \langle (a_{ij}^n(U) - \lambda\delta_{ij})\nabla e_j^n, \nabla e_i^{n+1}\rangle$$

$$+ \frac{1}{2}\lambda^2\Delta t\left[\|\frac{\partial^2}{\partial x\partial y} e_i^{n+1}\|_{L^2}^2 - \|\frac{\partial^2}{\partial x\partial y} e_i^n\|_{L^2}\right]$$

$$+ \langle \frac{e_i^{n+1}-e_i^n}{\Delta t}, \delta u_i^{n+1}\rangle$$

(3.62)

$$+ \lambda^2(\Delta t)^2 \langle \frac{\partial^2}{\partial x\partial y}\frac{e_i^{n+1}-e_i^n}{\Delta t}, \frac{\partial^2}{\partial x\partial y}\delta u_i^{n+1}\rangle$$

$$\leqslant \varepsilon\left[\|e^{n+1}\|_{H_0^1}^2 + \|e^n\|_{H_0^1}^2 + C\right]\|e^{n+1}\|_{L^2}^2$$

$$+ \|e^n\|_{L^2}^2 + (\Delta t)^2 + \|\delta u^{n+1}\|_{H_0^1}^2$$

$$i = 1,\cdots,K,$$

where $\|e^n\|^2 = \|e_1^n\|^2 + \cdots + \|e_K^n\|^2$, etc. Let $A^n(U)$ be the matrix $(a_{ij}^n(U))$ and assume that there exist constants $0 < a_{min} \leqslant a_{max} < \infty$ such that

(3.63) $$a_{min}|\xi|^2 \leqslant (A^n(U)\xi,\xi) \leqslant a_{max}|\xi|^2,$$

$$\xi \in R^K,$$

for all $(x,y) \in D$, $0 \leqslant t \leqslant T$, and all U. Then, for $x_1 = x$ and $x_2 = y$,

$$\left(\lambda I \frac{\partial e^{n+1}}{\partial x_k} + (A^n(U)-\lambda I)\frac{\partial e^n}{\partial x_k}, \frac{\partial e^{n+1}}{\partial x_k}\right) \geqslant \lambda \left|\frac{\partial e^{n+1}}{\partial x_k}\right|_2^2$$

$$- \frac{1}{2}\,|A^n(U)-\lambda I|_2 \left\{\left|\frac{\partial e^{n+1}}{\partial x_k}\right|_2^2 + \left|\frac{\partial e^n}{\partial x_k}\right|_2^2\right\} \quad ; \tag{3.64}$$

the norm $|\cdot|_2$ indicating the ordinary metric on R^K or its induced matrix norm. It follows from adding the inequalities (3.62) and the last remarks that

$$\frac{\|e^{n+1}\|_{L^2}^2 - \|e^n\|_{L^2}^2}{2\Delta t} + \{\lambda-\mu-\varepsilon\}\,\|e^{n+1}\|_{H_0^1}^2$$

$$- (\mu+\varepsilon)\,\|e^n\|_{H_0^1}^2 + \frac{1}{2}\lambda^2\Delta t\left[\,\left\|\frac{\partial^2}{\partial x\partial y} e^{n+1}\right\|_{L^2}^2\right.$$

$$\left.- \left\|\frac{\partial^2}{\partial x\partial y} e^n\right\|_{L^2}^2\right] + \sum_{i=1}^{K}\left\{< \frac{e_i^{n+1}-e_i^n}{\Delta t}, \delta u_i^{n+1}> \right. \tag{3.65}$$

$$\left.+ \lambda^2(\Delta t)^2 < \frac{\partial^2}{\partial x\partial y}\frac{e_i^{n+1}-e_i^n}{\Delta t}, \frac{\partial^2}{\partial x\partial y}\delta u_i^{n+1}> \right\}$$

$$\leqslant C\left[(\Delta t)^2 + \|e^{n+1}\|_{L^2}^2 + \|e^n\|_{L^2}^2 + \|\delta u^{n+1}\|_{H_0^1}^2\right],$$

where

$$\mu = \frac{1}{2}\sup_{x,y,t,u}|A^n(U)-\lambda I|_2 \,. \tag{3.66}$$

Again assume that

$$\lambda > \frac{1}{2} a_{max} . \tag{3.67}$$

The relation (3.65) is almost exactly the same as (3.25); consequently, the estimate (3.26) holds (with generalized meanings for the terms) for the approximation method (3.59).

The method specified by (3.39) can equally be generalized to systems; however, this shall be left to the reader. Only the definition of the right-hand side needs to be modified; again the arithmetic problem is decoupled. The only critical step in the proof is the analogue of (3.34), but this follows by the same type of argument as used in (3.64).

In the two-dimensional case of each of the techniques of Sections 2 and 3 an operator of the form

$$C_x \otimes C_y + \frac{\Delta t}{2} \left[A_x \otimes C_y + C_x \otimes A_y\right] \tag{3.68}$$

was perturbed to an operator

$$(C_x + \frac{\Delta t}{2} A_x) \otimes (C_y + \frac{\Delta t}{2} A_y) \tag{3.69}$$

in order to obtain a locally one-dimensional procedure. Operators of these forms can come from more general differential operators than the constant coefficient operators used. Consider the problem

$$a_1 b_1 \frac{\partial u}{\partial t} - \frac{\partial}{\partial x}\left(a_2 b_1 \frac{\partial u}{\partial x}\right) - \frac{\partial}{\partial y}\left(a_1 b_2 \frac{\partial u}{\partial y}\right) = f ,$$

$$(x,y) \in D , \quad t > 0 ,$$

$$(3.70)\qquad u = 0\ , \qquad (x,y) \in \partial D\ , \quad t \quad 0\ ,$$

$$u = u_0\ , \qquad (x,y) \in D\ , \quad t = 0\ ,$$

where the a_i's are functions of x and the b_i's are functions of y and $D = (0,1)^2$. Use a tensor product basis, and let

$$(3.71)\qquad \begin{aligned} C_x &= \langle a_1\alpha_i, \alpha_j\rangle_x\ , \quad C_y = \langle b_1\beta_i, \beta_j\rangle_y\ , \\ A_x &= \langle a_2\alpha_i, \alpha_j\rangle_x\ , \quad A_y = \langle b_2\beta_i, \beta_j\rangle_y\ . \end{aligned}$$

It is clear that the Crank-Nicolson alternating-direction procedure

$$(3.72)\qquad (C_x + \tfrac{\Delta t}{2} A_x) \otimes (C_y + \tfrac{\Delta t}{2} A_y)(\eta^{n+1} - \eta^n) = \Delta t \Phi\ ,$$

where

$$(3.73)\qquad \Phi_{pq} = \langle f, \alpha_p\beta_q\rangle - \left([A_x \otimes C_y + C_x \otimes A_y]\eta^n\right)_{pq}\ ,$$

can be analyzed in a fashion exactly parallel to that used in the case of the heat equation, provided that the a_i's and b_i's are reasonable. It is also clear that operators of this form can be used to build analogues of the Laplace-modified difference equations. An important special case of (3.70) is the radially symmetric heat equation on a three-dimensional annulus

$$r\frac{\partial u}{\partial t} - \frac{\partial}{\partial r}\left(r\frac{\partial u}{\partial r}\right) - r\frac{\partial^2 u}{\partial z^2}$$

$$= f, \quad 0 < r_0 < r < r_1, \quad z_0 < z < z_1, \tag{3.74}$$

$$t > 0 .$$

Lees [21] used a three-level finite-difference equation to obtain second-order accuracy in time for non-linear parabolic equations. The centered-in-time equation (3.4) could just as easily have been modified by some uniformly elliptic operator $(b_{ij}u_{x_j})x_i$ instead of the Laplace operator. All that was required in the argument (3.27)-(3.34) was a comparability between the original elliptic operator and the modifying operator. Thus, our discussion includes the generalization of Lees' method to a Galerkin procedure.

It is clear that higher order parabolic equations and systems can be treated by analogous alternating-direction Galerkin methods, but again this will be left to the reader.

Finally, the alternating-direction formulation of Douglas and Gunn [9] could just as easily have been utilized as the fractional step formulation. It is obvious that everything works for the (standard) heat eqation, since the commutativity of the x- and y-operators is trivial. Since the alternating-direction perturbations for the variable coefficient cases relate only to the Laplace operator, all the theorems proved have immediate analogues for the Douglas-Gunn approach. In fact, the Douglas-Gunn formulation will be employed for the elliptic problem in more than two space variables.

4. Hyperbolic Problems*

Consider the second-order hyperbolic problem

$$\frac{\partial^2 u}{\partial t^2} = \nabla\cdot(a(x,y)\nabla u) + f(x,y,t) , \qquad (x,y) \in D, \quad 0 < t \leq T ,$$

$$(4.1) \qquad u = 0 , \qquad (x,y) \in \partial D, \quad 0 < t \leq T ,$$

$$u = u_0(x,y), \quad u_t = u_1(x,y) , \qquad (x,y) \in D, \quad t = 0 .$$

An alternating-direction Galerkin procedure analogous to (3.39) can be defined by

$$(4.2) \qquad (C_x + \lambda(\Delta t)^2 A_x) \otimes (C_y + \lambda(\Delta t)^2 A_y)(\eta^{n+1} - 2\eta^n + \eta^{n-1}) = (\Delta t)^2 \Phi^n , \quad n \geq 1 ,$$

where

$$(4.3) \qquad (\Phi^n)_{pq} = -\langle a\nabla U^n, \nabla\alpha_p\beta_q\rangle + \langle f^n, \alpha_p\beta_q\rangle$$

Again a tensor product basis $M = \{\alpha_p\} \otimes \{\beta_q\} \subset H_0^1([0,1]) \otimes H_0^1([0,1])$, is assumed. It is obvious that the earlier

---

*The results of this section are due essentially to Dupont.

practical remarks apply to (4.2) as well.

The inner product form of (4.2) is

$$
\begin{aligned}
&< \frac{U^{n+1}-2U^{n}+U^{n-1}}{(\Delta t)^{2}},V> + \lambda <\nabla(U^{n+1}-2U^{n}+U^{n-1}),\nabla V \\
(4.4) \qquad &+ \lambda^{2}(\Delta t)^{2} < \frac{\partial^{2}}{\partial x\partial y}(U^{n+1}-2U^{n}+U^{n-1}), \frac{\partial^{2}}{\partial x\partial y} V> \\
&+ <a\nabla U^{n},\nabla V> = <f^{n},V> , \quad n \geqslant 1 , \quad V \in M .
\end{aligned}
$$

Both U^{o} and U^{1} must be specified to initiate the process. Assume that a solution $u \in C^{4}(\overline{D} \times [0,T])$ exist for (4.1). Then u satisfies (4.4) except for local errors that are $O((\Delta t)^{2})$. In the usual fashion, choose $v = V = e^{n+1} + \delta u^{n+1} - e^{n-1} - \delta u^{n-1} = (e^{n+1} - e^{n}) + (e^{n} - e^{n-1}) + 2\Delta t\partial\delta u^{n}$ and subtract (4.4) from the equation for u. After a small abount of manipulation, it follows that

$$
\begin{aligned}
&||\Delta_{t}e^{n}||^{2}_{L^2} - ||\Delta_{t}e^{n-1}||^{2}_{L^2} + \lambda\left[||e^{n+1}||^{2}_{H^{1}_{o}} - ||e^{n-1}||^{2}_{H^{1}_{o}}\right] \\
&+ 2 <(a-2\lambda)\nabla e^{n},\nabla\partial e^{n}> \Delta t \\
&+ \lambda^{2}(\Delta t)^{4}\left[||\frac{\partial^{2}}{\partial x\partial y} \Delta_{t}e^{n}||^{2}_{L^2} - ||\frac{\partial^{2}}{\partial x\partial y} \Delta_{t}e^{n-1}||^{2}_{L^2} \right] \\
&+ 2 < \frac{e^{n+1}-2e^{n}+e^{n-1}}{(\Delta t)^{2}}, \partial\delta u^{n}> \Delta t
\end{aligned}
$$

$$+ 2\lambda^2(\Delta t)^2 < \frac{\partial^2}{\partial x \partial y}(e^{n+1}-2e^n+e^{n-1}), \frac{\partial^2}{\partial x \partial y}\partial\delta u^n> \Delta t$$

(4.5)

$$\leqslant \varepsilon\Big[\|e^{n+1}\|^2_{H^1_0} + \|e^n\|^2_{H^1_0} + \|e^{n-1}\|^2_{H^1_0}\Big] \Delta t$$

$$+ C\Big[\|\Delta_t e^n\|^2_{L^2} + \|\Delta_t e^{n-1}\|^2_{L^2}$$

$$+ (\Delta t)^4 + \|\partial\delta u^n\|_{H^1_0}\Big]\Delta t + 2\lambda <\nabla\xi^n, \nabla(\partial e^n+\partial\delta u^n)> \Delta t$$

$$+ 2\lambda^2(\Delta t)^2 < \frac{\partial^2}{\partial x \partial y}\xi^n, \frac{\partial^2}{\partial x \partial y}(\partial e^n+\partial\delta u^n)> \Delta t \ ,$$

where $\Delta_t e^n = (e^{n+1}-e^n)/\Delta t$ and

(4.6) $$\xi^n = u^{n+1} - 2u^n + u^{n-1} \ .$$

Integration by parts implies that

$$|<\nabla\xi^n, \nabla(\partial e^n+\partial\delta u^n)>| = |<\Delta\xi^n, \partial e^n+\partial\delta u^n>|$$

(4.7)

$$\leqslant C\Big[(\Delta t)^4 + \|\Delta_t e^n\|^2_{L^2} + \|\Delta_t e^{n-1}\|^2_{L^2} + \|\partial\delta u^n\|^2_{H^1_0}\Big] \ .$$

Also,

$$|(\Delta t)^3 < \frac{\partial^2\xi^n}{\partial x \partial y}, \frac{\partial^2}{\partial x \partial y}(\partial e^n+\partial\delta u^n)>| \leqslant C\Big[(\Delta t)^4$$

(4.8) $$+ (\Delta t)^4 \|\frac{\partial^2}{\partial x \partial y}\Delta_t e^n\|^2_{L^2} + (\Delta t)^4 \|\frac{\partial^2}{\partial x \partial y}\Delta_t e^{n-1}\|^2_{L^2}$$

$$+ (\Delta t)^4 \left\|\frac{\partial^2}{\partial x \partial y} \partial\delta u^n\right\|^2_{L^2} \Big] \Delta t \, .$$

Relations (4.5)-(4.8) and addition on n from $n = 1$ to $n = N-1$ show that

$$\|\Delta_t e^{N-1}\|^2_{L^2} + \lambda\Big[\|e^N\|^2_{H^1_0} \; \|e^{N-1}\|^2_{H^1_0}\Big]$$

$$+ 2 \sum_1^{N-1} <(a-2\lambda)\nabla e^n, \nabla\partial e^n>\Delta t$$

$$+ \lambda^2(\Delta t)^4 \left\|\frac{\partial^2}{\partial x \partial y} \Delta_t e^{N-1}\right\|^2_{L^2}$$

$$+ 2 \sum_1^{N-1} < \frac{e^{n+1}-2e^n+e^{n-1}}{(\Delta t)^2} , \partial\delta u^n>\Delta t$$

$$+ 2\lambda^2(\Delta t)^4 \sum_1^{N-1} < \frac{\partial^2}{\partial x \partial y} \frac{e^{n+1}-2e^n+e^{n-1}}{(\Delta t)^2} , \frac{\partial^2}{\partial x \partial y} \partial\delta u^n>\Delta t$$

(4.9)

$$\lesssim \varepsilon \|e\|^2_{H^1_0 \times L^2(0,t_N)} + C \|\Delta_t e\|^2_{L^2 \times L^2(0,t_N)}$$

$$+ C(\Delta t)^4 \left\|\frac{\partial^2}{\partial x \partial y} \Delta_t e\right\|^2_{L^2 \times L^2(0,t_N)}$$

$$+ C\Big[(\Delta t)^4 + \|\partial\delta u\|^2_{H^1_0 \times L^2} + (\Delta t)^4 \left\|\frac{\partial^2}{\partial x \partial y} \partial\delta u\right\|^2_{L^2 \times L^2}\Big]$$

$$+ \ \|\Delta_t e^o\|^2_{L^2} + \lambda\left[\|e^1\|^2_{H^1_o} + \|e^o\|^2_{H^1_o}\right]$$

$$+ \lambda^2(\Delta t)^4 \left\|\frac{\partial^2}{\partial x \partial y} \Delta_t e^o\right\|^2_{L^2} \ .$$

Assume that

$$\lambda > \frac{1}{4} a_{max} \ , \tag{4.10}$$

where $0 < a_{min} \leqslant a(x,y) \leqslant a_{max} < \infty$. Then,

$$\left|2 \sum_1^{N-1} <\left(\frac{a}{\lambda} - 2\right)\nabla e^n, \nabla\partial e^n>\Delta t\right|$$

$$= \left|<\left(\frac{a}{\lambda} - 2\right)\nabla e^{N-1}, \nabla e^N> - <\left(\frac{a}{\lambda} - 2\right)\nabla e^1, \nabla e^o>\right|$$

$$\leqslant (1-q)\left[\|e^N\|^2_{H^1_o} + \|e^{N-1}\|^2_{H^1_o}\right] \tag{4.11}$$

$$+ C\left[\|e^1\|^2_{H^1_o} + \|e^o\|^2_{H^1_o}\right]$$

for some $q > 0$. Also, for $\tilde{D} = 1$ or $(\Delta t)^2\partial^2/\partial x\partial y$,

$$\left|\sum_1^{N-1} <\tilde{D}\ \frac{e^{n+1} - 2e^n + e^{n-1}}{(\Delta t)^2},\ \tilde{D}\partial\delta u^n>\Delta t\right|$$

$$= \left|- \sum_2^{N-1} <\tilde{D}\ \frac{e^{n+1} - e^n}{\Delta t},\ \tilde{D}\ \frac{\delta u^{n+2} - \delta u^{n+1} - \delta u^n + \delta u^{n-1}}{2(\Delta t)^2} >\Delta t\right.$$

$$+ \langle \tilde{D} \frac{e^N - e^{N-1}}{\Delta t}, \tilde{D}\partial\delta u^{N-1}\rangle - \langle \tilde{D} \frac{e^1 - e^0}{\Delta t}, \tilde{D}\partial\delta u^1\rangle \Big|$$

(4.12)

$$\leq \frac{1}{2} \|\tilde{D}\Delta_t e\|^2_{L^2 \times L^2(0,t_N)} + \frac{1}{2} \|\tilde{D}\Delta_t^2 \delta u\|^2_{L^2 \times L^2}$$

$$+ \varepsilon \|\tilde{D}\Delta_t e^{N-1}\|^2_{L^2} + C\Big[\|\tilde{D}\partial\delta u^{N-1}\|^2_{L^2}$$

$$+ \|\tilde{D}\Delta_t e^0\|^2_{L^2} + \|\tilde{D}\partial\delta u^1\|^2_{L^2} \Big] ,$$

where

(4.13) $$\Delta_t^2 g^n = (g^{n+2} - g^{n+1} - g^n + g^{n-1})/2(\Delta t)^2 .$$

It follows from (4.9)-(4.12) and the Gronwall lemma that

$$\|\Delta_t e\|^2_{L^2 \times L^\infty} + \|e\|^2_{H_0^1 \times L^\infty} + (\Delta t)^4 \|\frac{\partial^2}{\partial x \partial y} \Delta_t e\|^2_{L^2 \times L^\infty}$$

$$\leq C\Big[(\Delta t)^4 + \|e^0\|^2_{H_0^1} + \|e^1\|^2_{H_0^1} + \|\Delta_t e^0\|^2_{L^2}$$

(4.14)

$$+ (\Delta t)^4 \|\frac{\partial^2}{\partial x \partial y} \Delta_t e^0\|^2_{L^2} + \|\partial\delta u\|^2_{H_0^1 \times L^2}$$

$$+ \|\partial\delta u\|^2_{L^2 \times L^\infty} + \|\Delta_t^2 \delta u\|^2_{L^2 \times L^2}$$

$$+ (\Delta t)^4 \Big\{ \|\frac{\partial^2}{\partial x \partial y} \partial\delta u\|^2_{L^2 \times L^\infty}$$

$$+ \left\| \frac{\partial^2}{\partial x \partial y} \Delta_t^2 \delta u \right\|^2_{L^2 \times L^2} \Big\} \Big] .$$

Initial values U^0 and U^1 should be supplied so that $\|e^0\|^2_{H^1_0} + \|e^1\|^2_{H^1_0} + \|\Delta_t e^0\|^2_{L^2} + (\Delta t)^4 \| \partial^2/\partial x \partial y \, \Delta_t e^0\|^2_{L^2}$

is $O((\Delta t)^4)$ plus approximation errors. One way to do this is outlined below. First obtain U^0 from the following relation:

$$(4.15) \qquad (C_x + A_x) \otimes (C_y + A_y)\eta^0 = \phi^0 ,$$

where

$$(4.16) \qquad \begin{aligned} (\phi^0)_{pq} &= \langle u_0, \alpha_p \beta_q \rangle + \langle \nabla u_0, \nabla \alpha_p \beta_q \rangle \\ &+ \left\langle \frac{\partial^2}{\partial x \partial y} u_0, \frac{\partial^2}{\partial x \partial y} \alpha_p \beta_q \right\rangle . \end{aligned}$$

Then

$$(4.17) \qquad \begin{aligned} \|e^0\|^2_{H^1_0} + \left\| \frac{\partial^2}{\partial x \partial y} e^0 \right\|^2_{L^2} &\leqslant C \Big[\|\delta u^0\|^2_{H^1_0} \\ &+ \left\| \frac{\partial^2}{\partial x \partial y} \delta u^0 \right\|^2_{L^2} + (\Delta t)^4 \Big] . \end{aligned}$$

The initial data has been projected into the basis with greater smoothness than needed at a cost of the $\|(\partial^2/\partial x \partial y)\delta u^0\|_{L^2}$ term, but the ability to solve (4.15) must be present in the code for the general time step;

thus, (4.15) is easy to implement. Next, let

$$(4.18)\qquad [C_x \otimes C_y + \Delta t(A_x \otimes C_y + C_x \otimes A_y) + (\Delta t)^4 A_x \otimes A_y](\eta^1 - \eta^0) = \phi^1 ,$$

where

$$(4.19)\qquad \begin{aligned} (\phi^1)_{pq} &= \langle s, \alpha_p\beta_q\rangle + \Delta t \,\langle \nabla s, \nabla\alpha_p\beta_q\rangle \\ &\quad + (\Delta t)^4 \left\langle \frac{\partial^2 s}{\partial x \partial y}, \frac{\partial^2}{\partial x\partial y}\,\alpha_p\beta_q\right\rangle , \\ s &= \Delta t u_1 + \frac{(\Delta t)^2}{2}\,[\nabla\cdot(a\nabla u_0) + f^0] ; \end{aligned}$$

obviously, $s = u^1 - u^0 + O((\Delta t)^3)$. The test function $\Delta_t e^0 + \Delta_t \delta u^0$ leads to the estimate

$$(4.20)\qquad \begin{aligned} &\|\Delta_t e^0\|^2_{L^2} + (\Delta t)^{-1}\, \|e^1 - e^0\|^2_{H^1_0} \\ &\qquad + (\Delta t)^4 \left\|\frac{\partial^2}{\partial x\partial y}\,\Delta_t e^0\right\|^2_{H^1_0} \\ &\leqslant C\Big[\, \|\Delta_t \delta u^0\|^2_{H^1_0} \\ &\qquad + (\Delta t)^4 \left\|\frac{\partial^2}{\partial x \partial y}\,\Delta_t \delta u^0\right\|^2_{L^2} + (\Delta t)^4\Big] ; \end{aligned}$$

consequently, η^0 and η^1 are satisfactory initial conditions.

Note that (4.18) does not factor; however, it can be solved by iteration as follows:

(4.21)
$$(\eta^1 - \eta^0)^{(0)} \quad \text{arbitrary},$$
$$(C_x + \Delta t A_x) \otimes (C_y + \Delta t A_y)(\eta^1 - \eta^0)^{(q+1)}$$
$$= (\Delta t)^2(1-(\Delta t)^2)A_x \otimes A_y(\eta^1 - \eta^0)^{(q)} + \phi^1 .$$

If $\varepsilon^{(q)} = (\eta^1-\eta^0) - (\eta^1-\eta^0)^{(q)}$, then it is easy to see that

(4.22)
$$\|\varepsilon^{(q+1)}\|^2_{L^2} + \Delta t \|\varepsilon^{(q+1)}\|^2_{H^1_0}$$
$$+ \frac{1}{2} (\Delta t)^2(1+(\Delta t)^2) \left\|\frac{\partial^2}{\partial x \partial y} \varepsilon^{(q+1)}\right\|^2_{L^2}$$
$$\leqslant \frac{1}{2} (\Delta t)^2(1-(\Delta t)^2) \left\|\frac{\partial^2}{\partial x \partial y} \varepsilon^{(q)}\right\|^2_{L^2} ,$$

and the iteration converges. Again no major new piece of program is required.

Many of the additional developments given for the parabolic case have analogues in the hyperbolic case, but these results will be left to the reader.

5. Elliptic Problems

Galerkin methods are well-known to be very useful in obtaining highly accurate approximate solutions of elliptic problems [1, 2, 4, 5, 26]. The algebraic systems induced by Galerkin methods are similar in character to those arising from finite-difference methods; i.e., sparce matrices consisting of a (block) diagonal plus bands above and below the diagonal removed just far enough to make Gaussian elimination rather inefficient. Just as in the finite-difference case iterative procedures to solve the algebraic systems can be based on techniques used to solve the parabolic problem. The simplest problem, the Dirichlet problem for the Laplace equation on a rectangle, will be treated first. This analysis will then be used to derive a rapidly converging algorithm for a more general elliptic equation. The algorithm will be an extension of the multi-level iterative method of D'yakonov [17] and Gunn [19, 20] for treating elliptic difference equations. Finally, more than two space variables will be considered; there will be a distinct shift in point of view required between two space variables and more than two. Obviously, throughout the discussion the domain will remain rectangular. It is possible to generalize the three-level scheme of Dupont [14] using both the Gunn iteration and the Schwarz alternating principle [18, 23] to handle more general regions, but this treatment will be deferred.

Consider

$$(5.1)\qquad \begin{aligned} -\Delta u &= f , \quad (x,y) \in D = [0,1]^2 , \\ u &= 0 , \quad (x,y) \in \partial D . \end{aligned}$$

Inhomogeneous boundary values are assumed converted into a contribution to the function f. Let $M = M_x \otimes M_y =$

$= \{\alpha_k\}_1^{N_x} \otimes \{\beta_\ell\}_1^{N_y} \subset H_0^1([0,1]) \otimes H_0^1([0,1])$ and seek an approximate solution U,

(5.2) $$U(x,y) = \sum_{p,q} \eta_{pq}\alpha_p(x)\beta_q(y) ,$$

of the weak form

(5.3) $$<\nabla u,\nabla v> = <f,v> , \quad v \in H_0^1(D) ,$$

of (5.1). It is completely trivial to see that

(5.4) $$||u-U||_{H_0^1} \leq ||u-\tilde{u}||_{H_0^1} , \quad \forall \tilde{u} \in M .$$

Clearly, U is the H_0^1-projection of u into M; thus, the accuracy of the approximation is reduced solely to the study of approximation theory. Practically, the implementation of the Galerkin process with the tensor product basis means that an efficient procedure must be provided to solve the algebraic system

(5.5) $$(A_x \otimes C_y + C_x \otimes A_y)\eta = \phi ,$$
$$\phi_{pq} = <f,\alpha_p\beta_q> ;$$

the operators A_x, C_x, etc., retain their previous definitions.

An alternating-direction iterative procedure can be based on (2.28) as follows:

$$\eta^0 \in M \ , \quad \text{arbitrary} \ ,$$

(5.6) $$(\lambda_n C_x + A_x) \otimes (\lambda_n C_y + A_y)\eta^n$$

$$= (\lambda_n C_x - A_x) \otimes (\lambda_n C_y - A_y)\eta^{n-1} + z\lambda_n \phi \ , \quad n \geq 1 \ .$$

The parameters λ_n correspond to the usual alternating-direction iterative parameters for elliptic difference equations; consequently, one object of studying (5.6) is to exhibit a parameter sequence such that $O(\log N_x N_y)$ iterations suffice (for reasonable choices of M_x and M_y) so that $\|U-U^n\| \leq \delta \|U-U^0\|$. This will be done for both the L_2 and the H_0^1 norms for a wide class of bases, including all of the usual choices. Let

(5.7) $$e^n = U - U^n = \sum_{p,q} \varepsilon_{pq} \alpha_p \beta_q \ .$$

Let $|\cdot|_2$ indicate the usual Euclidean norm on $R^{N_x N_y}$ and let

(5.8) $$C = C_x \otimes C_y \ , \quad A = C_x \otimes A_y + A_x \otimes C_y \ .$$

Note that $C^{\frac{1}{2}} = C_x^{\frac{1}{2}} \otimes C_y^{\frac{1}{2}}$ and that

(5.9) $$|C^{\frac{1}{2}}\varepsilon^n|_2 = \|e^n\|_{L^2} \ , \quad |A^{\frac{1}{2}}\varepsilon^n|_2 = \|e^n\|_{H_0^1} \ .$$

It is easy to see that

$$\varepsilon^n = (\lambda_n I + C_x^{-1}A_x)^{-1}(\lambda_n I - C_x^{-1}A_x)$$

(5.10)

$$\otimes\, (\lambda_n I + C_y^{-1}A_y)^{-1}(\lambda_n I - C_y^{-1}A_y)\varepsilon^{n-1}$$

and that

$$(5.11)\qquad C^{\frac{1}{2}}\varepsilon^n = \left(\prod_{i=1}^{n} T_x(\lambda_i)\right) \otimes \left(\sum_{i=1}^{n} T_y(\lambda_i)\right) C^{\frac{1}{2}}\varepsilon^0 ,$$

where

$$T_x(\lambda) = (\lambda I + C_x^{-\frac{1}{2}}A_xC_x^{-\frac{1}{2}})^{-1}(\lambda I - C_x^{-\frac{1}{2}}A_xC_x^{-\frac{1}{2}}) ,$$

(5.12)

$$T_y(\lambda) = (\lambda I + C_y^{-\frac{1}{2}}A_yC_y^{-\frac{1}{2}})^{-1}(\lambda I - C_y^{-\frac{1}{2}}A_yC_y^{-\frac{1}{2}}) .$$

Let $\sigma(T)$ be the spectrum of T. Since the matrices involved in (5.11) are Hermitian and commute, then

$$\left|\prod_{i=1}^{n} T_x(\lambda_i)\right|_2 \leqslant \max_{x\,\epsilon\,\sigma(C_x^{\frac{1}{2}}A_xC_x^{-\frac{1}{2}})} \prod_{i=1}^{n} \left|\frac{\lambda_i - x}{\lambda_i + x}\right| .$$

Now

$$\sigma(C_x^{-\frac{1}{2}}A_xC_x^{-\frac{1}{2}}) \subset \left[\min_{\zeta\neq 0} \frac{(A_x\zeta,\zeta)_2}{(C_x\zeta,\zeta)_2},\ \max_{\zeta\neq 0} \frac{(A_x\zeta,\zeta)_2}{(C_x\zeta,\zeta)_2}\right]$$

$$(5.14)\qquad = \left[\min_{0\neq z\in M_X} \frac{\|z\|^2_{H^1_0}}{\|z\|^2_{L^2}},\ \max_{0\neq z\in M_X} \frac{\|z\|^2_{H^1_0}}{\|z\|^2_{L^2}}\right]$$

$$\subset \left|\pi^2,\ \max_{0\neq z\in M} \frac{\|z\|^2_{H^1_0([0,1])}}{\|z\|^2_{L^2([0,1])}}\right|$$

since the minimum on all of $H^1_0([0,1])$ is π^2. The upper bound is dependent on the choice of M_X.

Up to this point the authors have carefully avoided specializing the basis and it is our intention to maintain this independence to as great an extent as possible. Most commonly used subspaces employ a set of nodes and basis elements that have support in a fixed number of the subintervals generated by the nodes. Moreover, the basis elements generally are obtained by affine changes in coordinates. For some, such as the Hermite bases, unequal intervals cause the use of more than one affine map to define each basis element. For others, such as smooth splines, the basis elements associated with nodes near the ends of the interval are modified in shape to satisfy the boundary conditions, but still only a small number of fundamental functions can be used to define all the basis elements through affine transformations of the independent variable. Under these circumstances, it is clear that the homogeneity in the node spacing is -1 and +1, respectively, for the H^1_0 and L_2 norms of the individual basis elements. It is easy to see for subspaces having local bases with such homogeneity that

(5.15) $$\max_{0 \neq z \in M_x} \frac{\|z\|^2_{H^1_0}}{\|z\|^2_{L^2}} \leqslant ch^{-2} ,$$

where h is the minimum nodes spacing. Assume that the ratio of maximum to minimum spacing remains bounded. Since the number of basis elements associated with a single node is usually fixed, (5.15) usually is equivalent to

(5.16) $$\max_{0 \neq z \in M_x} \frac{\|z\|^2_{H^1_0}}{\|z\|^2_{L^2}} \leqslant cN_x^2$$

In particular, (5.16) holds for the Hermite bases of any (odd) degree, smooth splines, and the basis consisting of piecewise linear functions ("hat" functions) plus polynomials of some fixed degree defined on each subinterval and vanishing at the endpoints of the defining subinterval. It is also the case that the linear equations $(\lambda C_x + A_x)\xi = \theta$ can be solved by a careful use of Gaussian elimination in $O(N_x)$ operations for each of these bases. These remarks motivate the following definition.

Definition. The family of subspaces $M_x(h)$ of $H^1_0([0,1])$ is regular if

$$\max_{0 \neq z \in M_x(h)} \frac{\|z\|^2_{H^1_0}}{\|z\|^2_{L^2}} \leq c[\dim (M_x(h))]^2$$

and if the number of arithmetic operations necessary to solve the linear equations $(\lambda C_x(h) + A_x(h))\xi = \theta$ is not greater than $c \dim (M_x(h))$.

Assume from here on that M_x and M_y are regular. Then,

$$(5.17) \quad \sigma(C_x^{-\frac{1}{2}} A_x C_x^{-\frac{1}{2}}) \cup \sigma(C_y^{-\frac{1}{2}} A_y C_y^{-\frac{1}{2}}) \subset [\pi^2, c(N_x^2 + N_y^2)] .$$

Consequently, it is well known [24] that a geometric sequence of parameters $\{\lambda_n\}$, $n = 1, \cdots, O(\log N_x N_y) = K$, can be constructed so that

$$(5.18) \quad \left| \prod_{i\ 1}^{K} T_x(\lambda_i) \right|_2 , \quad \left| \prod_{i\ 1}^{K} T_y(\lambda_i) \right| \leq \delta^{\frac{1}{2}} .$$

Hence,

$$(5.19) \quad \|U - U^K\|_{L^2} \leq \delta \, \|U - U^0\|_{L^2} ,$$

$$K = O(\log N_x N_y) ,$$

and the great efficiency (in L_2) of alternating-direction iteration has carried over for the so-called "model problem."

The error bound (5.4) is in terms of the H_0^1 norm; thus, it would be desirable to obtain the analogue of (5.19) in the H_0^1 norm. Fortunately, this can be done simply. Let

$$\Lambda_K = C^{-\frac{1}{2}} \left(\prod_{i=1}^{K} T_x(\lambda_i)\right) \otimes \left(\prod_{i=1}^{K} T_y(\lambda_i)\right) C^{\frac{1}{2}} . \tag{5.20}$$

Then,

$$\varepsilon^K = \Lambda_K \varepsilon^0 , \tag{5.21}$$

and

$$A^{\frac{1}{2}}\varepsilon^K = (A^{\frac{1}{2}}\Lambda_K A^{-\frac{1}{2}})A^{\frac{1}{2}}\varepsilon^0 . \tag{5.22}$$

Thus

$$\|e^K\|_{H_0^1} \leq \delta \, \|e^0\|_{H_0^1} . \tag{5.23}$$

provided that $|A^{\frac{1}{2}}\Lambda_K A^{-\frac{1}{2}}|_2 \leq \delta$. Note that

$$\begin{aligned} C^{-\frac{1}{2}}AC^{-\frac{1}{2}} &= C_x^{-\frac{1}{2}}A_x C_x^{-\frac{1}{2}} \otimes I \\ &\quad + I \otimes C_y^{-\frac{1}{2}}A_y C_y^{-\frac{1}{2}} \end{aligned} \tag{5.24}$$

and $C^{-\frac{1}{2}}AC^{-\frac{1}{2}}$ commutes with $T_x(\lambda) \otimes T_y(\lambda)$. Now, $A^{\frac{1}{2}}\Lambda_K A^{-\frac{1}{2}}$ is Hermitian if $A^{\frac{1}{2}}(A^{\frac{1}{2}}\Lambda_K A^{-\frac{1}{2}})A^{\frac{1}{2}} = A\Lambda_K$ is; in turn $A\Lambda_K$ is Hermitian if

$$C^{\frac{1}{2}}(C^{-\frac{1}{2}}AC^{-\frac{1}{2}})\left(\prod_{i=1}^{K} T_x(\lambda_i)\right) \otimes \left(\prod_{i=1}^{K} T_y(\lambda_i)\right)C^{\frac{1}{2}}$$

is, and it is clear that this last product is Hermitian. Thus, $A^{\frac{1}{2}}\Lambda_K A^{-\frac{1}{2}}$ is Hermitian. Since $C^{\frac{1}{2}}\Lambda_K C^{-\frac{1}{2}}$ is Hermitian and has the same spectrum as $A^{\frac{1}{2}}\Lambda_K A^{-\frac{1}{2}}$, then (5.18) implies that $|A^{\frac{1}{2}}\Lambda_K A^{-\frac{1}{2}}|_2 \leq \delta$ and (5.23) holds.

Theorem 4. Let $U \in M$ be the Galerkin approximation solution of (5.1) and let U^n be the n-th iterate determined by the alternating-direction iteration (5.6). Let M_x and M_y be regular. Then a geometric sequence $\{\lambda_1, \cdots \lambda_K\}$. $K = O(\log N_x N_y)$, of iteration parameters can be constructed so that (simultaneously)

$$\|U-U^K\|_X \leq \delta \|U-U^0\|_X ,$$

$$X = L_2(D) \quad \text{or} \quad H_0^1(D) .$$

For any of the basis choices mentioned as examples above the number of arithmetic operations required to complete one step of (5.6) is $O(N_x N_y)$. Thus, $O(N_x N_y \log(N_x N_y))$ arithmetic operations are required to reduce the initial error in either norm by a fixed factor. This is the same estimate as is obtained for finite-differences. Since the truncation error tends to zero more rapidly with the node spacing for reasonable choices of M than for finite-differences, the algebraic system is much smaller for comparable accuracy for Galerkin and the iteration introduced above preserves the advantage.

The more general elliptic problem, as the more general parabolic problem, can be related to the simplest example. Consider (with $x = x_1$ and $x_2 = y$)

$$(5.25)\qquad \begin{aligned} -\frac{\partial}{\partial x_i}\left(a_{ij}(x,y)\frac{\partial u}{\partial x_j}\right) &= f(x,y) , && (x,y) \in D = (0,1)^2 , \\ u(x,y) &= 0 , && (x,y) \in \partial D , \end{aligned}$$

where

$$(5.26)\qquad \begin{gathered} 0 < \alpha_0|\xi|^2 \leqslant a_{ij}(x,y)\xi_i\xi_j \leqslant \alpha_1|\xi|^2 , \\ 0 \neq \xi \in R^2 . \end{gathered}$$

Let $M = M_x \otimes M_y$ as before and let U be the Galerkin approximation to u;

$$(5.27)\qquad \langle a_{ij}\frac{\partial U}{\partial x_j}, \frac{\partial V}{\partial x_i}\rangle = \langle f,V\rangle , \quad V \in M ,$$

or

$$\mathcal{B}\eta = \phi ,$$

$$(5.28)\qquad \mathcal{B} = (b_{rs}^{pq}) , \quad b_{rs}^{pq} = \langle a_{ij}\frac{\partial(\alpha_r\beta_s)}{\partial x_j}, \frac{\partial(\alpha_p\beta_q)}{\partial x_i}\rangle ,$$

$$\phi_{pq} = \langle f,\alpha_p\beta_q\rangle .$$

It is again easy to see that

$$(5.29)\qquad \|u-U\|_{H_0^1(D)} \leq c\,\|u-\tilde{u}\|_{H_0^1(D)}\,, \quad \forall u \in M\,.$$

A two-level iteration for the solution of (5.28) can be based on the ideas of D'yakonov [17] as improved by Gunn [19,20]. The outer iteration is of the form

$$(5.30)\qquad A\tilde{\eta}^{n+1} = A\eta^{n} - \rho_n(\mathcal{B}\eta^{n}-\phi)\,, \quad n \geq 0\,,$$

where A is given by (5.8) and ρ_n is a Chebyshev iteration parameter. The solution $\tilde{\eta}^{n+1}$ of (5.30) will not be obtained exactly. Let

$$(5.31)\qquad \psi^{n} = -\rho_n(\mathcal{B}\eta^{n}-\phi)$$

and solve

$$(5.32)\qquad A(\tilde{\eta}^{n+1}-\eta^{n}) = \psi^{n}$$

approximately by using K steps of the alternating-direction iteration (5.6) starting from initial values $(\tilde{\eta}^{n+1}-\eta^{n})^{(0)} = 0$. Then, (5.21) implies that

$$(5.33)\qquad \tilde{\eta}^{n+1} - \eta^{n+1} = \Lambda_K(\tilde{\eta}^{n+1} - \eta^{n})\;;$$

consequently, $\eta^{n+1} - \eta^{n} = (I-\Lambda_K)(\tilde{\eta}^{n+1}-\eta^{n})$ and

$$E\eta^{n+1} = E\eta^{n} - \rho_n(\mathcal{B}\eta^{n}-\phi) ,$$

(5.34)

$$E = A(I-\Lambda_K)^{-1} .$$

Note first that E is Hermitian. This follows trivially from the relation $A^{-\frac{1}{2}}E_K A^{-\frac{1}{2}} = (I-A^{\frac{1}{2}}\Lambda_K A^{-\frac{1}{2}})^{-1}$ and the fact that $A^{\frac{1}{2}}\Lambda_K A^{-\frac{1}{2}}$ is Hermitian. Next, E and $\mathcal{B}$ are comparable:

$$\frac{(\mathcal{B}\zeta,\zeta)_2}{(E\zeta,\zeta)_2} = \frac{(\mathcal{B}\zeta,\zeta)_2}{(C^{\frac{1}{2}}(C^{-\frac{1}{2}}AC^{-\frac{1}{2}})(I-C^{\frac{1}{2}}\Lambda_K C^{-\frac{1}{2}})^{-1}C^{\frac{1}{2}}\zeta,\zeta)_2}$$

(5.35) $$= \frac{(C^{-\frac{1}{2}}\mathcal{B}C^{-\frac{1}{2}}\theta,\theta)_2}{(C^{-\frac{1}{2}}AC^{-\frac{1}{2}})(I-C^{\frac{1}{2}}\Lambda_K C^{-\frac{1}{2}})^{-1}\theta,\theta)_2}$$

$$= \frac{(\mathcal{B}C^{-\frac{1}{2}}\theta,C^{-\frac{1}{2}}\theta)_2}{(AC^{-\frac{1}{2}}\theta,C^{-\frac{1}{2}}\theta)_2}\ \frac{(C^{-\frac{1}{2}}AC^{-\frac{1}{2}}\theta,\theta)_2}{((C^{-\frac{1}{2}}AC^{-\frac{1}{2}})(I-C^{\frac{1}{2}}\Lambda_K C^{-\frac{1}{2}})^{-1}\theta,\theta)_2}$$

for $0 \neq \zeta \in R^{N_x N_y}$ and $\theta = C^{\frac{1}{2}}\zeta$, C defined by (5.8). The first ratio in the product lies in the interval $[\alpha_0,\alpha_1]$ by the assumption (5.26). Since $|C^{\frac{1}{2}}\Lambda_K C^{-\frac{1}{2}}|_2 \leq \delta$, the second ratio falls in $[1-\delta,1+\delta]$. Hence,

(5.36) $$\frac{(\mathcal{B}\zeta,\zeta)_2}{(E\zeta,\zeta)_2} \in [(1-\delta)\alpha_0,(1+\delta)\alpha_1] .$$

Finally, note that if z is the function corresponding to the vector ζ, both $||a_{ij}z_{x_j}z_{x_i}||_{L^2}$ and $(E\zeta,\zeta)_2^{\frac{1}{2}}$ are uniformly comparable with $\|z\|_{H_0^1}$. It is then a standard argument [15, 16] that a sequence of parameters $\{\rho_n\}_1^N$, including as a special case the choice of constant ρ, can be chosen so that

$$(5.37)\qquad \|U-U^N\|_{H_0^1} \leqslant \delta' \|U-U^0\|_{H_0^1},$$

where $N = O(\log 1/\delta')$ is independent of M as a consequence of (5.36). The value of K, of course, does depend on M in exactly the fashion described earlier in this section. The above results can be summarized in the following theorem.

<u>Theorem 5</u>. Let U be the solution of (5.27). Assume that M_x and M_y are regular. Then the two-level Gunn iteration defined by (5.30) as the (theoretical) outer iteration and by the alternating-direction inner iteration (5.6) with error propagator Λ_K converges in $H_0^1(D)$. Moreover, the number of arithmetic operations required so that

$$\|U-U^n\|_{H_0^1(D)} \leqslant \gamma \|U-U^0\|_{H_0^1(D)}$$

is $O(N_xN_y\cdot\log(N_xN_y)\cdot\log \gamma^{-1})$.

It is relatively easy to extend the above results to include the midly nonlinear problem in which f depends on u with the usual requirement on the sign of $\partial f/\partial u$; however, this will be left to the reader.

As is the case for finite differences, the commutative case is more general than just the Laplace equation. The elliptic equation

$$\frac{\partial}{\partial x}\left(a_2 b_1 \frac{\partial u}{\partial x}\right) + \frac{\partial}{\partial y}\left(a_1 b_2 \frac{\partial u}{\partial y}\right) = f , \qquad (x,y) \in D ,$$

$$u = 0 , \qquad (x,y) \in \partial D ,$$

where the a_i's and b_i's are functions of x and y, respectively, gives a Galerkin problem of the form (5.5) with C_x, C_y, A_x, A_y defined by (3.71). The analysis for the Laplace operator carries through for operators of this form. Further, in the nested iteration the Laplacian can be replaced by operators of this more general form.

Consider next the question of more than two space variables. Clearly, the iterative procedure discussed above is strongly dependent on the ability to produce an efficient iterative solution of the Dirichlet problem on a rectangle. The straightforward generalization of (5.6),

$$(\lambda_n C_{x_1} + A_{x_1}) \otimes \cdots \otimes (\lambda_n C_{x_j} + A_{x_J})\eta^n$$

$$= (\lambda_n C_{x_1} - A_{x_1}) \otimes \cdots \otimes (\lambda_n C_{x_J} - A_{x_J})\eta^{n-1} + 2\lambda_n \phi ,$$

$$n \geqslant 1 ,$$

has the wrong (if any) fixed point; even with fixed λ the solution converges to solution of the wrong equation. One advantage of the Douglas-Gunn [9] formulation of alternating-direction methods over the fractional step formulation is that the proper fixed point is preserved; consequently, this approach will be used here. The notation introduced in (3.50)-(3.53) will be used again.

Consider the Dirichlet problem,

$$(5.38)\qquad \begin{aligned} -\Delta u = -\sum_{i=1}^{J} \frac{\partial^2 u}{\partial x_i^2} &= f(x) , \quad && x \in D = (0,1)^J , \\ u &= 0 , && x \in \partial D , \end{aligned}$$

and the associated Galerkin approximation

$$(5.39)\qquad \langle \nabla U, \nabla V\rangle = \langle f, V\rangle , \quad V \in M .$$

Let $U^0 \in M$ be arbitrary and define the iterates $U^{n,1}, U^{n,2}, \cdots, U^{n,J} = U^n$ by the equations

$$(5.40)\qquad \begin{aligned} &\lambda_n \langle U^{n,i} - U^{n-1}, V\rangle + \sum_{k=1}^{i} \left\langle \frac{\partial U^{n,k}}{\partial x_k}, \frac{\partial V}{\partial x_k} \right\rangle \\ &\quad + \sum_{k=i+1}^{J} \left\langle \frac{\partial U^{n-1}}{\partial x_k}, \frac{\partial V}{\partial x_k} \right\rangle = \langle f, V\rangle , \quad V \in M , \end{aligned}$$

for $i = 1,\cdots,J$. Obviously, a sum from $J+1$ to J is to be suppressed. For $i \geq 2$, the algebraic problem and the analysis can be simplified by differencing successive equations. Then, the system can be written in the form

$$\lambda_n \langle U^{n,1}-U^{n-1},V\rangle + \langle \frac{\partial(U^{n,1}-U^{n-1})}{\partial x_i}, \frac{\partial V}{\partial x_1}\rangle$$

(5.41)
$$+ \langle \nabla U^{n-1},\nabla V\rangle = \langle f,V\rangle \ , \quad i = 1 \ ,$$

$$\lambda_n \langle U^{n,i}-U^{n,i-1},V\rangle + \langle \frac{\partial}{\partial x_i}(U^{n,i}-U^{n-1}), \frac{\partial}{\partial x_i} V\rangle = 0 \ ,$$

$$i = 2,\cdots,J \ .$$

With $C = C_1 \otimes \cdots \otimes C_J$ and $A = Q_1 + \cdots + Q_J$, the equations above become

$$(\lambda_n C + Q_1)(\eta^{n,1} - \eta^{n-1}) + A\eta^{n-1} = \phi \ ,$$

(5.42)
$$(\phi)_k = \langle f,\alpha_k\rangle \ ,$$

$$(\lambda_n C + Q_i)\eta^{n,i} = \lambda_n C\eta^{n,i-1} + Q_i\eta^{n-1} \ ,$$

$$i = 2,\cdots,J \ .$$

Thus,

(5.43)
$$\eta^{n,i-1} - \eta^{n-1} = [I + \frac{1}{\lambda_n} I \otimes \cdots \otimes C_i^{-1}A_i \otimes \cdots \otimes I](\eta^{n,i} - \eta^{n-1})$$

and

$$\eta^{n,1} - \eta^{n-1} = \prod_{i=2}^{J} [I + \frac{1}{\lambda_n} I \otimes \cdots \otimes C_i^{-1} A_i \tag{5.44}$$
$$\otimes \cdots \otimes I](\eta^n - \eta^{n-1}) .$$

Substitute (5.44) into the first relation in (5.42):

$$\lambda_n \prod_{i=1}^{J} (P_i + \frac{1}{\lambda_n} Q_i)(\eta^n - \eta^{n-1}) + A\eta^{n-1} = \phi . \tag{5.45}$$

It is probably easier to implement the iteration in terms of (5.45) than of (5.42); since the equations are solved for the difference $\eta^n - \eta^{n-1}$, the rounding problem is also minimized. Let

$$\varepsilon^n = \eta - \eta^n , \quad A\eta = \phi . \tag{5.46}$$

Then,

$$\lambda_n \sum_{i=1}^{J} (P_i + \frac{1}{\lambda_n} Q_i)(\varepsilon^n - \varepsilon^{n-1}) + A\varepsilon^{n-1} = 0 . \tag{5.47}$$

Hence, with

$$\gamma^n = C^{\frac{1}{2}}\varepsilon^{\frac{1}{2}} , \tag{5.48}$$

$$\lambda_n(I + \frac{1}{\lambda_n} C_1^{-\frac{1}{2}}A_1C_1^{-\frac{1}{2}}) \otimes \cdots \otimes (I + \frac{1}{\lambda_n} C_J^{-\frac{1}{2}}A_JC_J^{-\frac{1}{2}})(\gamma^n-\gamma^{n-1}) \tag{5.49}$$
$$+ (C^{-\frac{1}{2}}AC^{-\frac{1}{2}})\gamma^{n-1} = 0 .$$

Since $C^{-1/2}AC^{-1/2}$ is Hermitian and commutes with each of the operators $I \otimes \cdots \otimes C_i^{-1/2}A_iC_i^{-1/2} \otimes \cdots \otimes I$, an eigenfunction expansion can be useful in analyzing (5.49). Let

$$(C_i^{-1/2}A_I C_I^{-1/2})\psi_{i,k_i} = \rho_{i,k_i}\psi_{i,k_i} , \qquad k_i = 1,\cdots,N_i ,$$

(5.50) $$\rho_k = \sum_{i=1}^{J} \rho_{i,k_i} , \qquad k = (k_1,\cdots,k_J) ,$$

$$(C^{-1/2}AC^{-1/2})\psi_k = \rho_k\psi_k , \qquad \psi_k = \bigotimes_{i=1}^{J} \psi_{i,k_i} .$$

Now, if each M_i is regular, then

(5.51)
$$\pi^2 \leqslant \rho_{i,k_i} = \frac{(A_i\theta,\theta)_2}{(C_i\theta,\theta)_2} \leqslant cN_i^2 ,$$
$$\theta = C_i^{-1/2}\psi_{i,k_i} .$$

If

(5.52) $$\gamma^o = \sum_k c_k^o \psi_k ,$$

then

(5.53)
$$\gamma^n = \sum_k \prod_{\ell=1}^{n} \left[1 - \frac{\rho_k}{\lambda \prod_{i=1}^{J} \left(1 + \frac{1}{\lambda_\ell}\rho_{i,k_i}\right)} \right] \cdot c_k^o \psi_k .$$

Note that $\lambda > 0$ implies that

$$(5.54)\qquad 0 < 1 - \frac{\rho_k}{\lambda_\ell \sum_{i=1}^{J} \left(1 + \frac{1}{\lambda_\ell} \rho_{i,k_i}\right)} < 1 ;$$

thus, the L^2 norm of the error is reduced on each iteration. That a set $\lambda_1, \cdots, \lambda_K$ of parameters can be chosen so that $K = O(\log(N_1 N_2 \cdots N_J))$ and $|C^{\frac{1}{2}}\varepsilon^K|_2 \leqslant \delta |C^{\frac{1}{2}}\varepsilon^0|_2$ is seen as follows. First.

$$(5.55)\qquad \frac{\rho_k}{\lambda e^{\rho_k/\lambda}} < \frac{\rho_k}{\lambda \prod_{i=1}^{J} \left(1 + \frac{1}{\lambda} \rho_{i,k_i}\right)} < \frac{\rho_k}{\lambda + \rho_k} .$$

Thus, if

$$(5.56)\qquad R_n(\lambda) = \max_k \prod_{\ell=1}^{n} \left[1 - \frac{\rho_k}{\lambda_\ell \prod_{i=1}^{J} \left(1 + \frac{1}{\lambda_\ell} \rho_{i,k_i}\right)}\right]$$

then

$$(5.57)\qquad R_n(\lambda) \leqslant \max_k \min_{1 \leqslant \ell \leqslant n} \left\{1 - \frac{\rho_k}{\lambda_\ell e^{\rho_k/\lambda_\ell}}\right\} .$$

Since $J\pi^2 \leqslant \rho_k \leqslant c(N_1^2 + \cdots + N_J^2)$, it is sufficient that

$$\max_{J\pi^2 \leqslant \tau \leqslant c(N_1^2+\cdots+N_J^2)} \ \min_{1 \leqslant \ell \leqslant K} \left\{1 - \frac{\tau}{\lambda_\ell} e^{-\tau/\lambda_\ell}\right\} \leqslant \delta \ . \tag{5.58}$$

Now, since $xe^{-x} \leqslant e^{-1}$ and there exists an interval about $x = 1$ such that $xe^{-x} \geqslant .2$ (for example), it is clear that a geometric sequence can be constructed of length $O(\log(N_1N_2\cdots N_J))$ so that

$$|C^{\frac{1}{2}}\varepsilon^K|_2 \leqslant .8|C^{\frac{1}{2}}\varepsilon^0|_2 \ . \tag{5.59}$$

Again it is desirable to obtain convergence in $H_0^1(D)$ as well as in $L^2(D)$. It is easy to see that

$$\varepsilon^K = C^{-\frac{1}{2}} \cdot \prod_{\ell=1}^{K} \left[I - \frac{1}{\lambda_\ell} (I + \frac{1}{\ell} C_1^{-\frac{1}{2}}A_1C_1^{-\frac{1}{2}})^{-1} \otimes \cdots \right.$$

$$\left. \otimes (I + \frac{1}{\lambda_\ell} C_J^{-\frac{1}{2}}A_JC_J^{-\frac{1}{2}})^{-1} \cdot (C^{-\frac{1}{2}}AC^{-\frac{1}{2}})\right] \cdot C\ \varepsilon^0 = \Lambda_K\varepsilon^0 \ . \tag{5.60}$$

Thus, $A^{\frac{1}{2}}\varepsilon^K = (A^{\frac{1}{2}}\Lambda_KA^{-\frac{1}{2}})A^{\frac{1}{2}}\varepsilon^0$, and the same steps as before indicate that

$$|A^{\frac{1}{2}}\varepsilon^K|_2 \leqslant .8|A^{\frac{1}{2}}\varepsilon^0|_2 \ . \tag{5.61}$$

The above argument shows that the direct generalization of Theorem 4 holds with the iteration (5.6) replaced by the iteration (5.45).

A slightly more efficient iteration could have been based on a Crank-Nicolson form of (5.40) rather than the backward form used. The argument is almost unchanged, except that the bounds on (5.54) become -1 and 1 and a slightly more complicated argument replaces the .8 in (5.59) by arbitrary $\delta > 0$. The iteration can be defined by halving Q_i in (5.45).

The iteration (5.45) can, of course, be combined with a Gunn iteration to treat the uniformly elliptic problem. The argument is modified so slightly that it will not be given; hence, the obvious generalization of Theorem 5 holds.

REFERENCES

1. Aubin, J. P., "Behavior of the Error of the Approximate Solutions of Boundary Value Problems for Linear Elliptic Operators by Galerkin's and Finite-difference Methods," Annali della Scuola Normale Superiore di Pisa 21 (1967), 599-637.
2. Birkhoff, G., Schultz, M. H., and Varga, R. S., "Piecewise Hermite Interpolation in One and Two Variables with Applications to Partial Differential Equations," Num. Math. 11 (1968), 232-256.
3. Brian, P. L. T., "A Finite Difference Method of High Order Accuracy for the Solution of Three-dimensional Heat Conduction Problems," A.I.Ch.E.J. 7 (1961), 367-370.
4. Ciarlet, P. G., Schultz, M. H., and Varga, R. S., "Numerical Methods of High-order Accuracy for Nonlinear Boundary Value Problems, I. One dimensional problem." Numer. Math. 9 (1967), 399-430.
5. ________, V. "Monotone Operators," Numer. Math. 13 (1969), 51-77.
6. Douglas, J., Jr., "On the Numerical Integration of $\frac{\partial^2 u}{\partial x^2} + \frac{\partial^2 u}{\partial y^2} = \frac{\partial u}{\partial t}$ by Implicit Methods," J. Sco. Ind. Appl. Math. 3 (1955), 42-65.
7. Douglas, J., Jr., and Rachford, H. H., Jr., "On the Numerical Solutions of Heat Conduction Problems in Two

and Three Space Variables," Trans. Amer. Math. Soc. 82 (1956), 421-439.
8. Douglas, J., Jr., "Alternating-direction Methods for Three Space Variables," Numer. Math. 4 (1962), 41-63.
9. Douglas, J., Jr., and Gunn, J. E., "A General Formulation of Alternating-direction Methods," Numer. Math. 6 (1964), 428-453.
10. Douglas, J., Jr., and Dupont, T., "The Numerical Solution of Waterflooding Problems in Petroleum Engineering," Studies in Numerical Analysis 2, SIAM, Philadelphia, 1970, 53-63.
11. Douglas, J., Jr., Dupont, T., and Rachford, H. H., Jr., "The Application of Variational Methods to Waterflooding Problems," J. Canadian Petroleum Technology 8 (1969), 79-85.
12. Douglas, J., Jr., and Dupont, T., "Galerkin Methods for Parabolic Equations," to appear in SIAM J. Numer. Anal. 7 (1969).
13. Douglas, J., Jr., Dupont, T., and Henderson, G.E., "Simulation of Gas Well Performance by Variational Methods," SPE 2891 and to appear.
14. Dupont, T., "On the Existence of an Iterative Method for the Solution of Elliptic Difference Equations with an Improved Work Estimate," Numerical Analysis of Partial Differential Equations, Edizioni Cremonese, Roma, (1968), 175-192.
15. Dupont, T., Kendall, R. P., and Rachford, H. H., Jr., "An Approximate Factorization Procedure for Solving Self-Adjoint Elliptic Difference Equations," SIAM J. Numer. Anal. 5 (1968), 559-573.
16. Dupont, T., "A Factorization Procedure for the Solution of Elliptic Difference Equations," SIAM J. Numer. Anal. 5 (1968), 753, 782.
17. D'yakonov, E. G., "On an Iterative Method for the Solution of a System of Finite-difference Equations," DAN SSSR 138 (1961), 522-525.
18. ________, "On a Method of Solution of Poisson's Equation," DAN SSSR 143 (1962), 21-24.
19. Gunn, J. E., "The Numerical Solution of $\nabla \cdot a \nabla u = f$ by a Semi-explicit Alternating-direction Iterative Method," Numer. Math. 6 (1964), 181-184.
20. ________, "The Solution of Elliptic Difference Equations by Semi-explicit Iterative Techniques," J. Soc. Ind. Appl. Math., Series B, Numer. Anal. 2 (1965), 24-45.

21. Lees, M., "A Linear Three-level Difference Scheme for Quasilinear Parabolic Equations," Math. of Comp., 10 (1966), 516-522.
22. Lions, J. L., and Magenes, E., "Problèmes aux limites non homogènes et applications," Vols. I-III, Paris, 1968-1969.
23. Miller, K., "Numerical Analogues to the Schwarz Alternating Procedure," Numer. Math. 7 (1965), 91-103.
24. Peacenan, D. W., and Rachford, H. H., Jr., "The Numerical Solution of Parabolic and Elliptic Differential Equations," J. Soc. Ind. Appl. Math. 3 (1955), 28-41.
25. Price, H. S. and Varga, R. S., "Error Bounds for Semi-discrete Galerkin Approximations of Parabolic Problems with Applications to Petroleum Reservoir Mechanics," Numerical Solution of Field Problems in Continuum Physics, Amer. Math. Soc., Providence, (1970), 74-94.
26. Schultz, M. H., "Rayleigh-Ritz-Galerkin Methods for Multidimensional Problems," SIAM J. Numer. Anal. 6 (1969), 523-538.
27. Swartz, B., and Wendroff, B., "Generalized Finite-Difference Schemes," Math. of Comp. 23 (1969), 37-49.
28. Temam, R., "Sur la stabilité et la convergence de la méthode des pas fractionnaires," Thèse, Paris, 1967.
29. Yanenko, N. N., "Méthode a pas fractionnaires, Librairie Armand Colin,"Paris, 1968. (This volume contains an extensive bibliography of Russian literature pertaining to alternating-direction methods.)

ON THE DIFFERENCE EQUATIONS FOR METEOROLOGICAL PREDICTION*

Alan J. Faller**

The equations used today for meteorological prediction are based upon a variety of physical approximations and inaccurate mathematical procedures, and they contain large finite-difference errors due to the sub-grid-scale motions. Because of these approximations and the fact that the Reynolds' stresses, if they are to be included, must be statistically related to the mean-flow variables, the equations used for numerical weather prediction have been called "empirical" equations.

An attempt is made here to rigorously derive a set of difference equations appropriate to the prediction of turbulent flows. The resultant equations contain a large number of terms analogous to the customary Reynolds' stresses, and the errors due to finite-difference approximations are explicit in the final equation. The variables to be observed and predicted are the space-time averaged

*The research reported here was carried out in part under Grant NSF - GA-4388 from the National Science Foundation.

**University of Maryland.

dependent variables, where the average is taken over the space-time mesh box used for the finite-difference approximations. Rigorously derived equations of this type are necessary for a complete and proper parameterization of the Reynolds' stresses, for a comparison of the relative magnitudes of the Reynolds' stresses and the finite-difference errors, and for estimates of the predictability of turbulent flow by finite-difference methods.

1. Introduction

In the prediction of atmospheric flow by the numerical integration of the equations of fluid motion it is necessary to select the time and space networks of grid points with rather large intervals. These intervals are dictated primarily by the following factors in combination: 1) The space-time density of observations, 2) The natural scales of atmospheric motion, 3) Numerical stability criteria, and 4) The capacity of existing computers. But no matter what reasonable density of grid points may eventually be attained there will always be sub-grid-scale motions, flows that cannot be resolved by the time-space finite-difference grid.

For the purposes of this paper we will often refer to these sub grid-scale motions as turbulence, although it should be recognized that many of these motions represent very-well organized flows. For example, if our horizontal grid had spacings of 100 km we could not resolve cumulus clouds or thunderstorm-scale motions which have typical scales of the order of 10 km. Thunderstorms and other convective systems have a considerable degree of regularity, yet if we could not observe them visually, i.e., by means other than instruments at our widely spaced network of observations, we probably would regard their effects as being due to some sort of turbulence.

It is fair to say that today the question of turbulence, or sub-grid-scale motions, constitutes the central problem of meteorological prediction. The capacity and speed of computers are continually increasing, and observations over the entire globe are now being obtained from satellites designed for the observation, collection and transmission of data. Nevertheless we never can expect to reduce the scale of observations or the speed of computation to the point where all turbulent motions are represented and calculated.

Some of the consequences of this lack of observational and computational resolution for the predictability of atmospheric flows have recently been examined by Lorenz (1969) who has classified deterministic systems into three categories. These are 1) Systems in which errors are bounded within limits close to their initial values so that the predictability of the system is essentially infinite; 2) Systems in which the errors of observation grow rapidly leading to a finite range of predictability but in which reduction of the initial errors leads to a correspondent increase in the range of predictability; and 3) Systems in which there is an upper limit to the range of predictability regardless of the magnitude of the initial errors. By means of a set of model equations Lorenz has given rather convincing evidence that many hydrodynamic flows, including the atmosphere, belong to the third category, although the precise limit of predictability (about 16 days in his model) may be sensitive to the particular model and its assumptions. On this basis it may be concluded that even if all turbulent motions were observable and computable we would still be limited in predictability by computer round off error, lack of information on the details of cloud physics processes, biological activity during the interval of prediction, variability of the sun, etc., and ultimately, molecular and atomic uncertainties.

The basic limitations on the range of predictability may be considered to be a consequence of 1) the nearly constant proportion between the temporal and spatial scales of the various motions, and 2) the inherent instabilities of the atmosphere on nearly all scales. Because of the hydrodynamical instabilities, small variations (or errors) in a particular scale of motion will grow exponentially in time in the numerical forecast. Because the time and space scales of the various motions are nearly proportional, the characteristic time for the growth of errors is very short for the small-scale motions. The errors on the smallest scale serve as perturbations upon the next largest scale which itself will grow exponentially but with a longer time scale. Thus, perturbations on even the smallest scales of motion cascade upward to eventually affect the global circulation. In this case "eventually" means a finite period of time because the time of growth of the smallest-scale perturbations adds imperceptibly to the total time in which the largest scales of motion are affected. A crude analogy may be drawn with a rapidly convergent infinite series where the nth term adds almost imperceptibly to the total sum when n is large.

Then one may well ask, why is it important to represent scales of motion smaller than those currently included in our space-time network of grid points? There are several reasons. First, the ultimate limits of predictability have not yet been approached and it is believed that the resolution of smaller scales than we now accommodate will add measureably to the range of predictability. Secondly, many of the sub grid-scale motions account for our interesting weather phenomena, frontal systems and hurricanes, for example, and it will be desirable to resolve these systems even though they may be predictable in detail for only a relatively short period of time. For particularly interesting phenomena of this type the predictions will be continually aided by new input data. Third, the various instabilities and the upward cascade of errors do not occur uniformly over the globe. There are regions such as the large high pressure areas where atmospheric conditions are rather stable and where the intermediate scales of motion are essentially missing. In these regions the upward cascade of energy will tend to be broken by the absence of intermediate scales.

More specifically, the widely quoted frequency spectrum of atmospheric energy at Brookhaven, L.I. (Van der Hoven, 1957) shows energy maxima of relative values 5, 3 and 2 units centered near the periods 4 days, 1 min, and 1 day, respectively. These maxima correspond to the large-scale cyclone waves, small-scale turbulence, and the diurnal oscillation. Between the maxima of 1 min and

1 day lies a broad minimum, the "spectral gap," centered at the period 1 hour with a minimum value of approximately 0.2 units. This broad minimum spans many interesting phenomena, such as fronts, thunderstorms, tornadoes, and numerous types of active cloud development. The lack of significant energy in these frequencies is attributable to the fact that large areas of the earth are covered by the relatively calm and stable high pressure systems. The active weather systems mentioned above are relatively local and sporadic, even though they may be violent and dramatic when they are present, so that the upward cascade of errors is limited to the local regions where these intermediate-scale instabilities are present. But while the broad outline of these problems has now emerged, many detailed questions concerning the sub-grid-scale motions, the "spectral gap," and their effects upon predictability remain for further theoretical and observational study.

One of the steps that might be taken to improve predictability, one that naturally follows from the above considerations, would be to introduce locally dense grid networks as suggested to this author by Dr. D. K. Lilly of NCAR. For example, one might wish to introduce a locally dense grid to follow the motion and development of a cold front. Such a grid would necessarily move with the front and mesh smoothly with the large-scale grid. Either it would have to be initiated automatically when a favorable large-scale situation presented itself or it could be introduced upon command when, from observational information, such a frontal system was suspected. It would necessarily dissolve itself when, by some criterion or other, the locally dense grid was no longer needed or when the computer capacity was preempted for some important feature of the circulation. Locally dense grids for fronts and smaller scale phenomena would necessarily be 3-dimensional as opposed to present atmospheric prediction models which deal with a set of weakly-coupled two-dimensional flows. Accordingly, it seems probable that just a few locally dense networks would require computer storage and time comparable to that necessary for the entire large-scale grid.

An alternative to locally-dense grids is the general process called parameterization, although these alternatives are not mutually exclusive. In so far as the sub-grid-scale motions are controlled by the macroscopic flow and the boundary conditions, it may be possible to statistically relate the unknown terms in the dynamical equations to those flows that are resolvable and predictable with the grid. The general process of parameterization involves three somewhat independent aspects. First, from the dynamical equations we must have a precise statement of the unknown terms that are to be parameterized. Second, we must select practical methods by which the unknown terms are to be related to the predictable variables. The necessary relations probably will emerge from some combination of theoretical and empirical procedures. Third, we must compare the results of the selected methods with observations from the real physical system with a view to improving the methods.

In this paper I wish to address only the first of the above aspects of parameterization, for it seems to be the case that mathematically correct and complete difference equations for the numerical prediction of turbulent flows have not clearly been set forth to date. In particular, the number of terms that represent the sub-grid-scale motions and that require parameterization is much larger than is commonly believed.

2. The "Empirical" Equations of Meteorological Prediction

At a symposium on the predictability of the atmosphere† Dr. G. D. Robinson referred to the equations that are generally used for meteorological prediction as "empirical" equations. This comment had two distinct bases in fact. First, the so called Reynolds' stresses are treated by empirical parameterization, if they are taken into account at all. Second, the equations that are customarily used have not been rigorously derived from the equations of fluid motion, and their principal justification is that they seem to work fairly well.

To adequately explain the basis of the latter statement it is necessary to briefly review the procedure used by Reynolds to derive the equations for a time-averaged flow.

We restrict our considerations to the Navier-Stokes equations modified by the Boussinesq approximation. Then the three momentum equations, the continuity equation, the equation of state, and the heat energy equation may be written, respectively:

---

†Washington meeting of the American Geophysical Union, April, 1969. Also, see Robinson (1967).

$$(1) \qquad \frac{\partial u}{\partial t} + \frac{\partial u^2}{\partial x} + \frac{\partial vu}{\partial y} + \frac{\partial wu}{\partial z} = fv - \frac{1}{\rho_0}\frac{\partial p}{\partial x} + \nu\nabla^2 u$$

$$(2) \qquad \frac{\partial v}{\partial t} + \frac{\partial uv}{\partial x} + \frac{\partial v^2}{\partial y} + \frac{\partial wv}{\partial z} = -\,fu - \frac{1}{\rho_0}\frac{\partial p}{\partial y} + \nu\nabla^2 v$$

$$(3) \qquad \frac{\partial w}{\partial t} + \frac{\partial uw}{\partial x} + \frac{\partial vw}{\partial y} + \frac{\partial w^2}{\partial z} = -\,\frac{\rho}{\rho_0}g - \frac{1}{\rho_0}\frac{\partial p}{\partial z} + \nu\nabla^2 w$$

$$(4) \qquad \frac{\partial u}{\partial x} + \frac{\partial v}{\partial y} + \frac{\partial w}{\partial z} = 0$$

$$(5) \qquad \rho = \rho_0(1-\varepsilon(T-T_0))$$

$$(6) \qquad \frac{\partial T}{\partial t} + \frac{\partial uT}{\partial x} + \frac{\partial vT}{\partial y} + \frac{\partial wT}{\partial z} = K\nabla^2 T$$

The equations have been expressed in a local right-handed Cartesian coordinate system with z opposed to the direction of gravity and with the x axis directed eastward. The dependent variables are: the three components of velocity u, v, and w in the x, y, and z directions, respectively; the pressure, p: the density ρ; and the temperature, T. Physical constants are: the mean density of the fluid, ρ_0; the kinematic viscosity, ν; the thermal diffusivity, K; the acceleration of gravity, g; the thermal expansion coefficient, ε; the mean temperature of the fluid, T_0; and the Coriolis parameter $f = 2\Omega \sin \phi$, where Ω is the rotation rate of the earth and ϕ is the latitude. Those familiar with hydrodynamics will realize that, in fact, these equations are not suitable for the prediction of large-scale atmospheric flows but, instead, are appropriate only for incompressible liquids with a linear relation between density and temperature.

However, the major arguments to follow apply equally well to all such systems and there is no need to use the more general equations.

Reynolds (1894) split the dependent variables into a time-averaged value and a deviation from the time average, e.g. $u = \bar{u} + u'$, where

$$\bar{u} = \frac{1}{\delta t}\int_{-\frac{\delta t}{2}}^{+\frac{\delta t}{2}} u \, dt \, .$$

Substituting this expansion into Eq. (1) and then averaging the equation term by term with respect to time over the same interval one obtains:

$$(7) \qquad \overline{\frac{\partial u'}{\partial t}} + \frac{\partial \bar{u}^2}{\partial x} + \frac{\partial \bar{v}\bar{u}}{\partial y} + \frac{\partial \bar{w}\bar{u}}{\partial z} = f\bar{v} - \frac{1}{\rho_0}\frac{\partial \bar{p}}{\partial x} + \frac{\partial}{\partial x}\left(\nu \frac{\partial \bar{u}}{\partial x} - \overline{u'^2}\right) + \frac{\partial}{\partial y}\left(\nu \frac{\partial \bar{u}}{\partial y} - \overline{v'u'}\right) + \frac{\partial}{\partial z}\left(\nu \frac{\partial \bar{u}}{\partial z} - \overline{w'u'}\right)$$

where the averaged perturbation product terms on the right are called the Reynolds' stresses. The leading term on the left may be expressed as

$$\overline{\frac{\partial u'}{\partial t}} = \frac{u'\left(\frac{\delta t}{2}\right) - u'\left(-\frac{\delta t}{2}\right)}{\delta t}$$

and for a stationary time series $\lim_{\delta t \to \infty} \overline{\frac{\partial u'}{\partial t}} = 0$. In this limit Eq. (7) is identical to Eq. (1) except that: 1) there is no time dependence, 2) the mean flow variables have replaced the instantaneous values, and 3) the

Reynolds' stresses appear on the right in combination with the viscous stresses. Similar results are obtained for Eqs. (2), (3), and (6).

Modifications of Eq. (7) are sometimes used for the prediction of turbulent flows with the following rationalizations: 1) The barred variables are thought to represent the larger-scale motions, those resolvable by a numerical grid, and the primed variables are thought to represent the remaining turbulent motions. 2) The viscous stresses are omitted since they generally are small compared to the Reynolds' stresses. 3) A time derivative of the average flow $\frac{\partial \bar{u}}{\partial t}$ is added since the flow is nonstationary with respect to a short averaging interval and since a time-dependent equation is required for prediction. The resultant equation is

$$\frac{\partial \bar{u}}{\partial t} + \frac{\partial \bar{u}^2}{\partial x} + \frac{\partial \bar{v}\bar{u}}{\partial y} + \frac{\partial \bar{w}\bar{u}}{\partial z} = + f\bar{v} - \frac{1}{\rho_0}\frac{\partial \bar{p}}{\partial x} - \left(\frac{\partial \overline{u'^2}}{\partial x} + \frac{\partial \overline{v'u'}}{\partial y} + \frac{\partial \overline{w'u'}}{\partial z} \right) . \tag{8}$$

Then 4) the Reynolds' stresses are either ignored or parameterized, and 5) the equations are approximated in a finite-difference form for numerical prediction.

These procedures contain a variety of physical assumptions, finite-difference approximations, and mathematically incorrect procedures which indeed do make the title "empirical" an appropriate one. This empirical approach, although not always spelled out in detail, has been rather typical of past applications of numerical prediction to the atmosphere. But despite the various uncertainties this procedure has been justifiable. With poor data coverage and the necessity of a rather gross parameterization of turbulence there has been little incentive to seek more accurate equations. Now, as observational techniques improve and as the possibilities for rapid computation grow, it is clear that a more rigorous development of the difference equations is desirable.

In a discussion of the representation and prediction of turbulent flows Lilly (1967) has introduced a somewhat different procedure. He defined the "mesh box average" by a relation of the form

$$\bar{u} = \frac{1}{\delta x \delta y \delta z} \int_{-\frac{\delta x}{2}}^{\frac{\delta x}{2}} \int_{-\frac{\delta y}{2}}^{\frac{\delta y}{2}} \int_{-\frac{\delta z}{2}}^{\frac{\delta z}{2}} u(\xi,\eta,\zeta,t)d\xi,\ d\eta,\ d\zeta$$

where ξ , η and ζ are dummy variables. The time-dependent equation that is obtained from Eq. (1) using the mesh box average and departures from this average is rather similar to Eq. (7) if one allows an interchange of the mesh box average and the spatial derivatives. However, it must be noted that for averages over short spatial and temporal intervals the motion cannot be considered stationary either in space or in time and the interchange of integration and differentiation without the introduction of additional terms may not be justified. Deardorff (1969) used this procedure for the numerical computation of turbulent shear flows. From comments in their respective papers both Lilly and Deardorff apparently were aware of the approximations made by their procedures but did not explicitly consider the additional terms that must arise nor the possible consequences of their assumptions. Accordingly, it is my purpose to develop here a complete set of equations so that meteorologists faced with the problem of parameterizing the sub-grid-scale motions may clearly see the task, and so that mathematicians who may wish to become involved in the numerical simulation of turbulent flow will be aware of the full problem.

3. Development of a Set of Complete Difference Equations

For the following work it is necessary to define averages and perturbations with respect to the four independent variables x, y, z, and t. In place of bars and primes the multiple-subscript notation introduced by Lorenze (1966) will be found convenient. Thus an expansion of the form $u - \bar{u} + u'$ is replaced by $u_0 = u_1 + u_2$ where the 0 subscript refers to the unmodified variable, 1 refers to an average, and 2 refers to a departure from that average. In the case of two independent variables, for example x and t in that order, the expansion becomes $u_{00} = u_{11} + u_{12} + u_{21} + u_{22}$, where u_{12} refers to time perturbations of an x average, u_{21} refers to x perturbations of a time average, and u_{22} refers to perturbations from the $x - t$ average. The term u_{11} is of course the space-time average, and in the later development such averages will be regarded as the measurable and predictable dependent variables. With the full set of independent variables the sequence of subscripts will be in the order x, y, z, t. It is apparent that u_{oooo} may be expanded into 16 terms.

For the initial discussion it is necessary to consider only the terms $\frac{\partial u}{\partial t} + \frac{\partial u^2}{\partial x}$ from Eq. (1). After the procedures for these terms are established the results are generalized to the full equations. The selected terms when expanded in only the two variables x and t become:

$$\frac{\partial u_{oo}}{\partial t} + \frac{\partial u_{oo}^2}{\partial x} \equiv \frac{\partial(u_{11}+u_{12}+u_{21}+u_{22})}{\partial t} + \frac{\partial(u_{11}+u_{12}+u_{21}+u_{22})^2}{\partial x} .$$

Then, following the Reynolds procedure, these terms are averaged with respect to x and t as indicated by the $[\ ]_{11}$. The terms that do not vanish are

$$(9)\quad \left[\frac{\partial u_{12}}{\partial t}\right]_{11} + \left[2u_{11}\frac{\partial u_{21}}{\partial x}\right]_{11} + \left[2u_{12}\frac{\partial u_{22}}{\partial x}\right]_{11} + \left[\frac{\partial u_{21}^2}{\partial x}\right]_{11} + \left[\frac{\partial u_{22}^2}{\partial x}\right]_{11}$$

Other terms are identically 0 either because they contain the x (or t) derivative of an x (or t) average or because the external average of an uncorrelated perturbation is zero. As examples, by the first criterion $\frac{\partial u_{11}^2}{\partial x} = 0$, and by the second $\left[u_{11}\frac{\partial u_{22}}{\partial x}\right]_{11} = 0$. Note that the external averages are applied over the same intervals of x and t that are used for the original expansions of the variables. Note also that these space-time averages are discontinuous functions of space and time as may be seen in Figure 1.†

Figure 1 illustrates a dependent variable $F_0(x)$ vs. x , the average $F_1(0)$ over the interval δx and centered at $x = 0$, and the perturbations F_2 at the locations $\pm \frac{\delta x}{2}$. The values of $F_2(\pm \frac{\delta x}{2})$ are perturbations from $F_1(0)$. It follows from the definition of the average that $\left[\frac{\partial F_2}{\partial x} \right]_1 = \frac{F_2(\frac{\delta x}{2}) - F_2(- \frac{\delta x}{2})}{\delta x}$.
Applying this type of relation to each of the terms in (9) we obtain:

$$\left[\frac{u_{12}(\frac{\delta t}{2}) - u_{12}(- \frac{\delta t}{2})}{\delta t} \right]_{00} \tag{10}$$

$$+ 2\left[u_{11}(0) \frac{u_{21}(\frac{\delta x}{2}) - u_{21}(- \frac{\delta x}{2})}{\delta x} \right]_{00}$$

$$+ 2\left[u_{12}(0) \frac{u_{22}(\frac{\delta x}{2}) - u_{22}(- \frac{\delta x}{2})}{\delta x} \right]_{01}$$

$$+ \left[\frac{u_{21}{}^2(\frac{\delta x}{2}) - u_{21}{}^2(- \frac{\delta x}{2})}{\delta x} \right]_{00}$$

$$+ \left[\frac{u_{22}{}^2(\frac{\delta x}{2}) - u_{22}{}^2(- \frac{\delta x}{2})}{\delta x} \right]_{01} ,$$

† The reader should note that the use of a continuous running average in time or in space, for example,

$$\bar{u}(t) = \frac{1}{\delta t} \int_{t-\frac{\delta t}{2}}^{t+\frac{\delta t}{2}} u(\tau)\, d\tau ,$$

where an external 0 subscript indicates that that external average is superfluous in that it has already been performed. As an example $\left[u_{22}{}^{2}\left(\frac{\delta x}{2}\right)\right]_{01}$ represents the square of the space-time perturbation evaluated at the edge of the box $+\frac{\delta x}{2}$ and then time averaged. No ambiguity should result from the fact that $u_{12}\left(\frac{\delta t}{2}\right)$ is to be evaluated at $x = 0$ and that terms like $u_{21}\left(\frac{\delta x}{2}\right)$ are to be evaluated at $t = 0$.

Clearly, if we wish to work with space-time averaged quantities the representation in (10) is inadequate since most of the terms are evaluated at the boundaries of the space-time mesh box. Referring again to Figure 1, it will be shown that finite differences logically arise when we approximate the perturbations at the boundaries in terms of the averaged variables. The mid points between $F_1(\delta x)$ and $F_1(0)$, and between $F_1(0)$ and $F_1(-\delta x)$ are indicated by crosses. These points are taken to be estimates of $F_0\left(\frac{\delta x}{2}\right)$ and $F_0\left(-\frac{\delta x}{2}\right)$, respectively, with errors of estimate $C_0(\delta x/2)$ and $C_0(-\delta x/2)$. Then it follows that

$$F_0\left(\frac{\delta x}{2}\right) \equiv F_1(0) + F_2\left(\frac{\delta x}{2}\right) = \frac{F_1(\delta x)+F_1(0)}{2} + C_0\left(\frac{\delta x}{2}\right)$$

and

$$F_0\left(-\frac{\delta x}{2}\right) \equiv F_1(0) + F_2\left(-\frac{\delta x}{2}\right) = \frac{F_1(0)+F_1(-\delta x)}{2} + C_0\left(-\frac{\delta x}{2}\right).$$

---

leads to serious difficulties of interpretation. Thus, in a term of the form $\overline{u^2} = \overline{\bar{u}^2} + \overline{2\bar{u}u'} + \overline{u'^2}$ it is not true that $\overline{\bar{u}^2} = \bar{u}^2$ and $\overline{\bar{u}u'} = 0$ if the bar indicates a running average.

Substituting these relations for $F_2(\pm \frac{\delta x}{2})$ expression (10) may be written

$$(11) \quad \left[\frac{u_{11}(\delta t) - u_{11}(-\delta t)}{2\delta t}\right]_{00}$$

$$+ 2\left[u_{11}(0)\ \frac{u_{11}(\delta x) - u_{11}(-\delta x)}{2\delta x}\right]_{00}$$

$$+ 2\left[u_{12}(0)\ \frac{u_{12}(\delta x) - u_{12}(-\delta x)}{2\delta x}\right]_{01}$$

$$+ \left[\frac{u_{21}{}^2(\delta x) - u_{21}{}^2(-\delta x)}{2\delta x}\right]_{10}$$

$$+ \left[\frac{u_{22}{}^2(\delta x) - u_{22}{}^2(-\delta x)}{2\delta x}\right]_{11}$$

$$+ \left[\frac{C_{10}(\frac{\delta t}{2}) - C_{10}(-\frac{\delta t}{2})}{\delta t}\right]_{00}$$

$$+ 2\left[u_{11}(0)\ \frac{C_{01}(\frac{\delta x}{2}) - C_{01}(-\frac{\delta x}{2})}{\delta x}\right]_{00}$$

$$+ 2\left[u_{12}(0)\ \frac{C_{02}(\frac{\delta x}{2}) - C_{02}(-\frac{\delta x}{2})}{\delta x}\right]_{01}$$

$$+ \left[\frac{f_{01}(\frac{\delta x}{2}) - f_{01}(-\frac{\delta x}{2})}{\delta x}\right]_{00}$$

$$+ \left[\frac{f_{02}(\frac{\delta x}{2}) - f_{02}(-\frac{\delta x}{2})}{\delta x}\right]_{00}$$

where the f terms are the errors that correspond to the

u^2 terms as illustrated in Figure 2. There it may be seen that, for example,

$$[u_{22}{}^2]_{01}\left(\frac{\delta x}{2}\right) = \frac{[u_{22}{}^2]_{11}(\delta x)+[u_{22}{}^2]_{11}(0)}{2} + f_{02}\left(\frac{\delta x}{2}\right)$$

and

$$[u_{22}{}^2]_{01}\left(-\frac{\delta x}{2}\right) = \frac{[u_{22}{}^2]_{11}(0)+[u_{22}{}^2]_{11}(-\delta x)}{2} + f_{02}\left(-\frac{\delta x}{2}\right) .$$

These relations then serve as definitions of the errors f. Now, by denoting $\dfrac{u_{11}(\delta t) - u_{11}(-\delta t)}{2\delta t}$ by $\dfrac{\Delta u_{11}}{\Delta t}$ and $\dfrac{C_{10}\left(\frac{\delta t}{2}\right) - C_{10}\left(-\frac{\delta t}{2}\right)}{\delta t}$ by $\dfrac{\delta C_{10}}{\delta t}$, etc., expression (11) may be simplified to

$$(12) \qquad \frac{\Delta u_{11}}{\Delta t} + 2u_{11}\frac{\Delta u_{11}}{\Delta x} + 2\left[u_{12}\frac{\Delta u_{12}}{\Delta x}\right]_{01} + \left[\frac{\Delta u_{21}{}^2}{\Delta x}\right]_{10}$$

$$+ \left[\frac{\Delta u_{22}{}^2}{\Delta x}\right]_{11} + \frac{\delta C_{10}}{\delta t} + 2u_{11}\frac{\delta C_{01}}{\delta x}$$

$$+ 2\left[u_{12}\frac{\delta C_{02}}{\delta x}\right]_{01} + \frac{\delta f_{01}}{\delta x} + \frac{\delta f_{02}}{\delta x}$$

where redundant external subscripts have been omitted.

The first two terms of expression (12) contain the space-time averaged values of u in an ordinary centered difference scheme. Since the next three terms contain averaged products of perturbations they correspond to the Reynolds' stresses. The last five terms are error terms that arise from the finite-difference approximations. Before generalizing the method, several interesting observations may be made:

1) It is apparent that in general development many more Reynolds' stresses will be generated than we are accustomed to deal with.

2) The finite-difference space-time averaged values arise from terms that were originally perturbation terms. Conversely, the space-time averages from the original expansion of the variables drop out when differentiated.

3) Both the error terms and the Reynolds' stresses depend upon the sub-grid-scale motions (or poorly resolved features) and essentially are indeterminate. (The possibility of a hierarchy of equations for the Reynolds' stresses and higher moments will not be considered here.) While there may be some possibility of parameterizing the Reynolds' stresses, we should first investigate the relative magnitude of the error terms.

4) The fact that <u>centered</u> space and time differences have emerged is simply a result of the manner in which the values on the edges of the space-time box have been approximated. Uncentered differences, higher order approximations,

or more complicated interlocking difference schemes may be obtained with consequent changes in the error terms.

5) The 4th and 5th terms of expression (12) can be combined to give $\left[\frac{\Delta u_{20}^{2}}{\Delta x}\right]_{11}$. In a similar way the 2nd and 3rd terms combine, but this combination is not profitable since the intent is to isolate the space-time averaged values from the perturbations.

The above procedures may be readily generalized by noting that all terms on the left hand sides of Eqs. (1-4) may be represented by the general term $\frac{\partial u_i u_j}{\partial x_j}$ where i and j run 1 - 4 representing x, y, z, and t and $u_i = dx_i/dt$. The expansion and averaging of this general term may be expressed by

$$\left[\frac{\partial u_{i0000} u_{j0000}}{\partial x_j}\right]_{iiii} = \sum_{m_k, n_k} \left[\frac{\partial u_{im_k} u_{jm_k}}{\partial x_j}\right]_{p_k} = \sum D$$

where i, j, and k run 1 - 4, m_k and n_k are either 1 or 2, and p_k is either 0 or 1. The rules for discarding or evaluating individual terms D are as follows:

1) If for any $k \neq j$, $m_k \neq n_k$, $D = 0$. This rule takes precedence.

2) If for $k = j$, $m_k = n_k = 1$, $D = 0$.

3) If for $k = j$, $m_k = 1$, $n_k = 2$,

$$D = \left[u_{im_k}\left(\frac{\Delta u_{jm_k}}{\Delta x_j} + \frac{\delta C_{jm_k}}{\delta x_j} \right) \right]_{p_k}$$

4) If for $k = j$, $m_k = 2$, $n_k = 1$,

$$D = \left[u_{jm_k} \left(\frac{\Delta u_{in_k}}{\Delta x_j} + \frac{\delta C_{in_k}}{\delta x_j} \right) \right]_{p_k}$$

5) If for $k = j$, $m_k = 2$, $n_k = 2$,

$$D = \left[\frac{\Delta u_{im_k} u_{jm_k}}{\Delta x_j} + \frac{\delta f_{m_k}}{\delta x_j} \right]_{p_k}$$

6) $p_k = 1$ for $k \neq j$, $p_k = 0$ for $k = j$. In δ terms $m_k = 0$ or $n_k = 0$ for $k = j$.

Applied to the entire right hand side of Eq. (1) the expansion and averaging gives the following expression that corresponds to $i = 1$:

(13)
$$\frac{\Delta \bar{u}}{\Delta t} + \bar{u} \frac{\Delta \bar{u}}{\Delta x} + \bar{v} \frac{\Delta \bar{u}}{\Delta y} + \bar{w} \frac{\Delta \bar{u}}{\Delta z}$$

$$+ \frac{1}{\Delta x} \left\{ 2[u_{1021} \Delta u_{1021}]_{0110} + 2[u_{1210} \Delta u_{1210}]_{0101} + 2[u_{1102} \Delta u_{1102}]_{0011} + 2[u_{1222} \Delta u_{1222}]_{0111} \right\}$$

$$+ \frac{1}{\Delta y} \left\{ [v_{1102} \Delta u_{1102} + u_{1102} \Delta v_{1102}]_{0011} + [v_{2110} \Delta u_{2110} + u_{2110} \Delta v_{2110}]_{1001} + [v_{0121} \Delta u_{0121} + u_{0121} \Delta v_{0121}]_{1010} + [v_{2122} \Delta u_{2122} + u_{2122} \Delta v_{2122}]_{1011} \right\}$$

$$+ \frac{1}{\Delta z} \Big\{ [w_{2110}\Delta u_{2110} + u_{2110}\Delta w_{2110}]_{1001}$$

$$+ [w_{0211}\Delta u_{0211} + u_{0211}\Delta w_{0211}]_{1100}$$

$$+ [w_{1012}\Delta u_{1012} + u_{1012}\Delta w_{1012}]_{0101}$$

$$+ [w_{2212}\Delta u_{2212} + u_{2212}\Delta w_{2212}]_{1101} \Big\}$$

$$+ \left[\frac{\Delta u_{2000}^{2}}{\Delta x} \right]_{1111} + \left[\frac{\Delta (v_{0200} u_{0200})}{\Delta y} \right]_{1111}$$

$$+ \left[\frac{\Delta (w_{0020} u_{0020})}{\Delta z} \right]_{1111}$$

$$+ \frac{\delta C_{1110}}{\delta t} - \left[\frac{\delta C_{0111}}{\delta x} + \frac{\delta d_{1011}}{\delta y} + \frac{\delta e_{1101}}{\delta z} \right]$$

$$+ 2 \left[u_{1000} \frac{\delta C_{0000}}{\delta x} \right]_{0111} + \left[\frac{u_{0100}\delta d_{0000} + v_{0100}\delta C_{0000}}{\delta y} \right]_{1011}$$

$$+ \left[\frac{u_{0010}\delta e_{0000} + w_{0010}\delta C_{0000}}{\delta z} \right]_{1101} + \frac{\delta f_{0111}}{\delta x}$$

$$+ \frac{\delta g_{1011}}{\delta x} + \frac{\delta h_{1101}}{\delta z}$$

where the bar now indicates the space-time average, the error terms C, d, and e correspond to errors in u, v, and w, and the error terms f, g, and h correspond to errors in the u^2, uv, and uw terms. Several of the Reynolds' stresses contain 0 subscripts indicating that they are contractions of more than one term. Thus the term $\left[\frac{\Delta (v_{0200} u_{0200})}{\Delta y} \right]_{1111}$ is a contraction of 8 separate terms that would occur in the general expansion. In the contracted form of Eq. (13) there are 23 Reynolds' stresses and in the fully expanded form there are 59 separate stress terms representing various combinations of averages and perturbations in the four independent variables. These may be contrasted with the 3 Reynolds' stresses of Eq. (7).

To obtain (13) use has been made of the expanded and averaged continuity equation which is simply:

$$\frac{\Delta\bar{u}}{\Delta x} + \frac{\Delta\bar{v}}{\Delta y} + \frac{\Delta\bar{w}}{\Delta z} + \frac{\delta C_{0111}}{\delta x} + \frac{\delta d_{1011}}{\delta y} + \frac{\delta e_{1101}}{\delta z} = 0 \; . \tag{14}$$

Other arrangements of the terms in Eq. (13) are possible although the given form appears to be the most compact without introducing further new notation. The Coriolis and pressure gradient terms from the right hand side of Eq. (1) reduce simply to the space-time averaged values plus (for the pressure gradient) the error terms. The viscous terms, involving second spatial derivatives, are also linear so that they reduce to a simple finite-difference representation of the average values plus errors, albeit in a somewhat more involved way. By a straightforward extension of the above methods a typical viscous term $\frac{\partial^2 u}{\partial x^2}$ is represented by $\frac{\bar{u}(2\delta x)-2\bar{u}(0)+\bar{u}(-2\delta x)}{\Delta x^2}$ + errors. Boundary conditions are handled in a manner very similar to the general development.

Finally, it is convenient to represent the difference terms in (13) in differential form, the form that would logically lead to the difference representation that we have found. The complete equation that corresponds to Eq. (1) then may be represented in differential form as:

$$\frac{\partial \bar{u}}{\partial t} + \bar{u}\frac{\partial \bar{u}}{\partial x} + \bar{v}\frac{\partial \bar{u}}{\partial y} + \bar{w}\frac{\partial \bar{u}}{\partial z} = f\bar{v} - \frac{1}{\rho_0}\frac{\partial \bar{p}}{\partial x}$$

$$+ \frac{\partial}{\partial x}\left(\nu \frac{\partial \bar{u}}{\partial x} - \overline{u_{m_k}^{2}}\right) + \frac{\partial}{\partial y}\left(\nu \frac{\partial \bar{u}}{\partial y} - \overline{v_{m_k} u_{m_k}}\right)$$

$$+ \frac{\partial}{\partial z}\left(\nu \frac{\partial \bar{u}}{\partial z} - \overline{w_{m_k} u_{m_k}}\right) \tag{15}$$

where the 4 subscripts m_k can assume all combinations of 1 and 2 except the combination 1111. The latter terms are already represented on the left hand side.

Conclusion

Equation (13) has been developed by a rigorous application of the Reynolds' method to a field of flow that is stationary neither in time nor in space with respect to the averaging intervals that are employed. A natural approximation of perturbations at the boundaries of the space-time mesh box has lead directly to a set of difference equations in the space-time averaged variables. These difference equations contain many more turbulent stresses than those originally derived by Reynolds for the case of a stationary time series. Each of the turbulent stresses can be thought of as representing a specific physical process on a sub-grid-scale level.

The fact that centered time and space differences were obtained is simply due to the manner in which the perturbations at the boundaries of the mesh box were related to the averaged values of the same variables. Other methods can lead to higher order approximations or non-centered difference schemes if these are desired. With other difference schemes the number and character of the Reynolds' stresses remain unchanged although the error terms necessarily are different.

An important issue for numerical weather prediction is the magnitude of the Reynolds' stresses in comparison with the error terms. If it should emerge that the Reynolds' stresses are neglible compared to the errors, then there is little point in attempting a detailed parameterization. If the Reynolds' stresses either singly or in combination make a significant contribution to Eq. (13), then the expanded form of the relevant terms may be particularly appropriate in isolating organized atmospheric processes for the purposes of parameterization.

Equation (15), although not rigorously derived from the differential equations of fluid motion, leads to Eq. (13) when properly interpreted and represented by finite differences. In the limiting case when the space-time mesh box is small enough to resolve all turbulent motions Eq. (15) reduce to Eq. (1). Equation (15) also may be used with other difference representations of the various terms with the understanding that the error terms will be different from those of Eq. (13).

To carry the error terms through the analysis may appear to be unnecessary and unwarrented. However, the relative magnitude of the errors and the Reynolds' stresses is of prime importance for the question of predictability and we must seek a means of estimating these effects. One method will be to construct simplified mathematical representations of small-scale atmospheric phenomena from which one may compute their contribution to the stresses and to the finite-difference errors. Estimates of this type have not yet been carried out in detail.

REFERENCES

1. Deardorff, J. W., "A Three-dimensional Numerical Study of Turbulent Channel Flow at Large Reynolds Numbers," NCAR Manuscript No. 69-19, National Center for Atmospheric Research, Boulder, Colorado, 1969.
2. Lilly, D. K., "The Representation of Small-scale Turbulence in Numerical Simulation Experiments," Proc. IBM Scientific Computing Symp. on Environmental Sciences, Nov. 14-16, Yorktown Heights, N.Y. IBM Form No. 320-1951, 195, 1966.
3. Lorenz, E. N., "A Multiple Index Notation for Describing Atmospheric Transport Processes," Sci. Report No. 2, Planetary Circulation Project, V. P. Starr, Director. Contract No. AF 19(628)-5816, AFCRL, 1966.
4. Lorenz, E. N., "The Predictability of a Flow which Possesses Many Scales of Motion," Tellus, 21, 289-307, 1969.
5. Reynolds, O., "On the Dynamical Theory of Incompressible Viscous Fluids and the Determination of the Criterion," Phil. Trans. Roy. Soc. London, A, 186, 123-164, 1894.
6. Robinson, G. D., "Some Current Projects for Global Meteorological Observation and Experiment," Q. J. Roy. Meteor. Soc., 93, 409-418, 1967.
7. Van der Hoven, Isaac, "Power Spectrum of Horizontal Wind Speed in the Frequency Range from .0007 to 900 Cycles Per Hour," J. Meteor., 14, 160-164, 1957.

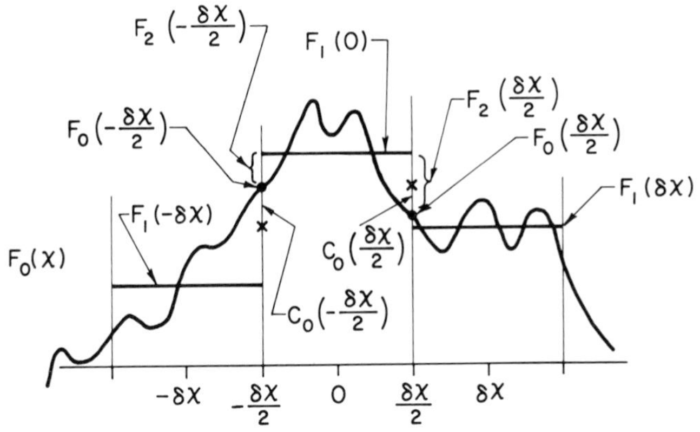

Fig. 1 The method of approximating values at the boundaries of the time-space mesh box. Only one spatial dimension is represented here. (See text.) Note that $F_2(\frac{\delta x}{2})$, $F_2(-\frac{\delta x}{2})$, and $C_o(\frac{\delta x}{2})$ have negative values in this example.

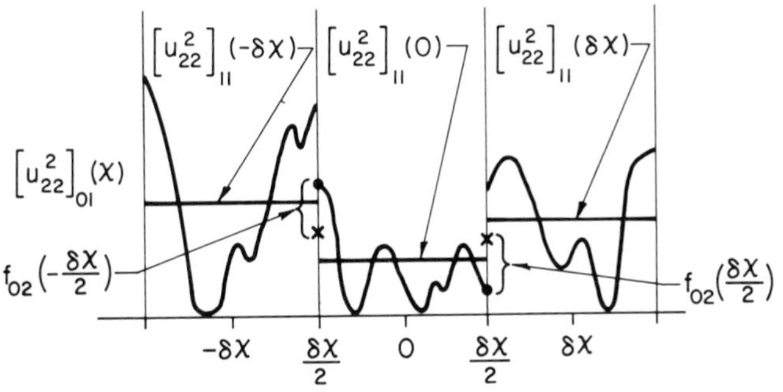

Fig. 2 In a manner similar to that illustrated in Fig. 1 the values $[u_{22}{}^2]_{01}(\pm\frac{\delta x}{2})$ (indicated by dots) are approximated by the averages of the spatial mean values $[u_{22}{}^2]_{11}$ with residuals denoted by $f_{02}(\pm\frac{\delta x}{2})$ (see text). Note that $f_{02}(\frac{\delta x}{2})$ as illustrated is a negative value.

FURTHER DEVELOPMENTS IN THE APPROXIMATION THEORY OF EIGENVALUES*

Gaetano Fichera**

This paper continues the one [1] which was presented at the Maryland Symposium on Numerical Solution of Partial Differential Equations in 1965. In [1] a method for getting lower bounds for the eigenvalues of a class of positive linear operators was given and denoted method of orthogonal invariants. It was shown how the method applies to the eigenvalue problems connected with boundary value problems of elliptic type. The application of the method to elliptic differential operators expounded in [1] relies on a structure theorem for the Green operator of the boundary value problem under consideration. On the other hand this structure theorem is founded on the explicit knowledge of a fundamental solution operator. That, in practice, restricts the application of the method to operators with constant coefficients. We wish now to show how the method can be applied to the general case of operators with

* This research has been sponsored in part by the Aerospace Research Laboratories through the European Office of Aerospace Research, OAR, United States Air Force, under Grant EOOAR 69-0066.

** University of Rome

variable coefficients. The proofs of the theorems we are going to discuss have been partly expounded in [2]. Additional proofs will be given in a forthcoming paper.

1. Eigenvalue Problems for Linear Elliptic Systems

Let A be a bounded domain of X^r which we suppose properly regular according to the definition given in [3] (see p. 21). Let $a_{pq}(x)$ $(0 \leqslant |p| \leqslant m,\ 0 \leqslant |q| \leqslant m)$ be an $\ell \times \ell$ complex matrix whose entries are supposed bounded and measurable in A. Set for $u, v \in H_m(A)$[†]

$$B(u,v) = \sum_{|p|,|q|}^{o,m} (-1)^{|p|} \int_A [a_{pq}(x)D^q u]D^p v \, dx .$$

Let V be a closed linear manifold of $H_m(A)$ such that $V \supset \overset{\circ}{H}_m(A)$. We suppose that

i) for any pair p,q: $a_{pq}(x) = (-1)^{|p|+|q|}\bar{a}_{qp}(x)$;

ii) the bilinear form $B(u,v)$ is coercive on V i.e.

$$(-1)^m B(v,v) \geqslant c \|v\|_m^2 \qquad (c > 0)$$

for every $v \in V$.

From the general theory of elliptic operators we have that, given arbitrarily $f \in H_0(A)$, there exists one and only one $u \in V$ such that

---

[†]For the definition of the spaces $H_m(A)$, $\overset{\circ}{H}_m(A)$, see [3] pp. 17-23.

$$B(u,v) = \int_A fv\, dx$$

for every $v \in V$.

Set $u = Gf$. The linear operator G has $H_0(A)$ as domain. Its range is contained in V. As an operator from $H_0(A)$ into $H_0(A)$, G is a positive compact operator. Since $H_0(A)$ is a complex Hilbert space, the operator G is symmetric. In the case that the boundary ∂A of A and the matrices $a_{pq}(x)$ are sufficiently smooth, the eigenvalue problem

$$Gf - \mu f = 0 , \qquad f \in H_0(A)$$

is equivalent to the differential eigenvalue problem

(1) $$(-1)^m D^p a_{pq}(x) D^q u - \lambda u = 0 , \qquad u \in V \qquad (\lambda = \mu^{-1})$$

with certain additional side conditions for u.

In the general case, the eigenvalue problem (1) can be only expressed in integral form

(2) $$B(u,v) - \lambda \int_A uv\, dx = 0 \qquad (v \in V) .$$

That is sufficient for our purposes.

Let us now consider a linear $\ell \times \ell$ -matrix operator of order $2m$ with constant coefficients

$$L_0(D) = \alpha_{pq} D^{p+q}$$

which we suppose formally self-adjoint, i.e.

$$\alpha_{pq} = (-1)^{|p|+|q|}\bar{\alpha}_{qp} .$$

Set

$$B_0(u,v) = \sum_{|p|,|q|}^{0,m} (-1)^{|p|} \int_A (\alpha_{pq} D^q u) D^p v \, dx$$

and suppose that for any $v \in V$

$$(-1)^m B_0(v,v) \geqslant c_0 \|v\|_m^2 \qquad (c_0 > 0)$$

and, moreover, $(-1)^m B_0(v,v) \geqslant 0$ for any $v \in H_m(A)$.

For instance we can assume

$$L_0(D) = \sum_{|p|}^{0,m} (-1)^{|p|+m} D^p D^p .$$

Concerning the functional equation

$$B_0(u,v) = \int_A f v \, dx , \qquad v \in V$$

and the eigenvalue problem

$$B_0(u,v) - \lambda \int_A u v \, dx = 0 , \qquad v \in V$$

the same results hold as in the case of the bilinear form $B(u,v)$. Let Γ_0 be the positive compact operator connected with $B_0(u,v)$, when considered in V. It is obvious that, without any restriction, we may assume

$$(-1)^m B(v,v) > (-1)^m B_o(v,v)$$

for any non-zero vector $v \in V$.

Set for $u, v \in V$

$$((u,v)) = (-1)^m [B(u,v) - B_o(u,v)]$$

Let $\{\psi_j\}$ $(j = 1, 2, \ldots)$ be a complete system of linearly independent vectors in the space $H_o(A)$. Set

$$\alpha_j = G\psi_j \qquad (j = 1, 2, \ldots) .$$

Let T and T_o be two linear bounded transformations from V into V such that, respectively,

$$B(u,v) = (Tu,v)_m , \qquad B_o(u,v) = (T_o u,v)_m .$$

Both T and T_o are symmetric.

Since the matrix

(3) $$\left\{((\alpha_i, \alpha_j))\right\} \qquad (i, j = 1, \ldots, n)$$

is non singular, we can consider in V the operator

$$T_n = T_o + b_{ij} ((\cdot,\alpha_i)) [T\alpha_j - T_o\alpha_j]$$

where b_{ij} is the inverse matrix of the matrix (3).

We claim that, for any given $f \in H_o(A)$, the problem

$$(4) \qquad (T_n u, v)_m = (f, v)_o \, , \quad v \in V$$

has one and only one solution.

Uniqueness is obvious. In fact $(T_n u, u)_m = 0$ implies

$$(T_o u, u) + b_{ij} ((u, \alpha_i)) ((u, \alpha_j)) = 0$$

and, in consequence, $u = 0$.

For proving existence, let us write (4) as follows

$$(T_o[u - b_{ij} ((u, \alpha_i)) \alpha_j], v)_m$$

$$= (f, v)_o - b_{ij} ((u, \alpha_i))(\psi_j, v)_o \, .$$

That implies

$$u = \Gamma_n f \equiv \Gamma_o f - \sigma_{jk} (f, \phi_k)_o \phi_j \, ,$$

where $\{\sigma_{jk}\}$ is the inverse matrix of the non singular $n \times n$ -matrix

$$(5) \qquad \{(\psi_j, \phi_k)_o\} \qquad j, k = 1, \ldots, n$$

and

$$\phi_k = \alpha_k - \Gamma_o \psi_k \ .$$

The matrix (5) is non-singular, because, otherwise, problem (4) should have more than one solution. By inspection we see that u satisfies (4).

Theorem 1. The operator Γ_n is a positive compact operator belonging to any class $\mathfrak{C}^p$ (see [1] p. 336) with

$$p > \frac{r}{2m} \ .$$

If we denote by $\{\mu_{n,k}\}$ the sequence of the eigenvalues of Γ_n and by $\{\mu_k\}$ the eigenvalues of G - ordered in the common way - we have

$$\mu_{o,k} \geqslant \mu_{n,k} \geqslant \mu_{n+1,k} \geqslant \mu_k \ .$$

Theorem 2. If for $s \geqslant 1$ and for $\nu \geqslant p$ we consider the orthogonal invariants of degree ν and order s (see [1] p. 335) we have

$$\mathfrak{J}_s^{\nu}(\Gamma_n) \geqslant \mathfrak{J}_s^{\nu}(\Gamma_{n+1}) \ , \qquad \lim_{n\to\infty} \mathfrak{J}_s^{\nu}(\Gamma_n) = \mathfrak{J}_s^{\nu}(G) \ .$$

Let $\{w_k\}$ be a sequence of linearly independent vectors complete in $H_o(A)$. Denote by W_m the manifold spanned by $w_1, \ldots, w_m$. Let P_m be the orthogonal projector of $H_o(A)$ onto W_m. Let $\mu_1^{(m)} \geqslant \ldots \geqslant \mu_m^{(m)}$ be the positive eigenvalues of the operator $P_m G P_m$. Denote by $w_k^{(m)}$ an eigenvector of the operator $P_m G P_m$

corresponding to the eigenvalue $\mu_k^{(m)}$. Let $W_{m,k}$ be the subspace of W_m formed by the vectors orthogonal to $w_k^{(m)}$. Let $P_{m,k}$ be the orthogonal projector of $H_o(A)$ onto $W_{m,k}$.

Theorem 3. For any fixed s and ν set

$$\mu_k^{(m,n)} = \left\{ \frac{\mathcal{J}_s^{\nu}(\Gamma_n) - \mathcal{J}_s^{\nu}(P_m G P_m)}{\mathcal{J}_{s-1}^{\nu}(P_{m,k} G P_{m,k})} + [\mu_k^{(m)}]^{\nu} \right\}^{1/\nu}$$

We have

$$\mu_k^{(m,n)} \geqslant \mu_k^{(\tilde{m},\tilde{n})} \qquad \text{if} \qquad m \leqslant \tilde{m}\ , \quad n \leqslant \tilde{n}$$

$$\lim_{\substack{m\to\infty \\ n\to\infty}} \mu_k^{(m,n)} = \mu_k \ .$$

Since $\mu_k^{(m)}$, $\mathcal{J}_s^{\nu}(P_m G P_m)$ and $\mathcal{J}_{s-1}^{\nu}(P_{m,k} G P_{m,k})$ are numerically computable, as it is shown in [1], the problem of the upper approximation of μ_k is reduced to construct upper bounds arbitrarily close to $\mathcal{J}_s^{\nu}(\Gamma_n)$.

2. Upper Approximation of $\mathcal{J}_s^{\nu}(\Gamma_n)$

Since $\mathcal{L}_o$ is an operator with constant coefficients, we may suppose that the following hypothesis be satisfied:

Hypothesis. There exists a linear operator R which maps $H_o(A)$ onto $H_m(A)$ such that for any $v \in V$ and any $f \in H_o(A)$

$$B_0(Rf, v) = (f, v)_0 .$$

R is a particular fundamental solution operator for $\mathcal{L}_0$. It is not difficult to construct explicitly, under very general hypotheses, the operator R.

Let K be the kernel of the quadratic form $B_0(v,v)$ in the space $H_m(A)$. Consider the quotient vector space $\mathcal{H} = H_m(A)/K$. Let us introduce in $\mathcal{H}$ the following scalar product:

$$[u,v] = (-1)^m B_0(u,v) .$$

Since V is a subspace of $\mathcal{H}$, we can consider the orthogonal complement Ω of V with respect to $\mathcal{H}$. Let P be the orthogonal projector of $\mathcal{H}$ onto Ω and denote by R^* the operator from $\mathcal{H}$ to $H_0(A)$ defined by the condition

$$[Rf, u] = (f, R^* u)_0 \qquad \left\{f \in H_0(A),\ u \in \mathcal{H}\right\} .$$

<u>Theorem 4</u>. The operator Γ_0 has the following structure:

$$\Gamma_0 = R^* R - R^* PR .$$

Let $\{\omega_s\}$ be a system of linearly independent vectors, complete in Ω. Let Ω_ρ be the manifold spanned by $\omega_1, \ldots, \omega_\rho$ and $P^{(\rho)}$ the orthogonal projector of $\mathcal{H}$ onto Ω_ρ. Set

$$\Gamma_o^{(\rho)} = R^*R - R^*P^{(\rho)}R$$

$$\Gamma_n^{(\rho)} = \Gamma_o^{(\rho)} - \sigma_{jk}(f,\phi_k)_o\phi_j \ .$$

Theorem 5. For any n we have

$$\mathcal{J}_s^\nu(\Gamma_n^{(\rho+1)}) \geqslant \mathcal{J}_s^\nu(\Gamma_n^{(\rho)}) \ ,$$

$$\lim_{\rho\to\infty} \mathcal{J}_s^\nu(\Gamma_n^{(\rho)}) = \mathcal{J}_s^\nu(\Gamma_n) \ .$$

From the theory developed in [1] it follows that $\mathcal{J}_s^\nu(\Gamma^{(\rho)})$ is numerically computable to any prescribed degree of accuracy, hence we obtain arbitrarily close upper bounds to $\mathcal{J}_s^\nu(\Gamma_n)$.

REFERENCES

1. Fichera, G., Approximation and Estimates for Eigenvalues - Numerical Solution of Partial Differential Equations, pp. 317-352, Academic Press, Inc., New York, 1966.
2. Fichera, G., "Upper Bounds for Orthogonal Invariants of Some Positive Linear Operators," Rend. Ist. Matem. Univ. Trieste, Vol. I, fasc. I, 1969.
3. Fichera, G., "Linear Elliptic Differential Systems and Eigenvalue Problems," Lecture Notes in Math., No. 8, Springer Verlag, Berlin, Heidelberg, New York, 1965.

NUMERICAL DESIGN OF TRANSONIC AIRFOILS*

P. R. Garabedian and D. G. Korn**

1. Introduction

Renewed interest in transonic flight has been stimulated by difficulties with the development of the SST and the desirability of speeding up the performance of subsonic aircraft like the Boeing 707. Recent experimental investigations [9,10] indicate that it is feasible to suppress wave drag and boost the force-break Mach number of airfoils by designing them to admit shock-free transonic flow at a specified cruising speed. The advent of Whitcomb's supercritical wing [13] challenges mathematicians to set up a purely theoretical procedure for calculating comparable airfoil shapes. With this in mind we shall describe here an inverse method of computing plane transonic flows past airfoils that are not only free of shocks, but also have adverse pressure gradients so moderate that no separation of the turbulent

* The work presented in this paper is supported by the AEC Computing and Applied Mathematics Center, Courant Institute of Mathematical Sciences, New York University, under Contract AT(30-1)-1480 with the U. S. Atomic Energy Commission.

** New York University.

boundary layer should take place. Up-to-date existence and uniqueness theorems combine with the experimental evidence to assure that these flows are physically realistic and will occur in practice.

We shall solve the partial differential equations governing steady two-dimensional flow of an inviscid compressible fluid by numerical analysis of characteristic initial value problems for the analytic continuation of the solution into the complex domain. The finite difference scheme we use was originally introduced [3] to describe the detached shock wave in front of a blunt body, but is actually better suited to the inverse problem of shaping airfoils so as to achieve shock-free transonic flow. It is related to Bergman's integral operator method [1] and does exploit simplifications associated with the linearity of the equations of motion in the hodograph plane, but it seems to be faster and more flexible than procedures such as that of Nieuwland [8] which involve combining known solutions found by separating the variables. Although the principal aim now is to construct smooth transonic flows, our method also generates solutions with limiting lines that can be interpreted as an approximation of weak shocks.

It is a pleasure to acknowledge the contribution to our work of Clinton Brown and Marshall Tulin, who made us aware of the physical importance of calculating shock-free airfoils whose adverse pressure gradient would not induce boundary layer separation. A more detailed account of the attack on this problem we shall describe is presented in an earlier report of Korn [4].

2. Transonic Controversy and the Inverse Problem

In the early days of breaking the sound barrier a controversy arose about shock-free transonic flow past airfoils with closed supersonic zones. The mathematical theorem stating that such flows do not in general exist for a given profile [6] has sometimes been interpreted erroneously to mean that they would be physically unstable even in the exceptional cases where they can be constructed. More modern results [5] suggest, however, that if an appropriate entropy inequality is imposed there should always exist a unique weak solution of the transonic boundary value problem for inviscid flow past an airfoil of reasonable shape, but that the flow can be expected to have at least weak shocks located presumably toward the back of the wing. In particular one can hope to prove that when a smooth solution is present, then by uniqueness it must coincide with the weak solution, which therefore has no shocks. Any slight perturbation in shape or Mach number should only lead to the appearance of small weak shocks. Moreover, we should be able to realize the flow experimentally if it does not exhibit an adverse pressure gradient that induces boundary layer separation effectively altering the profile in an essential way. Recently Cathleen Morawetz [7] has substantiated this analysis by showing that the Dirichlet problem becomes correctly set for the Tricomi equation when an appropriate singularity is permitted to occur at one end of the sonic line.

The contention that shock-free transonic flows are physically realistic is borne out by the experimental work of Pearcey and Spee. More specifically, Spee and Uijlenhoet [10] have made wind tunnel measurements on two-dimensional models built from Nieuwland's computation of smooth transonic flows by the hodograph method [8], and they obtain pressure distributions and schlieren photographs in excellent agreement with the theoretical predictions. Moreover, they establish that the flows are stable with respect to small changes in the angle of attack or the free stream Mach number. It therefore becomes evident that it would be useful to have more flexible and effective procedures for generating examples like Nieuwland's. Since the direct problem of determining smooth transonic flow past a given profile is not well posed, the question becomes how to calculate global solutions of the equations of motion that yield airfoil shapes admitting shock-free flow at a specified design Mach number. For this purpose our inverse method of complex characteristics provides an ideal tool.

Reasonable looking shock-free transonic airfoils are found with ease by the inverse method. Examination of the results reveals that limiting lines occurring inside the body penetrate the actual flow when we attempt to thicken the profile beyond a certain bound. This is the origin of the appealing but by now discredited limiting line hypothesis. By adjustment of parameters in the solution, the limiting line can be made to enter the flow at the back of the wing, representing a weak shock wave there, or it can be brought to the front, where it yields a physically favorable peaky pressure distribution of the type described by Pearcey [9]. Thus we are led in a natural way to Pearcey's conclusions by a new and quite different theory. Moreover, when a prescribed circulation causing lift is imposed on our solution, we come upon supercritical airfoils of the Whitcomb variety by choosing the parameters so as to delay the emergence of the forward limiting line at the upper surface of the wing.

By now the transonic controversy may be considered settled in favor of the physical importance of smooth transonic flow. However, we maintain only that at a given free stream Mach number shock-free flow can be achieved for certain exceptional airfoil shapes that do not induce boundary layer separation. The advantage of these shapes in practice is that they postpone the sharp increase in wave drag occurring at the force-break Mach number of the airfoil.

3. The Method of Complex Characteristics

We proceed to describe in broad terms the method of complex characteristics that enables us to calculate two-dimensional transonic flows without shocks. The essential element of our construction is a finite difference scheme that is based on analytic continuation into the complex domain and was first developed [3] to treat an elliptic Cauchy problem for the flow behind a detached shock wave, but is more effective in the present case both because what we are now dealing with is a genuinely inverse problem and because the equations for irrotational motion become linear in the hodograph plane. Instead of the Cauchy problem it is preferable to introduce characteristic initial value problems that convert analytic functions of a complex variable into solutions of the equations of gas dynamics. Such an analysis has features in common with Bergman's idea of integral operators [1] as well as with the more classical hodograph method.

The velocity potential ϕ for steady irrotational flow of a compressible inviscid fluid in the (x,y)-plane satisfies the second order partial differential equation

$$(c^2 - u^2)\phi_{xx} - 2uv\phi_{xy} + (c^2 - v^2)\phi_{yy} = 0 ,$$

where $u = \phi_x$ and $v = \phi_y$ are the velocity components, $c^2 = A\gamma\rho^{\gamma-1}$ is the square of the speed of sound given in terms of u and v by Bernoulli's law

$$\frac{q^2}{2} + \frac{c^2}{\gamma-1} = H , \quad q^2 = u^2 + v^2 ,$$

ρ is the density and γ , A and H are constants [2]. Both a stream function ψ and the transformation to the hodograph plane, or (u,v)-plane, can be introduced by means of the convenient formula

$$d\phi + i\,\frac{d\psi}{\rho} = (u - iv)(dx + idy) .$$

The characteristics of the equation for the velocity potential are defined by the ordinary differential equation

$$(c^2 - u^2)dy^2 + 2uv\;dydx + (c^2 - v^2)dx^2 = 0$$

in the physical plane, which takes the form

$$(c^2 - u^2)du^2 - 2uv\;dudv + (c^2 - v^2)dv^2 = 0$$

in the hodograph plane, where the equations of motion become linear so that the characteristics are fixed.

For supersonic flow, which means of course flow with Mach number $M = q/c > 1$, the equation for ϕ is hyperbolic and we can define characteristic coordinates ξ and η in terms of which the nontrivial canonical system

$$y_\xi + \lambda_+ x_\xi = 0 , \quad y_\eta + \lambda_- x_\eta = 0$$

for x and y is obtained, where

$$\lambda_\pm = \frac{uv \pm c\sqrt{q^2 - c^2}}{c^2 - u^2} .$$

The solution of the corresponding equations

$$u_\xi = \lambda_- v_\xi , \quad u_\eta = \lambda_+ v_\eta$$

for u and v has an explicit representation by means of epicycloids and will therefore be considered known. What we want to do is to solve the canonical equations for x and y numerically. For that purpose we shall set up characteristic initial value problems and apply to them the Massau finite difference scheme [4]

$$y_3 - y_1 + \lambda_+(x_3 - x_1) = 0 , \quad y_3 - y_2 + \lambda_-(x_3 - x_2) = 0 ,$$

which determines an approximation (x_3,y_3) for the coordinates of an unknown point that lies at the intersection of two characteristics through a given pair of points (x_1,y_1)

and (x_2,y_2) . If we think of u and v as specified functions of ξ and η on a rectangular grid, the coefficients λ_+ and λ_- may be approximated by average values such that the scheme becomes second order accurate in the mesh size.

The procedure we have outlined can be recast so that it becomes applicable to subsonic flow, too, even though λ_+ and λ_- are then complex because $M < 1$. In fact, if complex characteristic initial values of x are assigned along the coordinate axes $\xi = 0$ and $\eta = 0$ and if λ_+ and λ_- are complex, the method of Massau remains stable provided only that $\lambda_+ \neq \lambda_-$, for our canonical system merely reduces to a set of four instead of two simultaneous real partial differential equations of the hyperbolic type in the complex case. The same is true if the data are prescribed along any two paths connecting the origin to points ξ and η in the complex ξ-plane and the complex η-plane. However, the answer obtained will in general depend on the choice of the paths of complex initial data joining the two points where these planes $\xi = 0$ and $\eta = 0$ intersect the pair of complex characteristics through a third point at which the solution is sought. The important observation to be made is that when the data are specified by complex analytic functions of the complex variables ξ and η , then the answer becomes independent of the particular paths chosen and is, as a function of the end points ξ and η , an analytic solution of the canonical system for x and y . This result forms the foundation of our method. It is proved, as in the theory of ordinary differential equations, by using the Picard integral equations

$$y_3 + \lambda_+ x_3 = y_1 + \lambda_+ x_1 + \int x \frac{\partial \lambda_+}{\partial \xi} d\xi$$

and

$$y_3 + \lambda_- x_3 = y_2 + \lambda_- x_2 + \int x \frac{\partial \lambda_-}{\partial \eta} d\eta$$

for the solution to generate a convergent sequence of successive approximations which are all analytic and therefore are all independent of the paths of integration because of Cauchy's theorem [2].

After the analytic relation between u, v and ξ, η has been made explicit, we can arrive at real solutions of the linear flow equations in the elliptic region of the hodograph plane simply by taking real or imaginary parts of the complex answer. Our construction works when the desired solution has a global analytic continuation into the domain of the two independent complex variables ξ and η. The method is harder than the standard theory of ordinary differential equations in the complex domain only because the geometry we have to deal with is four-dimensional. The best way to visualize it is to consider the paths of initial data that are assigned initially in the two complex coordinate planes $\xi = 0$ and $\eta = 0$. These paths may be chosen more or less arbitrarily and, in particular, they need not be analytic even though the functions defining the initial values of x must be. The only place where the solution breaks down occurs when $\lambda_+ = \lambda_-$, so that the determinant of the Massau equations for x_3 and y_3 vanishes. That happens on the exceptional set $M = 1$ in four-dimensional complex space, which is a two-dimensional analytic surface intersecting the real domain along the sonic line. In the next section we shall give rules defining paths of data that circumvent the sonic line and lead to real results in both the subsonic and the supersonic zones [11].

In order to describe uniform flow at infinity in the hodograph plane, one must obtain a solution of the linear equations of motion there that has a certain elementary singularity. The answer can easily be represented by means of partial derivatives with respect to the parameters of a fundamental solution of the form

$$X(\xi,\eta;\xi_0,\eta_0) = R(\xi,\eta;\xi_0,\eta_0)\log(\xi-\xi_0)(\eta-\eta_0) + S(\xi,\eta;\xi_0,\eta_0),$$

where R and S are regular functions and ξ_0, η_0 are the characteristic coordinates of the singularity. It turns out that R is the Riemann function [2], which can be determined by the solving of a characteristic initial value problem with data prescribed at $\xi = \xi_0$, $\eta = \eta_0$. Afterwards the additional regular function S is found from another characteristic initial value problem with new data assigned elsewhere, say at $\xi = 0$, $\eta = 0$. In this way it is even possible to calculate uniform flows at infinity that are provided with circulation and sink terms leading to arbitrary coefficients of lift and drag C_L and C_D.

Because arbitrary analytic functions of the complex variables ξ and η define equally valid characteristic coordinates, our canonical equations are invariant under a conformal transformation of the ξ-plane and the η-plane.

Such a conformal mapping can be used to introduce branch points of the solution in the hodograph plane at points where the flow requires them. In this context a substitution that reduces to

$$w = 1 + 2B\eta - \eta^2$$

in the incompressible case, where $w = u-iv$ is a characteristic coordinate, is especially helpful in achieving a desired lift coefficient C_L related to the branch point parameter B .

Flow past a closed profile in the physical plane is obtained by choosing the characteristic initial data that define the regular part of our solution of the hodograph equations appropriately. One way of accomplishing this is to use analytic functions suggested by a study of the incompressible flow past some body such as an ellipse [8]. Corrections can be made later to change the shape of the profile or to improve the pressure distribution on it by adding small logarithmic terms to the original functions. To arrive at a tail with moderate adverse pressure gradient we assign its image point in the hodograph plane so as to bring the speed at the tail near that at infinity. We then put

$$\psi = \psi_u = \psi_v = 0$$

at this image point to simulate the Kutta-Joukowski condition. Our procedure is quite flexible owing to the large variety of analytic functions at our disposal, and it permits ample control of the subsonic portion of the flow. Since the characteristic coordinates ξ and η uniformize any limiting lines that may appear, these are accurately described, and it is not hard to design shock-free transonic airfoils of considerable physical interest.

4. Continuation Around the Sonic Locus

One of the principal difficulties in implementing the method of complex characteristics numerically is to fix the branches of the analytic functions that occur correctly. Another is to select paths of data along the initial complex characteristics that lead to real results for supersonic as well as subsonic flow. Both of these issues arise when we attempt to continue a two-dimensional slice of our solution of the equations of motion analytically around the sonic line $M = 1$, which is a surface in four-dimensional complex space where the coefficients λ_+ and λ_- become singular. The desired extension is obtained by using paths of integration that are defined in terms of a curve in the initial characteristic planes $\xi = 0$ and $\eta = 0$ that we shall call the sonic locus because it consists of points cut out by opposite characteristics coming from the real sonic line. The discovery by Eva Swenson [11] of the rules for analytic continuation associated with the sonic locus was a major breakthrough facilitating the computation of transonic flows by our finite difference scheme.

To calculate curves of real output in the region of subsonic flow we select paths of initial data in the ξ-plane and the η-plane that are complex conjugates of each other. The Schwarz principle of reflection shows that symmetric paths of this kind give real results along the diagonal of our grid bisecting the initial characteristics. The conjugate paths can be extended until they reach the sonic locus at a symmetric pair of points that project down onto the sonic line along the two outermost characteristics of the computation.

It is harder to determine real supersonic flow starting from complex analytic data because that necessitates retracing the real characteristics from the hyperbolic part of the hodograph plane back through the complex domain to the initial planes $\xi = 0$ and $\eta = 0$. To see where two such characteristics emerge we first follow them to the sonic line and then continue into the complex domain. It becomes clear that they must arrive ultimately at an asymmetric pair of points on the sonic locus which, together with their reflected images, project by our previous correspondence into the points on the sonic line where we left the real domain. Eva Swenson [11] has shown that if an S-shaped path of initial data is chosen which passes through the origin to connect these asymmetric points on the sonic locus, but which detours around their reflected images so as to fix the branch of the solution appropriately, then the last characteristics in our calculation will meet at a real supersonic point. Moreover, if the S-path terminates in two arcs of the sonic locus placed so that one of them lies in the ξ-plane and is the reflection of the other situated in the η-plane, then the outer triangle of our computation corresponding to mesh points on these arcs becomes meaningless because it involves singularities of λ_+ and λ_- , whereas the remaining inner triangle yields output in the real supersonic region of flow that appears to be, but is not, the result of solving a Cauchy problem with data assigned along the sonic line.

The method of complex characteristics generates solutions of the partial differential equations of gas dynamics that describe flows past profiles defined by a specific streamline which we shall assume to be $\psi = 0$. In the subsonic region real curves of data are computed that yield isolated points on the body. These can be tabulated conveniently by using paths of initial data that fork to cross several points where $\psi = 0$. According to the rules we have presented, the supersonic zone can be swept out by a single two-dimensional calculation based on one S-shaped path of integration terminating along arcs of the sonic locus. Inside the body limiting lines of both families of characteristics appear. At the front of the profile they cause a favorable peaky pressure distribution to prevail, whereas at the tail they must be interpreted as an approximation of weak shocks. It is in general easy to pick complex analytic initial data furnishing a reasonable airfoil shape in the transonic region or at the tail, but to achieve a closed streamline $\psi = 0$ around the nose more care must be exercised because at high Mach numbers or for large circulation the corresponding level curve $\psi = 0$ in the hodograph plane has to become quite elongated to reach the origin.

5. Construction and Properties of the Airfoils

A Fortran program [4] has been written for the CDC 6600 computer at N.Y.U. to calculate shockless transonic airfoils by the procedure we have outlined. An improved version of the program takes into account the effect of circulation so that any lift coefficient C_L can be assigned. We also plan to include an arbitrary sink at infinity that will yield a given drag coefficient C_D by creating a corresponding displacement thickness far downstream. This can be cancelled by making a reversed boundary layer correction at an appropriate finite Reynolds number. A variety of additional parameters defining the analytic functions from which the flow is computed enable us to control the shape of the airfoil. In particular, the adverse pressure gradient at the tail can be reduced enough to prevent separation of the turbulent boundary layer.

A typical run of the CDC 6600 program based on seven forked subsonic paths of integration plus one supersonic sweep, each of them subdivided by 30 or more mesh points, requires one minute of central processor machine time. This gives about sixty body points with an accuracy of two or three significant digits, which can be improved upon at will by refining the mesh when a desirable airfoil has been found. Elaborate parameter studies are usually needed to arrive at a reasonable shape whenever an appreciably larger free stream Mach number M or circulation Γ is imposed on the solution. Two examples that we have calculated in detail are presented in Figures 1 and 2.

Figure 1 shows the upper half of a symmetric airfoil

with a peaky pressure distribution of the Pearcey type near the nose. Mach lines are drawn in the supersonic zone and their physically meaningless analytic continuations inside the body are included to indicate where limiting lines occur. Only isolated points on the plot of the pressure coefficient C_p are given in the subsonic regime because of the restricted number of paths of integration used in the computation there.

In Figure 2 we present a lifting airfoil that is shock-free at $M = .75$ and $C_L = .63$. It has very much the appearance of the Whitcomb upside-down wing [13] because it is flat on top and has a concave lower surface near the trailing edge. This camber is achieved by moving the point in the hodograph plane corresponding to the tail sidewards toward the sonic line. An unusually gradual decay of the pressure coefficient along the rear of the upper surface should make the airfoil especially effective in suppressing turbulent boundary layer separation and consequently eliminating global perturbation by an unwanted wake. We hope to have experimental work done on this airfoil to test whether it really exhibits shock-free transonic flow in a wind tunnel and to see how it behaves at off-design conditions.

REFERENCES

1. Bergman, S., "Two-Dimensional Transonic Flow Patterns," Amer. J. Math., Vol. 70 (1948), 856-891.
2. Garabedian, P. R., Partial Differential Equations, Wiley, New York, 1964.
3. Garabedian, P. R., and Lieberstein, H. M., "On The Numerical Calculation of Detached Bow Shock Waves in

Hypersonic Flow," J. Aero. Sci., Vol. 25 (1958), 109-118.

4. Korn, D. G., "Computation of Shock-Free Transonic Flows for Airfoil Design," N.Y.U. Report NYO-1480-125, Courant Inst. Math. Sci., 1969.
5. Lax, P. D., "Hyperbolic Systems of Conservation Laws II," Comm. Pure Appl. Math., Vol. 10 (1957), 537-566.
6. Morawetz, C. S., "On The Non-Existence of Continuous Transonic Flows Past Profiles I," Comm. Pure Appl. Math., Vol. 9 (1956), 45-68.
7. Morawetz, C. S., "The Dirichlet Problem for the Tricomi Equation," Comm. Pure Appl. Math., Vol. 23 (1970).
8. Nieuwland, G. Y., "Transonic Potential Flow Around a Family of Quasi-Elliptical Aerofoil Sections," N.L.R. Report TR.T. 172, Amsterdam, 1967.
9. Pearcey, H. H., "The Aerodynamic Design of Section Shapes for Swept Wings," Advan. Aero. Sci., Vol. 3 (1962), 277-322.
10. Spee, B. M., and Uijlenhoet, R., "Experimental Verification of Shock-Free Transonic Flow Around Quasi-Elliptical Aerofoil Sections," N.L.R. Report MP. 68003U, Amsterdam, 1968.
11. Swenson, E. V., "Geometry of the Complex Characteristics in Transonic Flow," Comm. Pure Appl. Math., Vol. 21 (1968), 175-185.
12. Tomotika, S., and Tamada, K., "Studies in Two-Dimensional Transonic Flows of Compressible Fluid," Quart. Appl. Math., Vol. 7 (1950), 381-397.
13. Whitcomb, R. T., "The Upside-Down Wing," Science News Item in Time Magazine, Feb. 21, 1969, 66.

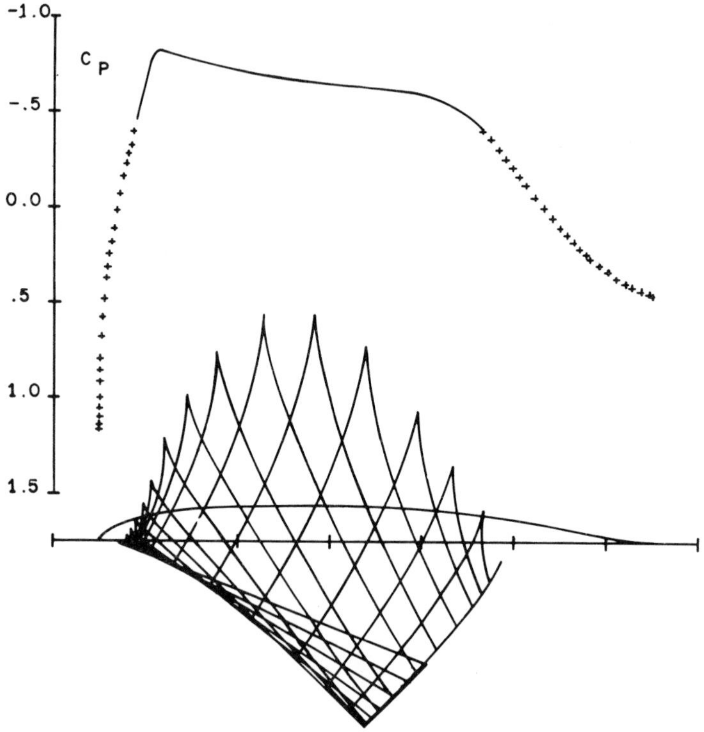

Fig. 1 Symmetric Transonic Airfoil at M = .8, T/C = .13

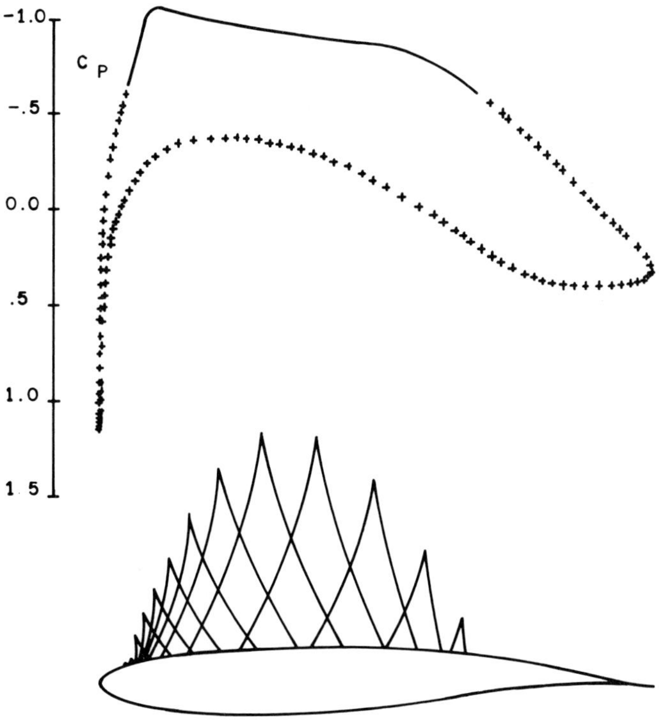

Fig. 2 Lifting Shockless Airfoil at $M = .75$, $C_L = .63$, $T/C = .12$

ON THE NUMERICAL TREATMENT OF PARTIAL DIFFERENTIAL EQUATIONS BY FUNCTION THEORETIC METHODS*

Robert P. Gilbert and David L. Colton**

1. Introduction

In the second section of this paper we shall show how constructive analytic methods may be used to solve boundary value problems for linear elliptic equations. We discuss in particular partial differential equations with analytic coefficients in two and three dimensions. The case in which the coefficients are only required to be smooth can also be handled if we first approximate these coefficients by polynomials.

In the third section we indicate how a direct numerical procedure is generated by an analytic procedure. This is done in detail for an n-dimensional equation with smooth, radially symmetric coefficients. Bounds are given on the discretization, and iterative errors. This approach

*This research was supported in part by the Air Force Office of Scientific Research through AF-AFOSR Grant 1206-67.

**University of Indiana, Bloomington, Indiana

is quite competative with the finite difference method, as was shown by the numerical example contained in the appendix of a previous paper [24].

In the fourth and fifth sections we investigate iterative methods for solving boundary value problems associated with semilinear equations in two dimensions. Our procedure depends on known methods for approximating the kernel function [8] and on the theory of generalized analytic functions [35]. Also in this regard see [24] page 347. It appears that these results can be extended to n-dimensions; however, we shall report on this at a later date.

2. Boundary Value Problems: Analytic Methods

In order to present our numerical schemes for solving boundary value problems we must first develop some analytic procedures for formulating these problems. In this section we do this for the following elliptic equations:

(2.1)

$$\underset{\sim}{\mathcal{E}}_2[u] \equiv \Delta_2 u + a(x,y)u_x + b(x,y)u_y - c(x,y)u = 0 ,$$

(2.2) $$\underset{\sim}{E}_3[u] \equiv \Delta_3 u - F(x,y,z)u = 0 ,$$

(2.3) $$\underset{\sim}{e}_n[u] \equiv \Delta_n u - B(r^2)u = 0 ,$$

$$\text{with} \quad r = |\underset{\sim}{x}| , \quad \underset{\sim}{x} \equiv (x_1, \ldots, x_n) .$$

Here $\Delta_n \equiv \frac{\partial^2}{\partial x_1^2} + \cdots + \frac{\partial^2}{\partial x_n^2}$, and the functions $c(x,y)$, $F(x,y,z)$, $B(r^2)$ are to be non-negative in their respective (real) domains of definition, D. For the purposes of simplicity of exposition we shall further assume, initially that these functions have entire extensions to $\mathbb{C}^2$, $\mathbb{C}^3$, and $\mathbb{C}$ respectively. Later we can relax these conditions to merely requiring that the coefficients be smooth.

The domains for which we shall pose the boundary value problems are to be bounded, star-like with respect to the origin, and have a Lyapunov boundary. We shall refer to such domains as being <u>appropriate</u>.

The approach we use is to derive a <u>general</u> integral representation for the solutions of each equation, and to then use this representation to formulate the respective boundary value problem as either a Fredholm integral equation, or to develop a complete family of functions for the purposes of approximating the solution. For Equation (2.1) the Fredholm integral equation method was developed by Vekua [34] via the theory of singular integral equations [29]. Our contribution is to develop an efficient iterative procedure for solving this equation by using Bergman's method for constructing [6] the integral representation. The approach to equations (2.2) and (2.3) is due to one of us [26], [22], [23] and is new.

Once the boundary value problems of Equations (2.1), (2.2), (2.3) have been reduced formally to a Fredholm equation, we are in the domain of numerical analysis. For instance, we are now faced with the various problems of numerically solving the integral equations which arise from the general integral representations. We shall discuss this in detail later for the case of equation (2.3).

A. The Equation $\underset{\sim}{\mathcal{E}}_2[u] = 0$:

As mentioned above we assume that the coefficinets of $\underset{\sim}{\mathcal{E}}_2[u] = 0$ have analytic continuation to $\mathbb{C}^2$ (or at least to $[D + \partial D] \times [D^* + \partial D^*]$), which permits us to transform (2.1) to a complex-valued hyperbolic equation,

$$(2.4) \qquad U_{ZZ^*} + AU_Z + BU_{Z^*} + CU = 0 \;, \quad B = \bar{A}$$

$$U(Z,Z^*) \equiv u\left(\frac{Z+Z^*}{2}, \frac{Z-Z^*}{2i}\right),$$

$$Z = x + iy \;, \quad Z^* = x - iy \;,$$

$(Z,Z^*) \in [D + \partial D] \times [D^* + \partial D^*]$; see [6], [20], and [21]. Bergman [6] has given the following integral representation for the solutions of (2.4),

(2.5)

$$U(Z,Z^*) = \exp\left[-\int_0^{Z^*} A(Z,s)ds\right] \cdot \left[g(Z) + \right.$$

$$\left. + \sum_{n\geq 1} \frac{Q^{(n)}(Z,Z^*)}{2^{2n}B(n,n+1)} \int_0^Z \int_0^{Z_1} \cdots \int_0^{Z_n} g(Z_n)dZ_n \ldots dZ_1 \right],$$

where $g(Z)$ is taken to be analytic in $D + \partial D$. The functions $Q^{(n)}(Z,Z^*)$ are defined by

(2.6) $$Q^{(n)}(Z,Z^*) = \int_0^{Z^*} P^{(2n)}(Z,s)ds ,$$

and the $P^{(2n)}(Z,Z^*)$ are defined recursively by the system,

(2.7) $$P^{(2)} \equiv -2F \equiv -2(A_Z - AB + C) ,$$

and

(2.8)
$$(2n+1)P^{(2n+2)} = -2\left[P_Z^{(2n)} + \left\{ \Phi' - \int_0^{Z^*} A_Z dZ^* + B \right\} P^{(2n)} + F \int_0^{Z^*} P^{(2n)} dZ^* \right] , \quad n \geq 1 .$$

Here $\Phi = \Phi(Z)$ is an arbitrary analytic function of one complex variable in $D + \partial D$. Another representation of solutions to (2.4) is given by

(2.9) $$U(Z,Z^*) = \exp\left[- \int_0^{Z^*} A(Z,\zeta)d\zeta \right] \left[g(Z) + \sum_{n\geq 1} \frac{Q^{(n)}(Z,Z^*)}{2^{2n}B(n,n+1)} \int_0^Z (Z-\zeta)^{n-1} g(\zeta)d\zeta \right] .$$

When $g(0) = 0$, this solution is identical to (2.5). It is straightforward to show that the representation (2.9) is identical with Vekua's representation, in terms of the complex Riemann function $R(t, t^*; Z, Z^*)$, namely

(2.10)

$$U(Z, Z^*) = R(Z, 0; Z, Z^*)g(Z) + \int_0^Z \left\{ -R_1(t, 0; Z, Z^*) + B(t, 0)R(t, 0; Z, Z^*) \right\} g(t)dt \,.$$

It is well known [29], [34] that if $g(Z)$ is holomorphic in D and also in the Hölder class, $\mathcal{H}^{\alpha}(D + \partial D)$, $0 < \alpha \leqslant 1$, then it has a representation as

$$G(Z) = \int_{\partial D} \frac{t\mu(t)ds}{t - Z} \,, \quad Z \in D \,, \tag{2.11}$$

where $\mu(t)$ is real valued and Hölder continuous. It is this device which Vekua exploits to rearrange (2.10) into a Fredholm equation for a density $\mu(t)$ such that $U(Z, \bar{Z}) = u(x, y)$ satisfies the Dirichlet data $u(x, y) = f(Z)$, $Z \in \partial D$. Since the Bergman solution (2.9) is identical to (2.10), we may exploit the already proved result of Vekua [34] that the resulting integral equation of the second kind is invertible. Furthermore, the Bergman formulation suggests an iterative procedure for solving this equation. Substituting (2.11) into (2.9) and inverting orders of integration, which is permissible, leads to [26]

(2.12)

$$u(x,y) = \mathrm{Re} \left\{ \hat{H}_0(Z) \int_{\partial D} t\mu(t) \left[\frac{1}{t-Z} + \sum_{n=1}^{\infty} \frac{Q^{(n)}(Z,\bar{Z})}{2^{2n} B(n,n+1)} \int_0^Z \frac{(Z-\zeta)^{n-1}}{t - \zeta} d\zeta \right] ds \right\}$$

where $\hat{H}_0(Z) \equiv \exp\left[-\int_0^{\bar{Z}} A(Z, \sigma)\, d\sigma\right]$. (Note, that in what follows, $\hat{H}_0(Z)$ etc. does not mean $\hat{H}_0$ is an analytic function of Z, rather it is a function of the point Z.) By computing the residue as $Z \to t_0 \in \partial D$, one obtains the singular integral equation for the density $\mu(t)$, namely [26]

(2.13)

$$f(t_0) = \mathrm{Re}\left\{\hat{H}_0(t_0)\left[\pi i t_0 \mu(t_0) \bar{t}'_0 + \int_{\partial D} t\mu(t)\left(\frac{1}{t-t_0} + \sum_{n=1}^{\infty} \frac{Q^{(n)}(t,t_0)}{2^{2n} B(n,n+1)} \int_0^{t_0} \frac{(t_0-\zeta)^{n-1}}{t-\zeta}\, d\zeta\right)\right] ds\right\}$$

where $t'(x) \equiv \frac{dt}{ds}$. An alternate form of this equation is

(2.14)

$$f(t_0) = A(t_0)\mu(t_0) + \frac{B(t_0)}{i\pi} \int_{\partial D} \frac{\mu(t)\,dt}{t-t_0} + \int_{\partial D} F_1(t_0,t)\mu(t)\,ds + \int_{\partial D} F_2(t_0,t)\mu(t)\,ds ,$$

(2.15) where $A(t_0) \equiv \mathrm{Re}\left[\pi i\; t_0\; \overline{t'_0}\; \hat{H}_0(t_0)\right]$,

$$i\pi \mathrm{Re}\left[t_0\; \overline{t'_0}\; \hat{H}_0(t)\right] ,$$

(2.16) $$F_1(t_0, t) = \mathrm{Re}\left\{\frac{t\hat{H}_0(t_0)}{t - t_0}\right\} - \frac{t' B(t_0)}{i\pi(t-t_0)}$$

and

(2.17)

$$F_2(t_0,t) = \mathrm{Re}\left\{ t\sum_{n=1}^{\infty} \frac{\hat{H}_0(t_0)Q^{(n)}(t_0,\bar{t}_0)}{2^{2n}B(n,n+1)} \int_0^{t_0} \frac{(t_0-\zeta)^{n-1}}{t-\zeta}\,d\zeta \right\}.$$

By using the Poincare-Bertrand formula [29] one may reduce (2.14) to the form of a Fredholm equation [26].

$$\mu(t_0) + \int_{\partial D} K(t_0,t)\mu(t)\,ds = F(t_0)\ , \tag{2.18}$$

with

$$F(t_0) \equiv \frac{1}{\pi^2|t_0|^2|\hat{H}_0(t_0)|^2}\left[A(t_0)f(t_0) - \frac{B(t_0)}{\pi i}\int_{\partial D} \frac{f(t)}{t-t_0}\,dt \right], \tag{2.19}$$

and

(2.20)

$$K(t_0,t) \equiv \frac{1}{\pi^2|t_0|^2|\hat{H}_0(t_0)|^2}\left[A(t_0)\left(\mathrm{Re}\left\{\frac{t\hat{H}_0(t_0)}{t-t_0}\right\} - \frac{t'B(t_0)}{i\pi(t-t_0)} \right) + A(t_0)F_2(t_0,t) - \frac{B(t_0)}{\pi i}\int_{\partial D} \mathrm{Re}\left\{\frac{t\hat{H}_0(t)}{t-\tau}\right\}\frac{d\tau}{\tau-t_0} - \frac{B(t_0)}{\pi i}\int_{\partial D} \frac{F_2(\tau,t)\,d\tau}{\tau-t_0} \right].$$

An iterative procedure may be obtained for solving (2.18) by replacing $F_2(\tau, t)$ in (2.20) by truncating its series representation (2.17). The series (2.17) converges rapidly in general, and this leads to a useful method. For equation (2.3) when $n = 2$, we have a special case of (2.1). We postpone to that case the numerical treatment of the corresponding integral equations. Numerical results have been published for this problem in [24].

B. The Equation $\underset{\sim}{E}_3[u] = 0$:

In her thesis Bwee Lan Tjong [33] obtained a generalization of the Bergman-Whittaker operator to the case of the equation $\underset{\sim}{E}_3[u] = 0$.[†] It is the following integral representation:

(2.21)

$$u(x,y,z) \equiv \psi(X,Z,Z^*) = \underset{\sim}{T}f$$

$$\equiv \int_{|\zeta|=1} \int_{\gamma} E(X,Z,Z^*,\zeta,t)f(w,\zeta) \frac{dt}{\sqrt{1-t^2}} \frac{d\zeta}{\zeta},$$

where γ is a rectifiable curve from $t = -1$ to $+1$. Here the variables X, Z, Z^*, and w are defined as $X = x$, $Z = \frac{1}{2}[y + iz]$, $Z^* = \frac{1}{2}[-y + iz]$

[†]It was subsequently shown by Colton and Gilbert [13] that it was possible to find an analogous operator for the four-dimensional equation,

$$\Delta_4\psi - F(\underset{\sim}{x})\psi = 0 , \qquad \underset{\sim}{x} = (x_1,x_2,x_3x_4) .$$

In this case this operator generalizes the integral operator $\underset{\sim}{G}_4$; see [22] for further details concerning $\underset{\sim}{G}_4$.

$$(2.22) \qquad w = (1-t^2)u \ , \quad \text{with} \quad u = X + \zeta Z + \zeta^{-1}Z^* \ .$$

Furthermore, the kernel $E(X,Z,Z^*,\zeta,t)$ satisfies the partial differential equation,

$$(2.23) \qquad ut\left\{\frac{\partial^2\hat{E}}{\partial\xi_1^2} + \frac{\partial^2\hat{E}}{\partial\xi_2^2} + \frac{\partial^2\hat{E}}{\partial\xi_3^2} + 2\frac{\partial^2\hat{E}}{\partial\xi_1\partial\xi_2} + 2\frac{\partial^2\hat{E}}{\partial\xi_1\partial\xi_3} - 2\frac{\partial^2\hat{E}}{\partial\xi_2\partial\xi_3} + \hat{F}\hat{E}\right\} + (1-t^2)\frac{\partial^2\hat{E}}{\partial\xi_1\partial t} - \frac{1}{t}\frac{\partial\hat{E}}{\partial\xi_1} = 0 \ ,$$

where $\hat{E}$ is obtained from E under the change of X, Z, Z^* - variables to $\xi_1 = X$, $\xi_2 = X + 2\zeta Z$, $\xi_3 = X + 2\zeta^{-1}Z^*$. She also gives a series representation for $\hat{E}$ of the form

$$(2.24) \qquad \hat{E}(\xi_1, \xi_2, \xi_3, \zeta, t) = 1 + \sum_{n\geq 1} t^{2n}u^n p^{(n)}(\underset{\sim}{\xi}; \zeta) \ ,$$

where the $p^{(n)}(\underset{\sim}{\xi}; \zeta)$ satisfy the differential equations

$$(2.25) \qquad p_1^{(n+1)} = -\frac{1}{2n+1}\left\{p_{11}^{(n)} + p_{22}^{(n)} + p_{33}^{(n)} + 2p_{12}^{(n)} + 2p_{13}^{(n)} - 2p_{23}^{(n)} + \hat{F}p^{(n)}\right\}$$

with $p^{(n+1)}(0, \xi_2, \xi_3, \zeta) = 0$, which gives rise to an alternate representation for solutions,

(2.26)

$$\psi = \frac{1}{2\pi i}\int_{|\zeta|=1} g(u,\zeta)\,\frac{d\zeta}{\zeta} + $$

$$+ \sum_{n\geq 1}\frac{1}{2\pi i B(n,\frac{1}{2})}\int_{|\zeta|=1}\left\{p^{(n)}(\xi,\zeta)\int_0^u (u-s)^{n-1} g(s,\zeta)\,ds\right\}\frac{d\zeta}{\zeta}\,,$$

where $g(u,\zeta) = \int_\gamma f(w,\zeta)\,\frac{dt}{\sqrt{1-t^2}}$, and γ is an arc from -1 to $+1$.

It was shown by Gilbert and Lo [26] that the solution having the representations (2.21, 2.26) was a general solution, i.e. any solution of (2.2) which is in $C^1(D+\partial D)$, where D is appropriate, has such a representation. This was done by making use of an inversion of the Bergman-Whittaker operator,

(2.27) $$H(X,Z,Z^*) = (\underset{\sim}{B}_3 g)(X,Z,Z^*)$$

$$\equiv \frac{1}{2\pi i}\int_{|\zeta|=1} g(u,\zeta)\,\frac{d\zeta}{\zeta}\,,$$

namely [26], $g(u,\zeta) = (B_3^{-1}H)(u,\zeta)$

(2.28) $$(B_3^{-1}H)(u,\zeta) \equiv \frac{1}{2\pi}\int_{\partial D_0} \rho(\underset{\sim}{Y})\ \frac{1}{\underset{\sim}{N}\cdot(\underset{\sim}{X}-\underset{\sim}{Y})}\,d\omega_y\ ;$$

here $\underset{\sim}{N}(\zeta)$ is the isotropic vector introduced earlier by the definition $u = \underset{\sim}{N}\cdot\underset{\sim}{X}$, and $\rho(\underset{\sim}{Y})$ is the single-layer desnity which generates the potential $H(\underset{\sim}{X})$ for $\underset{\sim}{X} \in D_0 \subset\subset D$. Using (2.28) in (2.26) yields

$$\psi(\underset{\sim}{X}) = H(\underset{\sim}{X}) + \int_{\partial D_0} \rho(\underset{\sim}{Y})K(\underset{\sim}{X},\underset{\sim}{Y})d\omega_y \,, \tag{2.29}$$

where

$$K(\underset{\sim}{X},\underset{\sim}{Y}) \equiv \frac{1}{2\pi i} \int_{|\zeta|=1} P(\underset{\sim}{X},\underset{\sim}{Y};\zeta) \frac{d\zeta}{\zeta} \,, \tag{2.30}$$

$$P(\underset{\sim}{X},\underset{\sim}{Y};\zeta) \equiv \sum_{n\geq 1} \frac{1}{B(n,\frac{1}{2})} p^{(n)}(\xi;\zeta)\Phi_n(u;\, \underset{\sim}{N}\cdot\underset{\sim}{Y}) \,, \tag{2.31}$$

and

$$\Phi_n(u;\, \underset{\sim}{N}\cdot\underset{\sim}{Y}) = [(\underset{\sim}{N}\cdot\underset{\sim}{Y}-u)^{n-1} - (\underset{\sim}{N}\cdot\underset{\sim}{Y})^{n-1}] \log(u-\underset{\sim}{N}\cdot\underset{\sim}{Y}) - \sum_{\nu=1}^{n-1} \binom{n-1}{\nu} \frac{(\underset{\sim}{N}\cdot\underset{\sim}{Y})^{n-\nu-1}}{\nu} \,. \tag{2.32}$$

The functions $\Phi_n(u;\, \underset{\sim}{N}\cdot\underset{\sim}{Y})$ are <u>universal functions</u>, and the kernel $K(\underset{\sim}{X},\underset{\sim}{Y})$ is <u>fixed for each particular differential equation</u> (2.2).

It is easy to show that the integral in (2.29) is a compact operator on the class of functions $\mathcal{C}[\partial D_0]$. Furthermore, if we formulate the Neumann problem for (2.2) with $\frac{\partial\psi}{\partial\nu_x} = f(\underset{\sim}{X})$, $\underset{\sim}{X} \in \partial D_0$, we are led to the Fredholm integral equation

$$f(\underset{\sim}{X}) = -\rho(\underset{\sim}{X}) + \frac{1}{2\pi} \int_{\partial D_0} \rho(Y) \frac{\partial}{\partial\nu_x} \left(\frac{1}{|\underset{\sim}{X}-\underset{\sim}{Y}|} \right) d\omega_y + \int_{\partial D_0} \rho(\underset{\sim}{Y}) \frac{\partial}{\partial\nu_x} K(\underset{\sim}{X},\underset{\sim}{Y})d\omega_y \,, \tag{2.33}$$

which we can show, following the discussion in Garabedian [20] for Laplace's equation, is uniquely soluble.

We now turn to the Dirichlet problem associated with (2.2), and we assume that the data is sufficiently smooth [20] page 347. Let us consider the class of all harmonic functions that are in $\mathcal{C}[D + \partial D]$, and let us designate this class by $\mathcal{H}[D]$. It has recently been shown by du Plessis [30] that the harmonic polynomials are complete for simply connected domains in $\mathbb{R}^n$ with respect to the uniform norm. With this in mind we wish to obtain a complete system of solutions for $\underset{\sim}{E}[u] = 0$, [14], [26]. One method of doing this is by means of the representation[†] [26] $\psi(\underset{\sim}{X}) = (\hat{\underset{\sim}{T}}H)(\underset{\sim}{X})$,

(2.34)
$$(\hat{\underset{\sim}{T}}H)(\underset{\sim}{X}) \equiv H(X) - \frac{a}{4\pi^3 i} \sum_{n\geq 1} \frac{1}{B(n,\frac{1}{2})} \int_0^{2\pi} d\phi' \int_0^{\pi} d\theta' \sin\theta'$$
$$\cdot \left\{ H(\underset{\sim}{X}' a^2/R^2) \int_{|\zeta|=1} p^{(n)}(\xi;\zeta) D(u,\hat{u};n) \frac{d\zeta}{\zeta} \right\},$$

where $\hat{u} \equiv X' + \zeta(1 - \frac{1}{\alpha})Z' + \zeta^{-1}(1 - \frac{1}{\alpha})^{-1} Z^{*\prime}$, and the coefficients $D(u, \hat{u}; n)$ are defined as

$$D(u, \hat{u}; n) \equiv \int_0^u (u-s)^{n-1} A(s,\hat{u}) \partial s , \tag{2.35}$$

[†]This representation comes about by making use of an inversion of $\underset{\sim}{B}_3^{-1}$ which is developed in [26]; see also [22] for another representation of $\underset{\sim}{B}_3^{-1}$ when the harmonic functions are regular at infinity.

with

$$(2.36) \quad A(s,\hat{u}) \equiv \int_0^1 d\alpha \int_0^1 \frac{d\beta\sqrt{\beta}}{\sqrt{1-\beta}} \left[\frac{12s\alpha\beta(1-\alpha)\hat{u}+a}{(4s\alpha\beta(1-\alpha)\hat{u}-a)^3} \right] .$$

(Here a is chosen so that D is contained in a sphere of radius a.) Indeed, we have the

<u>Thoerem 2.1</u>: Let $h_{n,m}(\underset{\sim}{X})$, $(m = 0, \pm 1, \pm 2, \ldots, +n;$ $n = 0, 1, 2, \ldots)$ represent the spherical harmonics. Then the functions given via (2.35) by

(2.37)

$$\psi_{n,m}(\underset{\sim}{X}) \equiv (\hat{\underset{\sim}{T}} h_{n,m})(\underset{\sim}{X}) , \quad (m = 0, \pm 1, \ldots, \pm n;\ n = 0, 1, \ldots)$$

are a complete system of solutions for $\underset{\sim}{E}[u] = 0$, with respect to uniform convergence in D.

<u>Proof</u>: To see that this is true, let $\hat{\psi}(\underset{\sim}{X}) \in \mathbf{E}(D)^{\dagger}$ be a solution in D which does not lie in the space spanned by the $\psi_{n,m}(\underset{\sim}{X})$. <u>In</u> D, $\psi(\underset{\sim}{X}) \in C^{\infty}$, and hence there exists a harmonic function $H(\underset{\sim}{X}) \in \mathcal{H}[D_0]$, where $D_0 \subset\subset D$ is an arbitrary compact <u>appropriate</u> domain, such that $\hat{\psi}(\underset{\sim}{X}) = (\underset{\sim}{T}H)(\underset{\sim}{X}) \in \mathbf{E}[D_0]$. Furthermore, $\hat{H}(\underset{\sim}{X})$ may be uniformly approximated in D_0 by the harmonic polynomials. Hence, since D_0 is arbitrary, $\hat{\psi}(\underset{\sim}{X})$ may be uniformly approximated in D by the $\psi_{n,m}(X)$, a contradiction.

Given a complete system of solutions $\{\psi_{\nu}(\underset{\sim}{X})\}$ there are various procedures we can use for approximating boundary

---

†The class of solutions of $\underset{\sim}{E}[u] = 0$ that are in $C[D+\partial D]$.

value problems. One such method is to approximate the boundary data by means of a linear combination,

$$f(\underset{\sim}{X}) \approx \sum_{j=1}^{N} c_j \psi_j(\underset{\sim}{X}) , \quad \underset{\sim}{X} \in \partial D \tag{2.38}$$

such that $\max_{X \in \partial D} |f - \sum_{j=1}^{N} c_j \psi_j| < \varepsilon$, for $\varepsilon > 0$ suitably small. The maximum principle for $\underset{\sim}{E}[u] = 0$, then says the solution is within an ε of the approximate solution in $D + \partial D$.

Another approach is to introduce the Dirichlet inner-product [7], [8], [20], for $\underset{\sim}{E}[u] = 0$, namely

$$(\psi,\Phi) \equiv \int_D [\nabla\psi \cdot \nabla\Phi + F\psi\Phi] d\underset{\sim}{x} , \tag{2.39}$$

and to obtain an orthonormal system $\{\Phi_j(\underset{\sim}{X})\}$, by means of the Gram-Schmidt process. One then expands the data as a Fourier series,

$$f(\underset{\sim}{X}) \simeq \sum_{j=1}^{N} a_j \Phi_j(\underset{\sim}{X}) , \quad \underset{\sim}{X} \in \partial D$$

$$a_j = (f,\Phi_j) = -\int_D f \frac{\partial \Phi_j}{\partial \nu} d\omega , \tag{2.40}$$

and this yields a solution, $\psi_N(\underset{\sim}{X}) = \sum_{j=1}^{N} a_j \Phi_j(\underset{\sim}{X})$, to $\underset{\sim}{E}[u] = 0$, which approximates ψ with respect to the Dirichlet norm. Furthermore, for complete systems, the theory of the kernel function tells us that the Fourier Series (2.40) converge uniformly <u>in</u> D as $N \to \infty$.

Yet another approach is to compute a truncated kernel function

$$(2.41) \qquad K_N(\underset{\sim}{X},\underset{\sim}{Y}) = \sum_{j=1}^{N} \Phi_j(\underset{\sim}{X})\Phi_j(\underset{\sim}{Y}) \;;$$

then $\psi_N(\underset{\sim}{X})$ may be written as

$$\psi_N(\underset{\sim}{X}) = (f(\underset{\sim}{Y}),\, K_N(\underset{\sim}{X},\underset{\sim}{Y})) \;,$$

and one has the estimate, for an arbitrary $\varepsilon > 0$ and a sufficiently large N,

$$(2.42) \quad |\psi(\underset{\sim}{X}) - \psi_N(\underset{\sim}{X})|^2 \leqslant \|\psi - \psi_N\|^2 \; K(\underset{\sim}{X},\underset{\sim}{X}) < \varepsilon K(\underset{\sim}{X},\underset{\sim}{X}) \;,$$

$$\underset{\sim}{X} \in D \;.$$

A similar procedure holds for the Neumann problem; however, here the Fourier coefficients are given as

$$a_j = (\Phi_j,\psi) = -\int_{\partial D} \Phi_j \frac{\partial \psi}{\partial \nu}\, d\omega \;,$$

where $\frac{\partial \psi}{\partial \nu}(\underset{\sim}{X}) = g(\underset{\sim}{X})$, $\underset{\sim}{X} \in \partial D$, is the Neumann data.

Integral operator techniques related to those above can also be used to construct global approximations to solutions of Cauchy's problem for elliptic and hyperbolic equations in three and four independent variables [13], [14].

C. The Equation $\underset{\sim}{e}_n[u] = 0$:

It may be shown by direct substitution that if $E(r, t; n)$ is a solution of

$$(2.43) \qquad (1-t^2)E_{rt} + (n-3)(t^{-1}-t)E_r + rt\left(E_{rr} + \frac{n-2}{r} E_r + BE\right) = 0 ,$$

which satisfies

$$(2.44) \qquad \lim_{t\to o^+} (t^{n-3}E_r)r^{-1} = 0 , \quad \lim_{t\to 1^-} ((1-t^2)^{1/2}E_r)r^{-1} = 0$$

$$\lim_{r\to o^+} E = 1 ,$$

then if $H(\underset{\sim}{x})$ is harmonic,

$$(2.45) \qquad u(\underset{\sim}{x}) = \int_0^1 t^{n-2} E(r,t;n)\, H(\underset{\sim}{x}[1-t^2]) \frac{dt}{(1-t^2)^{1/2}}$$

is a solution of $\underset{\sim}{e}_n(n) = 0$. The representation (2.45) may be reformulated in terms of a new harmonic function $h(\underset{\sim}{x})$ as

(2.46)

$$u(\underset{\sim}{x}) = (\underset{\sim}{I}+\underset{\sim}{G})h(\underset{\sim}{x}) \equiv h(\underset{\sim}{x}) + \int_0^1 \sigma^{n-1} G(r,1-\sigma^2) h(\underset{\sim}{x}\sigma^2)\, d\sigma$$

where

$$(2.47) \qquad h(\underset{\sim}{x}) \equiv \int_0^1 t^{n-2} H(\underset{\sim}{x}[1-t^2]) \frac{dt}{\sqrt{1-t^2}} ,$$

and $G(r, t)$ is a solution of the Goursat problem

$$(2.48) \qquad 2(1-t)G_{rr} - G_r + r(G_{rr} - BG) = 0$$

$$G(0,t) = 0 \ , \quad G(r,0) = \int_0^r rB(r^2)dr \ .$$

It is an interesting fact that $G(r,t)$ is independent of the dimension n of D. It is a more interesting fact that $G(r,t)$ is related to the Riemann function of

$$\frac{\partial^2 U}{\partial Z \partial Z^*} - \frac{1}{4} B(ZZ^*)U = 0 \ ,$$

when $B(\zeta)$ is analytic, namely by [22], [23]

$$(2.49) \qquad G(r,1-\sigma^2) \equiv -2r \, R_1(r\sigma^2,0;r,r) \ .$$

If $B(\zeta)$ is entire, then (2.45) is invertible for all appropriate domains in $\mathbb{R}^n$. This follows since (2.45) can be put in the form of a Volterra integral equation by a simple change of integration parameter. Actually, it is only necessary for $B(r^2) \in \mathcal{C}[0,a]$, where a is the radius of D, for (2.45) to be invertible. The following results were shown to be true in [24]:

Theorem 2.2. Let $\mathcal{H}[D]$ be the class of harmonic functions in $\mathcal{C}[D+\partial D]$, where $B(r^2) \geq 0$ and $B(r^2) \in \mathcal{C}[0,a]$ (i.e. $B(r^2) \in \mathcal{C}_+[0,a]$). Let $\mathcal{E}[D]$ be the class of solutions of (2.3) in $\mathcal{C}[D+\partial D]$. If D is appropriate, and if $u(\underset{\sim}{x}) \equiv (\underset{\sim}{I}+\underset{\sim}{G})h(\underset{\sim}{x})$, then $u \in \mathcal{E}[D]$ if and only if $h \in \mathcal{H}[D]$.

Theorem 2.3. If D is appropriate, and $B(r^2) \in \mathcal{C}_+^1[0,a]$, then the Dirichlet problem, $u \in \mathcal{E}[D]$, $u\big|_{\partial D} = f(\underset{\sim}{x}) \in \mathcal{H}[\partial D]$, has a unique solution, which has a representation of the form $u(\underset{\sim}{x}) = (\underset{\sim}{I}+\underset{\sim}{G})h(\underset{\sim}{x})$, where $h(\underset{\sim}{x}) \in \mathcal{H}[D]$ and

$$(2.50)\quad h(\underset{\sim}{x}) = \frac{\Gamma(n/2)}{(n-2)\pi^{n/2}} \int_{\partial D} \mu(\underset{\sim}{y}) \frac{\partial}{\partial \nu_y} \left(\frac{1}{|\underset{\sim}{x}-\underset{\sim}{y}|^{n-2}} \right) d\omega_y \, .$$

The density $\mu(\underset{\sim}{x})$ is a solution of the Fredholm integral equation

$$(2.51)\quad f(\underset{\sim}{x}) = \mu(\underset{\sim}{x}) + \int_{\partial D} K(\underset{\sim}{x},\underset{\sim}{y})\mu(\underset{\sim}{y})d\omega_y \, , \quad \underset{\sim}{x} \in \partial D \, .$$

with

$$(2.52)\quad K(\underset{\sim}{x},\underset{\sim}{y}) \equiv \frac{\Gamma(n/2)}{(n-2)\pi^{n/2}} \left\{ \frac{\partial}{\partial \nu_y} \frac{1}{|\underset{\sim}{x}-\underset{\sim}{y}|^{n-2}} + \int_0^1 \sigma^{n-1} G(r,1-\sigma^2) \cdot \frac{\partial}{\partial \nu_y} \left(\frac{1}{|\underset{\sim}{x}\sigma^2-\underset{\sim}{y}|^{n-2}} \right) d\sigma \right\}.$$

<u>Theorem 2.4</u>. Let D be appropriate, and $B(r^2) \in \mathcal{C}_+[0,a]$. Then the operator G is monotone in the sense of Collatz on $\mathcal{C}[D]$.

<u>Theorem 2.5</u>. Let $B(r^2) \in \mathcal{C}^1_+[0,a]$, and let $\psi(n;m_k;\underline{+};\underset{\sim}{x})$ be defined by the integrals,

(2.53)

$$\psi(n;m_k;\underline{+};\underset{\sim}{x}) = \int_0^1 t^{n-2} E(r,t;n) H(n;m_k;\underline{+},\underset{\sim}{x}[1-t^2]) \frac{dt}{\sqrt{1-t^2}} \, ,$$

where the $H(n;m_k;\underline{+};\underset{\sim}{x})$ are the homogeneous harmonic polynomials of degree n, $(n = 0, 1, 2, \ldots;\ 0 \leqslant m_{n-2} \leqslant \leqslant m_{n-3} \leqslant \ldots \leqslant m_1 \leqslant n)$, and

$$(2.54)\qquad E(r,t;n) \equiv 1 + t^2 \int_0^1 \sigma^{n-2}\, G(r,[1-\sigma^2]t^2)\,d\sigma \,.$$

Then the $\psi(n;m_k;\underline{+},\underset{\sim}{x})$ form a complete family of solutions of $\underset{\sim}{e}_n[u] = 0$, with respect to appropriate domains.

Proof: As mentioned before, the du Plessis theorem tells us the harmonic polynomials are complete in the uniform norm, for simply-connected domains. The remainder of the theorem comes about by realizing that the above representation is invertible. This follows from a formal identity involving $E(r,t;n)$, and $G(r,t)$. First, however, let us note that if $H(\underset{\sim}{x})$ is harmonic in an appropriate domain D, then so is $h(\underset{\sim}{x}) = (\underset{\sim}{J}H)(\underset{\sim}{x}) \equiv \int_0^1 t^{n-2} H(\underset{\sim}{x}[1-t^2]) \frac{dt}{\sqrt{1-t^2}}$.
From before, the function $u(\underset{\sim}{x}) = (\underset{\sim}{I}+\underset{\sim}{G})h(\underset{\sim}{x})$ is then seen to be harmonic in D also. Furthermore, if $u(\underset{\sim}{x}) \in \mathcal{E}[D]$, then $h(\underset{\sim}{x}) \in \mathcal{H}[D]$, which follows from Theorem 2.2. Actually, in [23] the inverse operator $(\underset{\sim}{I}+\underset{\sim}{G})^{-1}$ is given as a Volterra integral operator. That $H(\underset{\sim}{x})$ is also in $\mathcal{H}[D]$ follows from the claim: a representation for J^{-1} is given by

(2.55)

$$(J^{-1}h)(x) \equiv i^n \frac{(n-1)}{\pi} \int_{\mathcal{L}^{-1}}^{+1} t^{-n}\, h(\underset{\sim}{x}[1-t^2])(1-t^2)^{n/2-1}\, dt \,,$$

where $\mathcal{L}$ is a rectifiable arc from -1 to +1 which does not pass through the origin.

It follows if $u(\underset{\sim}{x}) \in \mathcal{E}[D]$, then the harmonic function $H(\underset{\sim}{x})$, which is the preimage of $u(\underset{\sim}{x})$ under the mapping $(\underset{\sim}{I}+\underset{\sim}{G})\underset{\sim}{J}$, is uniquely determined by the integral representation $H(\underset{\sim}{x}) = [\underset{\sim}{J}^{-1}(\underset{\sim}{I}+\underset{\sim}{G})^{-1}u](\underset{\sim}{x})$. Furthermore, $H(\underset{\sim}{x}) \in \mathcal{C}[D+\partial D]$. From this it follows directly that the functions

$$\Phi(n;m_k;\underline{+};\underset{\sim}{x}) \equiv (\underset{\sim}{I}+\underset{\sim}{G})\underset{\sim}{J}\ H(n;m_k;\underline{+};\underset{\sim}{x}) \tag{2.56}$$

form a complete family of solutions for $\underset{\sim}{e}_n[u] = 0$ for appropriate domains. It remains for us to identify these functions with the $\psi(n;m_k;\underline{+};\underset{\sim}{x})$. To this end we first show that $E(r,t)$ is given by the above integral identity (2.54). Substitution into equation (2.43) for the E-functions reveals, after a few manipulations plus integration by parts, that it is indeed a solution of this partial differential equation. We leave this for the reader to verify. We must next show that the function defined by (2.54) satisfies the boundary data (2.44). Since $E_r t^{n-3} r^{-1} = t^{n-1} r^{-1} \int_0^1 \sigma^{n-2} G_r(r,[1-\sigma^2]t^2)\,d\sigma$, we have that $\lim_{t\to 0^+} (E_r t^{n-3} r^{-1}) = \lim_{t\to 0^+} t^{n-1} \frac{B(r^2)}{n-1} = 0$ for $n \geqslant 2$. Likewise, one has

$$\lim_{t\to 1^-} [(1-t^2)^{1/2} E_r r^{-1}] = \lim_{t\to 1^-} (1-t^2)^{1/2} r^{-1} \int_0^1 \sigma^{n-2} G_r(r,1-\sigma^2)\,d\sigma = \tag{2.57}$$

$$= 2 \lim_{t \to 1^-} (1-t^2)^{\frac{1}{2}} \int_0^1 \sigma^{n-2} g_1(r^2, 1-\sigma^2) d\sigma = 0 ,$$

where $g(r^2,t) \equiv G(r,t)$ is clearly a $\mathcal{C}^1$ function of r^2. (See [24] for a proof that $G(r,t)$ is an even, $\mathcal{C}^1$-function of r.) Finally, using the initial condition $G(0,t) = 0$, it is immediate that $\lim_{r \to 0^+} E = 1$ from the representation (2.54). Consequently, the conditions (2.43-2.44) for the E-function are satisfied, and since (2.43-2.44) represents an over determined system, the E-function is then seen to be uniquely defined either by (2.54) and (2.48), or by (2.43-2.44).

3. Boundary Value Problems: Numerical Treatment

We next turn to the numerical evaluation of the G-function. There are several approaches to doing this. One such approach was mentioned earlier in a paper by one of us [24] for the special case when $B(r^2)$ was an analytic function of r^2. In this case it is known [23] that $G(r,t)$ may be found in the form

$$G(r,t) = \sum_{\ell=1}^{\infty} c_\ell(r^2) t^{\ell-1} , \tag{3.1}$$

where the expansion coefficients $c\ (r^2)$ satisfy the following recursion formulae

(3.2)

$$c_1'(r^2) = rB(r^2) , \qquad \left(\frac{d}{dr} \equiv {}' \right)$$

$$2\ell c'_{\ell+1}(r^2) = -rc''_{\ell}(r^2) + (2\ell-1)c'_{\ell}(r^2) + rP(r^2)c_{\ell}(r^2) \ .$$

$$\ell \geqslant 1 \ .$$

An approximate G-function may then be found by the truncated series

(3.3) $$G_N(r,t) \equiv \sum_{\ell=1}^{N} c_{\ell}(r^2)t^{\ell-1} \ .$$

It is easy to show that $G_N(r,t) \to G(r,t)$ uniformly on $[0,a] \times [0,1]$.

In order to compute $G(r,t)$ it is sometimes advantageous to introduce a change of variables, $w(\rho,t) = (1-t)G(r,t)$, $\rho = r\sqrt{1-t}$, $\tau = t$, and to extend the definition of $B(r^2)$ to $[0,\infty)$ by the scheme [24]

(3.4)

$$\tilde{B}(r^2) \equiv \begin{cases} B(r^2), & 0 \leqslant r \leqslant a \\ B(a^2)\exp\left[-\dfrac{\varepsilon^2}{\varepsilon^2-(r-a)^2}\right], & a \leqslant r \leqslant a+\varepsilon \\ 0\ , & r > a+\varepsilon\ , \end{cases}$$

where $\varepsilon > 0$ is taken arbitrarily small. The differential equation for $w(\rho,\tau)$ is then

(3.5) $$\frac{\partial^2 w}{\partial\rho\partial\tau} = \frac{1}{2}\,\frac{\rho\ \tilde{B}(\frac{\rho^2}{1-\tau})}{(1-\tau)^2}\,w\ ,$$

and it satisfies the data $w(0,\tau) = 0$, $w(\rho,0) = \int_0^{\rho} \rho B(\rho^2)d\rho$.

In the case where $B(r^2)$ is not analytic, but merely $C^1[0,a]$ various procedures are useful. One possibility is to approximate $B(r^2)$ by a polynomial and use the series procedure. Another such method is to use the finite difference Riemann function approach as developed in an interesting paper by Aziz and Hubbard [5]. We outline this latter method below. Let $R \equiv [0,a] \times [0,1]$ be a closed rectangle in the x, y plane, and let R_k be the set of grid points (mk, nk), m, n being positive integers. The real number k is the mesh constant. The "Goursat data" is to be given on the characteristic, mesh-point-surfaces, $(mk,0)$ and $(0,nk)$. Following Aziz and Hubbard we introduce a mesh Riemann function for the finite difference equation,

$$(3.6) \qquad \underset{\sim}{L}_k U \equiv U_{XY} + AU_X + BU_Y + CU = f(x,y) \ ,$$

where $x = mk - \frac{1}{2}k, \quad y = nk - \frac{1}{2}k,$

$$(3.7) \qquad U_X \equiv k^{-1}\left[U(x + \tfrac{1}{2}k, y) - U(x - \tfrac{1}{2}k, y) \right] ,$$

$$(3.8) \qquad U_Y \equiv k^{-1}\left[U(x, y + \tfrac{1}{2}) - U(x, y - \tfrac{1}{2}k) \right] ,$$

(3.9)

$$U_{XY} \equiv k^{-2}\left[U(x + \tfrac{1}{2}, y + \tfrac{1}{2}k) - U(x + \tfrac{1}{2}k, y - \tfrac{1}{2}k) - U(x - \tfrac{1}{2}k, y + \tfrac{1}{2}k) + U(x - \tfrac{1}{2}k, y - \tfrac{1}{2}k) \right] .$$

Here, the functions have been extended to the "half-mesh-points" by an averaging procedure, [5]. The mesh Riemann function is the solution of the adjoint finite difference problem

(3.10)

$$\underset{\sim}{L}_k^{*}V(x,y) \equiv V_{XY} - (AV)_X - (BV)_Y + CV = 0 \ ,$$

$$0 < x = mk - \frac{1}{2}k < \xi < a \ , \quad \text{and} \quad 0 < y = nk - \frac{1}{2}k < \eta < 1 \ ,$$

$$V(x,\eta;\xi,\eta)_X - B(x,\eta)V(x,\eta;\xi,\eta) = 0 \ , \quad 0 < x = mk - \frac{1}{2}k < \xi < a \ ,$$

(3.11)

$$V(\xi,y;\xi,\eta)_Y - A(\xi,y)V(\xi,y;\xi,\eta) = 0 \ , \quad 0 < y = nk - \frac{1}{2}k < \eta < 1 \ .$$

One may thereby solve the finite difference problem,

$$W(\rho,\tau)_{PT} = \frac{1}{2}\frac{\rho}{(1-\tau)^2}\tilde{B}\left(\frac{\rho^2}{1-\tau}\right)W(\rho,\tau) \ ,$$

(3.12)

$$W(0,\tau) = 0 \ , \quad W(\rho,0) = k\sum_{m=1}^{M}(mk - \frac{1}{k}k)\tilde{B}([mk - \frac{1}{2}k]^2) \ ,$$

$\rho = Mk$, M a positive integer, by means of the finite difference Riemann formula [5]. One obtains the following representation

(3.13)

$$W(\rho,\tau) = V(0,0;\rho,\tau)W(0,0) + \sum_{m=1}^{M}\Big\{(mk - \frac{1}{2}k)\tilde{B}([mk - \frac{1}{2}k]^2) \cdot$$

$$\cdot\ V(mk - \frac{1}{2}k, 0; \rho, \tau) \Big\} \quad , \quad \rho = Mk \ .$$

The dependence of this solution on the mesh constant k shall be indicated functionally by $W(\rho,\tau;k)$.

Recalling the definition $G(r,t) \equiv (1-t)^{-1}w(r\sqrt{1-t},t)$, we introduce the discrete function $G(r,t;k)$ by replacing $w(\rho,\tau)$ by $W(\rho,\tau;k)$. The function $G(r,t;k)$ is no longer defined on a rectangular grid; however, ignoring this point as being merely of technical interest, we represent the approximate operator $\underset{\sim}{G}_k$ by (for $n \geqslant 3$)

$$(3.14)\qquad (\underset{\sim}{G}_k h)(\underset{\sim}{x}) \equiv \int_0^1 \sigma^{n-3}\, W(r\sigma, 1-\sigma^2;k)h(\underset{\sim}{x}\sigma^2)\,d\sigma \ ,$$

where it is understood that the discrete function $W(r\sigma,1-\sigma^2;k)$ has been extended to the continuum by a smooth interpolation. This leads to a sequence of integral transformations

$$(3.15)\qquad u(\underset{\sim}{x};k) = (\underset{\sim}{I}+\underset{\sim}{G}_k)h(\underset{\sim}{x}) \ , \quad h(\underset{\sim}{x}) \in \mathcal{H}[D] \ .$$

The functions $u(\underset{\sim}{x};k)$ may be considered as belonging to a family of approximate solutions of (2.3). In order to obtain an approximate solution of the boundary value problem, we replace $h(x)$ in (3.15) by a double layer potential, as above, and compute the residue. We obtain a sequence of integral equations

$$(3.16)\qquad f(\underset{\sim}{x}) = (\underset{\sim}{I}+\underset{\sim}{K}_k)\mu(\underset{\sim}{x};k) \ ,$$

which may easily be shown to be of Fredholm type. The kernel $K(\underset{\sim}{x},\underset{\sim}{y};k)$ of $\underset{\sim}{K}_k$ is given by (2.52) with $G(r,t)$ replaced by the interpolated discrete function $G(r,t;k)$.

If we introduce the usual operator norm,

$$\|K\| \equiv \max_{\underset{\sim}{x}\in\partial D} \int_{\partial D} |K(\underset{\sim}{x},\underset{\sim}{y})|\,d\omega_y \quad , \tag{3.17}$$

then it is clear from the following

(3.18)

$$\frac{\Gamma(n/2)}{(n-2)\pi^{n/2}} \int_{\partial D} \frac{\partial}{\partial\nu_y}\left(\frac{1}{\|x-y\|^{n-2}}\right) d\omega_y = \begin{cases} 1 & , \ \underset{\sim}{x}\in D^0 \\ 1/2 & , \ \underset{\sim}{x}\in\Gamma \\ 0 & , \ \underset{\sim}{x}\in D' \ , \end{cases}$$

that $\|\underset{\sim}{K}\| < \infty$, and $\|\underset{\sim}{K}_k\| < \infty$ for $k > 0$ sufficiently small. Further more, we may show that $\|\underset{\sim}{K}-\underset{\sim}{K}_k\| \to 0$ as $k \to 0$. This will follow from the discretization error estimate on $w(\rho,\tau) - W(\rho,\tau;k)$ as given in [5]. This estimate applied to our case is,

(3.19)

$$|w(\rho,\tau) - W(\rho,\tau;k)| \leqslant \ell(\rho,\tau) \int_0^\rho d\tilde{\rho} \int_0^\tau d\tilde{\tau}\left[e_k^{K\{(\rho-\tilde{\rho})+(\tau-\tilde{\tau})\}}\right] ,$$

where

$$\ell(\rho,\tau) = \max_{\substack{0\leqslant\tilde{\rho}\leqslant\rho \\ 0\leqslant\tilde{\tau}\leqslant\tau}} |\underset{\sim}{L}_k\, w(\tilde{\rho},\tilde{\tau})| \ , \tag{3.20}$$

(3.21) $$2K^2 = \max_{\substack{0\leq\rho\leq a \\ 0\leq\tau\leq 1}} \left| \frac{\rho}{(1-\tau)^2} \tilde{B}\left(\frac{\rho^2}{(1-\tau)}\right) \right| ,$$

and $e_k^{f(x)}$ is a finite difference analogue of the exponential function $e^{f(x)}$. In [5] $e_k^{f(x)}$ is given as

(3.22)
$$e_k^{f(x)} = e_k^{f(0)} \prod_{m=1}^{M} \left[\frac{1 + \frac{1}{2}k\ f(mk - \frac{1}{2}k)_x}{1 - \frac{1}{2}k\ f(mk - \frac{1}{2}k)_x} \right] , \qquad x = mk .$$

We remark that the constant K given above is bounded because of the construction of $\tilde{B}\left(\frac{\rho^2}{1-\tau}\right)$ as a function which vanishes smoothly between $\frac{\rho^2}{1-\tau} = a$ and $a + \varepsilon$, $\varepsilon > 0$ arbitrarily small.

In [5] a bound on (ρ,τ) is given as Mk^2, with estimates on the constant M. In order to show that $\|\underset{\sim}{K}-\underset{\sim}{K}_k\| \to 0$ as $k \to 0$, it suffices to know that $|w(\rho,\tau) - W(\rho,\tau;k)| \leq \tilde{M}k^2$, which follows by an elementary computation. That $\|\underset{\sim}{K}-\underset{\sim}{K}_k\| \to 0$ can be seen directly from

(3.23)
$$K(\underset{\sim}{x},\underset{\sim}{y})-K(\underset{\sim}{x},\underset{\sim}{y};k) \equiv \frac{\Gamma(n/2)}{(n-2)\pi^{n/2}} \int_0^1 \sigma^{n-3} \left[w(r\sigma,1-\sigma^2)-W(r\sigma,1-\sigma^2;k)\right]$$
$$\cdot \frac{\partial}{\partial\nu_y} \left(\frac{1}{\|x\ {}^2-y\|^{n-2}}\right) d\sigma ,$$

which implies

(3.24)
$$\|\underset{\sim}{K}-\underset{\sim}{K}_k\| = \max_{x\in\partial D} \int_{\partial D} |K(\underset{\sim}{x},\underset{\sim}{y}) - K(\underset{\sim}{x},\underset{\sim}{y};k)|\, d\omega_y \leq$$

$$\leq \max_{x \in \partial D} \frac{\Gamma(n/2)}{(n-2)\pi^{n/2}} \int_{\partial D} \int_0^1 \sigma^{n-3} |w(r\sigma, 1-\sigma^2) - W(r\sigma, 1-\sigma^2; k)|$$

$$\cdot \left| \frac{\partial}{\partial \nu_y} \left(\frac{1}{\| x\sigma^2 - y \|^{n-2}} \right) \right| d\sigma \, d\omega_y$$

$$\leq k^2 \tilde{M} \max_{x \in \partial D} \frac{\Gamma(n/2)}{(n-2)\pi^{n/2}} \int_0^1 \sigma^{n-3} \int_{\partial D} \left| \frac{\partial}{\partial \nu_y} \left(\frac{1}{\| \underset{\sim}{x}\sigma^2 - \underset{\sim}{y} \|^{n-2}} \right) \right| d\sigma \, d\omega_y$$

$$\leq k^2 \tilde{M} \ , \quad \text{for} \quad n \geq 3 \ .$$

From this we conclude $\| \underset{\sim}{K} - \underset{\sim}{K}_k \| \approx O(k^2)$ as $k \to 0$. Using similar estimates we can show that

(3.25)

$$\lim_{\| \underset{\sim}{x}_1 - \underset{\sim}{x}_2 \| \to 0} \left\{ \int_{\partial D} |K(\underset{\sim}{x}_1, \underset{\sim}{y}; k) - K(\underset{\sim}{x}_2, \underset{\sim}{y}; k)| \, d\omega_y \right\} = 0$$

uniformly in $\underset{\sim}{x}_1$, $\underset{\sim}{x}_2$. This plus the fact that $\|K\| < \infty$, implies via the Arzela-Ascoli Theorem that the $\underset{\sim}{K}_k$ are compact. We already know that $\underset{\sim}{K}$ is compact, and furthermore, that $(\underset{\sim}{I} + \underset{\sim}{K})^{-1}$ exists. Two norm inequalities from Taylor [32] pg. 164, namely that when $\| \underset{\sim}{K} - \underset{\sim}{K}_k \| < \frac{1}{\| (I+K)^{-1} \|}$ then

$$\| (I + K_k)^{-1} \| \leq \frac{\| (\underset{\sim}{I} + \underset{\sim}{K})^{-1} \|}{1 - \| (\underset{\sim}{I} + \underset{\sim}{K})^{-1} \| \cdot \| \underset{\sim}{K} - \underset{\sim}{K}_k \|}$$

and

(3.27)

$$\| (\underset{\sim}{I}+\underset{\sim}{K}_k)^{-1} - (\underset{\sim}{I}+\underset{\sim}{K})^{-1} \| \leq \frac{\| (\underset{\sim}{I}+\underset{\sim}{K})^{-1} \|^2 \cdot \| \underset{\sim}{K}-\underset{\sim}{K}_k \|}{1 - \| (\underset{\sim}{I}+\underset{\sim}{K})^{-1} \| \cdot \| \underset{\sim}{K}-\underset{\sim}{K}_k \|}$$

tell us that $(\underset{\sim}{I}+\underset{\sim}{K}_k)^{-1}$ exists and that the unique solutions (for $k > 0$ sufficiently small) of

(3.28)
$$f(\underset{\sim}{x}) = (\underset{\sim}{I}+\underset{\sim}{K}_k)\mu(\underset{\sim}{x};k)$$

tend to the unique solution

(3.29)
$$f(\underset{\sim}{x}) = (\underset{\sim}{I}+\underset{\sim}{K})\mu(\underset{\sim}{x})$$

We next turn our attention to the numerical solution of the sequence of equations (3.28) above. The approach we use is an extension of the Nyström method to kernels with weak singularities and is due primarily to Anselone [1], [2]; see also in this regard Atkinson [3], [4].

We replace the integration in (3.28) by numerical quadrature. To this end, we first reparametrize (3.28) by introducing the following representation for ∂D, $\underset{\sim}{y} = \underset{\sim}{y}(\underset{\sim}{t})$, $\underset{\sim}{t} \in \mathcal{A}$, $\mathcal{A} \equiv \{\underset{\sim}{t} | 0 \leq \max t_i \leq 1; i = 1, 2, \ldots, (n-1)\}$, $\tilde{K}(\underset{\sim}{s},\underset{\sim}{t};k) \equiv K(\underset{\sim}{x}(\underset{\sim}{s}), \underset{\sim}{y}(\underset{\sim}{t});k)$. Then (3.28) becomes with $f(\underset{\sim}{x}(\underset{\sim}{s})) = F(\underset{\sim}{s})$, $\mu(\underset{\sim}{x}(\underset{\sim}{s});k) = \rho^{(k)}(\underset{\sim}{s})$

(3.30)
$$F(\underset{\sim}{s}) = \rho(\underset{\sim}{s}) + \int_0^1 dt_1 \int_0^1 dt_2 \cdots \int_0^1 dt_{n-1} \tilde{K}(\underset{\sim}{s},\underset{\sim}{t};k)\rho(\underset{\sim}{t}) ,$$

which may be then replaced by the numerical quadrature

(3.31)

$$\rho_{(m)}^{(k)}(\underset{\sim}{s}) + \sum_{j_1=1}^{m} \sum_{j_2=1}^{m} \cdots \sum_{j_{n-1}=1}^{m} \Phi_{j_1 \cdots j_{n-1}} \tilde{K}(\underset{\sim}{s},\underset{\sim}{t}_j;k)\ \rho_{(m)}^{(k)}(\underset{\sim}{t}_j) = F(\underset{\sim}{s}) .$$

The Nyström method is to now set the points $\underset{\sim}{s} = \underset{\sim}{t}_i$, $(i_\ell = 1, 2, \ldots, m)$ $(\ell = 1, 2, \ldots, n-1)$ and solve the linear system

(3.32)

$$z_i^{(k)} + \sum_{j_1=1}^{m} \cdots \sum_{j_{n-1}=1}^{m} \Phi_j\ K(\underset{\sim}{t}_i,\underset{\sim}{t}_j;k)\ z_j^{(k)} = F(\underset{\sim}{t}_i) ,$$

$(i_\ell = 1,2,\cdots,m)$, $(\ell = 1,2,\cdots,n-1)$, $i = (i_1,\cdots,i_{n-1})$.

The solution of (3.31) above is then given in terms of the solutions $z_j^{(k)}$ of (3.32) by

(3.33)

$$\rho_{(m)}^{(k)}(\underset{\sim}{s}) = \left[F(\underset{\sim}{s}) - \sum_{j_1=1}^{m} \cdots \sum_{j_{n-1}=1}^{m} \Phi_j\ \tilde{K}(\underset{\sim}{s},\underset{\sim}{t}_j;k) z_j^{(k)}\right] .$$

Let us put the above equations into a formal setting. Let $\mathcal{C}[\mathcal{A}]$ denote the Banach space of continuous functions on $\mathcal{A}$, with the uniform norm. Let $\underset{\sim}{\mathcal{K}}^{(k)}$ be the integral operator on $\mathcal{C}[\mathcal{A}]$ with the kernel $\tilde{K}(\underset{\sim}{s},\underset{\sim}{t};k)$:

$$(3.34) \qquad (\underset{\sim}{\mathcal{K}}^{(k)}\rho)(\underset{\sim}{s}) \equiv \int \tilde{K}(\underset{\sim}{s},\underset{\sim}{t};k)\rho(t)dt , \quad \underset{\sim}{s} \in \mathcal{A} .$$

Let $\underset{\sim}{\mathcal{K}}^{(k)}_{(m)}$ be the numerical quadrature operator on $C[\mathcal{A}]$ given by

$$(3.35)\qquad (\underset{\sim}{\mathcal{K}}^{(k)}_{(m)}\rho)(\underset{\sim}{s}) \equiv \sum_j \Phi_j\ \tilde{K}(\underset{\sim}{s},\underset{\sim}{t}_j;k)\rho(t_j)\ ,\quad \underset{\sim}{s}\in\mathcal{A}\ .$$

We have shown earlier that $\underset{\sim}{\mathcal{K}}^{(k)}$ is compact on $C[\mathcal{A}]$. Furthermore, since $\underset{\sim}{\mathcal{K}}^{(k)}_{(m)}$ is a finite rank operator it is also compact.

The equations (3.30) and (3.31) may now be written in the operator notation as

$$(3.36)\qquad (\underset{\sim}{I}+\underset{\sim}{\mathcal{K}}^{(k)})\rho^{(k)}(\underset{\sim}{s}) = F(\underset{\sim}{s})\ ,\quad \underset{\sim}{s}\in\mathcal{A},$$

and

$$(3.37)\qquad (\underset{\sim}{I}+\underset{\sim}{\mathcal{K}}^{(k)}_{(m)})\rho^{(k)}_{(m)}(\underset{\sim}{s}) = F(\underset{\sim}{s})\ ,\quad \underset{\sim}{s}\in\mathcal{A}.$$

As it was mentioned above, the work of Anselone [1], [2] concerning the solutions of (3.36) and (3.37) is influential to our approach. In particular, he has given estimates on the difference between the solutions of these equations. For instance, if $(\underset{\sim}{I}+\underset{\sim}{\mathcal{K}}^{(k)})^{-1}$ exists (which we showed was true), and if

$$(3.38)\qquad \|(\underset{\sim}{\mathcal{K}}^{(k)} - \underset{\sim}{\mathcal{K}}^{(k)}_{(\mathbf{m})})\underset{\sim}{\mathcal{K}}^{(k)}_{(m)}\| < \frac{1}{\|(\underset{\sim}{I}+\underset{\sim}{\mathcal{K}}^{(k)})^{-1}\|}\ ,$$

then $(\underset{\sim}{I}+\underset{\sim}{\mathcal{K}}^{(k)}_{(m)})^{-1}$ exists, and is bounded in norm by

$$(3.39)\qquad \|(\underset{\sim}{I}+\underset{\sim}{\mathcal{K}}^{(k)}_{(m)})^{-1}\| \leqslant$$

$$\leqslant \frac{1 + \|\underset{\sim}{\mathcal{K}}^{(k)}_{(m)}\| \cdot \|(\underset{\sim}{I}+\underset{\sim}{\mathcal{K}}^{(k)})^{-1}\|}{1 - \|(\underset{\sim}{\mathcal{K}}^{(k)}-\underset{\sim}{\mathcal{K}}^{(k)}_{(m)})\underset{\sim}{\mathcal{K}}^{(k)}_{(m)}\| \cdot \|(\underset{\sim}{I}+\underset{\sim}{\mathcal{K}}^{(k)})^{-1}\|} \ .$$

In addition, if $(\underset{\sim}{I}+\underset{\sim}{\mathcal{K}}^{(k)}_{(m)})^{-1}$ exists, say for m sufficiently large, then one has [1], [3],

(3.40)

$$\|\rho^{(k)}_{(m)}(\underset{\sim}{s}) - \rho^{(k)}(\underset{\sim}{s})\| \leqslant \|(\underset{\sim}{I}+\underset{\sim}{\mathcal{K}}^{(k)}_{(m)})^{-1}\| \cdot$$

$$\cdot \frac{\|(\underset{\sim}{\mathcal{K}}^{(k)}-\underset{\sim}{\mathcal{K}}^{(k)}_{(m)})F\| + \|\rho^{(k)}_{(m)}\| \cdot \|(\underset{\sim}{\mathcal{K}}^{(k)}-\underset{\sim}{\mathcal{K}}^{(k)}_{(m)})\underset{\sim}{\mathcal{K}}^{(k)}\|}{1 - \|(\underset{\sim}{\mathcal{K}}^{(k)}-\underset{\sim}{\mathcal{K}}^{(k)}_{(m)})\underset{\sim}{\mathcal{K}}^{(k)}\| \cdot \|(\underset{\sim}{I}+\underset{\sim}{\mathcal{K}}^{(k)}_{(m)})^{-1}\|} \ .$$

These estimates imply the following theorem:

Theorem 3.1. Let $D \subset \mathbb{R}^n$, $n \geqslant 3$, be an appropriate domain, $B(r^2) \in C^1_+[0,a]$, and $f(x) \in C^0[\partial D]$. Then the solution of the Dirichlet problem, $\Delta_n u - B(r^2)u = 0$, $u(\underset{\sim}{x}) = f(\underset{\sim}{x})$ for $\underset{\sim}{x} \in \partial D$, has the following approximate solution:

(3.41) $$u^k_m(\underset{\sim}{x}) = (\underset{\sim}{I}+\underset{\sim}{G}_k)h^k_m(\underset{\sim}{x}) \ , \quad \underset{\sim}{x} \in D+\partial D \ ,$$

where

(3.42)

$$h^k_m(\underset{\sim}{x}) = \frac{\Gamma(n/2)}{(n-2)\pi^{n/2}} \int_{\partial D} \mu^k_m(y) \frac{\partial}{\partial \nu_y} \left(\frac{1}{\|x-y\|^{n-2}}\right) d\omega_y \ ,$$

and $\mu^k_m(\underset{\sim}{y}) \equiv \rho^{(k)}_{(m)}(\underset{\sim}{t}(\underset{\sim}{y}))$, $\underset{\sim}{y} \in \partial D$. Furthermore, the error $|u(\underset{\sim}{x}) - u^k_m(\underset{\sim}{x})|$ is bounded by

(3.43) $$|u(\underset{\sim}{x}) - u_m^k(\underset{\sim}{x})| \leq C_1 \|\rho - \rho_{(m)}^{(k)}\| + C_2 k^2 ,$$

$$\underset{\sim}{x} \in D+\partial D ,$$

where C_1 and C_2 are constants and

(3.44)

$$\|\rho - \rho_{(m)}^{(k)}\| \leq \frac{\|(\underset{\sim}{I}+\underset{\sim}{\mathcal{K}}^{(k)})^{-1}\| \cdot \|\underset{\sim}{\mathcal{K}}^{(k)}-\underset{\sim}{\mathcal{K}}\| \cdot \|\underset{\sim}{\rho}^k\|}{1 - \|(\underset{\sim}{I}+\underset{\sim}{\mathcal{K}}^{(k)})^{-1}\| \cdot \|\underset{\sim}{\mathcal{K}}^{(k)}-\underset{\sim}{\mathcal{K}}\|}$$

$$+ \frac{\|(\underset{\sim}{\mathcal{K}}^{(k)}-\underset{\sim}{\mathcal{K}}_{(m)}^{(k)})f\| + \|\rho_{(m)}^{(k)}\| \cdot \|(\mathcal{K}^{(k)}-\mathcal{K}_{(m)}^{(k)})\mathcal{K}^{(k)}\|}{1 - \|(\underset{\sim}{\mathcal{K}}^{(k)}-\underset{\sim}{\mathcal{K}}_{(m)}^{(k)})\underset{\sim}{\mathcal{K}}^{(k)}\| \cdot \|(\underset{\sim}{I}+\mathcal{K}_{(m)}^{(k)})^{-1}\|} .$$

Proof: The bound on $\|\rho - \rho_{(m)}^{(k)}\|$ is found by the triangle inequality, equation (3.40), and an estimate on $\|\rho - \rho^{(k)}\|$. One has the identity (for k sufficiently small so that $(\underset{\sim}{I}+\underset{\sim}{\mathcal{K}}^{(k)})^{-1}$ exists)

(3.45)

$$(\underset{\sim}{I}+\underset{\sim}{\mathcal{K}})^{-1} - (\underset{\sim}{I}+\underset{\sim}{\mathcal{K}}^{(k)})^{-1} = (\underset{\sim}{I}+\underset{\sim}{\mathcal{K}}^{(k)})^{-1} (\underset{\sim}{\mathcal{K}}^{(k)}-\underset{\sim}{\mathcal{K}})$$

$$\cdot [\underset{\sim}{I}-(\underset{\sim}{I}+\underset{\sim}{\mathcal{K}}^{(k)})^{-1} (\underset{\sim}{\mathcal{K}}^{(k)}-\underset{\sim}{\mathcal{K}}]^{-1} (\underset{\sim}{I}+\underset{\sim}{\mathcal{K}}^{(k)})^{-1} .$$

The estimate for $\|\rho - \rho^{(k)}\|$ follows immediately from this. The estimate for $|u(x) - u_m^k(x)|$ follows from the maximum principal for solutions of $\Delta_n u - B(r^2)u = 0$, with $B(r^2) \geq 0$ for $\underset{\sim}{x} \in D$, the obvious inequality

$$|u-u_m^k| \leq |h-h_m^k| + |\underset{\sim}{G}(h-h_m^k)| + |(\underset{\sim}{G}-\underset{\sim}{G}_k)\, h_m^k| ,$$

and the fact that the operator $\underset{\sim}{G}$ is monotone in the sense of Collatz [10], [24]. The constants C_1 and C_2 may be estimated without difficulty.

An alternate procedure for computing an approximate G-function is to use the Cauchy-Euler Polygon Method. In this regard see the work of Diaz [D.1,2].

One chooses a sequence of subdivisions of the rectangle $\mathcal{R} \equiv [0,a] \times [0,1]$, i.e. for each (m,n), we form a subdivision.

$$(3.46) \qquad 0 = \rho_{0,m} < \rho_{1,m} < \cdots < \rho_{m,m} = a \, ,$$

$$0 = \tau_{0,n} < \tau_{1,n} < \cdots < \tau_{n,n} = a \, ,$$

and on each of the sub-rectangles $\mathcal{R}_{k,\ell} \equiv [\rho_{km}, \rho_{k+1,m}] \times [\tau_{\ell n}, \tau_{\ell+1,n}]$ we consider the "miniature problem"

$$\frac{\partial^2 w}{\partial\rho\partial\tau} = A_{k,\ell}\, w \; ; \quad A_{k,\ell} \equiv A(\rho_{k,m}, \tau_{\ell,n}) \, ,$$

where

$$(.47) \qquad A(\rho,\tau) \equiv \frac{1}{2} \frac{\rho}{(1-\tau)^2} \tilde{B} \left(\frac{\rho^2}{1-\tau}\right)$$

$$w(\rho,\tau_\ell) = D_{k\ell} + B_{k\ell}(\rho-\rho_k) \, , \quad \rho_k \leqslant \rho \leqslant \rho_{k+1} \, ,$$

$$w(\rho_k,\tau) = D_{k\ell} + C_{k\ell}(\tau-\tau_\ell) \, , \quad \tau_\ell \leqslant \tau \leqslant \tau_{\ell+1} \, .$$

Hence, in $\mathcal{R}_{k,\ell}$ we have,

(3.48) $$w(\rho,\tau) = A_{k,\ell}(\rho-\rho_k)(\tau-\tau_k) + B_{k\ell}(\rho-\rho_k) + C_{kk}(\tau-\tau_\ell) + D_{k\ell} .$$

On the rectangles having sides on $\tau = 0$ the initial data $w(\rho,0) = \int_0^\rho \rho\, B(\rho^2)d\rho$ is linearly approximated; on rectangles having sides on $\rho = 0$, the data is chosen to be zero. The general form of the solution in $\mathcal{R}_{k,\ell}$ has been given by Diaz in [18,19] and is for our case,

(3.49)

$$w_{mn}(\rho,\tau) = w_{k0} + w_{0\ell} - w_{00} + \frac{w_{k+1,0}-w_{k,0}}{\rho_{k+1}-\rho_k}(\rho-\rho_k)$$
$$+ \sum_{i=1}^{k}\sum_{j=1}^{\ell} A_{i-1,j-1}(\rho_i-\rho_{i-1})(\tau_j-\tau_{j-1})$$
$$+ \sum_{j=1}^{\ell} A_{k,j-1}(\rho-\rho_k)(\tau_j-\tau_{j-1})$$
$$+ \sum_{i=1}^{k} A_{i-1,\ell}(\rho_i-\rho_{i-1})(\tau-\tau_\ell)$$
$$+ A_{k\ell}(\rho-\rho_k)(\tau-\tau_\ell) .$$

Here we are using the notation $w_{ij} \equiv w(\rho_{im},\tau_{jn})$ and $w_{mn}(\rho,\tau)$ is the approximating solution computed by subdividing $\mathcal{R}$ into $m \times n$ smaller rectangles.

Such a construction is useful to use in order to obtain an ε-approximating solution as is done in the case of ordinary differential equations; i.e., given an $\varepsilon > 0$ we choose a subdivision such that in each $\mathcal{R}_{k\ell}$ the differential equation is almost satisfied by $w_{mn}(\rho,\tau)$. More precisely, we choose m and n such that $\left|\frac{\partial^2 w_{mn}}{\partial\rho\partial\tau} - Aw\right| < \varepsilon$. If this is the case we can obtain bounds on the difference between an actual solution and an ε-approximating solution.

Lemma 3.1. Let $K^2 = \max_{(\rho,\tau)\in\mathcal{R}} A(\rho,\tau)$, and $w_i(\rho,\tau)$, $(i=1,2)$ be ε_i-approximating solutions of (3.5), which satisfy the required data. Then for all $(\rho,\tau)\in\mathcal{R}$ one has the estimate

$$|w_1(\rho,\tau)-w_2(\rho,\tau)| \leqslant \varepsilon\rho\tau[1+\rho\tau K^2 e^{K(\rho+\tau)}] , \tag{3.50}$$

where $\varepsilon = \varepsilon_1 + \varepsilon_2$.

Proof: Our proof is modeled after the ordinary differential equation case in Coddington and Levinson [17], Chapter I. First we notice that the $w_i(\rho,\tau)$ satisfy

(3.51)

$$\left|w_i(\rho,\tau) - \int_0^\rho \rho\ \tilde{B}(\rho^2)d\rho - \frac{1}{2}\int_0^\rho\int_0^\tau \frac{\rho}{(1-\tau)^2}\tilde{B}\left(\frac{\rho^2}{1-\tau}\right) w_i(\rho,\tau)d\tau d\rho\right| \leqslant \varepsilon_i\ \rho\tau.$$

We notice that because of the extension of $\tilde{B}(r^2)$ to $[0,\infty)$, $\left|\frac{\rho}{2(1-\tau)^2}\tilde{B}\left(\frac{\rho^2}{1-\tau}\right)\right|$ is continuous on $\mathcal{R}$, and hence its maximum does exist on $\mathcal{R}$; we set it equal to K^2. If $c(\rho,\tau) \equiv |w_1(\rho,\tau)-w_2(\rho,\tau)|$, then by adding (3.51) for $i = 1,2$, we obtain

(3.52)

$$c(\rho,\tau) \leqslant \frac{1}{2}\int_0^\rho\int_0^\tau c(\xi,\eta)\,\frac{\xi}{(1-\eta)^2}\,\tilde{B}\left(\frac{\xi^2}{1-\eta}\right)d\eta d\xi + \varepsilon\rho\tau\ .$$

Defining $C(\rho,\tau) \equiv \int_0^\rho\int_0^\tau c(\xi,\eta)d\eta d\xi$, (3.52) becomes

$$\frac{\partial^2 C}{\partial\rho\partial\tau} \leqslant K^2 C(\rho,\tau) + \varepsilon\rho\tau\ . \tag{3.53}$$

Since the Riemann function for $C_{\rho\tau} - K^2 C = 0$ is $I_0(2K\sqrt{(\rho-\xi)(\tau-\eta)})$ we obtain the following estimate,

$$C(\rho,\tau) \leqslant \varepsilon\int_0^\rho\int_0^\tau \xi\eta\, I_0(2K\sqrt{(\rho-\xi)(\tau-\eta)})d\xi d\eta\ .$$

Since $I_0(x)$ is given by the series expansion

$$I_0(x) = \sum_{m\geqslant 0}\left(\frac{x}{2}\right)^{2m}\frac{1}{(m!)^2}\ ,$$

we have

$$C(\rho,\tau) \leqslant \varepsilon\rho^2\tau^2\, I_0(2K\sqrt{\rho\tau})\ ,$$

which upon substitution into (3.52) yields

$$\begin{aligned} c(\rho,\tau) &\leqslant \varepsilon\rho\tau[1 + \rho\tau K^2\, I_0(2K\sqrt{\rho\tau})] \\ &\leqslant \varepsilon\rho\tau[1 + \rho\tau K^2\, e^{K(\rho+\tau)}]\ . \end{aligned} \tag{3.54}$$

This is the stated result.

<u>Remark</u>: Since $A(\rho,\tau)$ is uniformly continuous on $\mathcal{R}$, it is possible to put a "uniform mesh" on $\mathcal{R}$ in order to obtain an ε-approximating solution.

We now introduce an approximate G-function $G_\varepsilon(r,t)$ by $G_\varepsilon(r,t) \equiv w_\varepsilon(r\sqrt{1-t},\, t)(1-t)^{-1}$. Then we have

(3.55)

$$|(\underset{\sim}{G}-\underset{\sim}{G}_\varepsilon)h(\underset{\sim}{x})| = \left|\int_0^1 \sigma^{n-3}[w(r\sigma,1-\sigma^2) - w_\varepsilon(r\sigma,1-\sigma^2)] \cdot h(x\sigma^2)\,d\sigma\right|$$

$$\leq \varepsilon r \int_0^1 \sigma^{n-2}(1-\sigma^2)[1+K^2 r\sigma(1-\sigma^2)e^{K(r\sigma+1-\sigma^2)}] \cdot |h(\underset{\sim}{x}\sigma^2)|\,d\sigma$$

$$\leq \varepsilon r\,\|h\| \left(\int_0^1 \sigma^{n-2}(1-\sigma^2)\,d\sigma\right)\cdot\left(1+k^2 r e^{K(r+1)}\right)$$

$$\leq \varepsilon\,\frac{2r}{n}\left(\frac{n-2}{n}\right)^{n-2}\left(1+K^2 r e^{K(r+1)}\right)\cdot \|h\|_{\partial D}\ ;$$

$$\|h\|_{\partial D} \equiv \max_{\partial D}|h|\ .$$

We next introduce the approximate kernel $K_\varepsilon(\underset{\sim}{x},\underset{\sim}{y})$ by replacing $G(r,t)$ in (2.52) by $G_\varepsilon(r,t)$, and then estimate $\|\underset{\sim}{K}-\underset{\sim}{K}_\varepsilon\|$ using the $\|\cdot\|$ defined by (3.17). We obtain

(3.56)

$$\left|\int_{\partial D} K(\underset{\sim}{x},\underset{\sim}{y})-K_\varepsilon(\underset{\sim}{x},\underset{\sim}{y})d\omega_y\right| = \frac{\Gamma(n/2)}{(n-2)\pi^{n/2}}\int_0^1 \sigma^{n-1}\left|\int_{\partial D}(G-G_\varepsilon)\right.$$

$$\left.\cdot\frac{\partial}{\partial\nu_y}\frac{1}{\|\underset{\sim}{x}\sigma^2-\underset{\sim}{y}\|^{n-2}}\right| d\omega_y\, d\sigma$$

$$\leqslant \frac{\Gamma(n/2)}{\pi^{n/2}}\int_0^1 \sigma^{n-1}\max_D|G-G_\varepsilon|\int_{\partial D}\frac{|\cos(\nu,\underset{\sim}{x}\sigma^2-\underset{\sim}{y})|}{\|\underset{\sim}{x}\sigma^2-\underset{\sim}{y}\|^{n-1}}d\omega_y\, d\sigma$$

$$\leqslant \int_0^1 \max_D |G-G_\varepsilon|\sigma^{n-1}\, d\sigma$$

$$\leqslant \varepsilon\,\frac{2r}{n}\left(\frac{n-2}{n}\right)^{n-2}(1+k^2re^{K(r+1)}) .$$

Hence, $\|K-K_\varepsilon\| \simeq O(\varepsilon)$ as $\varepsilon\to 0$.

Arguing as we did previously we may show that $(\underset{\sim}{I}+\underset{\sim}{K}_\varepsilon)^{-1}$ exists, and furthermore, that $\|(\underset{\sim}{I}+\underset{\sim}{K}_\varepsilon)^{-1} - (\underset{\sim}{I}+\underset{\sim}{K})^{-1}\| \to 0$ as $\varepsilon\to 0$; hence, it is sufficient for us to consider the sequence of integral equations

$$(\underset{\sim}{I}+\underset{\sim}{K}_\varepsilon)\,\mu_\varepsilon(\underset{\sim}{x}) = f(\underset{\sim}{x})$$

to obtain the approximate solutions $\mu_\varepsilon(\underset{\sim}{x})\to\mu(\underset{\sim}{x})$ as $\varepsilon\to 0$.

4. Nonlinear Equations in Two Independent Variables: The Dirichlet Problem

In this section we will obtain rapidly convergent analytic approximations to solutions of Dirichlet's problem for the equation

$$\Delta u = f(x,y,u, \frac{\partial u}{\partial x}, \frac{\partial u}{\partial y}) \tag{4.1}$$

defined in a simply connected domain D with Liapunov boundary C ([16]). Without loss of generality we assume $u = 0$ on C. We require $f(x,y,\xi_1,\xi_2,\xi_3)$ to satisfy the conditions

$$H_1: \quad f(x,y,0,0,0) \in L_p(D+C) \ , \quad p > 2 \ ,$$

$$H_2: \quad |f(x,y,\xi_1^o,\xi_2^o,\xi_3^o) - f(x,y,\xi_1^1,\xi_2^1,\xi_3^1)|$$
$$\leqslant f_o(x,y)\{|\xi_1^1-\xi_1^1| + |\xi_2^o-\xi_2^1| + |\xi_3^o-\xi_3^1|\}$$

where $f_o(x,y) \in L_p(D+C)$ and H_2 holds for $|\xi_1| + |\xi_2| + |\xi_3| < R$, R being a sufficiently large, but fixed, positive constant. In order to obtain the geometric convergence of our approximation sequence we rewrite equation (4.1) as

$$\begin{aligned} \Delta u - \alpha u &= f(x,y,u, \frac{\partial u}{\partial x}, \frac{\partial u}{\partial y}) - \alpha u \\ &= g(x,y,u, \frac{\partial u}{\partial x}, \frac{\partial u}{\partial y}) \ . \end{aligned} \tag{4.2}$$

where α is an arbitrary but fixed positive constant. It is easily seen that g satisfies the conditions H_1 and H_2 with f replaced by g. Now let $G(x,y;\xi,\eta)$ be the Green's function for $\Delta u - \alpha u$ in D and define the operators $\underset{\sim}{\Pi}_o$, $\underset{\sim}{\Pi}_1$, $\underset{\sim}{\Pi}_2$, and $\underset{\sim}{T}$ as follows:

$$(4.3)\quad u(x,y) = (\underset{\sim}{\Pi}_0\rho)(x,y) \equiv \iint_D G(x,y;\xi,\eta)\rho(\xi,\eta)d\xi d\eta$$

$$(4.4)\quad \frac{\partial u(x,y)}{\partial x} = (\underset{\sim}{\Pi}_1\rho)(x,y) \equiv \iint_D \frac{\partial}{\partial x} G(x,y;\xi,\eta)\rho(\xi,\eta)d\xi d\eta$$

$$(4.5)\quad \frac{\partial u(x,y)}{\partial y} = (\underset{\sim}{\Pi}_2\rho)(x,y) \equiv \iint_D \frac{\partial}{\partial y} G(x,y;\xi,\eta)\rho(\xi,\eta)d\xi d\eta$$

$$(4.6)\quad \Delta u - \alpha u = \rho(x,y) = (\underset{\sim}{T}_\rho)(x,y) = g(x,y,\Pi_0\rho,\Pi_1\rho,\Pi_2\rho)$$

where $\rho \in L_p(D+C)$ and the derivatives in equation (4.3)-(4.6) are to be interpreted in a generalized or Sobolev sense. In [16] it was shown that for D sufficiently small $\underset{\sim}{T}$ is a contraction mapping of a closed ball in $L_p(D+C)$ into itself and hence has a unique fixed point ρ in $L_p(D+C)$. The (generalized) solution of the Dirichlet problem is now given by

$$(4.7)\qquad u(x,y) = (\underset{\sim}{\Pi}_0\rho)(x,y)\ .$$

From the theory of integral operators whose kernel has a weak singularity and Sobolev's lemma we can conclude that $u(x,y)$ is continuously differentiable in $D + C$.

We will now construct a sequence of functions which converge geometrically to $u(x,y)$. Let $K(x,y;\xi,\eta) \equiv K(P,Q)$ be the kernel function ([8]) for $\Delta u - \alpha u$ in D. Then the Green's function $G(x,y;\xi,\eta) \equiv G(P,Q)$ can be represented as

$$(4.8)\quad G(P,Q) = \frac{1}{2\pi}\left[\log \overline{PQ} - \int_C \log \overline{PT}\,\frac{\partial K}{\partial n_T}(T,Q)\,ds_T\right]$$

where $\overline{PQ}$ denotes the distance from the point P to the point Q. Now let D_1 be a square containing D in its interior and let $S(P,Q)$ be Neumann's function for $\Delta u - \alpha u$ in D_1. Note that $S(P,Q)$ can be constructed in a variety of ways, including the method of images [16]. Following Bergman and Schiffer [8] we define $i^{(\nu)}(P,Q)$, $\nu = 1, 2, 3, \ldots,$ recursively by

$$(4.9) \qquad i^{(1)}(P,Q) = 4 \int_C S(Q,T) \frac{\partial S(T,P)}{\partial n_T} ds_T$$

$$(4.10) \quad i^{(\nu)}(P,Q) = -\int_C i^{(\nu-1)}(P,T) \frac{\partial i^{(1)}}{\partial n_T} (T,Q)\, ds_T \; ;$$

$$\nu \geqslant 2 \, .$$

We can then express the kernel function $K(P,Q)$ in terms of $i^{(\nu)}(P,Q)$ by ([8], p. 315)

$$(4.11) \qquad K(P,Q) = \sum_{m=0}^{\infty} \left[\sum_{\nu=0}^{m} (-1)^{\nu} \binom{m}{\nu} i^{(\nu+1)}(P,Q) \right]$$

and approximate Green's function by $G_N(X,y;\xi,\eta) \equiv G_N(P,Q)$ where

$$(4.12) \quad G_N(P,Q) = \frac{1}{2\pi} \left[\log \overline{PQ} - \int_C \log \overline{PT} \, \frac{\partial K_N(T,Q)}{\partial n_T} ds_T \right]$$

$$(4.13) \qquad K_N(P,Q) = \sum_{m=0}^{N} \sum_{\nu=0}^{m} (-1)^{\nu} \binom{m}{\nu} i^{(\nu+1)}(P,Q) \, .$$

Now define the operators $\underset{\sim}{\Pi}_0^{(N)}$, $\underset{\sim}{\Pi}_1^{(N)}$, $\underset{\sim}{\Pi}_2^{(N)}$, $\underset{\sim}{T}^{(N)}$ by

(4.14)

$$u^{(N)}(x,y) = (\underset{\sim}{\Pi}_0^{(N)}\rho)(x,y) \equiv \iint_D G_N(x,y;\xi,\eta)\rho(\xi,\eta)d\xi d\eta$$

(4.15)

$$\frac{\partial u^{(N)}(x,y)}{\partial x} = (\underset{\sim}{\Pi}_1^{(N)}\rho)(x,y) \equiv \iint_D \frac{\partial G_N(x,y;\xi,\eta)}{\partial x}\rho(\xi,\eta)d\xi d\eta$$

(4.16)

$$\frac{\partial u^{(N)}(x,y)}{\partial y} = (\underset{\sim}{\Pi}_2^{(N)}\rho)(x,y) \equiv \iint_D \frac{\partial G_N(x,y;\xi,\eta)}{\partial y}\rho(\xi,\eta)d\xi d\eta$$

(4.17)

$$\Delta u^{(N)} - \alpha u^{(N)} = \rho(x,y) = (\underset{\sim}{T}^{(N)}\rho)(x,y)$$

$$\equiv g(x,y,\underset{\sim}{\Pi}_0^{(N)}\rho,\underset{\sim}{\Pi}_1^{(N)}\rho,\underset{\sim}{\Pi}_2^{(N)}\rho)$$

where $\rho \in L_p(D+C)$ and the derivatives in equation (4.14)-(4.17) are to be interpreted in a generalized or Solobev sense. Again it can be shown ([23]) that if D is sufficiently small then $\underset{\sim}{T}^{(N)}$ is a contraction mapping of a closed ball in $L_p(D+C)$ into itself, and hence has a unique fixed point $\rho^{(N)}(x,y)$ in $L_p(D+C)$. Our candidate for an approximation to our original Dirichlet problem is now given by

$$u^{(N)}(x,y) = (\underset{\sim}{\Pi}_0^{(N)}\rho^{(N)})(x,y) . \tag{4.18}$$

We can again show that $u^{(N)}(x,y)$ is continuously differentiable in $D + C$ ([16]). Due to the particular choice of $S(P,Q)$ and the fact that $\alpha > 0$ it can in fact be shown ([16]) that the sequence $u^{(N)}(x,y)$ converges geometrically in $D + C$ to the solution of the Dirichlet problem for equation (4.1). More precisely we have the following theorem ([16]):

Theorem: Let $u(x,y)$ be the unique solution of the Dirichlet problem for equation (4.1), which exists for D sufficiently small. If the sequence $u^{(N)}(x,y)$, $N = 1, 2, 3, \ldots$ is defined by equations (4.9)-(4.18) then for D sufficiently small

$$\max_{(x,y)\in D+C} |u^{(N)}(x,y) - u(x,y)| = 0\left(\frac{1}{\lambda_1^{2N}}\right)$$

where $\lambda_1 > 1$ is the first eigenvalue of the Fredholm integral equation

$$\phi_\nu(P) = 2\lambda_\nu \int_C \frac{\partial S(P,Q)}{\partial n_Q} \phi_\nu(Q)\, ds_Q \;, \qquad P \in C \;.$$

An analogous theorem can be proved showing how to approximate solutions of Riquier's problem for higher order elliptic equations ([15]).

5. Nonlinear Equations in Two Independent Variables: The Cauchy Problem

In trying to solve free boundary problems by inverse methods it frequently becomes necessary to construct solutions to elliptic Cauchy problems (c.f. [20], Chapter 16). Garabedian and Lieberstein have used such an approach in a particularly elegant manner to study detached shock wave problems in fluid mechanics ([26]). Many problems in elasticity [31] also lend themselves to the use of inverse methods. Such problems in elasticity involve fourth order equations as opposed to the second order equations arising in fluid mechanics. With applications to elasticity in mind we outline below a constructive approach for solving the Cauchy problem

$$(5.1) \qquad \Delta^2 u = \tilde{g}(x,y,u,u_x,u_y,\Delta u, \frac{\partial \Delta u}{\partial x}, \frac{\partial \Delta u}{\partial y})$$

$$(5.2) \qquad \frac{\partial^\ell u(x,y)}{\partial n^\ell} = \tilde{\phi}_\ell(x+iy) \; ; \quad x + iy \in L \; ,$$

$$\ell = 0, 1, 2, 3$$

where L is a given analytic arc, n is the outward normal to L, and $\tilde{g}$, $\tilde{\phi}_\ell$, $\ell = 0, 1, 2, 3$, are assumed to have certain regularity properties to be described shortly. We first use a conformal mapping $z = f(\zeta)$ to map L onto a segment of the x-axis containing the origin. Since in the use of inverse methods the arc L is often assumed to be a portion of an algebraic curve, for example an ellipse, the mapping $z = f(\zeta)$ can be either calculated explicitly or approximated accurately by numerical methods [25]. Under this conformal transformation the Cauchy problem (5.1), (5.2) assumes the form

$$(5.3) \quad \Delta_1 \left(\frac{\Delta_1 u}{|f'(\zeta)|^2} \right) = g(\xi,\eta,u,u_\xi,u_\eta,\Delta_1 u, \frac{\partial \Delta_1 u}{\partial \xi}, \frac{\partial \Delta_1 u}{\partial \eta})$$

$$\frac{\partial^\ell u(\xi,0)}{\partial \eta^\ell} = \phi_\ell(\xi) \; ; \quad \ell = 0, 1, 2, 3$$

where $\zeta = \xi + i\eta$, $\Delta_1 = \frac{\partial^2}{\partial \xi^2} + \frac{\partial^2}{\partial \eta^2}$. In conjugate coordinates

$$\zeta = \xi + i\eta$$

$$\zeta^* = \xi - i\eta$$

equation (5.3), (5.4) become

(5.4)
$$\frac{\partial^2}{\partial\zeta\partial\zeta^*}\left(\frac{1}{|f'(\zeta)|^2}\frac{\partial^2 U}{\partial\zeta\partial\zeta^*}\right)$$
$$= G\left(\zeta,\zeta^*,U,\frac{\partial U}{\partial\zeta},\frac{\partial U}{\partial\zeta^*},\frac{\partial^2 U}{\partial\zeta\partial\zeta^*},\frac{\partial^3 U}{\partial\zeta^2\partial\zeta^*},\frac{\partial^3 U}{\partial\zeta\partial\zeta^{*2}}\right)$$
$$U(\zeta,\zeta^*) = \phi_0(\zeta) \;;\quad \zeta = \zeta^*$$

(5.5)
$$i^\ell\left(\frac{\partial}{\partial\zeta} - \frac{\partial}{\partial\zeta^*}\right)^\ell U(\zeta,\zeta^*) = \phi_\ell(\zeta) \;;\quad \zeta = \zeta^* \;;\quad \ell = 1,\,2,\,3$$

where $U(\zeta,\zeta^*) = u\left(\frac{\zeta+\zeta^*}{2},\frac{\zeta-\zeta^*}{2i}\right)$, $G\left(\zeta,\zeta^*,U,\ldots,\frac{\partial^3 U}{\partial\zeta^2\partial\zeta^*}\right)$ $=\frac{1}{16}g\left(\frac{\zeta+\zeta^*}{2},\frac{\zeta-\zeta^*}{2i},\ldots,4i\left(\frac{\partial}{\partial\zeta}-\frac{\partial}{\partial\zeta^*}\right)\frac{\partial^2 U}{\partial\zeta\partial\zeta^*}\right)$. We now assume that as a function of its first two arguments $G(\zeta,\zeta^*,Z_1,\ldots,Z_6)$ is holomorphic in a bicylinder $\theta\times\theta^*$ where $\theta^* = \{\zeta|\zeta^*\in\theta\}$, and as a function of its last six variables it is holomorphic in a sufficiently large ball about the origin. We further assume that θ is simply connected, contains the origin, is symmetric with respect to conjugation, i.e. $\theta = \theta^*$, and that $\phi_l(\zeta)$, $l = 0, 1, 2, 3$, are holomorphic for all $\zeta \in \theta$.

Now let

$$U^{(1)} = \frac{1}{|f'(\zeta)|^2} \frac{\partial^2 U}{\partial\zeta\partial\zeta^*} . \tag{5.6}$$

Then ([12]) we can define the operators $\underset{\sim}{A}_i$, $i = 1,2,3,4,5,6$, by

(5.7)

$$\underset{\sim}{A}_2(U^{(1)}) \equiv \frac{\partial U}{\partial \zeta} = \int_{\zeta}^{\zeta^*} |f'(\zeta)|^2 \, U^{(1)}(\zeta,\xi^*)d\xi^* + \frac{1}{2}\left[\frac{d\phi_0}{d\zeta}(\zeta) + i\phi_1(\zeta)\right]$$

(5.8)

$$\underset{\sim}{A}_3(U^{(1)}) \equiv \frac{\partial U}{\partial \zeta^*} = \int_{\zeta^*}^{\zeta} |f'(\zeta)|^2 U^{(1)}(\xi,\zeta^*)d\xi + \frac{1}{2}\left[\frac{d\phi_0}{d\zeta}(\zeta) + i\phi_1(\zeta)\right]$$

(5.9)

$$\underset{\sim}{A}_1(U)^{(1)}) \equiv U \equiv \int_{\zeta^*}^{\zeta} \left\{ \int_{\zeta}^{\zeta^*} |f'(\zeta)|^2 \, U^{(1)}(\xi,\xi^*)d\xi^* + \frac{1}{2}\left[\frac{d\phi_0^*}{d\xi}(\xi) - i\phi_1(\xi)\right] \right\} d\xi + \phi_0(\zeta) .$$

$$\underset{\sim}{A}_4(U^{(1)}) \equiv \frac{\partial^2 U}{\partial\zeta\partial\zeta^*} = |f'(\zeta)|^2 \, U^{(1)} \tag{5.10}$$

$$(5.11)\qquad \underset{\sim}{A}_5(U^{(1)}) \equiv \frac{\partial^3 U}{\partial\zeta^2\partial\zeta^*} = \frac{\partial}{\partial\zeta}\,[|f'(\zeta)|^2\, U^{(1)}] \quad ,$$

$$(5.12)\qquad \underset{\sim}{A}_6(U^{(1)}) \equiv \frac{\partial^3 U}{\partial\zeta\partial\zeta^{*2}} = \frac{\partial}{\partial\zeta^*}\,[|f'(\zeta)|^2\, U^{(1)}] \;.$$

The Cauchy problem (5.4), (5.5) now becomes

$$(5.13)\qquad \frac{\partial 2_U(1)}{\partial\zeta\partial\zeta^*} = G(\zeta,\zeta^*,\underset{\sim}{A}_1(U^{(1)}),\cdots,\underset{\sim}{A}_6(U^{(1)}))$$

$$U^{(1)}(\zeta,\zeta^*) = \phi_0^{(1)}(\zeta) \;; \qquad \zeta = \zeta^*$$

$$(5.14)\qquad i\left(\frac{\partial U^{(1)}}{\partial\zeta} - \frac{\partial U^{(1)}}{\partial\zeta^*}\right) = \phi_1^{(1)}(\zeta) \;; \qquad \zeta = \zeta^*$$

where $\phi_0^{(1)}$, $\phi_1^{(1)}$ can be computed from equation (5.6) and the original Cauchy data ϕ_ℓ, $\ell = 0, 1, 2, 3$ ([12]). We next define the operator $\underset{\sim}{B}$ by (also see [21], Chapter III)

$$(5.15)\qquad s(\zeta,\zeta^*) = \frac{\partial^2 U^{(1)}}{\partial\zeta\partial\zeta^*}$$

$$\underset{\sim}{B}(s) \equiv U^{(1)}(\zeta,\zeta^*) = \int_0^{\zeta}\int_0^{\zeta^*} s(\xi,\xi^*)\,d\xi^* d\xi + \int_0^{\zeta} \gamma(\xi)\,d\xi + \int_0^{\zeta^*} \psi(\xi^*)\,d\xi^* + \phi_0^{(1)}(0)$$

where [11]

(5.17)

$$\gamma(\zeta) = \frac{1}{2}\left[\frac{d\phi_0^{(1)}(\zeta)}{d\zeta} - i\phi_1^{(1)}(\zeta)\right] - \int_0^{\zeta} s(\zeta,\xi^*)\,d\xi^*$$

(5.18)
$$\psi(\zeta) = \frac{1}{2}\left|\frac{d\phi_0^{(1)}(\zeta)}{d\zeta} + i\phi_1^{(1)}(\zeta)\right| - \int_0^{\zeta} s(\xi,\zeta)d\xi \ .$$

Let $HB \equiv HB(\Delta\rho,\Delta\rho^*)$ be the Banach space of functions of two complex variables which are holomorphic and bounded in $\Delta\rho \times \Delta\rho^*$, $\Delta\rho = \{\zeta \mid |\zeta| < \rho\}$, $\Delta\rho^* = \{\zeta^* | \zeta^* \in \Delta\rho\}$, with norm

(5.19)
$$\|s\| = \sup_{\Delta\rho\times\Delta\rho} |s(\zeta,\zeta^*)| \ .$$

Finding a solution of the Cauchy problem (4.13), (4.14) is now equivalent to finding a fixed point in the Banach space HB of the operator $\underset{\sim}{T}:HB \rightarrow HB$ defined by

(5.20)
$$\underset{\sim}{T}s = G(\zeta,\zeta^*,\underset{\sim}{A}_1(B(s)),\cdots,\underset{\sim}{A}_6(B(s))) \ .$$

It was shown in [12] that due to the hypothesis imposed upon G that $\underset{\sim}{T}$ is a contraction mapping of a closed ball of HB into itself and hence has a unique fixed point $s(\zeta,\zeta^*)$. Equations (5.16) and (5.9) now allow us to construct $U(\zeta,\zeta^*)$, and use of the inverse mapping to $z = f(\zeta)$ yields the solution of our original Cauchy problem (5.1), (5.2). If equation (5.1) is linear, and one uses exponential majorization ([11], [21]), the above procedure yields global solutions.

Theorem: There exists a constructive procedure, suitable for analytic and numerical approximations, for solving the Cauchy problem (2.1), (2.2). Such a procedure is given explicitly by equations (5.3) - (5.20).

The Cauchy-Kowalewski Theorem also provides a method for constructing solutions to elliptic Cauchy problems. However, this method is not satisfactory for purposes of approximation theory of numerical computation. The difficulties which arise in devising efficient procedures for solving initial value problems for elliptic equations is due to the fact that such problems are improperly posed due to the lack of continuous dependence on the initial data ([20]). In our analysis this unstable dependence appears exclusively in the step where this data is extended to complex values of the independent variables ξ, η. When this can be done in an elementary way, for example by direct substitution via the conjugate-coordinate transformation, no instabilities will occur when one uses the contraction mapping operator $\underset{\sim}{T}$ to obtain approximations to the desired solution.

REFERENCES

1. Anselone, P. M., "Convergence and Error Bounds for Approximate Solutions of Integral and Operator Equations," Error in Digital Computation, Vol. 2, L. B. Rall, ed., John Wiley, New York, 1965, 231-252.
2. _______, "Uniform Approximation Theory for Integral Equations with Discontinuous Kernels," SIAM Journal on Numerical Analysis, Vol. 4, (2), 1967, 245-253.
3. Atkinson, K. E., "Extensions of the Nyström Method for the Numerical Solution of Linear Integral Equations of the Second Kind," MRC Technical Summary Report #686, August 1966.
4. _______, "The Numerical Solution of Fredholm Integral Equations of the Second Kind," SIAM Journal on Numerical Analysis, 4, (3), 337-348, 1967.
5. Aziz, A. K. and Hubbard, B., "Bounds on the Truncation Error by Finite Differences for the Goursat Problem," Math. Comp., 18, 1964, 19-35.

6. Bergman, S., Integral Operators in the Theory of Linear Partial Differential Equations, Springer, Berlin, 1961.
7. Bergman, S. and Herriot, J. G., "Numerical Solution of Boundary Value Problems by the Method of Integral Operators," Numer. Math., 7, 1965, 42-65.
8. Bergman, S. and Schiffer, M., Kernel Functions and Elliptic Differential Equations in Mathematical Physics, Academic Press, New York, 1953.
9. Bergman, S. and Schiffer, M., "Kernel Functions in the Theory of Partial Differential Equations of Elliptic Type," Duke Math. J., 15, 1948, 535-566.
10. Collatz, L., Functional Analysis and Numerical Mathematics, Academic Press, New York, 1966.
11. Colton, D., "Cauchy's Problem for Almost Linear Elliptic Equations in Two Independent Variables," J. Approx. Theory, 3, 1970, 66-71.
12. _______, "Cauchy's problem for Almost Linear Elliptic Equations in Two Independent Variables," II, to appear in J. Approx. Theory.
13. Colton, D. and Gilbert, R. P., "An Integral Operator Approach to Cauchy's Problem for $\Delta_{p+2}u(x)+F(x)u(x) = 0$," to appear in SIAM J. Math. Analysis.
14. _______, "New Results on the Approximation of Solutions to Partial Differential Equations: The Method of Particular Solutions," to appear in the proceedings of the Conference on the Analytic Theory of Differential Equations, Western Michigan University, May, 1970.
15. _______, "New Results on the Approximation of Solutions to Partial Differential Equations: Iterative Methods," to appear in the proceedings of the Conference on the Analytic Theory of Differential Equations, Western Michigan University, May, 1970.
16. _______, "Rapidly Convergent Approximations to Dirichlet's Problem for Semilinear Elliptic Equations," to appear.
17. Coddington, E. A. and Levinson, N., Theory of Ordinary Differential Equations, McGraw Hill, New York, 1955.
18. Diaz, J. B., "On an analogue of Euler-Cauchy Polygon Method for the Numerical Solution of $u_{xy} = f(x,y,u,u_x,u_y)$." Arch. Rational Mechanics Anal., 4, 1958, 357-390

19. Diaz, J. B., "On Existence, Uniqueness, and Numerical Evaluation of Solutions of Ordinary and Hyperbolic Differential Equations," Anali di matematica pura ed applicata, Serie IV, Tomo LII, (1960) 163-181.
20. Garabedian, P. R., Partial Differential Equations, John Wiley, New York, 1964.
21. Gilbert, R. P., Function Theoretic Methods in Partial Differential Equations, Academic Press, New York, 1969.
22. _______, "A Method of Ascent for Solving Boundary Value Problems," Bull. Amer. Math. Soc., 75, 1969, 1286-1289.
23. _______, "The Construction of Solutions for Boundary Value Problems by Function Theoretic Methods," SIAM J. Math. Anal., 1, 1970, 96-114.
24. _______, "Integral Operator Methods for Approximating Solutions of Dirichlet Problems," to appear in the proceedings of the conference "Numerische Methoden der Approximationationstheorie" at the Mathematisches Forchungsinstitut, Oberwolfach, Germany, June, 1969.
25. Gaier, D., Konstruktive Methoden der Konformen Abbildung, Springer-Verlag, Berlin, 1964.
26. Garabedian, P. R. and Lieberstein, H. M., "On the Numerical Calculation of Detached Bow Shock Waves in Hypersonic Flow," J. Aero. Sci., 25, 1958, 109-118.
27. Gilbert, R. P. and Lo, C. Y., "On the Approximation of Solutions of Elliptic Partial Differential Equations in Two and Three Dimensions," to appear.
28. Krzywoklocki, M. Z. v., "Bergman's and Gilbert's Operators in Elasticity, Electromagnetism, Fluid Dynamics, Wave Mechanics," Analytic Methods in Mathematical Physics, R. P. Gilbert and R. G. Newton eds., Gordon and Breach, New York, (1970), 207-247.
29. Muskhelishvili, N. I., Singular Integral Equations, Noordhoff, Groningen, 1953.
30. du Plessis, N., "Runge's Theorem for Harmonic Functions," J. London Math. Soc., 1, 1969, 404-408.
31. Sneddon, I. N. and Lowengrub, M., Crack Problems in the Classical Theory of Elasticity, John Wiley, New York, 1969.
32. Taylor, A. E., Introduction to Functional Analysis, John Wiley, New York, 1958.
33. Tjong, B. L., "Operators Generating Solutions of $\Delta_3\psi(x,y,z)+F(x,y,z)\psi(x,y,z) = 0$ and their properties," in Analytic Methods in Mathematical Physics, R. P. Gilbert

and R. G. Newton eds., Gordon and Breach, New York, (1970), 547-552.
34. Vekua. I. N., New Methods for Solving Elliptic Equations, John Wiley, New York, 1967.
35. _______, Generalized Analytic Functions, Pergamon Press, London, 1962.

A NEW DIFFERENCE SCHEME FOR PARABOLIC PROBLEMS*

Herbert B. Keller**

1. Introduction

We shall illustrate a new difference scheme for parabolic mixed initial-boundary value problems in one space dimension by applying it to the special case:

(1.1)
$$\frac{\partial U}{\partial t} = \frac{\partial}{\partial x}\left(a(x)\frac{\partial U}{\partial x}\right) + c(x)\,U + S(x,t)\ , \quad a(x) \geq a_0 > 0\ ;$$

(1.2)
$$U(x,0) = g(x)\ , \quad 0 < x < 1\ ;$$

(1.3)
$$\left.\begin{aligned} &a)\quad \alpha_0 U(0,t) + \alpha_1 a(0) U_x(0,t) = g_0(t) \\ &b)\quad \beta_0 U(1,t) + \beta_1 a(1) U_x(1,t) = g_1(t) \end{aligned}\right\}\ , \quad t > 0\ .$$

* This work was supported by the U. S. Army Research Office, Durham, under Contract DAHC 04-68-C-0006.

** Applied Mathematics, Firestone Laboratories, California Institute of Technology, Pasadena, California.

Our scheme has a number of very desirable features: i) it is simple, easy to program, and efficient, ii) it is unconditionally stable, iii) it has second order accuracy with nonuniform nets, iv) both $U(x,t)$ and $\partial U(x,t)/\partial x$ are approximated with the same accuracy, v) Richardson or $h \to 0$ extrapolation is valid and yields two orders of accuracy improvement per extrapolation (with nonuniform nets), vi) it is A-stable (i.e., if the exact solution decays in time so does the numerical scheme, with approximately the same rate), vii) the data, coefficients and solution need only be piecewise smooth and all the above remain valid. The method is also applicable to parabolic systems, to nonlinear parabolic equations and even to some hyperbolic systems with special properties. These more general applications will be reported elsewhere with the details of the convergence and A-stability proofs. Here we shall present the method, indicate the error estimates, $h \to 0$ extrapolation and discuss in some detail an efficient algorithm for applying it to the problem (1.1)-(1.3). The scheme is obviously motivated by a very powerful difference method for solving boundary value problems for systems of ordinary differential equations [3].

A crucial step in the numerical procedure is to reformulate the problem (1.1)-(1.3) in terms of a first order system of partial differential equations. In fact, with a slight gain in generality we consider the problem:

$$\text{(1.4)} \quad \begin{aligned} &\text{a)} \qquad a(x)\frac{\partial U}{\partial x} = V + R(x,t) \;, \\ &\text{b)} \qquad \frac{\partial V}{\partial x} = \frac{\partial U}{\partial t} - c(x)U - S(x,t) \;; \end{aligned}$$

(1.5) $$U(x,0) = g(x) \;;$$

(1.6)
a) $$\alpha_0 U(0,t) + \alpha_1 V(0,t) = g_0(t) \;,$$

b) $$\beta_0 U(1,t) + \beta_1 V(1,t) = g_1(t) \;.$$

With $R(x,t) \equiv 0$ this problem is equivalent to that in (1.1)-(1.3). Note that derivatives no longer occur in the boundary conditions.

2. The Box Scheme

On the rectangle $R(T)$: $0 \leqslant x \leqslant 1$, $0 \leqslant t \leqslant T$ we place a net $R_h(T)$ of points (x_j, t_n) subject only to the constraints that

(2.1) a) $$x_0 = 0, \quad x_J = 1, \quad t_0 = 0, \quad t_N = T \;.$$

The net spacings are otherwise arbitrary and are denoted by (2.1)

b) $$h_j = x_j - x_{j-1}, \quad 1 \leqslant j \leqslant J; \quad k_n = t_n - t_{n-1}, \quad 1 \leqslant n \leqslant N.$$

For net functions $\{\phi_j^n\}$ and coordinates of the net we employ the notation:

(2.2) a) $$x_{j \pm \frac{1}{2}} = \frac{1}{2}(x_j + x_{j \pm 1}) \;,$$

$$t_{n \pm \frac{1}{2}} = \frac{1}{2}(t_n + t_{n \pm 1}) \;,$$

b) $$\phi^n_{j \pm \frac{1}{2}} = \frac{1}{2}(\phi_j^n + \phi^n_{j \pm 1}) \;,$$

$$\phi_j^{n \pm \frac{1}{2}} = \frac{1}{2}(\phi_j^n + \phi_j^{n \pm 1}) \;,$$

$$\text{c)} \qquad D_x^- \phi_j^n = h_j^{-1}(\phi_j^n - \phi_{j-1}^n) ,$$

$$D_t^- \phi_j^n = k_n^{-1}(\phi_j^n - \phi_j^{n-1}) .$$

For functions $\psi(x,t)$ defined on $R(T)$ we employ the notation

$$(2.2) \quad \text{d)} \qquad \psi_j^n = \psi(x_j, t_n) , \quad \psi_{j\pm\frac{1}{2}}^n = \psi(x_{j\pm\frac{1}{2}}, t_n) ,$$

$$\psi_j^{n\pm\frac{1}{2}} = \psi(x_j, t_{n\pm\frac{1}{2}}) .$$

The numerical approximation to (1.4)-(1.6) is defined in terms of net functions $\{u_j^n\}$ and $\{v_j^n\}$ by employing the obvious centered difference approximations on $R_h(T)$. We get from (1.4):

(2.3)

$$\left.\begin{array}{ll} \text{a)} & a_{j-\frac{1}{2}} D_x^- u_j^n = v_{j-\frac{1}{2}}^n + R_{j-\frac{1}{2}}^n \\ & \\ \text{b)} & D_x^- v_j^{n-\frac{1}{2}} = D_t^- u_{j-\frac{1}{2}}^n - c_{j-\frac{1}{2}} u_{j-\frac{1}{2}}^{n-\frac{1}{2}} - S_{j-\frac{1}{2}}^{n-\frac{1}{2}} \end{array}\right\} 1 \leqslant j \leqslant J , \quad 1 \leqslant n \leqslant N .$$

The initial data are taken as, from (1.5) and (1.4a):

$$(2.4) \qquad \text{a)} \qquad u_j^0 = g(x_j) ,$$

$$\text{b)} \qquad v_j^0 = a_j \frac{dg(x_j)}{dx} - R_j^0 , \quad 0 \leqslant j \leqslant J .$$

The boundary conditions are simply

$$\text{(2.5)}\quad \left.\begin{aligned} &\text{a)}\quad \alpha_0 u_0^n + \alpha_1 v_0^n = g_0^n \\ &\text{b)}\quad \beta_0 u_J^n + \beta_1 v_J^n = g_1^n \end{aligned}\right\} 1 \leq n \leq N .$$

If we wish to allow piecewise smooth initial data then (2.4b) is altered to, say,

$$v_{j-\frac{1}{2}}^0 = a_{j-\frac{1}{2}} \frac{dg(x_{j-\frac{1}{2}})}{dx} - R_{j-\frac{1}{2}}^0 , \qquad 1 \leq j \leq J .$$

Here $v_{j-\frac{1}{2}}^0$ is not an average, as in (2.2b), but is a net function defined on $\{t=0;\ x=x_{j-\frac{1}{2}},\ 1 \leq j \leq J\}$. We must also modify (2.3b) for $n=1$ in an obvious way. Now jump discontinuities in $a(x)$, $d\,g(x)/dx$, $R(x,0)$ or their higher order derivatives are restricted to lie on net points $x=x_j$. Similarly if the boundary data are only piecewise smooth then (2.5) is replaced by

$$\left.\begin{aligned} \alpha_0 u_0^{n-\frac{1}{2}} + \alpha_1 v_0^{n-\frac{1}{2}} &= g_0^{n-\frac{1}{2}} \\ \beta_0 u_J^{n-\frac{1}{2}} + \beta_1 v_J^{n-\frac{1}{2}} &= g_1^{n-\frac{1}{2}} \end{aligned}\right\} 1 \leq n \leq N .$$

All jump discontinuities of $g_0(t)$, $g_1(t)$ or their derivatives must now occur for values $t = t_n$. In the usual diffusion problems with discontinuous diffusivity (i.e. $a(x)$ has a jump discontinuity) our difference equation (2.3a) with $R(x,t) \equiv 0$ implies the continuity of $a(x)\partial U/\partial x$ provided the jumps occur only at net points, $x = x_j$.

3. Error Estimates

Let $U(x,t)$ and $V(x,t)$ be the solution of (1.4)-(1.6) and have piecewise continuous derivatives of all orders $\leq M$ on $R(T)$. Further we assume that the at most finite number of jump discontinuties occur only on coordinate lines through our net $R_h(T)$. We summarize these continuity requirements by the notation: $U,\ V \in PC_M[R_h(T)]$. Introducing the error net functions

(3.1)
$$e_j^n \equiv U(x_j,t_n)-u_j^n, \quad f_j^n \equiv V(x_j,t_n)-v_j^n, \quad \begin{cases} 0 \leq j \leq J \\ 0 \leq n \leq N \end{cases} ;$$

we find, in the usual way, that they satisfy the difference equations:

(3.2)
$$\left.\begin{aligned} &\text{a)} \quad a_{j-\frac{1}{2}} D_x^- e_j^n = f_{j-\frac{1}{2}}^n + \rho_{j-\frac{1}{2}}^n \\ &\text{b)} \quad D_x^- f_j^{n-\frac{1}{2}} = D_t^- e_{j-\frac{1}{2}}^n - c_{j-\frac{1}{2}} e_{j-\frac{1}{2}}^{n-\frac{1}{2}} - \sigma_{j-\frac{1}{2}}^{n-\frac{1}{2}} \end{aligned}\right\} 1 \leq j \leq J, \quad 1 \leq n \leq N ;$$

(3.3) a) $e_j^0 = 0$, b) $f_j^0 = 0$, $0 \leq j \leq J$;

(3.4)
$$\left.\begin{aligned} &\text{a)} \quad \alpha_0 e_0^n + \alpha_1 f_0^n = 0 \\ &\text{b)} \quad \beta_0 e_J^n + \beta_1 f_J^n = 0 \end{aligned}\right\} \quad 1 \leq n \leq N .$$

The local truncation errors have been defined by

$$\text{(3.5a)}\quad \rho^n_{j-\frac{1}{2}} \equiv a_{j-\frac{1}{2}} \left\{ D_x^- U(x_j,t_n) - \frac{\partial U(x_{j-\frac{1}{2}},t_n)}{\partial x} \right\}$$

$$+ \left\{ V(x_{j-\frac{1}{2}},t_n) - \frac{1}{2}\left[V(x_j,t_n)+V(x_{j-1},t_n) \right] \right\}$$

and

(3.5b)

$$\sigma^{n-\frac{1}{2}}_{j-\frac{1}{2}} = \left\{ \frac{\partial V(x_{j-\frac{1}{2}},t_{n-\frac{1}{2}})}{\partial x} - \frac{1}{2} D_x^- \left[V(x_j,t_n)+V(x_j,t_{n-1}) \right] \right\}$$

$$+ \left\{ \frac{1}{2} D_t^- \left[U(x_j,t_n)+U(x_{j-1},t_n) \right] - \frac{\partial u(x_{j-\frac{1}{2}},t_{n-\frac{1}{2}})}{\partial t} \right\}$$

$$+ c_{j-\frac{1}{2}} \left\{ U(x_{j-\frac{1}{2}},t_{n-\frac{1}{2}}) - \frac{1}{4}\left[U(x_j,t_n)+U(x_{j-1},t_n) + U(x_j,t_{n-1})+U(x_{j-1},t_{n-1}) \right] \right\}$$

We now apply the usual Taylor expansions to estimate the dependence of ρ and σ on the net spacing. But since $U,\ V \in PC_M[R_h(T)]$ and <u>no difference or averages in (3.5) occur across net lines</u> we may proceed as if $U,\ V \in C_M[R(T)]$. Thus we get if $M \geqslant 2m + 2$:

$$\text{a)}\quad \rho^n_{j-\frac{1}{2}} \equiv \rho^n_{j-\frac{1}{2}}\{U,V\} = \sum_{\nu=1}^{m} \left(\frac{h}{2}\right)^{2\nu} R_\nu\{U,V;x_{j-\frac{1}{2}},t_n\} + \mathcal{O}(h^{2m+2}),$$

(3.6)

$$\text{b)}\quad \sigma^{n-\frac{1}{2}}_{j-\frac{1}{2}} \equiv \sigma^{n-\frac{1}{2}}_{j-\frac{1}{2}}\{U,V\} = \sum_{\nu=1}^{m} \left(\frac{h}{2}\right)^{2\nu} S_\nu\{U,V;x_{j-\frac{1}{2}},t_{n-\frac{1}{2}}\} + \mathcal{O}(h^{2m+2}).$$

Here we have introduced $h \equiv \max_j h_j$. Then for some fixed $r > 0$: $\max_n k_n = rh$. Now let $\theta(x)$ and $\phi(t)$ be piecewise smooth functions such that for some $\delta > 0$:

(3.7)

$$\text{a)} \quad h_j = \theta(x_{j-\frac{1}{2}})h \ , \quad 1 \leqslant j \leqslant J \ , \quad \delta \leqslant \theta(x) \leqslant 1 \ , \quad 0 \leqslant x \leqslant 1 \ ;$$

$$\text{b)} \quad k_n = \phi(t_{n-\frac{1}{2}})h \ , \quad 1 \leqslant n \leqslant N \ , \quad \delta \leqslant \phi(t) \leqslant r \ , \quad 0 \leqslant t \leqslant T \ .$$

With the above functions we have

(3.8)

$$\text{a)} \quad R_\nu\{U,V;x,t\} \equiv \frac{\theta^{2\nu}(x)}{(2\nu)!}\left\{ \frac{a(x)}{2\nu+1}\frac{\partial^{2\nu+1}U(x,t)}{\partial x^{2\nu+1}} - \frac{\partial^{2\nu}V(x,t)}{\partial x^{2\nu}} \right\} ;$$

$$\text{b)} \quad S_\nu\{U,V;x,t\} \equiv \sum_{\mu=0}^{\nu} \frac{\phi^{2\mu}(t)\theta^{2\nu-2\mu}(x)}{(2\mu)!(2\nu-2\mu)!}\left\{ \frac{\partial^{2\nu+1}U(x,t)}{\partial x^{2\mu}\partial t^{2\nu-2\mu+1}} - \frac{1}{2\nu-2\mu+1}\frac{\partial^{2\nu+1}V(x,t)}{\partial x^{2\nu-2\mu+1}\partial t^{2\mu}} - c(x)\frac{\partial^{2\nu}U(x,t)}{\partial x^{2\nu-2\mu}\partial t^{2\mu}} \right\} .$$

Note that the expressions in (3.8) are defined relative to some fixed but arbitrary net, $R_h(T)$, by means of the functions $\theta(x)$ and $\phi(t)$.

The error estimates are obtained by introducing, for arbitrary net functions $\{\phi_j^n\}$ and $\{\psi_j^n\}$:

$$(3.9)\qquad \text{a)}\quad (\phi^n,\psi^n)_h \equiv \sum_{j=1}^{J} \phi^n_{j-\frac{1}{2}}\,\psi^n_{j-\frac{1}{2}}\,h_j \;;$$

$$\text{b)}\quad \|\phi^n\|_h^2 \equiv (\phi^n,\phi^n)_h \,.$$

Employing the identities

$$(3.10)\qquad \text{a)}\quad (D_x^-\phi,\psi)_h = [\phi_J\psi_J-\phi_o\psi_o] - (\phi,D_x^-\psi)_h \,,$$

$$\text{b)}\quad (D_t^-\,\phi^n,\phi^{n-\frac{1}{2}})_h = \frac{1}{2k_n}\left(\|\phi^n\|_h^2 - \|\phi^{n-1}\|_h^2\right) ,$$

and discrete Sobelev inequalities to estimate boundary terms we find from (3.2)-(3.4) that there exist constants independent of the net spacing such that:

(3.11)

$$\left.\begin{aligned} \|e^n\|_h &\leq K_1 \sup_{\nu\leq n} \left(\|\rho^{\nu-\frac{1}{2}}\|_h + \|\sigma^{\nu-\frac{1}{2}}\|_h \right) \\ \|f^n\| &\leq K_2 \sup_{\nu\leq n} \left(\|\rho^{\nu-\frac{1}{2}}\|_h + \|\sigma^{\nu-\frac{1}{2}}\|_h \right) \end{aligned}\right\} 1 \leq n \leq N \,.$$

Similar error estimates follow from the general theory of H.-O. Kreiss as extended by J. Varah [5] to parabolic mixed problems.

Unfortunately convergence does not quite follow from the above estimates since $\|\cdot\|_h$ is, for our net functions, a semi-norm rather than a norm. However, it is not difficult to show that $\|\phi\|_h = \|\psi\|_h$ if and only if $\phi_j = \psi_j + (-1)^j p$ for some constant p. Thus if $\|e^n\|_h = \|\tilde{e}^n\|_h$ and $\|f^n\|_h = \|\tilde{f}^n\|_h$ then $\tilde{e}_j^n = e_j^n + (-1)^j p$ and $\tilde{f}_j^n = f_j^n + (-1)q$ for some constants p and q. Since $\{e_j^n\}$ and $\{f_j^n\}$ satisfy the homogeneous boundary conditions (3.4) in order that $\{\tilde{e}_j^n\}$ and $\{\tilde{f}_j^n\}$ also satisfy these conditions we must have:

$$\alpha_o p + \alpha_1 q = 0$$

$$\beta_o p + \beta_1 q = 0 \ .$$

Thus we deduce that:

i) $p = q = 0$ if $\alpha_o\beta_1 - \alpha_1\beta_o \neq 0$,

ii) $p = 0$ if $\alpha_1\beta_1 = 0$,

iii) $q = 0$ if $\alpha_o\beta_o = 0$.

In case i) our semi-norm is also a norm for net functions satisfying (3.4). In case ii) or iii) we may have an oscillatory error of constant amplitude in v_j^n or u_j^n , respectively. (These errors do occur in extreme situations but are easily observed.) To eliminate them we need only average neighboring values; that is we define

(3.12) a) $\bar{u}^n_{j-\frac{1}{2}} \equiv \frac{1}{2}(u^n_j+u^n_{j-1})$

or

b) $\bar{v}^n_{j-\frac{1}{2}} \equiv \frac{1}{2}(v^n_j+v^n_{j-1}), \quad 1 \leqslant j \leqslant J .$

Then clearly $\|\bar{u}^n\|_h = \|u^n\|_h$ and $\|\bar{v}^n\|_h = \|v^n\|_h$ but $\|\cdot\|_h$ is actually a norm for net functions defined on $\{x_{j-\frac{1}{2}}\}_1^J$ The oscillation is thus removed and the estimates in (3.11) yield the appropriate error bounds. If the solution of (1.4)-(1.6) is in $PC_4[R(T)]$ it follows that the error is at least $\mathcal{O}(h^2)$.

4. Richardson Extrapolation

The error in calculations with any fixed net, of arbitrary spacing, can be reduced by Richardson or $h \to 0$ extrapolation. The justification for such procedures is an asymptotic expansion of the errors $\{e^n_j, f^n_j\}$. To derive this expansion we introduce the notation

(4.1)

$$\text{a)} \quad \underset{\sim}{U}(x,t) \equiv \begin{pmatrix} U(x,t) \\ V(x,t) \end{pmatrix}$$

$$\mathcal{L}\underset{\sim}{U}(x,t) \equiv \begin{pmatrix} a(x)U_x(x,t)-V(x,t) \\ V_x(x,t)-U_t(x,t)+c(x)U(x,t) \end{pmatrix} \quad ;$$

b) $$\underset{\sim}{u}_j^n \equiv \begin{Bmatrix} u_j^n \\ v_j^n \end{Bmatrix}$$

$$\mathcal{L}_h \underset{\sim}{u}_j^n \equiv \begin{Bmatrix} a_{j-\frac{1}{2}} D_x^- u_j^n - v_{j-\frac{1}{2}}^n \\ D_x^- v_j^{n-\frac{1}{2}} - D_t^- u_{j-\frac{1}{2}}^n + c_{j-\frac{1}{2}} u_{j-\frac{1}{2}}^{n-\frac{1}{2}} \end{Bmatrix} ;$$

c) $$\underset{\sim}{R}_\nu\{\underset{\sim}{U}\} \equiv \begin{Bmatrix} R_\nu\{U,V;x,t\} \\ S_\nu\{U,V;x,t\} \end{Bmatrix}$$

Let the functions $\underset{\sim}{e}^{(\nu)}(x,t) \equiv \begin{Bmatrix} e^{(\nu)}(x,t) \\ f^{(\nu)}(x,t) \end{Bmatrix}$ be defined as the solutions on $(x,t) \in R(T)$ of the systems:

(4.2)

$$\mathcal{L}\underset{\sim}{e}^{(1)}(x,t) = \underset{\sim}{R}_1\{\underset{\sim}{U}\}$$

$$\mathcal{L}\underset{\sim}{e}^{(2)}(x,t) = \underset{\sim}{R}_2\{\underset{\sim}{U}\} - \underset{\sim}{R}_1\{\underset{\sim}{e}^{(1)}\} ,$$

$$\vdots$$

$$\mathcal{L}\underset{\sim}{e}^{(m)}(x,t) = \underset{\sim}{R}_m\{\underset{\sim}{U}\} - \underset{\sim}{R}_{m-1}\{\underset{\sim}{e}^{(1)}\} - \dots - \underset{\sim}{R}_1\{\underset{\sim}{e}^{(m-1)}\} ;$$

subject to the initial and boundary conditions:

$$\text{(4.3)}\quad \text{a)}\quad \underset{\sim}{e}^{(\nu)}(x,0) = 0 \qquad 0 < x < 1$$

$$\text{b)}\quad \left.\begin{aligned} \alpha_0 e^{(\nu)}(o,t)+\alpha_1 f^{(\nu)}(0,t) &= 0 \\ \beta_0 e^{(\nu)}(1,t)+\beta_1 f^{(\nu)}(1,t) &= 0 \end{aligned}\right\} \quad 0 \leqslant t \leqslant T$$

$$\nu = 1,2, \ldots, m .$$

A routine but messy calculation now reveals that

$$\text{(4.4)}\qquad \mathcal{L}_h \left[\underset{\sim}{U} - \underset{\sim}{u} - \sum_{\nu=1}^{m} \left(\frac{h}{2}\right)^{2\nu} \underset{\sim}{e}^{(\nu)} \right]_j^n = \mathcal{O}(h^{2m+2})$$

provided U, $V \in PC_{2m+2}[R_h(T)]$. Thus we obtain as in the derivation of (3.11) that

(4.5)

$$\begin{aligned} \| U(x_j,t_n)-u_j^n - \sum_{\nu=1}^{m} \left(\frac{h}{2}\right)^{2\nu} e^{(\nu)}(x_j,t_n) \|_h &= \mathcal{O}(h^{2m+2}) , \\ \| V(x_j,t_n)-v_j^n - \sum_{\nu=1}^{m} \left(\frac{h}{2}\right)^{2\nu} f^{(\nu)}(x_j,t_n) \|_h &= \mathcal{O}(h^{2m+2}) . \end{aligned}$$

The expansions in (4.5) justify several extrapolation procedures. First let the fixed net $R_h(T)$ be subdivided so that <u>every interval</u> $[x_j,x_{j-1}]$ <u>and</u> $[t_n,t_{n-1}]$ <u>is subdivided into</u> $\mu + 1$ <u>equal subintervals</u>, say for $\mu = 1,2, \ldots, q$. We denote the resulting sequence of nets by

$R_{h(\mu)}(T)$ and the maximum spacial meshlength of each such net is

$$(4.6) \qquad h^{(\mu)} = \frac{h}{\mu+1}, \qquad \mu = 0,1, \ldots, q .$$

On this set of nets the relations (3.7) are retained, when h, h_j are replaced by $h^{(\mu)}$, $h_j^{(\mu)}$ if:

$$\theta(x) \equiv h_j/h \quad \text{on} \quad x \in (x_{j-1}, x_j)$$

and

$$\phi(t) \equiv k_n/h \quad \text{on} \quad t \in (t_{n-1}, t_n) .$$

Clearly $R_{h^{(o)}}(T) \subset R_{h^{(\mu)}}(T)$ for $\mu = 1,2, \ldots, q$ and thus at each point of $R_h(T)$ we can make $\frac{q(q+1)}{2}$ extrapolations to get a final accuracy which is $\mathcal{O}(h^{2q+2})$. (Details of the extrapolation procedure can be found in [3].) Note that the net $R_{h^{(\mu)}}(T)$ contains $[(\mu+1)J+1]$ $[(\mu+1)N+1]$ points. Thus the total number of computations for all $(q+1)$ nets is essentially

$$(4.7) \qquad M_q \equiv \left(\frac{q+1}{2}\right) \left\{ (q+2) \ [(\tfrac{2}{3}\, q+1)JN+J+N]+2 \right\}$$

times the number of computations per net point.

On the other hand if <u>only the spatial net is subdivided</u> as above for $\mu = 0,1$ then one extrapolation can be performed and a glance at (3.8) and some thought reveals that the error is reduced to $\mathcal{O}(h^4 + k^2)$. Similarly a single <u>refinement only in the time</u> net yields, after one extrapolation, an accuracy that is $\mathcal{O}(h^2 + k^4)$. Further refinements of the space or time mesh alone in either of these cases cannot improve the order of the error. Previous considerations of Richardson extrapolation for parabolic difference equations have been made by Batten, see [1].

The tremendous effectiveness of these extrapolation procedures has been borne out in numerous test calculations including cases with only piecewise C_2 solutions. Rather than employ extrapolation to get great accuracy, which of course is easily done, we stress that with what seem to be unreasonably crude nets one obtains sufficient accuracy for most practical purposes.

Our analysis shows that it is theoretically possible to use nonuniform refinements of the initially nonuniform net and retain the validity of the extrapolation procedure. All that is required is that the relations (3.7) remain valid on the set of refined nets for some fixed functions $\theta(x)$ and $\phi(t)$ which are in $PC_{2m+2}[R_h(T)]$. Of course if each net of the set is to be used for extrapolation they must each have a different maximum mesh length, $h^{(\mu)}$. This further implies that each net subinterval must be subdivided in each refinement but the spacing and even the number of subdivisions need not be uniform. This added flexibility of our scheme is also shared by the difference scheme in [3]. It may be significant in current attempts to devise accurate and efficient procedures by means of which the net is altered, during the calculations, in a manner determined by the solution (i.e. "adaptive difference methods").

5. Solution of the Difference Equations

We indicate two methods for solving the difference equations (2.3)-(2.5). However, only the second procedure is advocated for practical calculations as it is more stable and more efficient. The first procedure easily reveals the nonsingularity of the linear systems to be solved.

Define the quantities:

(5.1)

a) $$R_j^n \equiv \begin{pmatrix} 1 & -b_{j-\frac{1}{2}}^n \\ -d_{j-\frac{1}{2}}^n & 1 \end{pmatrix},$$

$$L_j^n \equiv \begin{pmatrix} 1 & b_{j-\frac{1}{2}}^n \\ d_{j-\frac{1}{2}}^n & 1 \end{pmatrix}, \quad \begin{cases} b_{j-\frac{1}{2}}^n \equiv \dfrac{h_j}{2a_{j-\frac{1}{2}}^n} \\ d_{j-\frac{1}{2}}^n \equiv \frac{1}{2}h_j c_{j-\frac{1}{2}}^{n-\frac{1}{2}} + \dfrac{h_j}{k_n} \end{cases};$$

b) $$\underset{\sim}{y}_j^n \equiv \begin{pmatrix} u_j^n \\ v_j^n \end{pmatrix}, \qquad \underset{\sim}{q}_{j-\frac{1}{2}}^n \equiv \begin{pmatrix} r_{j-\frac{1}{2}}^n \\ s_{j-\frac{1}{2}}^n \end{pmatrix};$$

c) $$r_{j-\frac{1}{2}}^n \equiv 2b_{j-\frac{1}{2}}^n R_{j-\frac{1}{2}}^n$$

$$s_{j-\frac{1}{2}}^n \equiv (v_{j-1}^{n-1} - v_j^{n-1}) + [d_{j-\frac{1}{2}}^n - 2\frac{h_j}{k_n}](u_{j-1}^{n-1} + u_j^{n-1}) - 2h_j S_{j-\frac{1}{2}}^{n-\frac{1}{2}} .$$

Now the difference equations (2.3), generalized to allow time dependent coefficients $a(x,t)$ and $c(x,t)$, can be written as

$$(5.2) \qquad R_j^n \underset{\sim}{y}_j^n - L_j^n \underset{\sim}{y}_{j-1}^n = \underset{\sim}{q}_{j-\frac{1}{2}}^n , \qquad 1 \leqslant j \leqslant J , \quad 1 \leqslant n \leqslant N .$$

The boundary conditions (2.5) can be written as

$$(5.3) \qquad B_o \underset{\sim}{y}_o^n + B_J \underset{\sim}{y}_J^n = \underset{\sim}{g}^n$$

where we have introduced:

$$(5.4) \qquad B_o \equiv \begin{pmatrix} \alpha_o & \alpha_1 \\ 0 & 0 \end{pmatrix} , \qquad B_J \equiv \begin{pmatrix} 0 & 0 \\ \beta_o & \beta_1 \end{pmatrix} ,$$

$$\underset{\sim}{g}^n \equiv \begin{pmatrix} g_o^n \\ g_1^n \end{pmatrix}$$

The coefficient matrix of the system (5.2), (5.3) has the block matrix form:

$$\begin{pmatrix} B_o & 0 & 0 & B_J \\ -L_1^n & R_1^n & \ddots & 0 \\ 0 & \ddots & -L_J^n & R_J^n \end{pmatrix} .$$

A solution algorithm for such a system is presented in Lemma 3.2 of [3]. It is valid in the present circumstances provided that: the L_j^n are nonsingular and that an appropriate homogeneous two point boundary value problem has only the trivial solution. This latter condition now may be stated as: the eigenvalue problem:

(5.5)

a) $$(a(x,t_n)\phi_x)_x + [c(x,t_{n-\frac{1}{2}}) - \frac{2}{k_n}]\phi + \lambda\phi = 0$$

b) $$\alpha_o\phi(0) + \alpha_1 a(0,t_n)\phi_x(0) = 0$$

c) $$\beta_o\phi(1) + \beta_1 a(1,t_n)\phi_x(1) = 0$$

does not have $\lambda = 0$ as an eigenvalue. If, for example, we have $\alpha_o\alpha_1 \leqslant 0$, $\beta_o\beta_1 \geqslant 0$ and $c(x,t) \leqslant 0$ then $\lambda \neq 0$ for all k_n sufficiently small. Clearly the L_j^n are nonsingular under the mild conditions that

$$h_j^2 \left| \frac{1}{k_n} + \frac{1}{2} c_{j-\frac{1}{2}}^{n-\frac{1}{2}} \right| < 2a_{j-\frac{1}{2}}^n .$$

The system (5.2), (5.3) is more naturally treated if the separated endpoint boundary conditions are taken as the first and last equations of the system. That is we write the system as:

(5.6) a) $$A^n \underset{\sim}{Y}^n = \underset{\sim}{Q}^n ,$$

where

b)

$$A^n \equiv \begin{pmatrix} \alpha_o\alpha_1 & 0\ 0 & \cdots & & 0 \\ -L_1^n & \ddots & R_1^n & \ddots & \\ & \ddots & \ddots & \ddots & \\ & & -L_J^n & \ddots & R_J^n \\ 0 & \cdots & 0\ 0 & & \beta_o\beta_1 \end{pmatrix} ;$$

$$\underset{\sim}{Y}^n \equiv \begin{pmatrix} \underset{\sim}{y}_o^n \\ \underset{\sim}{y}_1 \\ \vdots \\ \underset{\sim}{y}_J^n \end{pmatrix} ; \qquad \underset{\sim}{Q}^n \equiv \begin{pmatrix} g_o^n \\ \underset{\sim}{q}_{1/2}^n \\ \vdots \\ \underset{\sim}{q}_{J-1/2}^n \\ g_1^n \end{pmatrix} .$$

Now we partition A^n into the block tridiagonal form

(5.7)

a)

$$A^n \equiv \begin{pmatrix} A_o^n & C_o^n & & & 0 \\ B_1^n & A_1^n & C_1^n & & \\ & \ddots & \ddots & \ddots & \\ & & \ddots & \ddots & C_{J-1}^n \\ 0 & & & B_J^n & A_J^n \end{pmatrix} ;$$

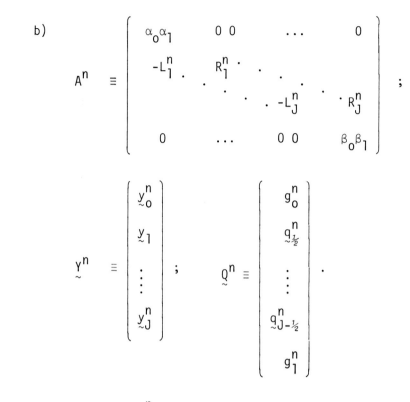

where:

$$A_o^n \equiv \begin{pmatrix} \alpha_o & \alpha_1 \\ -1 & -b_{1/2}^n \end{pmatrix} ; \qquad C_o^n \equiv \begin{pmatrix} 0 & 0 \\ 1 & -b_{1/2}^n \end{pmatrix} ;$$

b)

$$B_j^n \equiv \begin{pmatrix} -d_{j-1/2}^n & -1 \\ 0 & 0 \end{pmatrix} \qquad A_j^n \equiv \begin{pmatrix} -d_{j-1/2}^n & 1 \\ -1 & -b_{j+1/2}^n \end{pmatrix}$$

$$C_j^n \equiv \begin{pmatrix} 0 & 0 \\ 1 & -b_{j+1/2}^n \end{pmatrix} , \qquad 1 < j < J-1 ;$$

$$B_J^n \equiv \begin{pmatrix} -d_{J-1/2}^n & -1 \\ 0 & 0 \end{pmatrix} , \qquad A_J^n \equiv \begin{pmatrix} -d_{J-1/2}^n & 1 \\ \beta_o & \beta_1 \end{pmatrix} .$$

A standard algorithm (see [2], pp. 58-61) is now employed to solve (5.6a). Taking into account the zero elements in B_j^n and C_j^n simplifies this procedure further; the final formulas become in terms of some intermediate quantities $\{\xi_j,\eta_j\}$ and $\{w_{j,1}, w_{j,2}\}$:

<u>Forward sweep</u>:

(5.8)

$$\xi_0 = \alpha_0, \quad \eta_0 = \alpha_1$$

$$\left.\begin{aligned} \xi_{j+1} &= \Delta_j^{-1}(\xi_j - \eta_j d^n_{j+\frac{1}{2}}) - d^n_{j+\frac{1}{2}} \\ \eta_{j+1} &= -\Delta_j^{-1}(\xi_j - \eta_j d^n_{j+\frac{1}{2}}) b^n_{j+\frac{1}{2}} + 1 \end{aligned}\right\} \Delta_j \equiv \eta_j - \xi_j b^n_{j+\frac{1}{2}}, \quad 0 \leqslant j \leqslant J-1$$

$$\Delta_J = \beta_1 \xi_J - \beta_0 \eta_J$$

(5.9)

a)

$$w_{0,1} = -\Delta_0^{-1}(b^n_{\frac{1}{2}}\, g^n_0 + \eta_0 r^n_{\frac{1}{2}})$$

$$w_{0,2} = \Delta_0^{-1}(g^n_0 + \xi_0 r^n_{\frac{1}{2}})$$

b)

$$\left.\begin{aligned} w_{j,1} &= -\Delta_j^{-1}\Big[(b^n_{j+\frac{1}{2}} s^n_{j-\frac{1}{2}} + \eta_j r^n_{j+\frac{1}{2}}) \\ &\qquad + b^n_{j+\frac{1}{2}}(d^n_{j-\frac{1}{2}} w_{j-1,1} + w_{j-1,2})\Big] \\ w_{j,2} &= \Delta_j^{-1}\Big[(s^n_{j-\frac{1}{2}} + \xi_j r^n_{j-\frac{1}{2}}) \\ &\qquad + (d^n_{j-\frac{1}{2}} w_{j-1,1} + w_{j-1,2})\Big] \end{aligned}\right\} \quad 1 \leqslant j \leqslant J-1$$

c)

$$w_{J,1} = -\Delta_J^{-1}\Big[(-\beta_1 s^n_{J-\frac{1}{2}} + \eta_J g^n_1) - \beta_1(d^n_{J-\frac{1}{2}} w_{J-1,1} + w_{J-1,2})\Big]$$

$$w_{J,2} = \Delta_J^{-1}\Big[(-\beta_0 s^n_{J-\frac{1}{2}} + \xi_J g^n_1) - \beta_0(d^n_{J-\frac{1}{2}} w_{J-1,1} + w_{J-1,2})\Big]$$

Backward sweep:

(5.10)

a)

$$u_J^n = w_{J,1}$$

$$v_J^n = w_{J,2}$$

b)

$$\left.\begin{aligned} u_j^n &= w_{j,1} + \Delta_j^{-1}\eta_j(u_{j+1}^n - b_{j+\frac{1}{2}}^n v_{j+1}^n) \\ v_j^n &= w_{j,2} - \Delta_j^{-1}\xi_j(u_{j+1}^n - b_{j+\frac{1}{2}}^n v_{j+1}^n) \end{aligned}\right\} J-1 \geq j \geq 0 .$$

To insure the validity of the scheme (5.8)-(5.10) we need only show that all the principal minors, by 2 x 2 blocks, of A^n in (5.7) are nonsingular. Again this is assured if (J-1) appropriate eigenvalue problems do not have zero eigenvalues. Specifically the appropriate eigenvalue problems are (5.5a-c) and (5.5a,b) with the final boundary condition (5.5c), at $x = 1$, replaced by

(5.11) $$\phi(x_j) = 0 ; \quad j = 1,2, \ldots, J - 1 .$$

For many important classes of problems the monotone dependence of the eigenvalues on the interval length insure all of these conditions if only the problem (5.5a-c) has no zero eigenvalue. The details of the argument required above are contained in [4] and require in addition that the spacings h_j be sufficiently small. The idea is simply that if, for all sufficiently fine nets, the discrete problem is singular then the continuous problem has a zero eigenvalue.

To assess the efficiency of the scheme indicated in (5.8)-(5.11) with (5.1a,c) we compare with the operational count required in the Crank-Nicolson scheme, see [1], applied to the problem (1.1). Thus we set $R(x,t) \equiv 0$ and hence $r^n_{j+\frac{1}{2}} \equiv 0$ in (5.1c) and (5.9). It requires $[4J\mu + 4J\alpha]$ to evaluate all $\{\xi_j, \eta_j, \Delta_j\}$ in (5.8). Here μ is the time for a division or multiplication and α is the time for an addition. We have dropped odd operations not proportional to J. These are the "set-up" calculations which need be performed only once if $a(x,t)$ and $c(x,t)$ are independent of t and if the time steps, k_n, are constant. It requires $[7\,J\mu + 5\,J\alpha]$ to evaluate $\{u^n_j, v^n_j\}$ using (5.9) and (5.10). In addition $[2\,J\mu + 4\,J\alpha]$ are required to evaluate the source terms, $s^n_{j-\frac{1}{2}}$, in (5.1c).

The corresponding counts for Crank-Nicolson are: $[2J\mu + J\alpha]$ for the set-ups; $[3J\mu + 2J\alpha]$ to solve for $\{u^n_j\}$ and $[4J\mu + 3J\alpha]$ for the source terms. The totals are per time step:

$$[13J\mu + 13J\alpha] \text{ Box scheme; } [9J\mu + 6J\alpha] \text{ Crank-Nicolson.}$$

We hasten to recall however that C-N requires uniform spacing to retain its second order accuracy and does not approximate U_x to the same accuracy as U. Further if the set-ups are done only once (i.e. time independent coefficients and uniform time steps) the counts become:

$[9J\mu + 9J\alpha]$ Box scheme; $[7J\mu + 5J\alpha]$ Crank-Nicolson.

REFERENCES

1. Douglas, J., Jr., "Survey of Numerical Methods for Parabolic Differential Equations," Advances In Computers, Vol. II, Academic Press, N.Y. 1961, pp. 1-54.
2. Isaacson, E., and Keller, H. B., Analysis of Numerical Methods, John Wiley, New York, 1966.
3. Keller, H. B., "Accurate Difference Methods for Linear Ordinary Differential Systems Subject to Linear Constraints," SIAM Jour. on Num. Anal., 6 (1969), 8-30.
4. Kreiss, H.-O., "Numerical Solution of Ordinary Differential Equations," Lecture Notes, Computer Sci. Dept., Univ. of Uppsala, Uppsala, Sweden (1969).
5. Varah, J. M., "Stability of the Mixed Initial Boundary Value Problem for Parabolic Systems," to be published.

SINGULARITIES IN INTERFACE PROBLEMS*

R. B. Kellogg**

1. Introduction

In the first part of this paper we consider the interface problem for second order elliptic partial differential equations in two independent variables. The behavior of the solution is determined at the "singular" points of the problem: where two interface curves cross, where an interface curve meets the boundary, or where an interface or boundary has a discontinuous tangent (a corner). In the second part of the paper we apply these two dimensional results to analogous problems in three dimensions. We consider the Laplace operator on polyhedral domains with polyhedral interfaces.

Although there is a considerable literature on interface problems, there is little work on interfaces with singularities. Our methods are similar to those used by Birman and Skvorcov [1] in studying the effect of a corner on the

* This research was supported in part by National Science Foundation Grant NSF GP 9481 and in part by the Atomic Energy Commission under Grant AEC AT(40-1) 3443/3.

** University of Maryland.

solution of an elliptic boundary value problem. Some results on higher dimensional "corners" are contained in [2, 3].

Although the subject of the paper is not numerical analysis, the results of the paper have applications to the numerical solution of interface problems. Some of these will be considered in a later paper.

We will have occasion to use some of the usual Sobolev spaces $\overset{\circ}{W}{}^m_p(\Omega)$ and $W^m_p(\Omega)$, where Ω is a bounded open set in the plane or in Euclidean 3-space. If $m = 0$, we will write these spaces as $L_p(\Omega)$. We will write the norms on these spaces as $\|u\|_{W^m_p(\Omega)}$. If the domain Ω is clear from the context, we will write $\|u\|$ instead of $\|u\|_{L_2(\Omega)}$. We will let $C^\infty_0(\Omega)$ denote the set of infinitely differentiable functions u with compact support in Ω. Thus, $\overset{\circ}{W}{}^m_p(\Omega)$ is the closure in $W^m_p(\Omega)$ of the set $C^\infty_0(\Omega)$. We let ∇u and Δu denote respectively the gradient vector of u and the Laplacian of u. Throughout the paper c will denote a positive constant, not necessarily the same at each occurrence. Finally, all our functions will be real valued.

2. A Density Theorem in Two Dimensions

Let S be a disc of radius r_0 and center 0 in the (x_1,x_2) plane, let (r,θ) be polar coordinates in the (x_1,x_2) plane, and let p be a function on S which is a function of θ alone, and which takes on positive constant values in a finite number of intervals whose union is $[0,2\pi]$. It is well known that if $F \in L_2(S)$, the problem

$$(2.1) \qquad \int_S p\nabla u\cdot\nabla v\,dx = \int_S Fv\,dx\ , \quad \forall v \in \overset{\circ}{W}{}^1_2(S)$$

has a unique solution $u \in \overset{\circ}{W}{}^1_2(S)$. We let $\mathcal{D}(S,p)$ denote the set of all solutions u of (2.1) as F varies over $L_2(S)$, and we define the operator L_0 on $\mathcal{D}(S,p)$ by $L_0u = F$. It is not hard to see that L_0 is a densely defined, self adjoint, positive definite operator. Furthermore, $\mathcal{D}(S,p)$ is itself a Hilbert space with the norm $\|L_0u\|$, $u \in \mathcal{D}(S,p)$, and one has, for some $c > 0$,

$$(2.2) \qquad \|u\|_{W^1_2(S)} \leq c\|L_0u\| \ , \quad u \in \mathcal{D}(S,p) \ .$$

In this section we will determine $\mathcal{D}(S,p)$.

Let θ_j , $1 \leq j \leq J$, be the points of discontinuity of $p(\theta)$. The rays $\theta = \theta_j$ are called the <u>interfaces</u> of the problem (2.1). We denote by S_j the sector $\theta_{j-1} < \theta < \theta_j$, $0 < r < r_0$. Associated with (2.1) is the differential equation

$$(2.3) \qquad -p\Delta u = F\ , \quad \theta \neq \theta_j\ ,$$

the boundary condition

$$(2.4) \qquad u = 0 \quad \text{on} \quad r = r_0$$

and the interface conditions

$$(2.5) \qquad \left.\begin{aligned} u(r,\theta_j-0) &= u(r,\theta_j+0) \\ p(\theta-0)u_\theta(r,\theta_j-0) &= p(\theta+0)u_\theta(r,\theta_j+0) \end{aligned}\right\} \ 1 \leq j \leq J\ .$$

Let $\mathcal{W}_0(S,p)$ be the set of functions v in $\overset{\circ}{W}{}^1_2(S)$ which are continuous on S, which have all derivatives of order up to and including the third uniformly continuous in each region $\varepsilon \leq r \leq r_0$, $\theta_{j-1} \leq \theta \leq \theta_j$, for each $\varepsilon > 0$, which satisfy $rD^2v \to 0$ as $r \to 0$ for each second derivative D^2v of v, and which satisfy (2.5) for $u = v$ and $u = v_r$. With these assumptions, the following lemma is easily proved using Green's theorem.

<u>Lemma 2.1</u>. $\mathcal{W}_0(S,p) \subset \mathcal{D}(L_0,p)$; moreover, if $u \in \mathcal{W}_0(S,p)$, $L_0u = -p\Delta u$ in each sector S_j.

We let $\mathcal{W}(S,p)$ denote the closure of $\mathcal{W}_0(S,p)$ in the norm of $\mathcal{D}(S,p)$. The next lemma helps identify which functions are in $\mathcal{W}(S,p)$.

<u>Lemma 2.2</u>. If $u \in \mathcal{W}(S,p)$, the norm $\|L_0u\|$ is equivalent with the norm

$$\left\{\|u\|^2_{W^1_2(S)} + \sum_j \int_{S_j} (u^2_{x_1x_1} + u^2_{x_1x_2} + u^2_{x_2x_2})dx\right\}^{\frac{1}{2}} . \tag{2.6}$$

<u>Proof</u>. It suffices to prove the equivalence for $u \in \mathcal{W}_0(S,p)$. Letting A denote the expression (2.6), it is easy to see, using Lemma 2.1, that $\|L_0u\| \leq cA$. To see the converse, we use the integration formula

$$\int_{S_j}(\Delta u)^2 dx = \int_{S_j}(u^2_{rr} + 2r^{-2}u^2_{r\theta} + r^{-4}u^2_{\theta\theta})dx$$
$$+ 4\int_{S_j} r^{-3}u_ru_{\theta\theta}dx - 2\int_{\theta_{j-1}}^{\theta_j} r^{-1}u_ru_{\theta\theta}\Big]_0^{r_0} d\theta - 2\int_0^{r_0} r^{-1}u_ru_{r\theta}\Big]_{\theta_{j-1}}^{\theta_j} dr$$

Multiplying this by the constant value of p in the sector S_j, and summing over j, we get a formula for $\int_S p(\Delta u)^2 dx$. In this formula, the line integrals all cancel, by virtue of (2.4), (2.5) applied to u_r, and the fact that $r^{-2}u_{\theta\theta}$ is bounded at $r = 0$. Hence we obtain

$$\int_S p(\Delta u)^2 dx = \int_S p(u_{rr}^2 + 2r^{-2}u_{r\theta}^2 + r^{-4}u_{\theta\theta}^2 + 4r^{-2}u_r u_{\theta\theta}) r\,dr\,d\theta .$$

If the integrand on the right is written out in Cartesian coordinates, it is found to be equal to a positive definite quadratic form in the quantities $\partial^2 u/\partial x_j \partial x_k$, $1 \leqslant j$, $k \leqslant 2$, plus the expression

$$B = 2pr^{-2}u_\theta u_{r\theta} + 2pr^{-2}u_r u_{\theta\theta} - pr^{-3}u_\theta^2 - pr^{-1}u_r^2 .$$

Hence we obtain, also using (2.2),

$$A^2 \leqslant c\|L_0 u\|^2 + \int_S B\,dx . \tag{2.7}$$

Again using (2.5) we find that the integral of the first two terms in B vanishes, so

$$\int_S B dx \leqslant \int_S r^{-1}|\nabla u|^2 dx . \tag{2.8}$$

Now u is in W_2^2 on each sector S_j. Hence [4, pp. 106,

108], for any $\varepsilon > 0$, and $q < \infty$ there is a $c(\varepsilon) > 0$ such that

$$\left\{\int_S |\nabla u|^q \, dx\right\}^{1/q} \leqslant \varepsilon A^{\frac{1}{2}} + c(\varepsilon)\|u\|_{W_2^1} . \tag{2.9}$$

Using Hölder's inequality and (2.7), (2.8), (2.9), we deduce $A \leqslant c\|L_0 u\|^2$, which was to be shown.

Remark: It can be shown, conversely, that if $u \in \mathring{W}_2^1(S)$, $u \in W_2^2$ on each sector S_j, and if the traces of u satisfy (2.4) and (2.5), then $u \in \mathcal{W}(S,p)$.

We will see that the orthogonal complement of $\mathcal{W}(S,p)$ in the Hilbert space $\mathcal{D}(S,p)$ has finite dimension and that, using this, $\mathcal{D}(S,p)$ is easily determined. For this we consider the self adjoint Sturm-Liouville problem

$$(2.10) \quad \begin{cases} p\,\Theta'' + p\lambda\,\Theta = 0 , \quad \theta \neq \theta_j \\ \Theta(0) = \Theta(2\pi) \\ \Theta(\theta_j - 0) = \Theta(\theta_j + 0) \\ p(\theta_j - 0)\,\Theta'(\theta_j - 0) = p(\theta_j + 0)\,\Theta'(\theta_j + 0) , \quad 1 \leqslant j \leqslant J . \end{cases}$$

One may extend the usual results on Sturm-Liouville problems to the system (2.10). In particular, the following result is easily obtained.

Lemma 2.3. The problem (2.10) has a countable number of eigenvalues $0 = \lambda_0 < \lambda_1 \leqslant \lambda_2 \leqslant \cdots$, and eigenfunctions

$\Theta_n(\theta)$. The eigenfunctions are complete in $L_2(0,2\pi)$ and may be taken to satisfy

$$\int_0^{2\pi} p(\theta)\, \Theta_m(\theta)\, \Theta_n(\theta) d\theta = \begin{cases} 0 , & m \neq n \\ 1 , & m = n . \end{cases}$$

The eigenvalues are stationary values of the Rayleigh quotient

$$\frac{\int_0^{2\pi} p\, \Theta'^2\, d\theta}{\int_0^{2\pi} p\, \Theta^2\, d\theta} .$$

The nth eigenvalue $\lambda_n \to \infty$ and satisfies

$$c_1 n^2 \leqslant \lambda_n \leqslant c_2 n^2 , \quad n = 1,2, \cdots ,$$

where $c_1 > 0$, $c_2 > 0$ depend only on $p(\theta)$.

We set $a_n = \lambda_n^{\frac{1}{2}}$, $v_n = r^{a_n} \Theta_n(\theta)$, and we let $\zeta(r)$ be a C^∞ function which is $\equiv 1$ near $r = 0$ and $\equiv 0$ near $r = r_0$. Since $\Delta v_n = 0$ for $\theta \neq \theta_j$, it is not hard to see that $\zeta v_n \in \mathcal{D}(S,p)$, and $\zeta v_n \in \mathcal{W}_0(S,p)$ provided $a_n \geqslant 1$. We let $\mathcal{H}(S,p)$ be the linear span of the functions ζv_n with $a_n < 1$. Then we have

<u>Theorem 2.1</u>. $\mathcal{D}(S,p) = \mathcal{H}(S,p) + \mathcal{W}(S,p)$.

<u>Proof</u>. Since $\mathcal{H}(S,p)$ is finite dimensional, it is closed in the topology of $\mathcal{D}(S,p)$. Hence $\mathcal{H}(S,p) + \mathcal{W}(S,p)$ is a

closed subspace of $\mathcal{D}(S,p)$. To prove the theorem we will let $v \in \mathcal{D}(S,p)$ be perpendicular to $\mathcal{H}(S,p) + \mathcal{W}(S,p)$, and we will show that $v = 0$. Setting $f = L_0 v$, we will show that

$$(2.11) \qquad \int_S f\, L_0 u \, dx = 0 \ , \quad \forall u \in \mathcal{H}(S,p) + \mathcal{W}(S,p) \ , \Rightarrow f = 0 \ .$$

From (2.11) and Weyl's lemma [5, p. 182] it is seen that f is harmonic in each sector S_j . Also, from an interface form of Weyl's lemma [6, Lemma 2.3], it is seen that f has uniformly continuous second derivatives in each subset of S_j bounded away from $r = 0$ and $r = r_0$, and f satisfies the interface conditions (2.5). As in [6, Theorem 4.1], or in Theorem 5.1 of this paper in the more complicated 3-dimensional case, we may obtain the expansion

$$(2.12) \qquad f = \sum_0^\infty R_m(r)\, \Theta_m(\theta) \ ,$$

$$R_0(r) = \alpha_0 + \beta_0 \,\ell n r \ ,$$
$$R_m(r) = \alpha_m r^{a_m} + \beta_m r^{-a_m} \ , \quad m = 1,2, \cdots \ .$$

The expansion is convergent in $L_2(S)$. Set $u = \psi(r)\, \Theta_m(\theta)$ in (2.11), where $\psi(r)$ is a smooth function which is $\equiv 0$ near $r = 0$ and $\equiv r_0 - r$ near $r = r_0$. Then $u \in \mathcal{W}_0(S,p)$ and from the formula

$$p\Delta u = p\, \Theta_m(\theta)\, [r^{-1}(r\psi')' - \lambda_m r^{-2}\, \psi] \ ,$$

the expansion (2.12), and Lemma 2.3, we obtain

$$0 = \int_0^{r_0} (\alpha_m r^{a_m} + \beta_m r^{-a_m})((r\psi')' - \lambda_m r^{-1} \psi)dr$$

$$= \alpha_m r_0^{1+a_m} + \beta_m r_0^{1-a_m} , \quad m = 1,2, \cdots .$$

On the other hand, since $\zeta v_m \in \mathcal{H}(S,p) + \mathcal{W}(S,p)$, we have with $\psi = \zeta r^{a_m}$,

$$0 = \int_0^{r_0} (\alpha_m r^{a_m} + \beta_m r^{-a_m}) [(r\psi')' - \lambda_m r^{-1} \psi]dr$$

$$= -\beta_m(1+a_m) , \quad m = 1,2, \cdots .$$

Hence $\alpha_m = \beta_m = 0$, $m = 1,2, \cdots$, and we have $f = \alpha_0 + \beta_0 \ell n\, r$. Now setting $u = \zeta(r)$,

$$0 = \int_0^{r_0} (\alpha_0 + \beta_0 \ell n r)(r\zeta')'dr = \beta_0 ,$$

and setting $u = \psi(r) \equiv 0$ near $r = 0$ and $\equiv r_0 - r$ near $r = r_0$, $0 = -\alpha_0 r_0$. Hence $f = 0$ and the theorem is proved.

<u>Remark</u>. From the above proof one may see that the solution u of (2.1) is in $\mathcal{W}(S,p)$ if and only if F satisfies the orthogonality conditions

(2.13) $$\int_S v_n F dx = 0 , \quad \text{if } 0 < \lambda_n < 1 .$$

The above theorem may be obtained in the case that S is replaced by a sector S_0. Let S_0 be given by the inequalities $0 < \theta < \theta_J$, and let S_0 be divided into sectors $S_j: \theta_{j-1} < \theta < \theta_j$, $1 \leq j \leq J$, in each of which p is a constant. Then with S replaced by S_0, we may define $\mathcal{D}(S_0,p)$, $L_0: \mathcal{D}(S_0,p) \to L_2(S_0)$, and $\mathcal{W}(S_0,p)$. We may note that functions in $\mathcal{D}(S_0,p)$ now vanish on $\theta = 0$, $\theta = \theta_J$, as well as the curved part of S_0. In place of (2.10) we consider the Sturm-Liouville problem

$$(2.14)\quad \begin{cases} p\Theta'' + p\lambda\Theta = 0 , & \theta \neq \theta_j , \\ \Theta(0) = \Theta(\theta_J) = 0 , & \\ \Theta(\theta_j-0) = \Theta(\theta_j+0) , & 1 \leq j < J \\ p(\theta_j-0)\,\Theta'(\theta_j-0) = p(\theta_j+0)\,\Theta(\theta_j+0) , & 1 \leq j < J , \end{cases}$$

with eigenvalues $\lambda_1 \leq \lambda_2 \leq \cdots$, $\lambda_1 > 0$, and we let $\mathcal{H}(S_0,p)$ be spanned by the finite set of functions $\zeta(r)r^{a_n}\,\Theta_n(\theta)$ with $0 < \lambda_n < 1$. As above, we may prove the following result.

<u>Theorem 2.2</u>. $\mathcal{D}(S_0,p) = \mathcal{H}(S_0,p) + \mathcal{W}(S_0,p)$. If $u \in \mathcal{W}(S_0,p)$, the norm $\|L_0u\|$ is equivalent with the expression

$$(2.15)\quad \left\{\|u\|^2_{W^1_2} + \sum_{j=1}^{J} \int_{S_j} (u^2_{x_1x_1} + u^2_{x_1x_2} + u^2_{x_2x_2})dx\right\}^{\frac{1}{2}} .$$

<u>Remark</u>. The eigenvalue problems (2.10) and (2.14) are studied in more detail in [6]. In particular it is found

that λ_1 may be arbitarily close to 0, but, in the case (2.10) with only two sectors, S_1, S_2, or in the case (2.14) with p = constant, one has $\lambda_1 > \frac{1}{2}$. In the latter case, if $\theta_J \leqslant \pi$, then $\lambda_1 \geqslant 1$ and $\mathcal{H}(S_0,p)$ is empty.

The space $\mathcal{H}(S,p)$ is not uniquely determined by the problem. By modifying it, one may for example replace $\mathcal{H}(S,p)$ by the orthogonal complement of $\mathcal{W}(S,p)$ in the inner product of $\mathcal{D}(S,p)$.

3. The Method of Continuation

Let there be given on the closed disc $\bar{S}$ a continuously differentiable, symmetric, positive definite second order matrix function $(a_{jk}(x))$, and a bounded nonnegative function $a_0(x)$. Let p be the piecewise constant function considered in Section 2. We consider the problem

$$(3.1) \qquad \int_S \left[\sum_{j,k=1}^{2} p a_{jk} u_{x_j} v_{x_k} + a_0 \, uv\right] dx = \int_S Fv dx \,, \quad \forall v \in \mathring{W}_2^1(S) \,.$$

As in Section 2, it is easily seen that (3.1) has, for each $F \in L_2(S)$, a unique solution $u \in \mathring{W}_2^1(S)$. The collection of all such weak solutions is denoted by $\mathcal{D}(S,L)$, and an unbounded, linear operator $L: \mathcal{D}(S,L) \to L_2(S)$ is defined by $Lu = F$ where u and F are related by (3.1). As in Section 2, $\mathcal{D}(S,L)$ is dense in $L_2(S)$, L is closed and self adjoint, and for some $c > 0$,

(3.2) $$\|u\|_{W_2^1(S)} \leqslant c\|Lu\| \ , \quad u \in \mathcal{D}(S,L) \ .$$

Also, $\mathcal{D}(S,L)$ is a Hilbert space with the norm $\|Lu\|$, $u \in \mathcal{D}(S,L)$.

In this section we assume that the $a_{jk}(x)$ satisfy

(3.3) $$a_{jk}(x) = \delta_{jk} \quad \text{on the rays} \quad \theta = \theta_j \ , \ r > 0 \ ,$$

and we show that $\mathcal{D}(S,L) = \mathcal{D}(S,p)$. The condition (3.3) will be removed in the next section.

Our proof will involve the method of continuation. For this we define $a_{jk}(x,t) = ta_{jk}(x) + (1-t)\delta_{jk}$, $a(x,t) = ta(x)$, and we consider (3.1) with the functions $a_{jk}(x,t)$ and $a(x,t)$ instead of $a_{jk}(x)$ and $a(x)$. The corresponding operator and domain are denoted L_t and $\mathcal{D}_t$. These functions satisfy (3.3), and we have for some $c > 0$,

(3.4) $$\|u\|_{W_2^1(S)} \leqslant c\|L_t u\| \ , \quad u \in \mathcal{D}_t \ , \ 0 \leqslant t \leqslant 1 \ .$$

The same argument that gave Lemma 2.1 gives

<u>Lemma 3.1</u>. $\mathcal{W}_0(S,p) \subset \mathcal{D}_t$; moreover, if $u \in \mathcal{W}_0(S,p)$, then

(3.5) $$F = L_t u = -p \sum_{j,k} \frac{\partial}{\partial x_j} \left(a_{jk}(x,t) \frac{\partial u}{\partial x_k}\right) + a(x,t)u$$

in each sector S_j.

Also, an argument similar to that of Lemma 2.2 gives

Lemma 3.2. $\mathcal{W}(S,p) \subset \mathcal{D}_t$. Also, if $u \in \mathcal{W}(S,p)$, the norm $\|L_t u\|$ is equivalent with the norm (2.6) with a constant independent of $t \in [0,1]$.

Lemma 3.3. $\mathcal{H}(S,p) \subset \mathcal{D}_t$.

Proof. For $u = \zeta v_n$, let F be given by (3.5) in each sector S_j . Since $a_{jk}(x,t)$ is continuously differentiable and tends to δ_{jk} as $x \to 0$, it is seen by a calculation that $F \sim r^{a_n - 1}$ as $x \to 0$; hence $F \in L_2(S)$. Since u satisfies the interface conditions (2.5), use of Green's theorem and (3.3) shows that u satisfies (3.1), so $u \in \mathcal{D}_t$ as was to be shown.

The following simple lemma will be used several times.

Lemma 3.4. Let $t_m \in [0,1]$ with $t_m \to t$. Let $u_m \in \mathcal{D}_{t_m}$ with $L_{t_m} u_m \rightharpoonup F \in L_2(S)$, the one-sided arrow denoting weak convergence in $L_2(S)$. Suppose $u_m \to u \in \mathring{W}_2^1(S)$. Then $u \in \mathcal{D}_t$ and $L_t u = F$.

Proof. Setting $F_m = L_{t_m} u_m$, we have for each $v \in \mathring{W}_2^1(S)$,

$$\int_S \left[\sum p a_{jk}(x,t_m) u_{m,x_j} v_{x_k} + a(x,t) u_m v\right] dx = \int_S F_m v dx .$$

The right side tends to $\int_S Fv\,dx$. Using the continuity of the coefficients with respect to t and the convergence of u_m to u , the left side is seen to converge to the corresponding expression with the m's deleted. Hence $u \in \mathcal{D}_t$ and $L_t u_t = F$, as asserted.

Lemma 3.5. There is a γ , $0 < \gamma < 1$, such that

$$|(L_t w, L_t v)| \leq \gamma \|L_t w\| \cdot \|L_t v\| , \tag{3.6}$$

for all $w \in \mathcal{H}(S,p)$, $v \in \mathcal{W}(S,p)$, and $t \in [0,1]$.

Proof. The inequality (3.6) certainly holds with $\gamma = 1$, and since $\mathcal{H}(S,p)$ and $\mathcal{W}(S,p)$ are disjoint closed subspaces of $\mathcal{D}_t$, it certainly holds, for t fixed, with some $\gamma < 1$. Assume the lemma is false. Then there are sequences of numbers $t_m \in [0,1]$, operators $L_m = L_{t_m}$, functions $w_m \in \mathcal{H}(S,p)$ with $\|L_m w_m\| = 1$, functions $v_m \in \mathcal{W}(S,p)$ with $\|L_m v_m\| = 1$, and numbers $\gamma_m < 1$ with $\gamma_m \to 1$, such that

$$(L_m w_m, L_m v_m) = \gamma_m . \tag{3.7}$$

By picking subsequences we may assume that $t_m \to t \in [0,1]$, $L_m v_m \rightharpoonup G \in L_2(S)$, and $L_m w_m \rightharpoonup H \in L_2(S)$. Since $\mathcal{H}(S,p)$ is a finite dimensional subspace, we have $w_m \to w$ in $W_2^1(S)$, $L_m w_m \to H$ in $L_2(S)$, and $w \in \mathcal{H}(S,p)$. By Lemma 3.4, $L_t w = H$. By Lemma 3.2, v_m is a bounded sequence in $W_2^2(S_j)$ for each sector S_j . Hence, picking subsequences if necessary, the v_m form a Cauchy sequence in $W_2^1(S_j)$ for each j. Hence, the v_m form a Cauchy sequence in $\overset{\circ}{W}{}_2^1(S)$, so $v_m \to v$ in $\overset{\circ}{W}{}_2^1(S)$. Now $\mathfrak{W}(S,p)$ with the norm (2.15) forms a Hilbert space. Since v_m converges weakly in this space, $v \in \mathfrak{W}(S,p)$. From Lemma 3.4, $L_t v = G$.

Now as m increases, the left side of (3.7) tends to (Lw, Lv), for

$$
\begin{aligned}
&|(L_m w_m, L_m v_m) - (Lw, v)| \\
&\leqslant |(L_m w_m - Lw\ L_m v_m)| \\
&\quad + |(Lw, L_m v_m - Lv)| \\
&\leqslant \|L_m w_m - Lw\| + |(Lw, L_m v_m - Lv)| \\
&\to 0 .
\end{aligned}
$$

Hence $(L_t w, L_t v) = 1 = \|L_t w\| \cdot \|L_t v\|$, which implies that $w = cv$. This is impossible, and the lemma is proved.

Lemma 3.6. $\mathcal{D}_0 \subset \mathcal{D}_t$, and for fixed $u \in \mathcal{D}_0$, the map $t \to L_t u$ is a continuous map of $[0,1] \to L_2(S)$.

Proof. From Theorem 2.1, Lemma 3.2, and Lemma 3.3, $\mathcal{D}_0 = \mathfrak{W}(S,p) + \mathcal{H}(S,p) \subset \mathcal{D}_t$. Now for $u \in \mathcal{D}_0$ and using (3.5), we have in each sector S_j,

$$L_s u - L_t u = (s-t)\left\{p \sum_{j,k} (a_{jk}(x)-\delta_{jk})u_{x_j x_k}(x) + a(x)u(x)\right\}$$

$$= (s-t)\left\{L_1 u - L_0 u\right\} .$$

Since $L_1 u \in L_2(S)$, $L_0 u \in L_2(S)$, the result follows.

We are now able to give the main theorem of the section.

Theorem 3.1. For $t \in [0,1]$, $\mathcal{D}_t = \mathcal{H}(S,p) + \mathcal{W}(S,p)$. Furthermore, there is a $c > 0$ such that if $u = \mathcal{D}_t$, so $u = v + w$, $v \in \mathcal{H}(S,p)$, $w \in \mathcal{W}(S,p)$, then

$$\|L_t v\|^2 + \|w\|^2_{W_2^1} + \sum_j \int_{S_j} \left[(w^2_{x_1x_1} + w^2_{x_1x_2} + w^2_{x_2x_2})\right] dx \tag{3.8}$$

$$\leq c\|L_t u\|^2 .$$

Proof. Let $A \subset [0,1]$ be the set of t such that $\mathcal{D}_t = \mathcal{D}_0$. Then $0 \in A$. We will show that $A = [0,1]$ by showing that A is both open and closed. To show that A is open, let $t \in A$, and suppose, to the contrary, that there is a sequence of $t_m \notin A$, $t_m \to t$. Since $\mathcal{H}(S,p)$ is finite dimensional, for $s \in [0,1]$ $L_s\mathcal{H}(S,p)$ is a closed subspace of $L_2(S)$. Using Lemma 3.2, it may be seen that $L_s\mathcal{W}(S,p)$ is a closed subspace of $L_2(S)$. Hence $L_s\mathcal{D}_0$ is a closed subspace of $L_2(S)$, for each $s \in [0,1]$. Hence there are functions $z_m \in L_2(S)$, $\|z_m\| = 1$, such that

$$\int_S z_m L_{t_m} u dx = 0 , \quad \forall u \in \mathcal{D}_0 . \tag{3.9}$$

By choosing a subsequence if necessary, we may suppose that $z_m \rightharpoonup z \in L_2(S)$. Let $u \in \mathcal{D}_0$ satisfy $L_t u = z$. Then

$$1 = \int_S |z|^2 \, dx = \int_S z \, L_t u \, dx$$

$$= \int_S (z - z_m) L_t u \, dx + \int_S z_m (L_t u - L_{t_m} u) dx$$

$$+ \int_S z_m \, L_{t_m} u \, dx \quad .$$

The first term on the right tends to 0 by weak convergence of the z_m to z . The second term tends to 0 by Lemma 3.6. The last term is zero for each m by (3.9). This gives a contradiction and hence A is open.

We now show that A is closed. Let $t_m \in A$, $t_m \to t$. To show that $t \in A$, we will show that $L_t u = F$ may be solved for $u \in \mathcal{D}_0$, for each $F \in L_2(S)$. Since $t_m \in A$, we know that there is $u_m \in \mathcal{D}_0$ with $L_m u_m = F$, where $L_m = L_{t_m}$. Write $u_m = v_m + w_m$, $v_m \in \mathcal{H}(S,p)$, $w_m \in \mathcal{W}(S,p)$. Then using Lemma 3.5,

$$\|F\|^2 = \|L_m v_m\|^2 + 2(L_m v_m, L_m w_m) + \|L_m w_m\|^2 \tag{3.10}$$

$$\geq (1-\gamma)\left\{\|L_m v_m\|^2 + \|L_m w_m\|^2\right\} \quad .$$

Hence $\|L_m v_m\|$ and $\|L_m w_m\|$ are bounded sequences. As in the proof of Lemma 3.5 we may, by picking subsequences, assume that for some $v \in \mathcal{H}(S,p)$, $w \in \mathcal{W}(S,p)$,

$$\|v-v_m\|_{W_2^1} \to 0 \ , \quad \|w-w_m\|_{W_2^1} \to 0 \ ,$$

and the sequences $L_m v_m$ and $L_m w_m$ converge weakly in $L_2(S)$. From Lemma 3.4, we obtain $L_m v_m \rightharpoonup Lv$, $L_m w_m \rightharpoonup Lw$, so $L(v+w) = F$. Since $v + w \in \mathcal{D}_0$, this shows that A is closed, so $A = [0,1]$.

It remains to derive (3.8). Pick $u \in \mathcal{D}_0$, so $u = v + w$, $v \in \mathcal{H}(S,p)$, $w \in \mathcal{W}(S,p)$. As in (3.10), we find that

$$\|L_t v\|^2 + \|L_t w\|^2 \leqslant c\|L_t u\|^2 \ .$$

Inequality (3.8) then follows from Lemma 3.2.

. Theorem 3.1 may also be obtained if the disc S is replaced by a sector S_0 defined by the inequalities $0 < \theta < \theta_J$, $0 < r < r_0$. As before, we suppose that S_0 is divided into sectors S_j on which p is a constant. We denote by $\mathcal{D}(S_0,L)$ the set of all solutions of (3.1), with S replaced by S_0, and with $F \in L_2(S_0)$. Then we have

Theorem 3.2. $\mathcal{D}(S_0,L) = \mathcal{H}(S_0,p) + \mathcal{W}(S_0,p)$. Furthermore, there is a $c > 0$ such that if $u \in \mathcal{D}(S_0,L)$, so $u = v+w$, $v \in \mathcal{H}(S_0,p)$, $w \in \mathcal{W}(S_0,p)$, then (3.8) holds, where the norms are taken over S_0.

4. The General Equation in the Plane

We now consider a general domain Ω whose boundary consists of a finite number of C'' arcs which meet at non-zero angles. Let Ω be partitioned into a finite number of subdomains, Ω_j, by a finite number of curves, called interfaces. It is assumed that the interfaces meet one another and the boundary $\partial\Omega$ at non-zero angles. Also, it is assumed that each interface consists of a finite number of C'' arcs which meet at non-zero angles. By a <u>singular point</u> we will mean a point where two interfaces meet, or where an interface meets the boundary, or a point where two of the C'' arcs defining an interface or the boundary $\partial\Omega$ meet. The singular points will be denoted $\bar{x}^1, \cdots, \bar{x}^N$.

Let $a_{jk}(x)$, $1 \leq j, k \leq 2$, be a family of C' functions on $\bar{\Omega}$ which form a symmetric, positive definite matrix, and let $a_0(x)$ be a bounded nonnegative function on $\bar{\Omega}$. Let p be a function on Ω which takes a positive constant value in each subdomain Ω_j. We consider the problem

$$(4.1) \quad \sum_{j,k} \int_\Omega [p a_{jk} u_{x_j} v_{x_k} + a_0 uv]dx = \int_\Omega Fv dx , \quad \forall v \in \overset{\circ}{W}{}_2^1(\Omega) .$$

As in Section 2, it is easily seen that (4.1) has, for each $F \in L_2(\Omega)$, a unique solution $u \in \overset{\circ}{W}{}_2^1(\Omega)$. The collection of all such weak solutions is denoted by $\mathcal{D}(\Omega,L)$, and the operator

$$L: \mathcal{D}(\Omega,L) \to L_2(\Omega)$$

is defined by $Lu = F$, if u satisfies (4.1). Again as in Section 2, we have

$$(4.2) \qquad \|u\|_{W_2^1(\Omega)} \leq c\|Lu\| \ , \quad u \in \mathcal{D}(\Omega,L) \ .$$

Associated with (4.1) is the differential equation

$$(4.3) \qquad -p \sum_{j,k} \frac{\partial}{\partial x_k} (a_{jk} \frac{\partial u}{\partial x_j}) + a_0 u = F \ , \quad \text{in each } \Omega_j \ ,$$

the boundary condition

$$(4.4) \qquad u = 0 \qquad \text{on } \partial\Omega \ ,$$

and the interface conditions

$$(4.5) \qquad \begin{cases} u \ \text{continuous} & \text{across each interface} \\ p \sum a_{jk} n_k u_{x_j} \ \text{continuous} & \text{across each interface,} \end{cases}$$

where n_k are the components to the normal vector of the interface.

Let $\mathcal{W}_0(\Omega,L)$ be the set of functions $v \in \overset{\circ}{W}{}_2^1(\Omega)$ which are continuous on Ω ; which are of class C''' on each compactly contained subregion of each subdomain Ω_j ; which satisfy $|x-\overline{x}^n| D^2 v(x) \to 0$ as $x \to \overline{x}^n$, where $\overline{x}^n$ is any singular point, and $D^2 v$ is any second derivative of v ; and which satisfy (4.5). As in Section 2, we obtain the following lemma whose proof is omitted.

Lemma 4.1. $\mathcal{W}_0(\Omega,L) \subset \mathcal{D}(\Omega,L)$. If $u \in \mathcal{W}_0(\Omega,L)$, then (4.3) is satisfied in each subdomain Ω_j . Also, the norm $\|Lu\|$ is equivalent with the expression

$$\|u\|_{W_2^1(\Omega)} + \sum_j \|u\|_{W_2^2(\Omega_j)} .$$

We let $\mathcal{W}(\Omega,L)$ denote the closure, with respect to the norm $\|Lu\|$, of the set $\mathcal{W}_0(\Omega,L)$. Then $\mathcal{W}(\Omega,L)$ is a closed subspace of the Hilbert space $\mathcal{D}(\Omega,L)$.

To proceed with our analysis we must construct a mapping of the x-plane which straightens out the interfaces. This is done in the following lemma. As Lemma 5.1 of [6] contains a similar construction, we shall not give a proof of this lemma.

Lemma 4.2. There is, for each point $\bar{x}$ on one of the interfaces or on $\partial\Omega$, a number $\delta(\bar{x}) > 0$, an open set $U(\bar{x})$ containing $\bar{x}$, and a (1-1) C' map $y = y(x)$ of $U(\bar{x})$ onto the disc $|y| < \delta(\bar{x})$, such that $y(\bar{x}) = 0$; $y(x)$ is of class C'' at all points of $U(\bar{x})$ except $\bar{x}$; the second derivatives of $y(x)$ are uniformly bounded in $U(\bar{x})\setminus\{\bar{x}\}$; the image of each of the interface and boundary curves which meet at $\bar{x}$ is a straight line in the y-plane; if $\bar{a}_{mn}(y)$ are defined by

$$\bar{a}_{mn} = \frac{\partial(x_1,x_2)}{\partial(y_1,y_2)} \sum_{j,k} a_{jk} \frac{\partial y_m}{\partial x_j} \frac{\partial y_n}{\partial x_k} ,$$

then $\bar{a}_{mn} = \delta_{mn}$ along the images in the y-plane of the interface and boundary curves which meet at $\bar{x}$.

With this change of variables it is convenient to define

$$\bar{a}_0 = a_0 \frac{\partial(x_1,x_2)}{\partial(y_1,y_2)} .$$

Evidently, in Lemma 4.2 the numbers $\delta(\bar{x})$ may be taken so small that either $\bar{x}$ is a singular point or $U(\bar{x})$ contains no singular points. Let $\mathscr{A}$ be the union of all the points on the interface curves and the boundary. Then $\mathscr{A}$ is a compact set and is covered by a finite number of the sets $U'(\bar{x})$ which map onto the discs $|y| < \delta(\bar{x})/2$. The corresponding points, which must include all the singular points, will be labeled $\bar{x}^n$, $1 \leqslant n \leqslant N'$, $N' \geqslant N$. Now let $\tilde{\zeta}_n(x)$ be a function satisfying

$$\tilde{\zeta}_n(x) = \begin{cases} 1 , & x \in U'(\bar{x}^n) , \\ 0 , & x \notin U(\bar{x}^n) , \end{cases}$$

and which, in the variables $y(x)$ defined by the mapping of Lemma 4.2 at the point $\bar{x}^n$, is a function only of $|y|$. Define

$$\zeta_n(x) = \frac{\tilde{\zeta}_n(x)}{\pi(1-\tilde{\zeta}_j(x))+\Sigma\tilde{\zeta}_j(x)} , \quad 1 \leqslant n \leqslant N' ,$$

$$\zeta_0(x) = 1 - \Sigma\zeta_n(x) .$$

Then $\zeta_0(x) = 0$ in each of the sets $U'(\bar{x}^n)$. Furthermore, using the fact that $(\bar{a}_{jk})$ is the identity matrix on the interfaces and that $\tilde{\zeta}_n$ is a radial function in the y-plane, it may be seen that $\tilde{\zeta}_n$, and hence ζ_n, satisfies the interface conditions (4.5).

Now let n satisfy $1 \leqslant n \leqslant N'$, and let S_n denote the image in the y-plane of the set $U(\bar{x}^n) \cap \Omega$. Then either S_n is the disc of radius $\delta(\overline{x}^n)$, or a sector of this disc. The latter possibility occurs if $\overline{x}^n \in \partial\Omega$. Then Theorem 3.1 or 3.2 gives a finite dimensional set $\mathcal{H}(S_n,p)$. Note that the set $\mathcal{H}(S_n,p)$ is empty if $n > N$; that is, if $\overline{x}^n$ is not a singular point. It may happen that $\mathcal{H}(S_n,p)$ is empty for some singular points $\overline{x}^n$. Let $1 \leqslant n \leqslant M$ be the indices such that $\mathcal{H}(S_n,p)$ is not empty. For such n, let $\mathcal{H}_n$ denote the set of functions $u = v\zeta_n$ where $v \in \mathcal{H}(S_n,p)$. Let $\mathcal{H} = \Sigma\mathcal{H}_n$. Then $\mathcal{H} \cap \mathcal{W}(\Omega,L) = \{0\}$. We have

<u>Lemma 4.3</u>. $\mathcal{H}_n \subset \mathcal{D}(\Omega,L)$.

<u>Proof</u>. By Lemma 3.3 or the corresponding result for sectors, each $u \in \mathcal{H}(S_n,p)$ satisfies an equation of the form

$$\sum_{j,k} \int_{S_n} p\overline{a}_{jk} v_{y_j} w_{y_k} dy + \int_{S_n} \overline{a}_0 vwdy = \int Fwdy\ , \quad \forall w \in \overset{\circ}{W}{}^1_2(S_n)\ .$$

Hence $u = v\zeta_n$ satisfies a similar equation with F replaced by

$$G = \zeta_n F - p \sum \Big\{ \overline{a}_{jk,y_k} \zeta_{n,y_j} v + \overline{a}_{jk}(v_{y_j} \zeta_{n,y_k} + v_{y_k} \zeta_{n,y_j}) + \overline{a}_{jk} v \zeta_{n,y_j y_k} \Big\}\ .$$

Thus, $G \in L_2(S_n)$. Finally, changing to the x-plane, we see that u satisfies (4.1) with F replaced by

$$H = G \frac{\partial(y_1, y_2)}{\partial(x_1, x_2)} .$$

Hence $u \in \mathcal{D}(\Omega, L)$.

With the set $\mathcal{H}$, we are able to state our desired result.

Theorem 4.1. $\mathcal{D}(\Omega,L) = \mathcal{H} + \mathcal{W}(\Omega,L)$. Furthermore, there is a $c > 0$ such that if $u \in \mathcal{D}(\Omega,L)$, so $u = v + w$, $v \in \mathcal{H}$, $w \in \mathcal{W}(\Omega,L)$, then

$$\|Lv\| + \|w\|_{W_2^1(\Omega)} + \sum_j \|w\|_{W_2^2(\Omega_j)} \leq c \|Lu\| .$$

Proof. Let $u \in \mathcal{D}(\Omega,L)$, and let $u_n = \zeta_n u$, $0 \leq m \leq N'$. Then

$$v = \sum_1^M u_n \in \mathcal{H} , \quad w = u_0 + \sum_{M+1}^{N'} u_n \in \mathcal{W}(\Omega,L) ,$$

and $u = v + w$. The inequality then follows from the corresponding inequalities of Theorems 3.1 and 3.2.

The following result easily follows from Theorem 4.1 and the results of Section 2.

Theorem 4.2. There are numbers $\varepsilon_n \in (0,1)$, $1 \leq n \leq M$, and $c > 0$ such that for each $u \in \mathcal{D}(\Omega,L)$,

$$\sum_j \|\omega D^2 u\|_{L_2(\Omega_j)} \leq c \|Lu\| ,$$

where D^2u is any second derivative of u, and where

$$\omega(x) = \min\{|x-\bar{x}_n|^{1-\varepsilon_n}, 1 \leqslant n \leqslant M\} .$$

5. A Density Theorem in Three Dimensions

Let $x = (x_1,x_2,x_3) \in$ Euclidean 3 space, and let B be the ball $r = (x_1^2 + x_2^2 + x_3^2)^{\frac{1}{2}} < r_0$. Let B be divided into open subregions B_j by a finite number of planes passing through 0. Let there be given a positive function p on B which takes on the constant value $p_j > 0$ in each B_j. It is well known that if $F \in L_2(B)$, the problem

$$(5.1) \qquad \int_B p\nabla u\cdot\nabla v dx = \int_B Fv dx , \quad \forall v \in \overset{\circ}{W}{}_2^1(B) ,$$

has a unique solution $u \in \overset{\circ}{W}{}_2^1(B)$. We let $\mathcal{D}(B,p)$ denote the set of all solutions u of (5.1) as F varies over $L_2(B)$, and we define the operator M on $\mathcal{D}(B,p)$ by $Mu = F$. It is not hard to see that M is a densely defined, self-adjoint, positive definite operator. Furthermore, $\mathcal{D}(B,p)$ is itself a Hilbert space with the norm $\|Mu\|$, $u \in \mathcal{D}(B,p)$, and one has, for some $c > 0$,

$$(5.2) \qquad \|u\|_{W_2^1(B)} \leqslant c\,\|Mu\| , \quad u \in \mathcal{D}(B,p) .$$

In this section we shall determine $\mathcal{D}(B,p)$ and use the result to obtain some weighted inequalities for second derivatives of solutions of (5.1).

Each of the subregions B_j is a truncated polyhedral cone whose boundary consists of a spherical polygon on the sphere $r = r_0$, plus a lateral boundary consisting of a number of flat sides and a number of edges. The flat sides will be called interfaces of the problem (5.1). By $\rho = \rho(x)$ we will mean the minimum distance from x/r to any one of the edges of any of the subregions B_j. In other words, $\rho(x)$ represents the smallest angular distance from x to an edge.

Associated with (5.1) is the equation

$$-p\Delta u = F \,, \quad x \in \text{some } B_j \,, \tag{5.3}$$

the boundary condition

$$u = 0 \quad \text{on} \quad r = r_0 \,, \tag{5.4}$$

and the interface conditions

$$\left\{ \begin{array}{l} u(x) \text{ and } p(x)u_n(x) \text{ continuous} \\ \text{as } x \text{ passes through an interface,} \end{array} \right. \tag{5.5}$$

where u_n signifies the directional derivative of u in the direction normal to the interface.

We use the usual multi-index notation for a derivative $D^\alpha u = \partial^{|\alpha|} u/\partial x_1^{\alpha_1}\, \partial x_2^{\alpha_2}\, \partial x_3^{\alpha_3}$ of order $|\alpha| = \alpha_1+\alpha_2+\alpha_3$. Let $\mathcal{W}_0(B,p)$ be the set of functions v in $\overset{\circ}{W}{}_2^1(B)$ satisfying the following conditions: (a) v is of class C''' in each subregion B_j; (b) each derivative of order ≤ 2 of v is continuous at each boundary point of B_j except possibly the edges, and each derivative of order 1 of v is continuous at all points of $\overline{B}_j$ except possibly $x = 0$; (c) one has for some $\varepsilon > 0$,

$$(5.7)\quad \begin{cases} r^{\frac{3}{2}-\varepsilon}\rho^{1-\varepsilon}D^{\alpha}v \quad \text{bounded in } B_j\,, \quad |\alpha| = 2 \\ r^{\frac{1}{2}-\varepsilon}D^{\alpha}v \quad \text{bounded in } B_j\,, \quad |\alpha| = 1\,; \end{cases}$$

(d) the interface conditions (5.5) are satisfied for $v = u$; (e) the quantities pv_{tn} and v_{tt} are continuous across interfaces, where n and t denote respectively the normal derivative and any tangential derivative at the interface plane.

The conditions (5.7) may seem an artifical requirement. However we shall deal not with $\mathcal{W}_0(B,p)$ but with its closure in the norm of $\mathcal{D}(B,p)$. Any set of functions which gives the same closure would suffice for our purposes.

With this definition the following lemma is easily proved.

<u>Lemma 5.1</u>. $\mathcal{W}_0(B,p) \subset \mathcal{D}(B,p)$; moreover, if $u \in \mathcal{W}_0(B,p)$, then $Mu = -p\Delta u$ in each subregion B_j.

<u>Proof</u>. From (5.7) we see that $-p\Delta u$ is square integrable in each subregion B_j. Let $\eta > 0$ be arbitrarily small, and let $B_j(\eta) = B_j \setminus \{x : r \leq \eta \text{ or } \rho \leq \eta\}$. Then for $v \in C_0^\infty(B)$,

$$\int_{B_j(\eta)} vMudx = \int_{\partial B_j(\eta)} pvu_n dS + \int_{B_j(\eta)} p\nabla v\cdot\nabla udx\,.$$

Adding this over j and using (5.5), we see that all the boundary integrals vanish or cancel except those around the $x = 0$ or the edges. Using (5.7), it is seen that these tend to 0 with η. Hence in the limit we obtain (5.1) for $v \in C_0^\infty(B)$. Since this is dense in $\overset{\circ}{W}{}_2^1(B)$, we obtain our result.

We now define $\mathcal{W}(B,p)$ to be the closure of $\mathcal{W}_0(B,p)$ in the norm $\|Mu\|$. The next lemma helps identify which functions are in $\mathcal{W}(B,p)$.

Lemma 5.2. If $u \in \mathcal{W}(S,p)$, the norm $\|Mu\|$ is equivalent with the norm

$$(5.8) \qquad |||u||| = \{\|u\|^2_{W_2^1(B)} + \sum_j \sum_{|\alpha|=2} \int_{B_j} |D^\alpha u|^2 dx\}^{\frac{1}{2}} .$$

Proof. As in the proof of Lemma 5.1, for each subregion B_j, let

$$B_j(\eta) = B_j \setminus \{x:\ r \leq \eta \ \text{ or } \ \rho \leq \eta\} .$$

Then, using the formula

$$u_{x_i x_i} u_{x_k x_k} = (u_{x_i x_i} u_{x_k})_{x_k} - (u_{x_i x_k} u_{x_k})_{x_i} + (u_{x_i x_k})^2 ,$$

we see that

$$(5.9) \qquad \int_{B_j(\eta)} p(\Delta u)^2 dx = \sum_{|\alpha|=2} \int_{B_j(\eta)} p(D^\alpha u)^2 dx + I(u) ,$$

where $I(u)$ is an integral over $\partial B_j(\eta)$ of products of the form $pn_k D^{\alpha}u \cdot D^{\beta}u$, $|\alpha| = 2$, $|\beta| = 1$, (n_k) being the normal vector to $\partial B_j(\eta)$. We may write this surface integral as a sum of four terms , $I = I_1 + I_2 + I_3 + I_4$, where:

$I_1(u)$ is the integral over the part of $\partial B_j(\eta)$ with $r = r_0$;

$I_2(u)$ is the integral over the part of $\partial B_j(\eta)$ with $r = \eta$;

$I_3(u)$ is the integral over the part of $\partial B_j(\eta)$ with $\rho = \eta$;

$I_4(u)$ is the integral over the part of $\partial B_j(\eta)$ which lies on the interfaces.

Because the surface $r = r_0$ is convex, it may be shown that $I_1(u) \geq 0$. When the integrand of $I_4(u)$ is written out, it is seen to consist of quantities of the form $pu_n u_{tt}$ or $pu_t u_{nt}$, where the subscripts n and t refer respectively to directional derivatives normal and tangent to the interface plane. Hence, using condition (e) in the definition of $\mathcal{W}_0(B,p)$, we see that when the sum

$$\sum_j \int_{B_j(\eta)} p(\Delta u)^2 dx$$

is formed, the contributions to $I_4(u)$ from the different subregions B_j all cancel one another out. To finish the proof it therefore suffices to show that

$$I_2(u) \to 0 \quad \text{as} \quad \eta \to 0$$

$$I_4(u) \to 0 \quad \text{as} \quad \eta \to 0 \ .$$

Using (5.7) and the fact that $\eta\rho = \tilde{\rho}$ is, in the neighborhood of an edge, equivalent to the distance on the sphere $r = \eta$, we have

$$I_2(u) \leq c\eta^{2\varepsilon-2} \int_{\tilde{\rho}=0}^{c\eta} \rho^{\varepsilon-1} \tilde{\rho} d\tilde{\rho}$$

$$\leq c\eta^{2\varepsilon}$$

so $I_2(u) \to 0$ as $\eta \to 0$. Similarly, integrating over the conical surface $\rho = \eta$, we have

$$I_3(u) \leq c\eta^{\varepsilon-1} \int_0^{r_0} r^{2\varepsilon-2} r\eta dr$$

$$\leq c\, r_0^{2\varepsilon} \eta^{\varepsilon} ,$$

so $I_3(u) \to 0$ as $\eta \to 0$. This completes the proof of the lemma.

To proceed further we introduce spherical coordinates $x_1 = r \sin\phi \cos\theta$, $x_2 = r \sin\phi \sin\theta$, $x_3 = r\cos\phi$, and write the Laplace operator as

$$\Delta u = r^{-2}(r^2 u_r)_r + r^{-2}\Lambda u ,$$

$$\Lambda u = \csc\phi[(u_\theta \csc\phi)_\theta + (u_\phi \sin\phi)_\phi] .$$

Notice that in spherical coordinates, p is a function of ϕ and θ alone. As in (2.10), we consider the following eigenvalue problem for the Laplace-Beltrami operator with discontinuous coefficients:

$$(5.10) \qquad \int_{\partial B} p[\Psi_\theta \Phi_\theta \csc\phi + \Psi_\phi \Phi_\phi \sin\phi] d\theta d\phi = \lambda \int_{\partial B} p\Psi\Phi \sin\phi \, d\theta d\phi ,$$

$$\forall \Phi \in W_2^1(\partial B) .$$

The following lemma may be proved by following the usual development of properties of eigenvalue problems ([4], p. 50). We shall not give the proof here.

Lemma 5.3. The problem (5.10) has a countable number of eigenvalues $0 = \lambda_0 < \lambda_1 \leq \lambda_2 \leq \cdots$, and eigenfunctions $\Psi_n \in W_2^1(\partial B)$. The eigenfunctions are complete in $L_2(\partial B)$ and may be taken to satisfy

$$\int p\Psi_m \Psi_n \sin\phi \, d\theta d\phi = \begin{cases} 0 , & m \neq n \\ 1 , & m = n . \end{cases}$$

Inside each of the spherical polygons $\overline{B}_j \cap \partial B$, we have

$$(5.11) \qquad \Lambda\Psi_n + \lambda_n \Psi_n = 0 .$$

The nth eigenvalue $\lambda_n \to \infty$ as $n \to \infty$.

For a fixed eigenvalue λ_n and eigenfunction Ψ_n, (5.10) may be regarded as an inhomogeneous interface problem on the sphere with right hand side $p\lambda_n \Psi_n$. With the two dimensional results of the preceding sections, we can deduce interface conditions that Ψ_n satisfies on the sides of the spherical polygons $\overline{B}_j \cap \partial B$, and integrability conditions that Ψ_n and its derivatives satisfy at the vertices of these spherical polygons. In particular, it may happen that the second derivatives of Ψ_n are not integrable in the neighborhood of a vertex. We let J denote the set of all indices m such that $\Psi_m \notin W_2^2(\partial B \cap \overline{B}_j)$ for some subregion B_j. Also, we let K denote the set of all indices $m \notin J$ such that $\lambda_m \leq \frac{3}{4}$. Obviously, K is a finite set, but J may be an infinite set of indices.

It is easily seen from the interface conditions satisfied by Ψ_m, that if $R(r)$ is a smooth function, then the function $R(r)\Psi_m(\phi,\theta)$ satisfies the interface conditions (5.5). Also, from (5.11) we see that if a satisfies

$$a^2 + a - \lambda_m = 0 \, , \tag{5.12}$$

then $r^a\Psi_m(\phi,\theta)$ is a harmonic function in each subregion B_j.

For $m \in J$, we let $\mathcal{H}_m(B,p)$ denote the set of all functions $v = R(r)\Psi_m(\phi,\theta)$ such that $v \in \overset{\circ}{W}{}_2^1(B)$, such that v has weak square integrable second derivatives in each open set R with $\overline{R} \subset B_j$, for each subregion B_j, and such that

$$\int_{B_j} (\Delta v)^2 dx < \infty \, , \quad \text{each subregion } B_j \, .$$

In terms of $R(r)$, these conditions mean that

$$(5.13)\quad \begin{cases} \displaystyle\int_0^{r_0} [r^2R'^2 + R^2 + r^{-2}(r^2R'' + 2rR' - \lambda_m R)^2]dr < \infty , \\ R(r_0) = 0 . \end{cases}$$

From the definition of J, $v \notin W_2^2(B_j)$ for at least one j, so $|||v||| = \infty$, $v \notin \mathcal{W}(B,p)$.

Now we select a fixed C^∞ function $\xi(r)$ which is $\equiv 1$ in a neighborhood of $r = 0$ and $\equiv 0$ in a neighborhood of $r = r_0$. We let $\mathcal{K}(B,p)$ be the linear span of the set of functions (finite in number)

$$(5.14)\qquad z_m = \xi(r)r^{a_m}\,\Psi_m(\phi,\theta) , \quad m \in K ,$$

where a_m is the positive root of (5.12). Note that

$$\int_{B_j} pz_{m,rr}^2\, dx \geq c \int_{r=0}^{\varepsilon} r^{2a_m-2}\, dr = \infty$$

since, by (5.12), $a_m \leq \frac{1}{2}$. Hence $z_m \notin \mathcal{W}(B,p)$.

We have seen that $\mathcal{D}(B,p)$ is a Hilbert space with the norm $\|Mu\|$. It is convenient to use instead the equivalent norm $\|p^{-\frac{1}{2}}Mu\|$ and the corresponding inner product. Then we have

<u>Lemma 5.5.</u> The collection of sets consisting of $\mathcal{K}(B,p)$ and $\mathcal{H}_m(B,p)$, $m \in J$, are a family of closed, mutually orthogonal subspaces of $\mathcal{D}(B,p)$.

<u>Proof.</u> Select a smooth function $R(r)$ satisfying (5.13)

and set $v = R\Psi_m$, $m \in J$. An argument similar to the proof of Lemma 5.1 shows that $v \in \mathcal{D}(B,p)$, with

$$Mv = -p\Delta v = -p\, r^{-2}[(r^2R')' - \lambda_m R]\Psi_m$$

in each subregion B_j . Approximating any function R satisfying (5.13) by smooth functions, it is seen that $\mathcal{H}_m(B,p)$ is a closed subspace of $\mathcal{D}(B,p)$. Using Lemma 5.3, we see that the different $\mathcal{H}_m(B,p)$ are mutually orthogonal. A similar proof shows that $\mathcal{K}(B,p) \subset \mathcal{D}(B,p)$ and is orthogonal to the $\mathcal{H}_m(B,p)$.

We can now state the main theorem of this section. Set $\mathcal{H}(B,p) = \Sigma\,\mathcal{H}_m(B,p)$, the sum being taken in the sense of orthogonal subspaces of a Hilbert space. Then we have

Theorem 5.1. $\mathcal{D}(B,p) = \mathcal{W}(B,p) + \mathcal{H}(B,p) + \mathcal{K}(B,p)$.

Proof. To prove the theorem, we let $u \in \mathcal{D}(B,p)$ be in the orthogonal complement of $\mathcal{W}(B,p) + \mathcal{H}(B,p) + \mathcal{K}(B,p)$ and we show that $u = 0$. Setting $f = p^{-1}Mu$, we have

$$\int_B f\, M\, v\, dx = 0 \ , \quad v \in \mathcal{W}(B,p) + \mathcal{H}(B,p) + \mathcal{K}(B,p) \ . \tag{5.15}$$

Choosing $v \in C_0^\infty(B_j) \subset \mathcal{W}(B,p)$, we see from Weyl's lemma [5, p. 182] that f is infinitely differentiable in each subregion B_j and satisfies $\Delta f = 0$ there. Also, by Fubini's theorem, for almost every (a.e.) $r \in [0,r_0]$, $f(r,\phi,\theta) \in L_2(\partial B)$. For such r , set

(5.16) $$f(r,\phi,\theta) = \sum_0^\infty R_m(r)\,\Psi_m(\phi,\theta)\ ,$$

so the functions $R_m(r)$ are defined a.e. Applying Fubini's theorem to the convergent integral $\int_B f p \Psi_m dx$, we see that for a.e. r, the integral

$$\int_{\partial B} f p \Psi_m \sin\phi \, d\theta d\phi$$

is finite, and the integral is a measurable function of r. Since this integral is $R_m(r)$ for a.e. r, $R_m(r)$ is measurable. The same argument also shows that

(5.17) $$\int_B f^2 p dx = \sum_0^\infty \int_0^{r_0} r^2 R_m^2 \, dr < \infty\ .$$

Let $\zeta(r)$ be a C^∞ function which is $\equiv 0$ near $r = 0$ and $r = r_0$. Then if $m \in J$, $v = \zeta\Psi_m \in \mathcal{H}(B,p)$, and if $m \notin J$, $v \in \mathcal{W}(B,p)$. Hence we may in any event apply (5.15) and obtain, using (5.16) and Fubini's theorem,

$$\int_0^{r_0} R_m(r)\{(r^2\zeta')' - \lambda_m \zeta\}dr = 0\ , \quad \forall\zeta \in C_0^\infty([0,r_0])\ .$$

By Weyl's lemma for the one dimensional differential operator appearing in the braces, we deduce that $R_m(r)$ is twice continuously differentiable and satisfies

$$(r^2 R_m')' - \lambda_m R_m = 0\ .$$

Hence

$$R_m(r) = \alpha_m r^{a_m} + \beta_m r^{b_m} , \quad m = 0,1, \cdots . \tag{5.18}$$

where a_m , b_m are the larger and smaller roots of (5.12). We finish the proof by showing that $\alpha_m = \beta_m = 0$ for each $m \geqslant 0$.

From (5.12), we see that $a_m \geqslant 0 > b_m$, and $b_m \to -\infty$ as $m \to \infty$. Hence, from (5.17), $\beta_m = 0$ for $2 + 2b_m \leqslant -1$, or, $\lambda_m \geqslant \frac{3}{4}$. Let $\zeta(r)$ vanish in a neighborhood of $r = 0$, and satisfy $\zeta(r_0) = 0$. Then $v = \zeta\Psi_m \in \mathcal{W}(B,p) + \mathcal{H}(B,p)$, so from (5.15) and the formula

$$\int_\varepsilon^{r_0} R_m[(r^2\zeta')' - \lambda_m \zeta]\, dr = r^2(R_m\zeta' - R_m'\zeta)\Big|_\varepsilon^{r_0} , \tag{5.19}$$

we find that $R_m(r_0) = 0$. For $\lambda_m \geqslant \frac{3}{4}$, this gives $\alpha_m = \beta_m = 0$. For $\lambda_m < \frac{3}{4}$, $m \in J \cup K$. In any event, $v = \xi(r)r^{a_m}\Psi_m(\phi,\theta)$ is in $\mathcal{H}(B,p) + \mathcal{K}(B,p)$, where $\xi(r)$ is the function used in (5.14). Hence from (5.15) and (5.19), we get

$$-\beta_m(a_m - b_m) \lim_{\varepsilon \to 0} \varepsilon^{a_m + b_m + 1} = 0 .$$

From (5.12), $a_m + b_m = -1$. Hence, $\beta_m = 0$, so from $R(r_0) = 0$, $\alpha_m = 0$ in this case. Thus $R_m \equiv 0$ for all m , and $f = 0$, proving the theorem.

The above theorem may be obtained also in the case that B is replaced by $B_0 = B \cap P$, where P is a polyhedron with vertex at 0 and where r_0 is small enough so that B_0 contains only those edges and sides of P that meet at 0 . Again we suppose B_0 to be divided into subregions B_j by planes passing through 0 , and we suppose given a function p which is a constant $p_j > 0$ on each B_j . Then with B replaced by B_0 we may define $\mathcal{D}(B_0,p)$, $M: \mathcal{D}(B_0,p) \to L_2(B_0)$, and $\mathcal{W}(B_0,p)$. We let σ denote the spherical polygon $\partial B \cap \partial B_0$. We may note that functions in $\mathcal{D}(B_0,p)$ now vanish on the flat sides of ∂B_0 as well as on σ . In place of (5.10) we consider the eigenvalue problem

$$(5.20)\quad \int_\sigma p[\Psi_\theta \Phi_\theta \csc\phi + \Psi_\phi \Phi_\phi \sin\phi]d\theta d\phi = \lambda \int_\sigma p\Psi\Phi \sin\phi\, d\theta d\phi,$$

$$\forall \Phi \in \mathring{W}_2^1(\sigma)$$

with eigenvalues λ_m satisfying $0 < \lambda_1 \leqslant \lambda_2 \leqslant \cdots$, and eigenfunctions $\Psi_m \in \mathring{W}_2^1(\sigma)$ which are complete in $L_2(\sigma)$ and satisfy

$$\int_\sigma p\Psi_m\Psi_n \sin\phi\; d\theta d\phi = \begin{cases} 1\,, & m = n \\ 0\,, & m \neq n\,. \end{cases}$$

With these eigenvalues, we define the index sets J and K exactly as above, and with these eigenfunctions we define the sets $\mathcal{H}(B_0,p)$, $\mathcal{K}(B_0,p)$ exactly as above. Then all the preceding lemmas, and the preceding theorem, are true, with exactly the same proofs, when B is replaced by B_0 .

In the case p = constant in B_0 , so there are no interfaces, Theorem 5.1 gives a new result about solutions of the Poisson equation $\Delta u = F$ with $u = 0$ on ∂B_0. If B_0 is convex, then $\mathcal{H}(B_0,p) = \mathcal{K}(B_0,p) = \{0\}$, and

$$\mathcal{W}(B_0,p) = W_2^2(B_0) \cap \overset{\circ}{W}{}_2^1(B_0) . \tag{5.21}$$

In the non-convex case, $\mathcal{W}(B_0,p)$ will be a proper subset of $\mathcal{D}(B_0,p)$. By a result of Hanna and Smith [3], (5.21) still holds in this case, but this fact is irrelevant for our analysis and in fact (5.21) is false for some non-polyhedral domains. (An example is given in [2], p. 311.)

Theorem 5.1 may be used to obtain inequalities for weighted norms of second derivatives of the solution u of (5.1). For this we let D_r denote the directional derivative in the radial direction, and we let D_t denote any directional derivative in a direction perpendicular to the radial direction. Then we have

Theorem 5.2. Let $\Omega = B$ or B_0. Then there are constants $\varepsilon > 0$ and $c > 0$ depending only on Ω and p such that if $u \in \mathcal{D}(\Omega,p)$,

$$\sum_j \|r^{\frac{1}{2}-\varepsilon} D_r^2 u\|_{L_2(B_j)} \leq c\, \|Mu\|_{L_2(\Omega)} , \tag{5.22}$$

$$\sum_j \|r^{\frac{1}{2}-\varepsilon} D_r D_t u\|_{L_2(B_j)} \leq c\, \|Mu\|_{L_2(\Omega)} , \tag{5.23}$$

$$\sum_j \|r^{\frac{1}{2}-\varepsilon} \rho^{1-\varepsilon} D_t^2 u\|_{L_2(B_j)} \leq c\, \|Mu\|_{L_2(\Omega)} . \tag{5.24}$$

Before giving the proof of this theorem, it is

convenient to give an inequality for the solution $R(r)$ of the ordinary differential equation

$$rR'' + 2R' - \lambda r^{-1}R = f \tag{5.25}$$

with

$$R(r_0) = 0 , \tag{5.26}$$

$$\int_0^{r_0} (r^2R'^2 + R^2)dr < \infty . \tag{5.27}$$

<u>Lemma 5.6</u>. Let $f \in L_2([0,r_0])$. Then for $\lambda > 0$ there is a unique solution, $R(r)$, of (5.25), (5.26), (5.27). Furthermore, if $\lambda \geq \frac{3}{4} + \varepsilon$, there is a $c > 0$ depending only on r_0 and $\varepsilon > 0$ (but not on λ) such that

$$\int_0^{r_0} [r^2(R'')^2 + \lambda(R')^2 + \lambda^2 r^{-2}R^2]dr \leq c \int_0^{r_0} f^2 dr . \tag{5.28}$$

For $0 < \lambda < \frac{3}{4}$, there is an $\varepsilon > 0$ and a $c > 0$, depending on λ and r_0, such that

$$\int_0^{r_0} r^{1-\varepsilon} [r^2(R'')^2 + \lambda R'^2 + \lambda^2 r^{-2}R^2]dr \leq c \int_0^{r_0} f^2 dr . \tag{5.29}$$

<u>Proof</u>. The existence and uniqueness of the solution $R(r)$ follows from the method of variation of parameters. The solution is given by the formulas

$$R(r) = R_1(r) + R_2(r) ,$$

$$R_1(r) = \frac{1}{a-b} \int_0^{r_0} k(r,t) f(t) dt ,$$

$$k(r,t) = \begin{cases} -t^{-b} r^b , & 0 < t < r \\ -t^{-a} r^a , & r < t < r_0 , \end{cases}$$

$$R_2(r) = \alpha r^a ,$$

$$\alpha = \frac{r_0^{b-a}}{a-b} \int_0^{r_0} t^{-b} f(t) dt ,$$

$$a = \frac{-1+\sqrt{1+4\lambda}}{2} , \quad b = \frac{-1-\sqrt{1+4\lambda}}{2} .$$

The inequalities (5.28) and (5.29) also follow from these formulas for $R(r)$. To illustrate this we prove

$$\text{(5.30)} \qquad \lambda^2 \int_0^{r_0} r^{-2} R^2 dr \leqslant c \int_0^{r_0} f^2 dr , \quad \lambda \geqslant \tfrac{3}{4} + \varepsilon .$$

It suffices to prove (5.30) with R replaced by R_1 and R_2. For $R = R_2$, we have, noting that

$$\text{(5.31)} \qquad \lambda > \tfrac{3}{4} \Longrightarrow a > \tfrac{1}{2} , \quad b < -\tfrac{3}{2} ,$$

$$\lambda^2 \int_0^{r_0} r^{-2} R_2^2 dr = \frac{\lambda^2 \alpha^2}{2a-1} r_0^{2a-1} .$$

From Schwartz's inequality,

$$\alpha^2 \leqslant \frac{r_0^{1-2a}}{(a-b)^2(-2b+1)} \int_0^{r_0} f^2 dr .$$

Since $|a| \leqslant c\lambda^{\frac{1}{2}}$, $|b| \leqslant c\lambda^{\frac{1}{2}}$, we obtain (5.30) with R replaced by R_2 . To finish the proof of (5.30) we must consider

$$(5.32) \qquad \int_0^{r_0} r^{-2} R_1(r)^2 dr \leqslant \frac{1}{(a-b)^2} \int_0^{r_0}\int_0^{r_0} K(s,t) f(s) f(t) ds dt$$

where

$$K(s,t) = \int_{r=0}^{r_0} r^{-2} k(r,s) k(r,t) dr$$

is a positive symmetric kernel which is given, for $s < t$, by

$$(5.33) \qquad \begin{aligned} K(s,t) &= \frac{1}{2a-1} s^{a-1} t^{-a} + \tfrac{1}{2} s^{-b} t^{-a} (s^{-2} - t^{-2}) \\ &+ \frac{1}{1-2b} s^{-b} t^{-b} (t^{-1+2b} - r_0^{-1+2b}) . \end{aligned}$$

Using the formulas

$$|a| \leqslant c\lambda^{\frac{1}{2}} , \quad |b| \leqslant c\lambda^{\frac{1}{2}} , \quad a + b = -1 ,$$

we may show by a calculation that

$$(5.34) \qquad \int_{s=0}^{r_0} K(s,t)ds \leqslant c\lambda^{-1} , \quad 0 \leqslant t \leqslant r_0 ,$$

by showing (5.34) separately for each of the three terms on the right side of (5.33). From (5.32), (5.34), and Schwartz's inequality we have

$$\int_0^{r_0} r^{-2} R_1(r)^2 dr \leqslant c\lambda^{-2} \int_0^{r_0} f^2 dr ,$$

as was desired. This completes the proof of (5.30). The other inequalities of the lemma are proved in the same way.

Proof of Theorem 5.2. Write $u \in \mathcal{D}(\Omega,p)$ as $u = v + w + z$ with $v \in \mathcal{W}(\Omega,p)$, $w \in \mathcal{H}(\Omega,p)$, and $z \in \mathcal{K}(\Omega,p)$. Since these are closed subspaces of $\mathcal{D}(\Omega,p)$, we have for some $c_1 > 0$,

$$\|Mv\| + \|Mw\| + \|Mz\| \leqslant c_1 \|Mu\| .$$

Hence it suffices to prove the inequalities with u replaced by v , w , or z . For $u = v$, the results follow from Lemma 5.2. Let $u = w \in \mathcal{H}(\Omega,p)$. Then we may write $w = \Sigma\, w_m$ with $w_m = R_m \Psi_m \in \mathcal{H}_m$. To establish (5.22) we write, using Lemma 5.6 and 5.2,

$$\sum_j \|r^{\frac{1}{2}-\varepsilon} D_r^2 w\|_{L_2(B_j)}^2$$

$$= \sum_j \int_{B_j} r^{1-2\varepsilon} |D_r^2 w|^2 p r^2 \sin\phi \; dr d\phi d\theta$$

$$\leq \sum_m \int_0^{r_0} r^{3-2\varepsilon} (R_m'')^2 dr$$

$$\leq c \sum_m \int_0^{r_0} r^{-2} [r^2 R_m'' + 2rR_m' - \lambda_m R_m]^2 dr$$

$$\leq c \sum_m \int_\Omega |Mw_m|^2 p r^2 \sin\phi \; dr d\phi d\theta$$

$$= c \int_\Omega |Mw|^2 \; r^2 \sin\phi \; dr d\phi d\theta$$

$$= c \; \|Mw\|^2 \quad .$$

Next, we consider (5.23). We have

$$\|r^{\frac{1}{2}-\varepsilon} \; D_r \; D_t \; w\|^2_{L_2(B_j)}$$

$$= \int_{B_j} r^{3-2\varepsilon} \; |D_t w_r|^2 \sin\phi \; d\phi d\theta dr$$

$$\leq c \int_\Omega r^{1-2\varepsilon} \; [w_{r\theta}^2 \csc^2\phi + w_{r\theta}^2] p \sin\phi \; d\phi d\theta dr$$

$$\leq c \sum_m \int_\Omega r^{1-2\varepsilon} \; (R_m')^2 \Psi_m \Lambda \Psi_m \; p \sin\phi \; d\phi d\theta dr$$

$$= c \sum_m \lambda_m \int_0^{r_0} r^{1-2\varepsilon} (R_m')^2 dr$$

$$\leq c \sum_m \int_0^{r_0} r^{-2}[r^2 R_m'' + 2rR_m' - \lambda_m R_m]^2 dr$$

$$\leq c \|Mw\|^2 .$$

To prove (5.24), we first note that as a consequence of the work in Section 4, if $w(r,\phi,\theta)$ is in the domain of Λ, for fixed r,

$$r^4 \int_{r=\text{const.}} \rho^{2-\varepsilon} |D_t^2 w|^2 \sin\phi \, d\phi d\theta \leq c \int_{r=\text{const.}} |\Lambda w|^2 \sin\phi \, d\phi d\theta .$$

Hence

$$\| r^{\frac{1}{2}-\varepsilon} \rho^{1-\varepsilon} D_t^2 w \|^2_{L_2(B_j)}$$

$$= \int_{B_j} r^{3-2\varepsilon} \rho^{2-2\varepsilon} |D_t w|^2 \sin\phi \, d\phi d\theta dr$$

$$\leq c \int_B r^{-1-2\varepsilon} |\Lambda w|^2 p \sin\phi \, d\phi d\theta dr$$

$$\leq c \sum_m \lambda_m^2 \int_0^{r_0} r^{-1-2\varepsilon} R_m^2 \, dr$$

$$\leq c \sum_m \int_0^{r_0} r^{-2}[r^2 R_m'' + 2rR_m' - \lambda_m R_m]^2 \, dr$$

$$\leq c \|Mw\|^2 .$$

Thus, the inequalities (5.22), (5.23), (5.24) are satisfied for $u = w \in \mathcal{H}$. If $u = z \in \mathcal{K}$, the proof of the inequalities may be carried out in the same way. This completes the proof of the theorem.

6. The Interface Problem in a Polyhedron

We now consider a polyhedron Ω in Euclidean 3-space, which we suppose is partitioned by a finite number of planes into a finite number of polyhedral subdomains Ω_j. The faces of the polyhedra Ω_j which do not lie on $\partial\Omega$ will be called interfaces of the problem. The collection of all vertices of all the Ω_j will be called the vertices, and will be denoted $\overline{x}^n$, $1 \leqslant n \leqslant N$.

Let p be a function on Ω which takes on a positive constant value on each subdomain Ω_j. We consider the problem

$$\text{(6.1)} \qquad \int_\Omega p\nabla u\cdot\nabla v \, dx = \int Fv dx , \quad \forall v \in \overset{\circ}{W}{}^1_2(\Omega) .$$

Equation (6.1) has a solution $u \in \overset{\circ}{W}{}^1_2(\Omega)$ for each $F \in L_2(\Omega)$. The set of all solutions will be denoted by $\mathcal{D}(\Omega,p)$, and a map $M: \mathcal{D}(\Omega,p) \to L_2(\Omega)$ will be defined by $Mu = F$. The set $\mathcal{D}(\Omega,p)$ is a Hilbert space with the norm $\|Mu\|$, and, for some $c > 0$,

$$\text{(6.2)} \qquad \|u\|_{W^1_2(\Omega)} \leqslant c \, \|Mu\| , \quad u \in \mathcal{D}(\Omega,p) .$$

Associated with (6.1) is the differential equation

(6.3) $$-p\Delta u = F \quad \text{in } \Omega_j ,$$

the boundary condition

(6.4) $$u = 0 \quad \text{on } \partial\Omega ,$$

and the interface conditions

(6.5) $$\begin{cases} u(x) \text{ and } p(x)u_n(x) \text{ continuous} \\ \text{as } x \text{ passes through an interface,} \end{cases}$$

where u_n means the directional derivative of u in the direction normal to the interface.

Let there be given, for each vertex $\bar{x}^n$, an open set U_n containing $\bar{x}^n$ such that the U_n form an open cover of $\bar{\Omega}$, and such that $\bar{x}^n \notin U_m$ for $n \neq m$. Also suppose that the only faces of the polyhedra $\partial\Omega_j$, or of $\partial\Omega$, with which $\bar{U}_n$ has a non-empty intersection, are faces which contain the point $\bar{x}^n$. Let there be given a nonnegative function $\tilde{\zeta}_n \in C_0^\infty(U_n)$ such that $\sum_n \tilde{\zeta}_n(x) \neq 0$ in $\bar{\Omega}$, and such that near each interface, $\tilde{\zeta}_n(x)$ is a function of $|x-\bar{x}^n|$. Let

$$\zeta_n(x) = \frac{\tilde{\zeta}_n(x)}{\Sigma\, \tilde{\zeta}_m(x)} .$$

Then $\Sigma\, \zeta_n(x) = 1$ in $\bar{\Omega}$, and each ζ_n satisfies the interface conditions (6.5).

Let r_0 be the diameter of the polyhedron Ω. Let $\bar{x}^n$ be a given vertex, and let $\varepsilon < \min\{|\bar{x}^n-\bar{x}^m| , \ m \neq n\}$.

Let

$$B_n = \{x\colon \bar{x}^n + \frac{\varepsilon}{|x-\bar{x}^n|}(x-\bar{x}^n) \in \Omega \text{ and } |x-\bar{x}^n| < r_0\} .$$

Thus, if $\bar{x}^n \notin \partial\Omega$, B_n is a ball of radius r_0 , and if $\bar{x}^n \in \partial\Omega$, B_n is part of the ball of radius r_0 with polyhedral sides. Define a function $p_n(x)$, $x \in B_n$, by

$$p_n(x) = p(\bar{x}^n + \frac{\varepsilon}{|x-\bar{x}^n|}(x-\bar{x}^n)) .$$

With these definitions we consider the problem

$$(6.6) \qquad \int_{B_n} p_n \nabla v \cdot \nabla z dx = \int_{B_n} F z dx , \quad \forall z \in \overset{\circ}{W}{}_2^1(B_n) ,$$

and the associated domain

$$\mathcal{D}(B_n,p_n) = \mathcal{H}(B_n,p_n) + \mathcal{K}(B_n,p_n) + \mathcal{W}(B_n,p_n) .$$

We have

<u>Lemma 6.1</u>. If $u \in \mathcal{D}(\Omega,p)$, then $u\zeta_n \in \mathcal{D}(B_n,p_n)$.

<u>Proof</u>. Let $u \in \mathcal{D}(\Omega,p)$, and let $z \in \overset{\circ}{W}{}_2^1(B_n)$. Then $\zeta_n z \in \overset{\circ}{W}{}_2^1(\Omega)$, and, since u satisfies (6.1) for some $F \in L_2(\Omega)$,

$$\int_\Omega p \nabla u \cdot \nabla (z\zeta_n) dx = \int_\Omega z\zeta_n F dx .$$

Since $\zeta_n \in C_0^\infty(U_n)$, the integrals may be taken over B_n , but the integrands are non-zero only on $\Omega \cap U_n$. On this set, $p_n(x) = p(x)$. Hence we have

$$\int_{B_n} p_n \nabla(u\zeta_n)\cdot\nabla z dx = \int_{B_n} z(\zeta_n F - p\nabla u\cdot\nabla\zeta_n)dx + \int_{B_n} pu\nabla\zeta_n\cdot\nabla z dx \ .$$

Writing the last integral on the right as a sum of integrals over the subregions Ω_j , using Green's theorem and the fact that ζ_n satisfies the interface conditions, we find that

$$\int_{B_n} pu\nabla\zeta_n\cdot\nabla z dx = - \int_{B_n} p(\nabla u\cdot\nabla\zeta_n + u\Delta\zeta_n)z dx \ .$$

Hence $u\zeta_n$ satisfies (6.6) with F replaced by

$$\zeta_n F - 2\nabla u\cdot\nabla\zeta_n - pu\Delta\zeta_n \ .$$

Our object is to obtain a weighted inequality for second derivatives of the problem (6.1) analogous to Theorem 5.2. For this we define, for $x \in \Omega$,

$$r(x) = \min\{|x-\bar{x}^n| \ , \ 1 \leq n \leq N\} \ ,$$

$$\rho_n(x) = [\text{minimum distance from } x \text{ to any of the edges containing } \bar{x}^n \text{ of any of the polydedra } \Omega_j]/|x-\bar{x}^n| \ ,$$

$$\rho(x) = \min\{\rho_n(x) \ , \ 1 \leq n \leq N\} \ .$$

Then we have

Theorem 6.1. Each $u \in \mathcal{D}(\Omega,p)$ has locally square integrable second derivatives at each $x \in \Omega_j$, $1 \leq j \leq N$. Furthermore, there are constants $\varepsilon > 0$ and $c > 0$ depending only on Ω and p such that if $u \in \mathcal{D}(\Omega,p)$ and if D^2u represents any second derivative of u , then

$$\sum_j \|r^{\frac{1}{2}-\varepsilon} \rho^{1-\varepsilon} D^2u\|_{L_2(\Omega_j)} \leq c \, \|Mu\|_{L_2(\Omega)} .$$

Proof. Since $u = \Sigma\, u\zeta_n$ and $u\zeta_n \in \mathcal{D}(B_n,p_n)$, we have $D^2(u\zeta_n)$ locally square integrable at each $x \in \Omega_j$. Hence, using (6.2), D^2u is locally square integrable at each $x \in \Omega_j$. Also, using Lemma 6.1 and Theorem 5.2,

$$\begin{aligned} &\|r^{\frac{1}{2}-\varepsilon} \rho^{1-\varepsilon} D^2u\|_{L_2(\Omega_j)} \\ &\leq \sum_n \|r^{\frac{1}{2}-\varepsilon} \rho^{1-\varepsilon} D^2(\zeta_n u)\|_{L_2(\Omega_j)} \\ &\leq c \sum_n \|\zeta_n F - 2\nabla u \cdot \nabla \zeta_n - pu\Delta\zeta_n\| \\ &\leq c[\|F\|_{L_2(\Omega)} + \|u\|_{W_2^1(\Omega)}] \\ &\leq c\, \|F\|_{L_2(\Omega)} , \end{aligned}$$

proving the theorem.

We remark that this theorem could be made more precise, in the manner of Theorem 5.2, by giving different inequalities for derivatives tangent and normal to a given edge. Also, this theorem could be made more precise, in the manner of Theorem 4.2 by giving different values of ε for each vertex and each edge.

REFERENCES

1. Birman, M. Š., and Skvorcov, G.E., "On the Quadratic Integrability of the Highest Derivatives of the Solution of the Dirichlet Problem in a Domain with Piecewise Smooth Boundary," Izv. Vysš. Ucben. Zaved. Matematika, Vol. 30 (1962), 11-21.
2. Kondrat'ev, V. A., "Boundary Problems for Elliptic Equations with Conical or Angular Points," Trans. Moscow Math. Soc. Vol. 16 (1967), translated by Am. Math. Soc., 1968.
3. Hanna, Martin S., and Smith, Kennan T., "Some Remarks on the Dirichlet Problem in Piecewise Smooth Domains," Comm. Pure Appl. Math. Vol. 20 (1967), 575-593.
4. Nečas, J., Les methodes directes en théorie des equations elliptiques, Masson et Cie, Paris, 1967.
5. Hellwig, G., Partial Differential Equations, Blaisdell, New York, 1964.
6. Kellogg, R. B., "On the Poisson Equation with Intersecting Interfaces," Technical Note BN-643, University of Maryland, Feb., 1970.
7. Ladyzhenskaya, O. A., and Ural'tseva, N. N., Linear and Quasilinear Elliptic Equations, Academic Press, New York, 1968.

INITIAL BOUNDARY VALUE PROBLEMS FOR PARTIAL DIFFERENTIAL AND DIFFERENCE EQUATIONS IN ONE SPACE DIMENSION

by

Heinz-Otto Kreiss*

1. Differential Equations

Consider a system of partial differential equations

$$\partial u(x,t)/\partial t = P(\partial/\partial x)\, u + F(x,t) \ , \tag{1.1}$$

$$P(\partial/\partial x) = \sum_{\nu=0}^{m} A_\nu \, \partial^\nu/\partial x^\nu$$

in the quarter plane $x \geqslant 0$, $t \geqslant 0$. Here $u(x,t) = (u^{(1)}(x,t), u^{(2)}(x,t), \ldots , u^{(n)}(x,t))'$ † is a vector function of the real variables x,t and A_ν are constant square matrices of order n . We make always

Assumption 1.1. A_m is nonsingular.

The solution of (1.1) is determined only if we specify initial conditions:

---

* Uppsala University, Sweden.

† If y is a vector then y' denotes its transpose, y^* its adjoint and $|y|^2 = \Sigma |y_i|^2$ its Euclidean norm. Similar notations hold for matrices.

(1.2) $$u(x,0) = f(x) \ , \quad x \geq 0$$

and boundary conditions for $x = 0$:

(1.3a) $$B_{jj}\, \partial^j u/\partial x^j + \sum_{i=0}^{j-1} B_{ij}\, \partial^i u/\partial x^i = g_j \ , \ j = 0,1,2,m-1,$$

which we shall also write as

(1.3b) $$Bu = g \ , \qquad g = (g_0, g_1, \ldots, g_{m-1})' \ .$$

Here B_{ij} are rectangular matrices and without restriction we can assume that the rows of B_{jj} are linearly independent.

Let $L_2(R)$ denote the space of all vector functions which are quadratically integrable over $-\infty < x < \infty$ and use the notation

$$(u(\cdot),\ v(\cdot))_R = \int_{-\infty}^{\infty} u(x)^* v(x)dx \ ,$$

$$(u(\cdot),\ u(\cdot))_R = \|u(\cdot)\|^2_R \ .$$

We make always

<u>Assumption 1.2</u>. The Cauchy problem for the system of differential equations (1.1) is well posed i.e. there are constants K_0 , α_0 such that if $F \equiv 0$ then for every $f \in L_2(R)$, there is a solution with

(1.4) $$\|u(\cdot,t)\|_R \leq K_0 e^{\alpha_0 t} \|f(\cdot)\|_R \ .$$

There are a number of ways to define when the initial boundary value problem is well posed. One way is to use essentially the same definition as for the Cauchy problem:

Let $L_2(0,\infty)$ denote the space of all vector functions which are quadratically integrable for $0 \leqslant x < \infty$ and use the notation

$$(u,v) = \int_0^\infty u^* v \, dx \, , \quad \|u\|^2 = (u,u) \, .$$

Then we get:

<u>Definition 1.1</u>. Assume that the differential equations (1.1) and the boundary conditions (1.3) are homogeneous. $(F \equiv g \equiv 0)$. The problem is well posed if there are real constants α_0, K_0 such that for every $f \in L_2(0,\infty)$ there is a solution with

$$\|u(\cdot,t)\| \leqslant K_0 e^{\alpha t} \|f(\cdot,t)\| \, , \; \alpha > \alpha_0 \tag{1.5}$$

for all $\alpha > \alpha_0$.

The disadvantage of this definition is that it is rather difficult to develop a general theory. This is much easier if we instead use one of the following three definitions:

<u>Definition 1.2</u>. Consider the initial boundary value problem (1.1)-(1.3) with homogeneous initial conditions $(f \equiv 0)$. The problem is well posed if there are real constants α_0, K_0 such that for every smooth F and all $\alpha > \alpha_0$ there is a solution with:

$$(1.6) \qquad \int_0^\infty e^{-2\alpha t}\left((\alpha-\alpha_0)^2\,\|u(\cdot,t)\|^2 + (\alpha-\alpha_0)\,|u(0,t)|^2\right)dt$$
$$\leqslant K_0^2 \int_0^\infty e^{-2\alpha t}\left(\|F(\cdot,t)\|^2 + (\alpha-\alpha_0)\,|g(t)|^2\right)dt\ .$$

<u>Definition 1.3</u>. Consider the initial boundary value problem (1.1)-(1.3) with homogeneous initial and boundary conditions $(f \equiv g \equiv 0)$. The problem is well posed if instead of (1.6)

$$(1.7) \qquad \int_0^\infty e^{-2\alpha t}\|u(\cdot,t)\|^2 dt \leqslant \left(\frac{K_0}{\alpha-\alpha_0}\right)^2 \int_0^\infty e^{-2\alpha t}\|F(\cdot,t)\|^2 dt$$

holds.

<u>Definition 1.4</u>. Assume that the differential equations (1.1) and the initial conditions are homogeneous $(F \equiv f \equiv 0)$. The problem is well posed if instead of (1.7)

$$(1.8) \qquad \int_0^\infty e^{-2\alpha t}\|u(\cdot,t)\|^2 dt \leqslant \left(\frac{K_0}{\alpha-\alpha_0}\right)^2 \int_0^\infty e^{-2\alpha t}|g(t)|^2\, dt$$

holds.

The reason why the last three definitions are more suited for developing a general theory is that they are equivalent with certain estimates of the resolvent equation. If we Laplace Transform the above problem then we get a system of ordinary differential equations:

$$(1.9) \qquad (sI - P(d/dx)\,\hat{u}(x,s) \quad \hat{F}(x,s)\ , \quad B\hat{u}(x,s) = \hat{g}(s)\ ,$$

and there is no difficulty to prove the following

Lemma 1.1. The problem is well posed according to definition 1.2 if and only if (1.9) has for every complex s with $\text{real}\, s > \alpha_0$ and every $\hat{F}, \hat{g}$ a unique solution with:

$$(1.10a) \qquad \left(\text{Real}(s-\alpha_0)\right)^2 \|\hat{u}(\cdot,s)\|^2 + \text{Real}(s-\alpha_0)|u(0,s)|^2$$

$$\leqslant K_0^2\left(\|\hat{F}(\cdot,s)\|^2 + \text{Real}(s-\alpha_0)|\hat{g}(s)|^2\right) .$$

It is well posed according to definition 1.3 if $\hat{g} \equiv 0$ and (1.10a) is replaced by

$$(1.10b) \qquad \text{Real}(s-\alpha_0)\|u(\cdot,s)\| \leqslant K_0\|\hat{F}(\cdot,s)\| .$$

Finally it is stable according to definition (1.4) if $\hat{F} \equiv 0$ and (1.10b) is replaced by

$$(1.10c) \qquad \text{Real}(s-\alpha_0)\|\hat{u}(\cdot,s)\| \leqslant K_0|\hat{g}| .$$

The question then is which of the three definitions shall we adopt? Here there is no problem because they are all equivalent. In fact we believe that they are also equivalent with the first definition. Furthermore we get a rather simple algebraic criteria for well posedness. Consider the following reduced equation

$$(1.11) \qquad \begin{aligned} &(sI - A_m\, d^m/dx^m)\,\hat{u} = 0 , \\ &B_{jj} d^j\, \hat{u}/dx^j = g_j \qquad j = 0,1,2, \ldots , m-1 . \end{aligned}$$

Then the following theorem holds:

Theorem 1.1. The problem is well posed according to any of the definitions 1.2-1.4 if and only if (1.11) has for every s with Real $s > 0$ and every g a unique solution.

For hyperbolic first order systems this leads to the usual condition, namely that the variables corresponding to the "ingoing" characteristics can be expressed by the variables corresponding to the "outgoing" variables. For second order parabolic systems we get that

$$\text{Det} \begin{vmatrix} B_{00} & \\ B_{11} & A_2^{1/2} \end{vmatrix} \neq 0$$

is necessary and sufficient for well posedness.

I believe that the same is true for parabolic systems in more than one space dimension. Let us now discuss strictly hyperbolic systems in more than one space dimension. Then these definitions are not equivalent any more. One can show that definition 1.2 is a stronger requirement than definition 1.3 which in turn is stronger than definition 1.4. It can furthermore be shown that definition 1.2 is stronger than definition 1.1 but that definition 1.4 is weaker than definition 1.1.

In this case we have thus to decide which definition we want to choose. There are two requirements for a sound definition of well-posedness.

1. The algebraic criteria shall be simple (or better: as simple as possible).

2. It must be possible to apply these criteria "pointwise" to equations with variable coefficients.

We have shown that both these conditions are fulfilled if we choose definition 1.2. For definition 1.3 the algebraic criteria become much more involved while they are again reasonable for definition 1.4. It is not known whether for the last two definitions the second condition is fulfilled. However there are strong indications that this is so.

Let us now return to the one dimensional problems, and let us for simplicity use definition 1.3. To estimate the solution of (1.9) we always use the energy-method: We write the equations (1.9) as a first order system by introducing new variables

$$s^{j/m} w_j = d^j \hat{u}/dx^j \;, \quad j = 0,1,2, \ldots , m-1 \;.$$

Then (1.9) gets the form:

$$(1.12) \qquad \frac{d\tilde{w}}{dx} = M(s)\tilde{w} + \tilde{G} = s^{1/m}\left(M_0 + \sum_{\nu=1}^{m-1} s^{-\nu/m} M_\nu\right) + \tilde{G} \;.$$

Here $\tilde{w} = (w_0, w_1, \ldots , w_{m-1})'$, $\tilde{G} = s^{-(m-1)/m}(0,\ldots 0,,-\hat{F})'$

and

$$M_0 = \begin{pmatrix} 0 & I & 0 & \cdot & \cdot & \cdot & 0 \\ 0 & 0 & I & 0 & \cdot & \cdot & \cdot & 0 \\ \cdot & \cdot & \cdot & \cdot & \cdot & \cdot \\ 0 & \cdot & \cdot & \cdot & 0 & I \\ 0 & \cdot & \cdot & \cdot & 0 & A_m^{-1} & 0 \end{pmatrix} \;.$$

The boundary conditions get the form:

$$(1.13) \qquad s^{j/m} B_{jj} w_j + \sum_{i=0}^{j-1} s^{i/m} B_{ij} w_i = 0 ,$$

$$j = 0,1,2, \ldots , m-1 .$$

The main theorem then is

<u>Theorem 1.2.</u> Assume that the problem (1.11) has for every g and every s with $\text{Real } s > 0$ a unique solution. Then there are constants α_0, K and $\delta_j > 0$, $j = 1,2,3$ such that for $\text{Real } s > \alpha_0 \geq 0$ there is a matrix $H = H(s)$ with the following properties:

1) H is Hermitian and $|H| < K$,

2) $\tilde{w}^* H \tilde{w} \geq 0$ for all vectors $\tilde{w}$ which fulfill the boundary conditions (1.13),

3) $|s|^{(m-1)/m}(HM + M^*H) \geq \delta_3 \text{ Real}(s-\alpha_0)I$.

From theorem 1.2 the estimate (1.10b) follows without difficulties because (1.12) implies:

$$-\text{Real}(\tilde{w}, Hd\tilde{w}/dx) + \text{Real}(\tilde{w}, HM\tilde{w}) = -\text{Real}(\tilde{w}, H\tilde{G}) .$$

Now

$$\text{Real}(\tilde{w}, Hd\tilde{w}/dx) = \tilde{w}^*(0,s)H\tilde{w}(0,s) \geq 0 ,$$

and

$$|s|^{(m-1)/m}\ \text{Real}(\tilde{w},HM\tilde{w}) \geqslant \delta_3\ \text{Real}(s-\alpha_0)\|\tilde{w}\|^2 .$$

Therefore

$$\delta_3\ \text{Real}(s-\alpha_0)\|w\|^2 \leqslant K\|F\| \cdot \|w\|$$

and the estimate (1.10b) follows without difficulty.

For hyperbolic and parabolic systems it is easy to generalize the estimates to equations with variable coefficients because the matrix H depends smoothly on the coefficients of the differential equations. The following result holds:

Theorem 1.3. Consider hyperbolic and parabolic systems and assume that the coefficients $A_j(x)$ are Lipschitz continuous functions of x. Replace $A_j(x)$ by $A_j(0)$ and assume that this problem with constant coefficients is well posed in the sense of definition 1.2. Then the problem with variable coefficients is also well pos d in the same sense.

It should be pointed out that this theorem is also true for strictly hyperbolic systems in more than one space variable. Furthermore the obvious generalization to the case with two boundaries is also valid.

2. Difference Approximations for Hyperbolic Systems

Consider now a first order system of partial differential equations

(2.1) $\partial u(x,t)/\partial t = A\partial u/\partial x + F(x,t)\ ,\quad x \geqslant 0\ ,\quad t \geqslant 0\ ,$

with initial values

$$(2.2) \qquad u(x,0) = f(x) \ ,$$

and boundary conditions

$$(2.3) \qquad u^{I}(0,t) = Su^{II}(0,t) + g(t) \ .$$

Here

$$(2.4) \quad A = \begin{pmatrix} A_1 & 0 \\ 0 & A_2 \end{pmatrix}, \quad A_1 = \begin{pmatrix} a_1 & 0 & \cdots & 0 \\ 0 & a_2 & 0\cdots & 0 \\ \cdots & & & \cdots \\ 0 & \cdots & 0 & a_\ell \end{pmatrix} > 0 \ ,$$

$$A_2 = \begin{pmatrix} a_{\ell+1} & 0 & \cdots & 0 \\ 0 & a_{\ell+2} & 0\cdots & 0 \\ \cdots & & & \cdots \\ 0 & \cdots & 0 & a_n \end{pmatrix} < 0 \ ,$$

S is a rectangular matrix and $u^{I} = (u^{(1)}, \ldots u^{(\ell)})'$, $u^{II} = (u^{(\ell+1)}, \ldots , u^{(n)})'$ correspond to the partition of A .

We want to solve the above problem by difference approximation. For that reason we introduce a time-step $k > 0$, a mesh width $h = 1/N$, N natural number and divide the x-axis into intervals of length h . As usual we assume that $k/h = \lambda = \text{const.}$ Let $p, q, r,$ and s be natural numbers and use the notation:

$$v_\nu(t) = v(x_\nu,t) \ , \ x_\nu = \nu h \ , \ \nu = -r+1,-r+2, \ \ldots \ ,0,1,\ldots \ .$$

We approximate (1.1) for $\nu = 1,2,3, \ \ldots$, and $t = t_\tau = \tau k$, $\tau = s,s+1, \ \ldots$ by a multistep method

$$(2.5) \qquad Q_{-1} \, v_\nu(t+k) = \sum_{\sigma=0}^{s} Q_\sigma v_\nu(t-\sigma k) + k \, F_\nu(t) \quad .$$

Here

$$Q = \sum_{j=-r}^{r} A_{j\sigma} E^j \ , \quad E v_\nu = v_{\nu+1}$$

are difference operators with matrix coefficients. For the solution of (1.5) to be uniquely determined it is necessary to specify initial values

$$(2.6) \qquad v_\nu(\sigma k) = f_\nu(\sigma k) \ , \ \sigma = 0,1,2,\ldots,s; \ \nu = -r+1,-r+2,\ldots$$

and for $t = t_\tau \geqslant sk$ boundary conditions

$$(2.7) \quad v_\mu(t+k) = \sum_{\sigma=-1}^{s} S_\sigma^{(\mu)} v_1(t-\sigma k) + k \, F_\mu(t) \ , \ \mu = -r+1,\ldots,0.$$

Here

$$(2.8) \qquad S_\sigma^{(\mu)} = \sum_{j=0}^{q} C_{j\sigma}^{(\mu)} \ E^j$$

are one sided difference operators i.e. (2.7) expresses the solution at the boundary points $x_\mu \leqslant 0$ in terms of the solution at interior points.

We shall always assume that the approximation is stable for the Cauchy problem. Let us explain this assumption in detail: If we Fourier transform the homogeneous equations (2.5) with respect to x we get

$$(2.9) \qquad \hat{Q}_{-1}(i\xi)\, \hat{v}(t+k) = \sum_{\sigma=0}^{s} \hat{Q}_{\sigma}(i\xi)\, \hat{v}(t-\sigma k) \,,$$

$$\hat{Q}_{\sigma} = \sum A_{j\sigma}\, e^{ij\xi} \,,$$

and we make

Assumption 2.1. The von Neuman condition is fulfilled, i.e. the eigenvalue problem

$$(2.10) \qquad \left(\hat{Q}_{-1}(i\xi)\, z^{s+1} - \sum_{\sigma=0}^{s} \hat{Q}_{\sigma}(i\xi)\, z^{s-\sigma}\right) y = 0$$

has no solution z with $|z| > 1$.

It is well known that the von Neuman condition is not sufficient for stability, therefore we shall make also one of the two following assumptions:

Assumption 2.2. Let z_j, $j = 1,2, \ldots (s+1)n$ be the eigenvalues of (2.10). If $z_i = z_j$ for some $\xi = \xi_0$ and $|z_i| = 1$ then z_i, z_j are continuously differentiable and $\partial z_\nu(\xi_0)/\partial\xi \neq 0$, $\nu = i,j$.

This assumption is for example fulfilled if the approximation is dissipative. We can replaced the last assumption by:

Assumption 2.3. The matrices $A_{j\sigma}$ can be transformed to diagonal form by the same transformation.

Let $\ell_2(h)$ denote the space of all grid functions G_ν , $\nu = -r+1,-r+2, \ldots , 0,1, \ldots$ with $\|G\|_h < \infty$ and use the notation

$$(G,F)_h = \sum_{\nu=-r+1}^{\infty} G_\nu^* F_\nu h , \quad \|G\|_h^2 = (G,G)_h .$$

Corresponding to the definitions 1.2-1.4 for the continuous problem we have now three different definitions for stability. Here all these definitions lead to different algebraic criteria and we shall only discuss the weakest one:

Definition 2.1. Consider the difference approximation (2.5)-(2.7) with $f(\sigma k) \equiv 0$, $\sigma = 0,1,2, \ldots , s$. The approximation is stable if there are constants $\alpha_0 \geq 0$, $K > 0$ such that for all F with $F_\nu \equiv 0$, $\nu = 1,2, \ldots$ and all $\alpha > \alpha_0$ and all h

$$\text{(2.11)} \quad \sum_{\tau=0}^{\infty} \|e^{-\alpha\tau k} v(\tau k)\|_h^2 k \leq \left(\frac{K_0}{\alpha-\alpha_0}\right)^2 \sum_{\tau=0}^{\infty} e^{-\alpha\tau k} \|F(t)\|_h^k$$

holds. (Observe that $\|F\|_h^2 = \sum_{\mu=-r+1}^{0} |F_\mu|^2 h)$.

In the samy way as in the continuous case this definition is equivalent with an estimate of the resolvent equation

$$\text{(2.12)} \quad \left(Q_{-1} - \sum_{\sigma=0}^{s} z^{-\sigma-1} Q_\sigma\right) \hat{v}_\nu = 0 \qquad \nu = 1,2,\ldots .$$

$$\text{(2.13)} \quad \hat{v}_\mu - \sum_{\sigma=-1}^{s} z^{-\sigma-1} S_\sigma^{(\mu)} \hat{v}_1 = G_\mu , \ \mu = -r+1, \ldots , 0 .$$

The following theorem holds.

Theorem 2.1. The approximation is stable according to definition 2.1 if and only if (2.12), (2.13) has for every G_μ and every z with $|z| > 1$ a unique solution belonging to $\ell_2(h)$ with:

$$(2.14)\quad (|z| - 1)^2 \|\hat{w}\|_h^2 \leqslant |z|^2 K_o^2 \|G\|_h^2 , \quad \|G\|_h^2 = \sum_{\mu=-r+1}^{0} |G_\mu|_h^2$$

The proof of the theorem is again done by the energy method. Therefore it can be generalized to equations with variable coefficients and to problems with two boundaries.

We shall now derive explicit algebraic conditions for (2.14) to hold. The following lemma is essential.

Lemma 2.1. Consider the solutions $\kappa = \kappa(z)$ of the characteristic equation

$$(2.15)\quad \mathrm{Det} \left| \hat{Q}_{-1}(\kappa) - \sum_{\sigma=0}^{s} z^{-\sigma-1} \hat{Q}_\sigma(\kappa) \right| = 0 ,$$

$$\hat{Q}_\sigma(\kappa) = \sum_{j=-r}^{p} A_{j\sigma}{}^{j}$$

for $|z| > 1$. There is a constant $\delta > 0$ such that

$$(2.16)\qquad |z| \Big| |\kappa(z)| - 1 \Big| \geqslant \delta (|z| - 1) .$$

Lemma 2.1 implies immediately:

Theorem 2.2. Let $\xi > 0$ be any constant. An estimate of

type (2.14) holds for all $|z| > 1 + \varepsilon$ if and only if (2.12), (2.13) has no eigenvalue z with $|z| > 1$.

This is the well known Ryabenkii-Gudonow condition. We need thus only consider a neighbourhood of $|z| = 1$. Here the following lemma is essential.

Lemma 2.2. Consider the solutions of the characteristic equation (2.15) in a neighbourhood of a fixed point $z = z_0$ with $|z_0| = 1$. Then the $\kappa_j(z)$ with $|\kappa_j| < 1$ for $|z| > 1$ can be developed into Puisseux series

$$\kappa_{j\tau}(z) = \kappa_j(z_0) + a(z - z_0)^{1/r_j} + \cdots \qquad \tau = 1,2,\ldots\ell_j$$

and there are constants $\delta_j > 0$ such that either

$$|\kappa_j(z_0)| < 1 - \delta_1 \quad \text{or} \quad |\kappa_j(z_0)| = 1 .$$

Therefore the general solution of (2.12) which for $|z| > 1$ lies in $\ell_2(h)$ can be written as:

$$\hat{v}_\nu = \sum_j \sum_{\tau=1}^{\ell_j} \lambda_{j\tau}\, y_{j\tau}(z,\nu) .$$

Here $y_{j\tau}(z,\nu)$ are also Puisseux series which depend linearly on $\kappa_{j1}^{\nu}(z)$, ... $\kappa_{j\ell_j}^{\nu}(z)$, are linearly independent for $z = z_0$ and for which an estimate

$$|y_{j\tau}| \leqslant \text{const.}(1 + |\nu|^{\tau-1})\ |\kappa_{j\tau}|^{\nu}$$

holds.

To determine the $\lambda_{j\tau}$ we introduce (2.3) into the boundary conditions (2.3) thus deriving a linear system for them. Then we can for example show

<u>Theorem 2.3</u>. The approximation is stable in the sense of definition 2.1 if the equations (2.12) and (2.13) have no eigenvalue z with $|z| > 1$ and in the neighbourhood of every z_0 with $|z_0| = 1$ estimates

$$\frac{|\lambda_{j\tau}|^2}{\sum_{\mu=-r+1}^{0} |G_\mu|^2} \leq \begin{cases} \dfrac{\text{const}}{|z-z_0|^2} & \text{if } |\kappa_j(z_0)| < 1 \\[2ex] \dfrac{\text{const}}{|z-z_0|^{2-2/r_j}}, & \text{if } |\kappa_j(z_0)| = 1 \end{cases}$$

hold.

<u>Remark</u>. If the approximation is dissipative and $|\kappa_j(z_0)| = 1$ then necessarily $r_j = 1$. Therefore the conditions of the last theorem are much simpler in that case. Furthermore in most applications all $\lambda_{j\tau}$ are bounded. Extensive examples are discussed in [2].

REFERENCES

1. Kreiss, H. O., "Initial Boundary Value Problems for Partial Differential Equations in One Space Dimension," to appear.
2. Gustavsson, B., Kreiss, H. O., and Sundström, A., "Stability Theory of Difference Approximations for Mixed Initial Boundary Value Problems II," to appear.

NUMERICAL SOLUTION OF CONDITIONALLY PROPERLY POSED PROBLEMS

M. M. Lavrentiev*

1. The problems of mathematical analysis to be discussed here are usually called in literature improperly posed problems.

The problems are said to be "improperly posed" if the principle of continuous dependence of their solution on the initial data is violated. However, in formulating "improperly posed" problems [1], [5] it is a priori assumed that the solution belongs to some set (usually to a compact set). In this case the solution on the given set continuously depends on the data. Therefore it seems to us more natural to call the problems under consideration "conditionally properly posed problems" (for the first time this term was used in Ref [4]).

Investigation of the methods of the numerical solution of conditionally properly posed problems involves additional difficulties (as compared with classical properly posed problems). The theory of the numerical solution of such problems is just developing.

*Computer Center, Siberian Branch of the Academy of Sciences of the U.S.S.R., Novosibirsk.

In the present paper we will report two results on the theory of the numerical solution of conditionally properly posed problems.

We will consider only the following linear operator equations:

$$(1.1) \qquad Ax = f, \quad x \in X, \quad f \in F,$$

where X, F are Banach spaces, A a linear completely continuous operator.

In studying conditionally properly posed problems of great importance is the notion of regularization [2], [3]. Instead of Eq.(1.1) we consider a family of equations:

$$(1.2) \qquad A_\alpha x = f$$

depending on the parameter $\alpha > 0$ and satisfying the following conditions:

1) The operator A_α has a continuous inverse A_α^{-1}.
2) For any $x \in X$

$$\lim_{\alpha \to 0} A_\alpha^{-1} A x = x .$$

A family of such operators $A_\alpha^{-1} = R_\alpha$ is called a regularizer. The following is one of the approaches to the construction of the numerical solution of the conditionally properly posed problems. A conditionally properly posed problem is a regularized (a family A_α or R_α is built), then a numerical method is built for the classical properly posed problem of Eq.(1.2) (or numerical realization of the operator R_α). We present V. G. Vasiliev's results devoted to the development of the above approach [6].

2. Consider a linear operator equation

$$(2.1) \qquad Ax = f\,, \quad x \in X\,, \quad f \in F\,,$$

where X is a Banach, F a hilbert space, $A[X \to F]$ a linear continuous operator whose inverse operator A^{-1} exists but is not bounded.

Let the solution x_0 of Eq. (2.1) exist at $f = f_0$ and belong to the correctness class M to be determined as in [5].

Denote by Z a Hilbert space and consider a continuous operator $B[Z \to X]$ such that B^{-1} exists.

The image of a sphere $\|z\| \leqslant r$ in Z under the mapping B will be regarded as the correctness class M.

We introduce now the notations

$$(2.2) \qquad C = AB\,, \quad C[Z \to F]\,, \quad x = Bz\,, \quad z \in Z\,,$$

then Eq. (2.1) can be written as

$$(2.3) \qquad Cz = f\,,$$

and, under the above assumption,

$$(2.4) \qquad Cz_0 = f_0\,, \quad x_0 = Bz_0 \in M\,.$$

Let us bring Eq. (2.4) to correspond to

(2.5) $$C_1 z^\alpha + \alpha z^\alpha = C^* f_o \ , \quad C_1 = C^* C \ ,$$

$$\alpha > 0 \ , \quad x^\alpha = B z^\alpha \ .$$

If B is continuous, then for any $x_o \in B[Z]$ the relations

(2.6) $$\lim_{\alpha \to 0} \| z^\alpha - z_o \|_Z = 0 \ , \quad \lim_{\alpha \to 0} x^\alpha = x_o$$

are valid, i.e. the operator $R_\alpha = B(C_1 + \alpha E)^{-1} C^*$ is a regularizer, see [7].

Next, let X_h, F_h, $Z_h[h = (h_1, h_2, \ldots, h_n) \in R_n$, $h_i > 0]$ be the total subspaces corresponding to the spaces X, F, Z and $P_h^{(1)}$, $P_h^{(2)}$, $P_h^{(3)}$ linear operators of the projection of X onto X_h, F onto F_h and Z onto Z_h, respectively, so that

$$X_h = P_h^{(1)}[X] \ , \quad F_h = P_h^{(2)}[F] \ , \quad Z_h = P_h^{(3)}[Z] \ .$$

In addition, let us introduce into consideration the linear continuous operator $B_h[Z_h \to X_h]$, where

(2.7) $$\gamma_1(h) = \| B - B_h \|_{Z_h \to X} \to 0$$

if $h \to h_o$, $h^o = (h_1^o, h_2^o, \ldots, h_n^o)$, $h_i > 0$.

The following

(2.8) $$C_{1h} z_h^\alpha + z_h^\alpha = C_h^* f_h \ , \quad C_{1h} = C_h^* C_h \ ,$$

$$f_h \in F_h \ , \quad z_h^\alpha \in Z_h$$

will be taken as an "approximate" equation, where

(2.9) $$\alpha = \alpha(h) , \quad \lim_{h\to h^o} \alpha(h) = 0 ,$$

and

(2.10) $$\gamma_2(h) = \frac{1}{\alpha(h)} \|f_o - f_h\|_F \to 0 \quad \text{if} \quad h \to h^o .$$

Here $C_h[Z_h \to F_h]$ is a linear continuous operator satisfying the conditions

(2.11) $$\gamma_3(h) = \frac{1}{\alpha(h)} \|C^* - C_h^*\|_{F_h \to Z} \to 0 \quad \text{if} \quad h \to h^o ,$$

and

$$\gamma_4(h) = \frac{1}{\alpha^2(h)} \|C_{1h} - C_1\|_{Z_h \to Z} \to 0 \quad \text{if} \quad h \to h^o .$$

From (2.5) and (2.8) it follows that

(2.12)

$$\begin{aligned} z^\alpha - z_h^\alpha &= (C_1+\alpha E)^{-1} C^* f_o - (C_{1h}+\alpha E)^{-1} C_h^* f_h \\ &= (C_1+\alpha E)^{-1}\{C^* f_o - C_h^* f_h \\ &+ [E - (C_1+\alpha E)(C_{1h}+\alpha E)^{-1}] C_h^* f_h\} \\ &= (C_1+\alpha E)^{-1}\{C^*(f_o - f_h) + (C^* - C_h^*) f_h \\ &+ (C_{1h} - C_1)(C_{1h}+\alpha E)^{-1} C_h^* f_h\} . \end{aligned}$$

In [5] we see that

$$\|(C_1+\alpha E)^{-1}\| = \frac{1}{2}, \quad \|(C_{1h}+\alpha E)^{-1}\| = \frac{1}{2}$$

and hence from (2.12) we get the inequality

(2.13)

$$0 \leqslant \|z^\alpha - z_h^\alpha\|_Z \leqslant \gamma_2(h)\,\|C^*\| + \gamma_3(h)(\,\|f_o\| + \|f_o - f_h\|_F\,)$$

$$+\ \gamma_4(h)(\,\|C^*\| + \|C^* - C_h^*\|_{F_n \to Z})(\,\|f_o\| + \|f_o - f_h\|_F\,)\ .$$

It is evident that under the above assumptions the equalities

$$\lim_{h\to h^o} \|f_o - f_h\|_F = 0\ ,$$

$$\lim_{h\to h^o} \|C^* - C_h^*\|_{F_h \to Z} = 0$$

are valid.

Therefore, if we use equalities (2.10)-(2.12) and inequality (2.13), we get

$$\lim_{h\to h^o} \|z^\alpha - z_h^\alpha\| = 0\ . \tag{2.14}$$

Then, the inequality

$$0 \leqslant \|z_o - z_h^\alpha\| \leqslant \|z_o - z^\alpha\| + \|z^\alpha - z_h^\alpha\| \tag{2.15}$$

is valid. But, according to (2.6),

$$\lim_{h\to h^o} \|z_o - z^\alpha\| = \lim_{\alpha\to 0} \|z_o - z^\alpha\| = 0\ . \tag{2.16}$$

Hence, from (2.14), (2.15) and (2.16) we have

(2.17) $$\lim_{h\to h^o} z_h^\alpha = z_o \ .$$

As C_{1h} is a self-adjoint positive operator and $\alpha > 0$, the solution z_h^α of Eq. (2.8) exists and is unique for any f_h from F_h.

Therefore, specifying $f_h \in F_h$, we find the unique solution

(2.18) $$x_h^\alpha = B_h z_h^\alpha$$

which can be taken as an approximate solution of the initial equation (2.1).

Indeed, from the inequality

$$0 \leq \|x_o - x_h^\alpha\| = \|B z_o - B_h z_h^\alpha\|$$

$$\leq \|B\| \cdot \|z_o - z_h^\alpha\| + \|B - B_h\| \cdot \|z_h^\alpha\|$$

and equalities (2.7) and (2.17) it follows that

(2.19) $$\lim_{h\to h^o} x_h^\alpha = x_o \ .$$

In this case, if the operator B is completely continuous, then it is shown in [7] that

$$\|x_o - x_h^\alpha\| \leq \omega(\alpha)\, \|z_o\| \ , \quad \lim_{\alpha\to 0} \omega(\alpha) = 0 \ ,$$

where

$$\omega(\eta) = \sup \|Bz\| \quad \text{if} \quad \|z\| \leqslant 1 , \quad \|C_1 z\| \leqslant \eta .$$

Therefore

$$(2.20) \qquad \|x_o - x_h^\alpha\| \leqslant \|x_o - x^\alpha\| + \|x^\alpha - x_h^\alpha\|$$

$$\leqslant \omega(\alpha) \|z_o\| + \|B\| \{\gamma_2(h) \|C^*\|$$

$$+ [\gamma_3(h) + \gamma_4(h) \|C_h^*\|] \|f_n\|\}$$

$$+ \gamma_1(h) \|z_h\| .$$

If the upper bound for $\|z_o\|$ is known, the last inequality gives an estimation of the deviation of the approximate solution x_h^α from the exact x_o.

In this way we have proved the following theorem.

Theorem. If the operator B is continuous and the limit equalities (2.7), (2.9)-(2.11) are valid, then

$$\lim_{h \to h^o} x_h^\alpha = x_o .$$

If, in addition, the operator B is completely continuous, then the estimation (2.20) of the deviation of the approximate solution x_h^α from the exact x_o holds.

Similar results can be obtained also for other regularizations of Eq. (2.1), for example, for the solution of (2.1) by the iterative method when the regularization parameter is the number of iterations.

3. An important class of operator equations of the first kind is the class of the Volterra-type equations of the first kind. In constructing the numerical solution of these equations one can use universal regularizations (e.g. regularizer (2.6)). However, in this case R_α is not an operator of Volterra-type which makes its numerical realization difficult.

We present V. O. Sergeev's results on constructing a Volterra-type regularizer for integral Volterra equations of the first kind.

Thus consider the integral equation

$$\int_0^x K(x,t)\phi(t)dt = f(x) , \quad x, t \in [0,x_o] . \tag{3.1}$$

Let the kernel $K(x,t)$ vanish on the diagonal together with its derivatives with respect to x up to the order $n-1$ inclusive, $K_n(x,t) = \frac{\partial^n K(x,t)}{\partial x^n}$ be continuous with respect to x and t and

$$\max_{x,t} |K_n(x,t)| \leq K_n , \quad K_n(x,x) = 1 .$$

After n differentiations of both parts of the equation we get

$$\phi(x) + \int_0^x K_n(x,t)\phi(t)dt = f^{(n)}(x) . \tag{3.2}$$

If the function $f \in C_o^{(n)}(0,x_o) = \{v | v \in C^{(n)}(0,x_o),$ $\frac{d^i v(0)}{dx^i} = 0, \ i = 0,\dots,n-1\}$ Eq. (3.1) has the unique solution $\phi^* \in C^o$. It is a priori assumed that $\phi^* \in C^{(m)}$.

First, it will be assumed that the values

$$\left.\frac{d^k \phi^*(x)}{dx^k}\right|_{x=o} = \phi_k^* , \qquad k = 0,1,\cdots,m-1 ,$$

are known. Consider the function

$$W_\varepsilon(x) = \sum_{k=o}^{n+1} \frac{1}{\varepsilon^k} C_{m+k-1}^{m-1} e^{-\frac{x}{\varepsilon}} \frac{x^{m+k-1}}{(m+k-1)!}$$

such that $x = 0$: $W_\varepsilon^{(j)}(0) = 0$, $j = 0,1,\ldots,m-2,m,\ldots,m+n-1$; $W_\varepsilon^{(m-1)}(0) = 1$.

Determine, on the space $C^o(0,x_o)$, the operator

$$B_\varepsilon \in L(C^o(0,x_o) \to C_o^{(n)}(0,x_o)) : \ B_\varepsilon u = y ,$$

$$y(x) = u(x) - \frac{d^m}{dx^m} \int_0^x W_\varepsilon(x-t)u(t)dt = - \int_0^x W_\varepsilon^{(m)}(x-t)u(t)dt .$$

If $\varepsilon > 0$, B_ε has an inverse. Indeed, its kernel on the diagonal vanishes together with all the derivatives with respect to x up to the order n, and the $(n+1)$-th derivative: $W_\varepsilon^{(m+n+1)}(x-t)$ if $\varepsilon > 0$ is bounded and $W_\varepsilon^{(m+n+1)}(0) = - \frac{1}{\varepsilon^{n+2}} C_{m+n+1}^{n+2}$.

Then, if $\varepsilon > 0$, Eq. (3.1) and

$$(3.3) \qquad \phi + B_\varepsilon K_n \phi - f_o = B_\varepsilon f^{(n)} - f_o + \phi - B_\varepsilon \phi$$

are equivalent and $f_o(x)$ is the function of the boundary layer type:

$$f_o(x) = \sum_{k=0}^{m-1} \phi_k^* W_\varepsilon^{(m-k-1)}(x)$$

Integrating by parts, we get

$$\phi^* - f_o - B_\varepsilon\phi^* = - f_o + \int_0^x W_\varepsilon^{(m-1)}(x-t)(\phi^*)'dt$$

$$+ \phi_o^* W_\varepsilon^{(m-1)}(x) = \cdots = \int_0^x W_\varepsilon(x-t)\phi^{*(m)}(t)dt .$$

Defining $\max\limits_{x\in[0,x_0]} |\phi^{*(m)}(x)| = \Phi_m$, we get the estimation

$$\text{(3.4)} \quad \|\phi^* - B_\varepsilon\phi^* - f_o\|_{C_o} \leqslant \Phi_m \varepsilon^m C_{m+n+1}^m = \Phi_m \varepsilon^m A ,$$

$$A = C_{m+n+1}^m .$$

Consider now the kernel of the integral operator $B_\varepsilon K_n$:

$$u = B_\varepsilon K_n \phi ,$$

$$u(x) = - \int_0^x \phi(t) \int_t^x W^{(m)}(x-\tau)K_n(\tau,t)d\tau dt .$$

The kernels modulus can be estimated as follows:

$$\text{(3.5)} \quad \left|\int_t^x W_\varepsilon^{(m)}(x-\tau)K_n(\tau,t)d\tau\right| \leqslant K_n \int_0^\infty |W_\varepsilon^{(m)}(t)|dt$$

$$\leqslant 2^m C_{m+n+1}^m K_n = B .$$

In a similar way we estimate the norm $\|B_\varepsilon f^{(n)}\|_{C^0}$:

$$\|B_\varepsilon f^{(n)}\|_{C^0} \leqslant \max_{(0,x_0)} |f^{(n)}(x)| \cdot 2^m C^m_{m+n+1} . \tag{3.6}$$

Consider the equation

$$\phi + B_\varepsilon K_n \phi = B_\varepsilon f^{(n)} + f_0(x) . \tag{3.7}$$

If $\varepsilon > 0$ the right-hand side is bounded and, by estimations (3.5) and (3.6), we get the existence and uniqueness of the solution $\phi_\varepsilon(x)$ of this equation. The difference $w_\varepsilon(x) = \phi^*(x) - \phi_\varepsilon(x)$ satisfies the equation

$$w_\varepsilon + B_\varepsilon K_n w_\varepsilon = \phi^* - B_\varepsilon \phi^* - f_0$$

hence, according to (3.4)

$$\|\phi^* - \phi_\varepsilon\|_{C^0} \leqslant \varepsilon^m \Phi_m A e^{Bx_0} . \tag{3.8}$$

Thus, if $\varepsilon \to 0$, $\phi_\varepsilon \to \phi^*$ uniformly on $[0,x_0]$.

Integrating by parts and taking into account that $f^{(k)}(0) = 0$, $k = 0,1,\dots,n-1$, we get for the function $B_\varepsilon f^{(n)}$ the following representation:

(3.9)

$$- \int_0^x W_\varepsilon^{(m)}(x-t) f^{(n)}(t)dt = - \int_0^x W_\varepsilon^{(m+1)}(x-t) f^{(n-1)}(t)dt$$

$$= \cdots = - \int_0^x W_\varepsilon^{(m+n)}(x-t) f(t)dt .$$

Extend the domain of definition of the right-hand side of Eq. (3.7) according to formula (3.9). Then estimate the value $\|B_\varepsilon f^{(n)}\|_{C^0}$ in terms of $\|f\|_{C^0}$. We have

$$\|B_\varepsilon f^{(n)}\|_{C^0} \leqslant \|f\|_{C^0} \cdot \frac{2^{n+m}}{\varepsilon^n} C^m_{m+n+1}$$
$$= \|f\|_{C^0} \frac{2^{n+m}}{n} \cdot A \; .$$

Now it will be assumed that, instead of the function $f(x)$ in the right-hand side of Eq. (3.1), we know the function $f^\delta(x) = f(x) + \delta(x)$, where $\delta(x)$ is continuous and $\|\delta(x)\| \leqslant \delta$. Consider the equation

$$\phi(x) - \int_0^x \phi(t) \int_t^x W_\varepsilon^{(m)}(x-\tau)K_n(\tau,t)d\tau dt$$
$$= f_0(x) - \int_0^x W_\varepsilon^{(m+n)}(x-t)f(t)dt \tag{3.10}$$

whose solution will be denoted as $\phi_\varepsilon^\delta(x)$. Then the difference $w_\varepsilon^\delta(x) = \phi_\varepsilon(x) - \phi_\varepsilon^\delta(x)$ satisfies the equation

$$w_\varepsilon^\delta(x) - \int_0^x \phi(t) \int_t^x W_\varepsilon^{(m)}(x-\tau)K_n(\tau,t)d\tau dt$$
$$= - \int_0^x W_\varepsilon^{(m+n)}(x-t)\delta(t)dt \; .$$

Hence,

$$\|\phi_\varepsilon - \phi_\varepsilon^\delta\|_{C^0} \leqslant \frac{\delta}{\varepsilon^n} A \cdot e^{Bx_0} \cdot 2^{n+m} \; . \tag{3.11}$$

Joining estimations (3.8) and (3.11) we get

$$\|\phi^* - \phi_\varepsilon^\delta\|_{C^0} \leqslant \varepsilon^m \Phi_m A e^{Bx_0} + \frac{\delta}{\varepsilon^n} 2^{n+m} A e^{Bx_0} .$$

The minimum in ε of this expression is obtained if

$\varepsilon(\delta) = 2\left(\frac{n}{m\Phi_m}\right)^{\frac{1}{m+n}} \delta^{\frac{1}{m+n}}$ and it equals

$$(3.12) \qquad 2^m A e^{Bx_0} \left(1+ \frac{m}{n}\right) \Phi_m^{\frac{n}{n+m}} \left(\delta \frac{n}{m}\right)^{\frac{m}{n+m}} .$$

Now compare the order in δ of this estimation with a highest attainable order of error under our assumptions concerning the error of the right-hand side of Eq. (3.1) and the smoothness of the exact solution.

For this purpose let us consider the functions $\delta(x) \in C^{(n+m)}$, $\|\delta\|_{C^0} < \delta$ whose derivatives up to the order m vanish if $x = 0$. Turning to Eq. (3.2) with the right-hand side $f^{(n)}(x) + \delta^{(n)}(x)$ we get for its solution, $\tilde{\phi}_\delta$ the estimation

$$\|\phi^* - \tilde{\phi}_\delta\|_{C^0} \leqslant e^{k_n x_0} \|\delta^{(n)}\|_{C^0}$$

and, since $\delta^{(n)} \in C_0^{(m)}$ then it is seen in [8] that

$$\|\delta^{(n)}\|_{C^0} \leqslant C \cdot \|\delta\|_{C^0}^{\frac{m}{n+m}} , \qquad C = \text{const} ,$$

$$\|\phi^* - \tilde{\phi}_\delta\|_{C^0} \leqslant C_1 \cdot \delta^{\frac{m}{n+m}} , \qquad C_1 = \text{const} .$$

Thus, the order of estimation (3.12) coincides with the highest attainable order of the error under our assumptions.

If the values ϕ_k^*, $k = 1,\dots,m-1$, are not known, but ϕ_0^* is known, then, letting $f_0(x) = \phi_0^* W_\varepsilon^{(m)}(x)$ in Eq. (3.10) we get that the difference $\tilde{w}_\varepsilon$ between the solution $\tilde{\phi}_\varepsilon$ of such an equation and the solution ϕ_ε^δ satisfies the equation

$$\tilde{w}_\varepsilon(x) - \int_0^x \tilde{w}_\varepsilon(t) \int_t^x W_\varepsilon^{(m)}(x-\tau) K_n(\tau,t)\, d\tau\, dt$$

$$= \sum_{k=1}^{m} \phi_k^* W_\varepsilon^{(m-k)}(x) = \tilde{f}_0(x)$$

and then

$$\tilde{w}_\varepsilon = \tilde{f}_0 + B_\varepsilon K_n \tilde{f}_0 + B_\varepsilon K_n (B_\varepsilon K_n \tilde{f}_0) + B_\varepsilon K_n)^2 (B_\varepsilon K_n \tilde{f}_0) + \cdots .$$

But

$$\| B_\varepsilon K_n \tilde{f}_0 \|_{C^0} \leqslant \max_x \int_0^x \left| W_\varepsilon^{(m)}(x-t) \right| \cdot \left| \int_0^t K_n(t,\tau) \tilde{f}_0(\tau)\, d\tau \right| dt$$

$$\leqslant K_n \cdot \int_0^\infty \sum_{k=1}^{m} \phi_k^* \, |W_\varepsilon^{(m-k)}(t)| \, dt \cdot 2^m C_{m+n+1}^m \leqslant \varepsilon \cdot C_3 ,$$

$$C_3 = \text{const} .$$

Thus, $\tilde{\phi}_\varepsilon - \phi_\varepsilon^\delta$ is the sum of a function of the type of the boundary value and of a function which decreases with ε.

REFERENCES

1. Tychonov, A. N., "On the Stability of Inverse Problems" (Russian), Dokl.Akad.Nauk SSSR, 39, No. 5, 1944.
2. Tychonov, A. N., "Incorrectly Posed Problems and the Method of Regularization" (Russian), DAN SSSR, 151, No. 3, 1963.
3. Tychonov, A. N., "On the Regularization of Improperly Posed Problems" (Russian), DAN SSSR, 153, No. 1, 1963.
4. Krein, S. G., "On Correctness Classes for Some Boundary-Value Problems" (Russian), DAN SSSR, 114, No. 6, 1957.
5. Lavrentiev, M. M., "On Some Improperly Posed Problems of Mathematical Physics"(Russian), Izd.SO AN SSSR, Novosibirsk, 1962.
6. Vasiliev, V. G., "On Approximate Methods of Solving Operator Equations of the First Kind" (Russian), Sb. "Matematicheskie Problemy Geofiziki," Novosibirsk, 1969.
7. Ivanov, V. K., "On the Regularization of Linear Operator Equations of the First Kind" (Russian), Izv.Vuzov, Matematika, No. 10, 1967.
8. Krein, S. G., Petunin Yu.I., "Scales of Banach Spaces." Russian Math. Surveys, Vol. 21, No. 2, March-April, 1966.

ON GENERAL PURPOSE PROGRAMS FOR FINITE ELEMENT ANALYSIS, WITH SPECIAL REFERENCE TO GEOMETRIC AND MATERIAL NONLINEARITIES

Pedro V. Marcal*

Abstract

A summary of the state-of-the-art in nonlinear finite element analysis is made by describing a nonlinear theory and presenting some case studies. The formulation is applicable to problems of large displacement and small strains.

The paper then focuses on the general purpose program. The concept and development of a general purpose program is described. A discussion is then made of the different sizes of problems which can be solved by such a program. These sizes are dependent on the available computer core. The conclusion is made that the general purpose program is a powerful means of implementing finite element analysis over a wide spectrum of problems in structural mechanics.

---

* Associate Professor of Engineering, Brown University, Providence, Rhode Island.

PEDRO V. MARCAL

Introduction

In recent years we have seen an increasing use of the finite element method for both research and development. This paper briefly summarizes the method and traces the reasons for its widespread use.

The finite element method is dependent on the combination of two basic ideas. The first is the recognition that problems in continuum mechanics may be solved by complete satisfaction of only one of the two requirements of equilibrium and compatibility if the other condition is also satisfied in an integral sense. This approximate solution of the remaining condition in the integral sense is brought about by the use of the principle of virtual work and theorems resulting from it. The second idea is that the function for a whole domain may be better approximated by local functions assumed within subdomains which also maintain continuity of the functions across the subdomains. The undetermined parameters for the assumed local subdomain functions are then related to physical quantities of displacement [1] or force [2] degrees of freedom at points or nodes on the boundaries of the subdomain. This then allows the definition of equations which define either stiffness or flexibility matrices for the subdomain (or discrete element). The combination of the relaxation of the requirement of either equilibrium or compatibility and the localized functions whose unknowns are represented by quantities at nodes results in a considerable easing of the problems of geometry and boundary conditions. Different subdomains may be modeled with different functions and these may be used simultaneously for an analysis.

The finite element theory is usually cast in matrix theory since this allows the large background of matrix theory to be exploited. Its development occurred at the same time as the order of magnitude increase in computer speeds and core size. This happy confluence of all the factors discussed above has given rise to the widespread development and use of the finite element method. Initially, its implementation took place in the form of specialized programs written for specific purposes. Then as the method developed it became obvious that a more general approach could be adopted in which the common tasks for every finite element could be programmed once and for all. This has resulted in the development of the general purpose finite element programs. These efforts have the same overall strategy as the SPADE projects adopted for partial differential equations. At present, the continued development of the general purpose programs appears to be the best means of implementing finite element theory. Yet notwithstanding this, little has appeared in the literature which specifically concerns itself with the features and underlying philosophy of the general purpose program. It is the purpose of the current paper to discuss the development of a general purpose program.

Review of Literature

The present paper will trace developments from the original paper by Turner et al. [1], and attention will be confined solely to the direct stiffness method of finite element analysis.

Initially, work was concerned with developing elements [3-8] with compatible displacements at the boundaries. This phase can now be said to be complete, and elements exist to cover any two- or three-dimensional solids (including shell structures). We may classify elements by the type of displacement modes assumed and with this classification three types of elements can be recognized. In its two-dimensional form these three may be referred to as the triangular, the orthogonal and the piecewise patching type of element.

In the triangular type of element, the displacement modes are assumed to take the form of complete polynomials [1-4]. In the orthogonal type of element, the displacement modes are assumed to take the form of either Lagrange Polynomials [3,5] or Hermitian Polynomials [6]. The Lagrange Polynomials [5] are also used to extend the element formulation to quadrilaterals and curve shaped elements by isoparametric techniques. In the patching type of element [7, 8], usually used for shells and referred to as the DeVeubeke element, the displacement modes are assumed to be made up of a compatible patching of complete polynomials. Three-dimensional equivalents also exist for the first two types of elements.

At the same time work [9,10] was reported which established a framework by which the finite element method could be related to the methods used in continuum mechanics. The finite element method is now recognized as a special case of the Rayleigh-Ritz method where generalized modes are assumed over subdomains where the generalized modes give rise to inter-element compatibility of displacements.

With the establishment of the method, attention was turned to extensions for nonlinear analysis. In the area of material nonlinearity, two methods were developed for elastic-plastic analysis. The method of initial strains is based on the idea of modifying the equations of equilibrium so that the elastic equations can be used throughout on the left-hand side of the equations. Modifications are introduced on the right-hand side of the equation to compensate for the fact that the plastic strains do not cause any change in the stresses. On the other hand, the tangent modulus method is based on the linearity of the incremental laws of plasticity and approaches the problem in a piecewise linear fashion. The load is applied in increments, and at each increment a new set of coefficients is obtained for the equilibrium equations. The matrix equations for the finite element analysis using the method of initial strains were developed by Padlog et al. [11], Argyris et al. [12] and Jensen et al. [13]. A recent paper by Witmer [14] summarizes the latest application of the method. The equations for the tangent modulus method were developed by Pope [15], Swedlow and Yang [16] and Marcal and King [17]. The two methods were compared by Marcal [18] and a close similarity was found between them.

Progress has also been made in the area of geometric nonlinearity. Large displacement analysis by the finite element method was first proposed by Turner et al. [19]. Initial stress stiffness matrices were developed to account for the effect of initial stress in truss and plane stress assemblies. Subsequent work on the derivation of the initial stress matrices for other elements was reported by Argyris et al. [20], Gallagher et al. [21] and Kapur and Hartz [22]. Martin [23] placed the derivation of the initial stress matrix on a firm foundation by using a potential energy formulation together with the nonlinear strain displacement relation (for Green's strain). The above papers were concerned with forming matrices which account for geometric changes in the solid during an increment of load. These matrices were then either used in a piecewise linear manner or used in an eigenvalue analysis of the Euler type. Recent papers [24,25] have drawn attention to the fact that certain important terms were neglected in finite element large displacement analysis. These neglected terms result in what was called the initial displacement matrix and is a result of the coupling between the quadratic and the linear terms in the strain displacement expressions.

Other workers solved the nonlinear equations of the finite element method directly. Bogner et al. [26] performed a direct minimization of the potential energy without explicitly forming the matrix stiffness equations. The large displacement behavior was followed into the post-buckling region. Mallett and Berke [27] applied this method to frameworks and Bogner et al. to plates and shells [28]. Oden [29] and Oden and Kubitza [30] developed nonlinear stiffness relations for the nonlinear elasticity problem. The equations were solved by a Newton-Raphson method. This series of works is perhaps best placed in a separate category. It concerns itself with large strain, large displacement analysis. The other papers reviewed previously have all implicitly assumed a large displacement small strain theory. In addition, the constitutive equations used there are in terms of an energy potential which is appropriate for a rubber-like material. A similar formulation with appropriate assumptions of constitutive behavior gives rise to equations for large strain large displacement analysis of metal structures [31]. A recent survey by Oden [32] brings out the very general nature of the finite element formulation.

Hence, we have seen that progress has been made in both nonlinear material and geometric behavior. The two nonlinear formulations do not depend on each other so that they may be profitably combined. In the present paper we shall restrict our attention to a small strain large displacement theory appropriate to shells and other solid metal structures.

In the area of elastic analysis by the finite element method, general purpose computer programs exist which are written with a view to covering the whole area of stress analysis. These general purpose programs exploit the generality of the matrix formulation of the finite ele- method. The programs have a library of elements which can be used for the modeling of most structures in service. Melosh et al. [33] have summarized the more recent general purpose programs.

Technical Considerations

The theory outlined here has been developed previously in [34]. It is included here for completeness. The displacement method of finite element analysis will be used throughout. The structure to be analyzed is divided into a number of elements. The behavior of each element is lumped into a number of nodal point displacements. Conforming displacement modes or simply conforming elements are modes which maintain displacement compatibility between adjacent elements at the element boundaries. The principle of virtual work is then used to effect the lumping of the equivalent forces. The principle of virtual work is of course applicable to large displacement as well as nonlinear material behavior.

A brief outline of the method is now given. A displacement mode is first chosen for the type of element being used,

$$u = \sum_{i=1}^{n} a_i f_i(x) = [f(x)]\{a\} \tag{1}$$

where u is the displacement at position x

x is used to represent the coordinates of the element

a_i are the generalized displacements (also written $\{a\}$)

n is the number of terms in the summation.

By substituting for x at the nodes obtain

$$\{a\} = [\alpha]\{u\} \tag{2}$$

where $\{u\}$ is the displacement at the nodal points (note that there is a distinction between the bracketed and unbracketed u).

$[\alpha]$ is the nodal point to generalized displacement transformation matrix.

By the assumption of small strains we also have

(3) $$\Delta\{a\} = [\alpha]\Delta\{u\}$$

where the prefix Δ denotes an increment of the quantity immediately following it.

We now define the so-called differential operator [B] which transforms an increment of generalized displacement to an increment of strain.

(4) $$\Delta\{e\} = [B]\Delta\{a\}$$

The differential operator [B] is a function of position x and of current displacement u. It is defined by writing the nonlinear strain displacement equations (Green's strain) in incremental form.

The strain increment can be written as a stress increment for an elastic-plastic material in the manner of Marcal and King for solids [17] and Marcal [35] for plates and shells.

(5) $$\Delta\{\sigma\} = [D]\Delta\{e\}$$

The stress increments $\Delta\{\sigma\}$ are accumulated at representtative points within each element. Each representative point is given a set of reference axes which deform with the element and so take the same direction as that defined by an increment of Green's strain. Thus stresses and strain increments are automatically aligned and the non-linear equations of equilibrium can be set up with ease.

We now use the principle of virtual work to define equivalent forces $\{P\}$ at the nodes for a virtual displacement $\delta\lfloor u \rfloor$.

$$(6) \qquad \Delta\lfloor u \rfloor\{P\} = \int_V \delta\lfloor e \rfloor\{\sigma\}dV = \int_V \delta\lfloor u \rfloor[\alpha]^T[B]^T\{\sigma\}dV$$

where $\lfloor\ \rfloor$ denotes a row vector, and integration is performed over the volume V.

Cancelling the non-zero virtual displacements from both sides and writing equation (6) in incremental form with aid of equation (5), obtain

$$(7) \ \Delta\{P\} = \int_V [\alpha]^T\Delta[B]^T\{\sigma\}dV + \int_V [\alpha]^T[B]^T[D][B][\alpha]dV\Delta\{u\}$$

The last term on the right of equation (7) can be divided into a matrix which is dependent on the current displacement and one which is not. With some rearrangement, we obtain the element stiffness matrices

$$(8) \qquad \Delta\{P\} = \left([k^{(1)}] + [k^{(2)}] + [k^{(0)}]\right) \Delta\{u\}$$

where $[k^{(1)}]$ is the initial stress matrix and is obtained from the first term on the right of equation (7)

$[k^{(2)}]$ is the initial displacement matrix of Marcal [25].

$[k^{(0)}]$ is the small displacement stiffness matrix.

The element stiffness matrices and the nodal equivalent forces are then summed in the usual direct stiffness manner to obtain master stiffness equations represented by equations in capitals

$$\Delta\{P\} = \left([K^{(1)}] + [K^{(2)}] + [K^{(0)}]\right) \Delta\{u\} \tag{9}$$

General Purpose Programs

The stiffness equations developed are quite general and are not restricted to a particular type of element. It is therefore possible to write a general purpose program which implements the theory. This program will then serve as a basic and common program to which subroutines may be added to account for specific characteristics belonging to the particular type of element (or element combinations) being used. Two general approaches to programming language have been adopted. The one most favored by the developers of ASKA, FORMAT, MAGIC, NASTRAN, SAMIS and STRUDL is to make use of some type of matrix interpretive language. Here the intention is to develop machine independent concepts and also lay the foundation for easy implementation of further theoretical developments. However, most of these programs have been developed under the influence of particular computers and programming languages, so that the aims of complete machine independence and freedom from bookkeeping requirements have not been fully achieved. On the other hand, there have been other programs (ELAS, MARC2) which were developed with FORTRAN as the programming language and which made use of matrix support packages.

The one common feature to both approaches is the attempt to implement the matrix manipulation required by the theory in as general a form as possible. We see from (7) and (8) that the matrix operations required to form the stiffness matrices do not change. Similarly, the assembly of the element stiffness to form the master stiffness matrix does not change with different elements.

In order to focus on the advantages of general purpose programs, we shall now focus on the program MARC2 developed at Brown University. Most of the features found in this program can readily be included in other programs so that the points to be discussed can be thought to apply equally to all programs in general. This program was developed with the intention of carrying out the common matrix operations required to solve finite element problems with nonlinear material and/or geometric behavior. Because it was meant to be used in a research environment, it was organized with a view to minimize the coding required to implement new elements. This is made easier by the use of numerical integration to form the element stiffness matrices. Only four user subroutines are required to form the $[\alpha][B]$ quantities and specify the weighting functions required to perform the numerical integration. The general purpose program carries out the rest of the calculations based on input data. In particular, a subroutine has been developed to implement the incremental Prandtl-Reuss relations. Another subroutine integrates these relations through the thickness for a plate or shell when required. Various subroutines enable the assembled matrix equations to be solved by either the direct or iterative approach, as well as giving the option of an in-core assembly and out-of-core solution. This program simplifies combined elastic-plastic or creep and large displacement analysis by reducing the amount of additional programming required from a user. The nonlinear problem is converted to a series of piecewise linear problems. There is now an increase of an order of magnitude in computing time required compared to a linear elastic solution since it takes about ten steps to trace the load history of a structure to its buckling or limit load.

Figure 1 shows the flow chart for the general purpose program MARC2. The procedures depicted in the main flow are the control, assembly, application of boundary conditions and the solution of the master stiffness equation. Two subroutines interface with the user subroutines and form the element stiffnesses and the initial stress and strain vectors respectively. Their purpose is to organize and perform the numerical integration to obtain the required quantities. In turn, these subroutines draw on the subroutines which form the linear incremental strain to stress transformation matrix [D] referred to above.

It is of interest to note here the various types of problems that can be handled by the program. It is noted that these can be performed with any combinations of elements and any combination of the following classes of problems:

1. Elastic
2. Elastic-plastic
3. Creep

4. Thermal strains
5. Large displacement
6. Large strains
7. Buckling (eigenvalue analysis at any load level)

We see immediately the advantages of using a general purpose program. Any feature implemented in the program can be combined with all previous developments. As illustrations of this we give examples of two recent additions to the general purpose program. The first was the implementation of a buckling analysis. Once this feature was checked it meant that it was possible to make use of all previously developed elements with large displacement capabilities and perform buckling analysis of beams, plates and shells. Conversely, as an example of using the common features in the program, a new arbitrary, doubly curved shell element [36] was recently developed. It was then possible to use the element in the solution of all the seven classes of problems outlined above. This ability to preserve and exploit all previous developments is the main reason behind the impetus towards development of general purpose programs in recent years. This generalization is in accord with the development and use of computer programs in other areas. One other advantage of the general purpose program is that the main flow of this program will be used frequently and confidence in such a program will be more readily established.

There are also some drawbacks in such a general approach which are perhaps not so evident. First of all, there is its slower running time because of the many conditional statements in the program. This slower speed is particularly noticeable in the larger computer systems where parallel computing devices are employed. Such a program also tends to become large, particularly if there is a large team working on it, and the limits of computer storage are quickly reached. Another problem to be overcome is that of verification documentation and dissemination of such a program. Because MARC2 was intended to be used in a research environment, the coding has been kept to a minimum. Even so, it has grown to about 6000 FORTRAN statements and already makes severe demands on new users. It is interesting to note here that it takes a new user about a month and a half to learn the program and begin to contribute to its development by modifying it. One major disadvantage in developing such a program is the difficulty in keeping changes made by one worker from interfering with the programming work of others.

Note on the Cost of Computing

In this note we shall examine the relationship between the size of a problem and the cost of computing. The actual cost of computing depends on the system configuration and it is possible to obtain differences in costs of up to factors of 2 by simply choosing different machines of the same nominal speed, as well as by choosing the same make of computer but using it at different installations. The important point to be recognized is that core is now required on the faster machines in such large sizes that its cost is as much as that of the central processor (C.P.U.). Thus realistic accounting procedures recognize this and, since particular system configurations are designed with certain functions in mind, its use for other purposes may cause a certain penalty. We shall use here the total system time as a measure of the computing required to solve each problem and not merely the C.P.U. time used. The size of a problem is dependent on the system configuration and will differ from one machine to the next. Thus the following discussion is an attempt to measure the relative cost of solving relatively large to small size problems on a particular computer. The ability to handle large problems is dependent on the size of core available to the user and, at the same time, it is also dependent on the core left for simultaneous use by other users. Because the core requirements for a finite element problem are determined by its master stiffness equation, we shall measure the size of the problem by the product of the number of degrees of freedom with the half-bandwidth. With this definition the size of a problem can be divided into three distinct categories which are determined by the peripheral storage required to solve the master stiffness equations. These solutions fall into the following categories, viz., the in-core assembly and solution, in-core assembly and out-of-core solution, and the out-of-core assembly and solution. In the first category the complete solution can be effected in-core. In the second, the core requirements are such that assembly of the matrix can be performed in core by packing the matrix while the solution is carried out with the aid of magnetic discs or tapes. This course of action usually doubles the size that can be handled by an in-core solution. Finally, the third category is one in which the problem is so large that both assembly and solution must be performed out of core. Figure 2 shows a plot of total system time used against the size of problem. The limits of the in-core solution and the partial in-core solution are shown. Because of the relative speeds of C.P.U. to I./O. operations the total system time is shown increasing with increasing computer speeds. No attempt is made to show relative computing costs. Figure 2 also shows a curve for computing on a time-shared machine which is able to simulate practically unlimited core size for the user, the so-called virtual machines. It appears that operation on such a computer does not penalize a user unduly for working out of core since this is the normal mode of operation for which the system is planned. A distinct advantage in cost can be gained by solving the larger problems on such a machine. The writer's experience does bear this out.

The demand on the computer is not the only cost involved in a finite element analysis since the effort required for coding and the preparation of data must also be taken into account. The question of overall economics deserves further attention. It would be interesting, too, to combine this with further study of future hardware and software development of computer systems.

Case Studies

In this section we present a series of case studies which illustrate various facets of the current state-of-the-art of finite element analysis. It is hoped that these studies taken together will give an overall picture of the progress made in this area. The writer has mainly drawn from results obtained in conjunction with his colleagues. Other choices could also have been made, but the present ones simply reflect greater familiarity with the results, as well as ready access to it.

1. Substructural Analysis of the 747 Aircraft Wing-Body Intersection [37]

This example of an elastic analysis is included to show the large-scale problems that are handled by the finite element method. The method of substructures or matrix partitioning is found to be the best way of reducing the problem to manageable proportions in both the data handling and the equation solving aspects of the problem. The substructures are shown in schematic form in Figure 3. These were idealized by a combination of rod, beam, shear and constant strain elements. The whole problem resulted in 13,870 degrees of freedom. The substructuring reduced the largest bandwidth that had to be handled at any one stage. In connecting the substructures there were a total of 709 degrees of freedom that interacted at the interface. The effort that is involved in performing the analysis of the problem is described in [37]. The problem is restricted to linear elastic behavior; however, with the current rate of progress in the area, it is not difficult to envisage the same problem being solved with nonlinear material and geometric behavior.

It is of interest to note that about a hundred man-months of effort stretching over seven months was required. Much of the model idealization and work on the substructures proceeded in parallel. Twenty-eight hours of CDC 6600 C.P.U. time and one hundred and twenty hours of residency time was required for an error-free pass through the system.

2. Elastic-Plastic Analysis of Tensile Specimen with Semi-Elliptic Crack [38]

The next example is one of the analysis of a semi-elliptic crack in a tensile specimen using a combination of 648 three-dimensional cubic and isoparametric elements [38] and 900 nodal points or 2,700 degrees of freedom. Eight layers of elements were used to model the problem. Figure 4 shows the bottom layer of elements and the layout adopted to represent the semi-elliptic crack. The solution was carried out by first obtaining the load to cause the most highly stressed element to yield. Four increments, each equal to 0.07 of the elastic load were added to study the elastic-plastic behavior. The progress of plastic yielding in the second and fourth load increment is shown in Figure 5. The elastic solution took 45 minutes of C.P.U. time on the IBM 360-91, and subsequent elastic-plastic increments took about 15 minutes per increment. A method of successive over-relaxation was used.

3. Analysis of Shell-Nozzle Junction with Combined Shell and Triangular Ring Elements [39]

This example is included to show a combination of shell and solid elements by the method of linear constraints [39]. A mild steel shell nozzle junction under pressure was studied experimentally by Dinno and Gill [40]. This same problem was analyzed using the mesh in Figure 6. Triangular ring elements are used in and around the weld section, and shell elements are used throughout the main body of the shell and nozzle. Comparison of the finite element results with experimental data is shown in Figure 4. The actual differences between the peak stresses can be seen in Table 1. The hybrid finite element results show considerable improvement over a previous modified shell theory approach using a band of pressure for the junction [41]. That theory was itself a large improvement over the simple shell theory.

	LIMIT OF PROPORTIONALITY (lb/in^2)
Experimental [40]	800
Simple shell theory [41]	340
Band theory [41]	630
Hybrid analysis	793

TABLE 1. First Yield of Shell-Nozzle Junction with Internal Pressure

4. Imperfect Hemisphere under External Pressure [42]

This example shows the combined effect of nonlinear geometric and material behavior. Both of these act together to drastically weaken the load resistance of the structure. The oblate spherical shell of Figure 8 was analyzed by Bushnell [43] in the nonlinear elastic region. It was there observed that high "elastic" stresses were observed at the crown of the shell prior to collapse. This shell was analyzed with a dimensionless yield stress of 0.00666 E together with a linear work-hardening curve with a slope of 0.05 E. This corresponds roughly to the stress strain curve of an Aluminum Alloy.

Figure 9 gives a comparison between the buckling pressure of the elastic shell of [43] and the present elastic-plastic results. The results are plotted in terms of the parameters used in [43]. The classical buckling load ρ_c is defined by

$$\rho_c = 1.21(2H/R)^2 E$$

where E is the Young's Modulus
R is the radius of the sphere
and H is the half-thickness.
The geometric parameter λ is defined by

$$\lambda = \sqrt{12(1-\nu^2)} \left(\frac{R}{2H}\right)^{1/2} \frac{R}{R_{imp.}}$$

where $R_{imp.}$ = mean radius of the oblate portion of the sphere, ν = Poisson's Ratio.

Plastic yielding has a considerable effect on the behavior of the oblate shells under external pressure. This effect increases with the thickness to radius ratio.

It is noted that the failures at the higher thickness to radius ratio ($\lambda \leqslant 1.5$) are due to membrane yield.

Reasonable agreement is also obtained with the elastic results of Bushnell [43] for the thinner shells which do not yield before buckling.

5. Infinite Incompressible Log [44]

The infinite log under symmetric line loading is shown in Figure 10. This problem was solved by Oden [44] using the generalized Newton-Raphson method. The problem was reduced to 22 simultaneous equations by taking advantage of symmetry. For illustration purposes, the line loading P was taken to be 200 lb/in and for the Mooney constants $C_1 = 43.75$ lb/in^2 and $C_2 = 6.25$ lb/in^2 were used. Results in the form of the deformed profile are indicated in Figure 11.

Conclusions and Future Work

In this paper we have examined the formulation and the implementation of a theory for nonlinear finite element analysis. The general purpose program was shown to be a versatile and flexible method of implementing the basic theory. It was found possible to classify three basic sizes of problems which were dependent on the ability of the computer to either assemble or solve the master stiffness matrix in core.

Case studies were given to illustrate representative applications of the theory. If, as is argued here, the general purpose program is the key to widespread applications of the theory, then much more remains to be learned about its development and organization. Little is known about the best way to match programs with particular computer system configurations. Even less is known about the impact of future hardware developments. Finally, rigorous procedures have yet to be developed for verifications of these programs.

REFERENCES

1. Turner, J. L., Clough, R. W., Martin, H. C., and Topp, L. J., "Stiffness and Deflection Analysis of Complex Structures," J. Aero. Science, Vol. 23, No. 9, Sept. 1956, pp. 805-825.
2. De Veubeke, B. F., "Displacement and Equilibrium Models in the Finite Element Method," Chapter 9 of Stress Analysis, ed. O. C. Zienkiewicz and G. S. Holister, Wiley, 1965.
3. Clough, R. W., "The Finite Element in Structural Mechanics," Chapter 7 of Stress Analysis, ed. O. C. Zienkiewicz and G. S. Holister, Wiley, 1965.
4. Argyris, J. H., "Triangular Elements with Linearly Varying Strain for the Matrix Displacement Method," J. Roy. Aero. Soc., Tech. Note 69, Oct. 1965, pp. 711-713.
5. Ergatoudis, I., "Quadrilateral Elements in Plane Analysis," M.Sc. thesis, University of Wales, Swansea, 1966.
6. Bogner, F. K., Fox, R. L., and Schmit, L. A., "The Generation of Inter Element-compatible Stiffness and Mass Matrices by the Use of Interpolation Formulae," Proc. 1st Conf. on Matrix Methods in Struct. Mech., AFFDL-TR-66-80, 1966, pp. 397-444.
7. De Veubeke, B. F., "Bending and Stretching of Plates," Proc. 1st Conf. on Matrix Methods in Struct. Mech., AFFDL-TR-66-80, 1966, pp. 863-886.
8. Clough, R. W., and Tocher, J. L., "Finite Element Stiffness Matrices for Analysis of Plates in Bending," Proc. 1st Conf. on Matrix Methods in Struct. Mech., AFFDL-TR-66-80, 1966, pp. 515-546.

9. Szmelter, J., "Energy Method of Networks of Arbitrary Shape in Problems of the Theory of Elasticity," Proc. I.U.T.A.M., Symposium on Non-Homogeneity in Elasticity and Plasticity, ed. W. Olszak, Pergamon Press, 1959.
10. Melosh, R. J., "Basis for Derivation of Matrices for the Direct Stiffness Method," AIAA Journal, Vol. 1, No. 7, 1963, pp. 1631-1637.
11. Padlog, J., Huff, R. D., and Holloway, G. F., "The Unelastic Behavior of Structures Subjected to Cyclic, Thermal and Mechanical Stressing Conditions," Bell Aerosystems, Co., Report WPADD-TR-60-271, 1960.
12. Argyris, J. H., Kelsey, S., and Kamel, W. H., "Matrix Methods of Structural Analysis, A Precis of Recent Developments," Proc. 14th Meeting of Structures and Materials Panel, AGARD, 1963.
13. Jensen, W. R., Falby, W. E., and Prince, N., "Matrix Analysis Methods for Anisotropic Inelastic Structures," AFFDL-TR-65-220, 1966.
14. Witmer, E. A., and Kotanchik, J. J., "Progress Report on Discrete-Element Elastic and Elastic-Plastic Analysis of Shells of Revolution Subjected to Axisymmetric and Asymmetric Loading," Proc. 2nd Conf. on Matrix Methods in Struct. Mechanics, AFFDL, October 1968.
15. Pope, G., "A Discrete Element Method for Analysis of Plane Elastic-Plastic Stress Problems," Royal Aeronautical Establishment TR 65028, 1965.
16. Swedlow, J. L., and Yang, W. H., "Stiffness Analysis of Elastic-Plastic Plates," Graduate Aeronautical Lab., California Institute of Technology SM 65-10, 1965.
17. Marcal, P. V., and King, I. P., "Elastic-Plastic Analysis of Two-Dimensional Stress Systems by the Finite Element Method," Intern. J. Mech. Sci., Vol. 9, No. 3, 1967, pp. 143-155.
18. Marcal. P. V., "Comparative Study of Numerical Methods of Elastic-Plastic Analysis," AIAA Journal, Vol. 6, No. 1, 1967, pp. 157-158.
19. Turner, M. J., Dill, E. H., Martin, H. C., and Melosh, R. J., "Large Deflection Analysis of Complex Structures Subjected to Heating and External Loads," Journal Aero Space Sciences, Vol. 27, Feb. 1960, pp. 97-106.
20. Argyris, J. H., Kelsey, S., and Kamel, H., Matrix Methods of Structural Analysis, AGARDOGRAPH No. 72, Pergamon Press, 1964, pp. 105-120.

21. Gallagher, R. H., Gellatly, R. A., Padlog, J., and Mallett, R. H., "Discrete Element Procedure for Thin-Shell Instability Analysis," AIAA Journal, Vol. 5, No. 1, 1967, pp. 138-144.
22. Kapur, K. K., and Hartz, B. J., "Stability of Plates Using the Finite Element Method," Journal Eng. Mech. Div., ASCE, Vol. 92, EM2, 1966, pp. 177-195.
23. Martin, H. C., "Derivation of Stiffness Matrices for the Analysis of Large Deflection and Stability Problems," Proc. 1st Conf. on Matrix Methods in Struct. Mech., AFFDL-TR-66-80, 1966, pp. 697-715.
24. Purdy, D. M., and Przemieniecki, I. S., "Influence of Higher-Order Terms in the Large Deflection Analysis of Frameworks," Proc., ASCE, Joint Specialty Conf., Optimization and Nonlinear Problems, April 1968, pp. 142-152.
25. Mallett, R. H., and Marcal, P. V., "Finite Element Analysis of Nonlinear Structures," J. Struct. Div., ASCE, Vol. 94, No. ST 9, Sept. 1968, pp. 2081-2105.
26. Bogner, F. K., Mallett, R. H., Minich, M. D., and Schmit, L. A., "Development and Evaluation of Energy Search Methods of Nonlinear Structural Analysis," AFFDL-TR-65-113, 1965.
27. Mallett, R. H., and Berke, L., "Automated Method for the Finite Displacement Analysis of Three Dimensional Truss and Frame Assemblies," AFFDL-TR-102, 1966.
28. Bogner, F. K., Fox, R. L., and Schmit, L. A., "A Cylindrical Shell Discrete Element," AIAA Journal, Vol. 5, No. 4, 1967, pp. 745-750.
29. Oden, J. T., "Numerical Formulation of Nonlinear Elasticity Problems," Journal Struct. Div., ASCE, Vol. 93, No. ST 3, June 1967, pp. 235-255.
30. Oden, J. T., and Kubitza, W. K., "Numerical Analysis of Nonlinear Pneumatic Structures," Proc. 1st International Colloquium on Pneumatic Structures, Stuttgart, May 1967.
31. Hibbitt, H. D., Marcal, P. V., and Rice, J. R., "Finite Element Formulation for Problems of Large Strain and Large Displacement," Brown University Engineering Report N00014-0007/2, June 1969.
32. Oden, J. T., "General Theory of Finite Elements," Int. Journal Numerical Methods in Engineering, Parts I and II, 1, 1969, p. 205 and p. 247.
33. Melosh, R., Lang, T., Schmele, L., and Bamford, R., "Computer Analysis of Large Structural Systems," Proc. AIAA 4th Annual Meeting, Paper 67-955, 1967.

34. Marcal, P. V., "Finite Element Analysis of Combined Problems of Nonlinear Material and Geometric Behavior," Proc. ASME Computer Conference on Computational Approaches in Applied Mechanics, June 1969, p. 133. Also, Brown University Engineering Report N00014-0007/1, March 1969.
35. Marcal, P. V., "Elastic-Plastic Analysis of Pressure Vessel Components," Proc. 1st Pressure Vessel and Piping Conf., ASME Computer Seminar, Dallas, Sept. 1968.
36. Dupuis, G., and Goel, J. J., "A Curved Finite Element for Thin Elastic Shells, "Brown University Engineering Report N00014-0007/4, December 1969.
37. Anderton, G. L., Connacher, N. E., Dougherty, C. S., and Hansen, S. D., "Analysis of the 747 Aircraft Wing-Body Intersection," Proc. Second Conference on Matrix Methods in Structural Mechanics, AFFDL-TR-68-150, 1968, p. 743.
38. Zienkiewicz, O. C., <u>The Finite Element Method in Structural and Continuum Mechanics</u>, McGraw-Hill, 1967, p. 82.
39. Hibbitt, H. D., and Marcal, P. V., "Hybrid Finite Element Analysis with Particular Reference to Axisymmetric Structures," Proc. AIAA 8th Aerospace Sciences Meeting, Paper No. 70-137, 1970.
40. Dinno, K. S., and Gill, S. S., "An Experimental Investigation into the Plastic Behavior of Flush Nozzles in Spherical Pressure Vessels," Int. J. Mech. Sci., <u>7</u>, 1965, pp. 817-839.
41. Marcal, P. V., and Turner, C. E., "Elastic-plastic Behavior of Flush Nozzles in Spherical Pressure Vessels," 9, 3, Journal Mech. Eng. Sci., 1967, p. 182.
42. Marcal. P. V., "Large Deflection Analysis of Elastic-Plastic Shells of Revolution," AIAA/ASME 10th Structures, Structural Dynamics and Materials Conference, April 1969.
43. Bushnell, D., "Nonlinear Axisymmetric Behavior of Shells of Revolution," AIAA Journal, Vol. 5, No. 3, 1967, pp. 432-439.
44. Oden, J. T., "Finite Plane Strain of Incompressible Elastic Solids by the Finite Element Method," The Aeronautical Quarterly, 19, August 1968, p. 254.

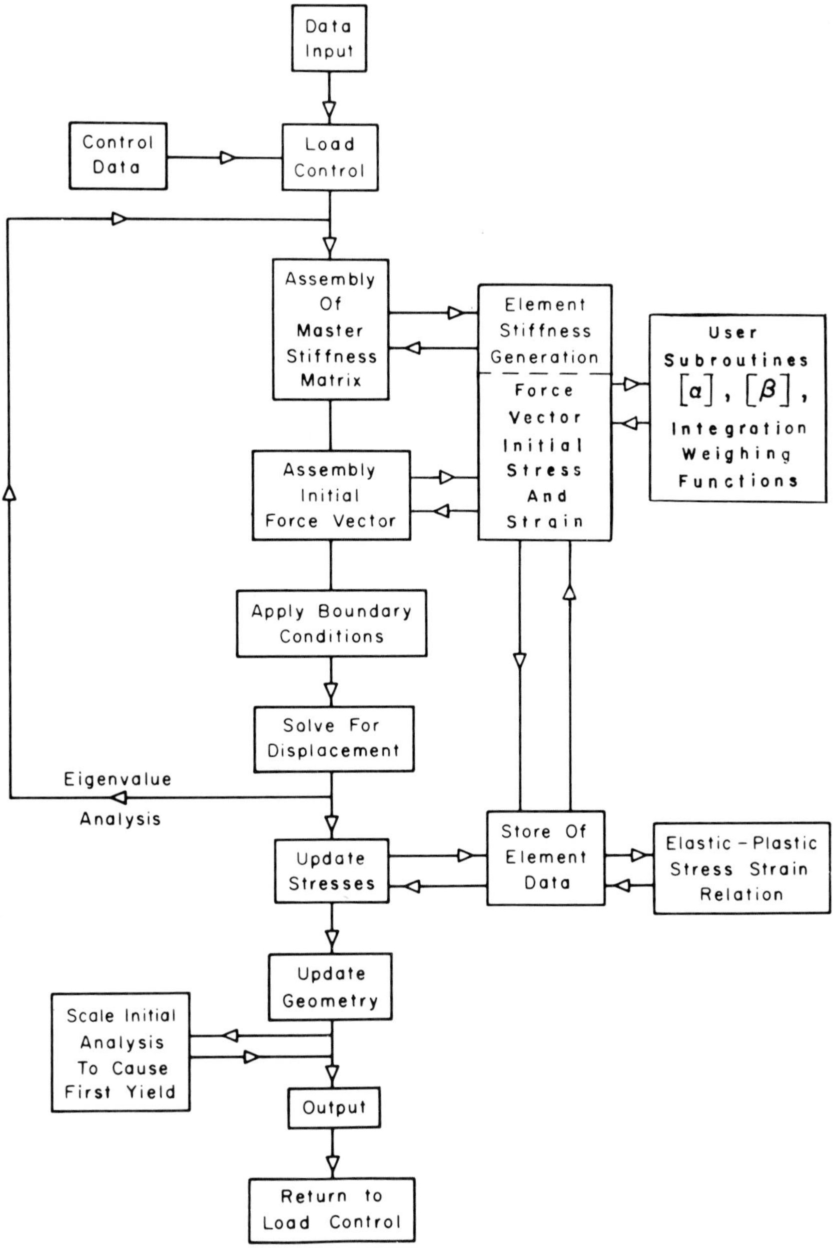

Fig. 1 Flow Chart for Computer Program

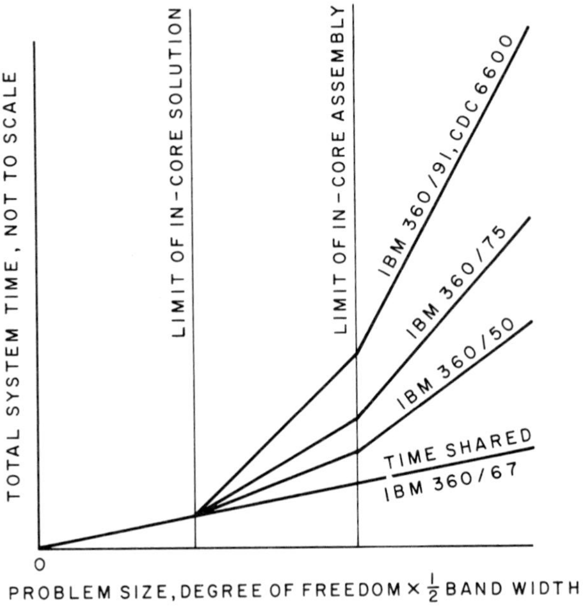

Fig. 2 Computing Time-Problem Size

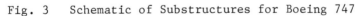

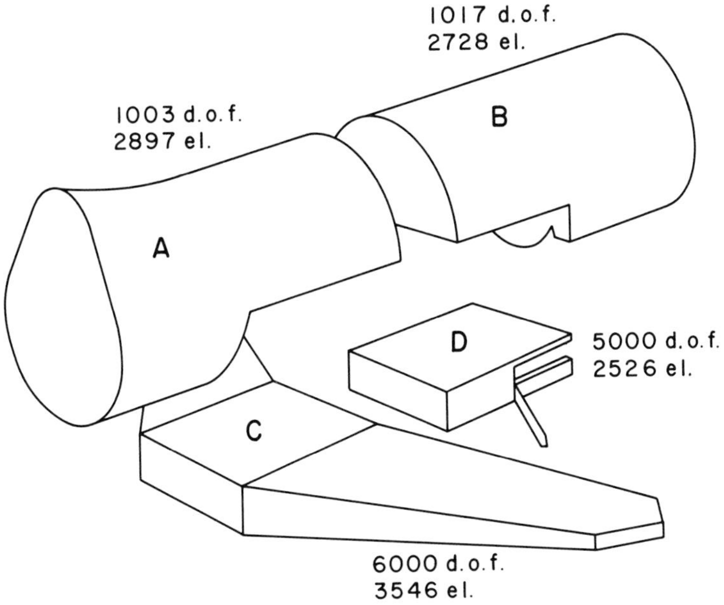

Fig. 3 Schematic of Substructures for Boeing 747

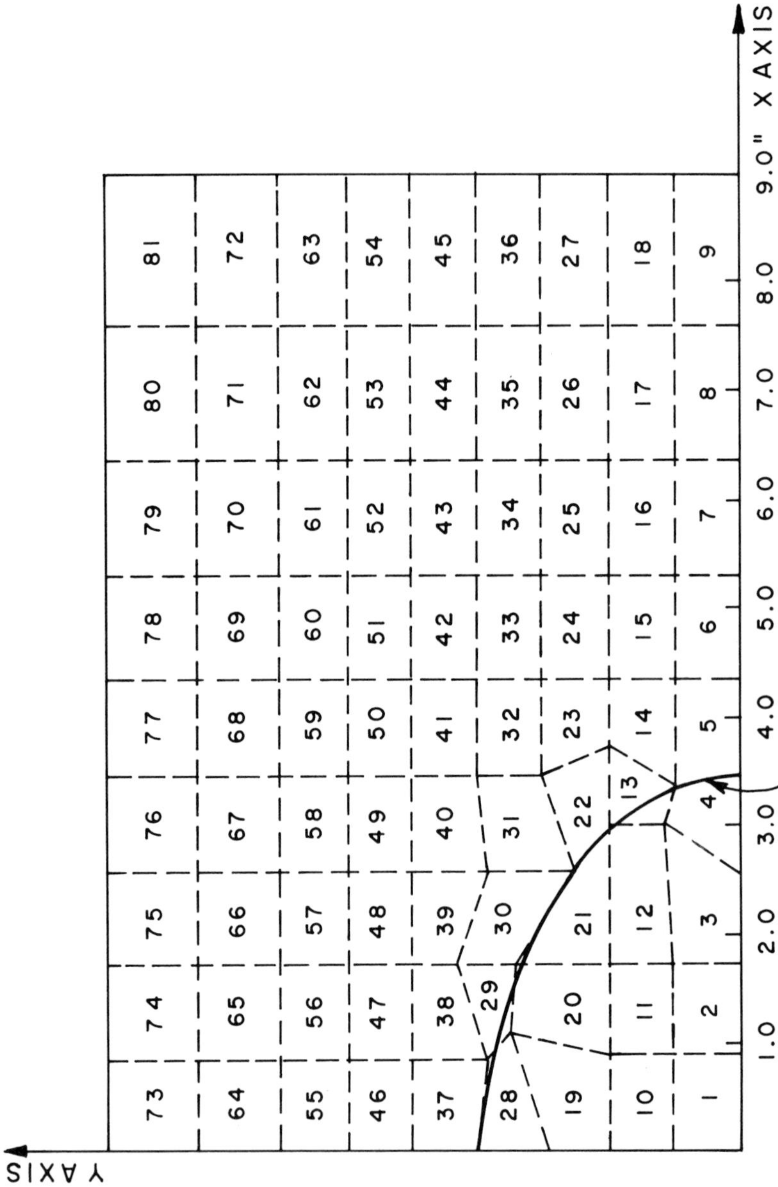

Fig. 4 Elements Identification in the Crack Plane

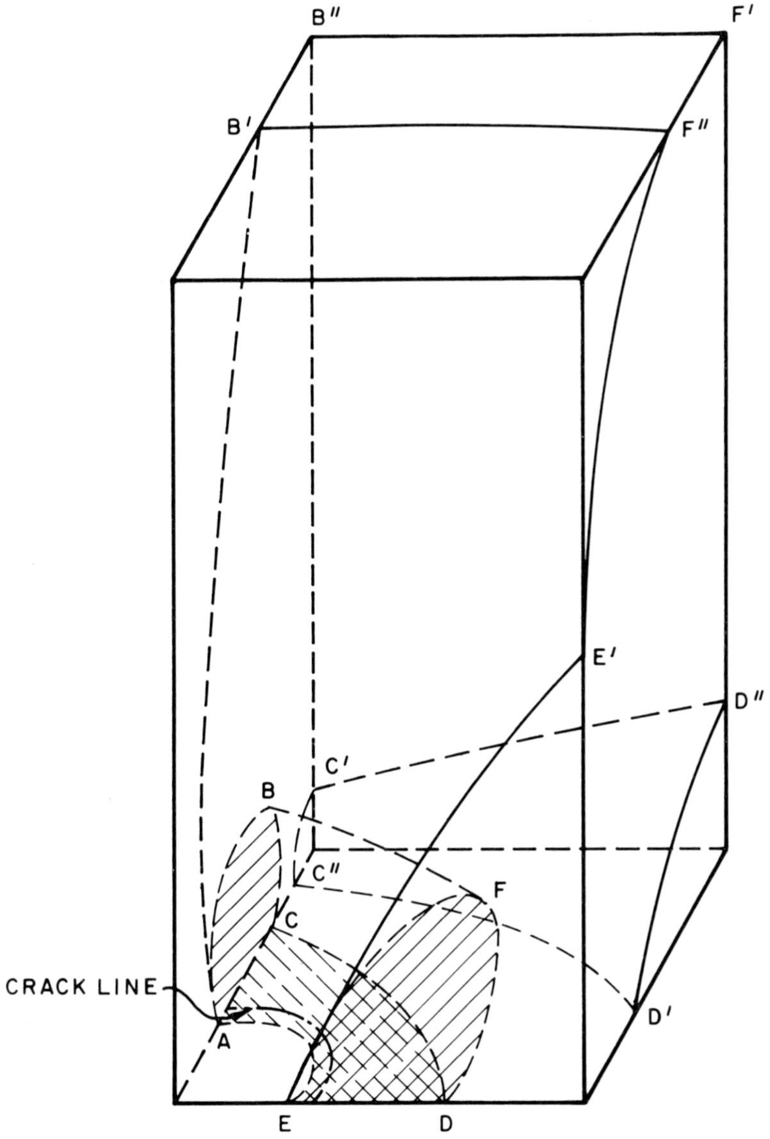

1) PLASTIC ZONE ABCDEF IS AFTER TWO INCREMENTS
2) PLASTIC ZONE AB′B″C′C″D′D″F′F″E′E IS AFTER 4TH INCREMENT

Fig. 5 Progress in Plastic Yielding Tensile Plate with Semi-Elliptic Crack

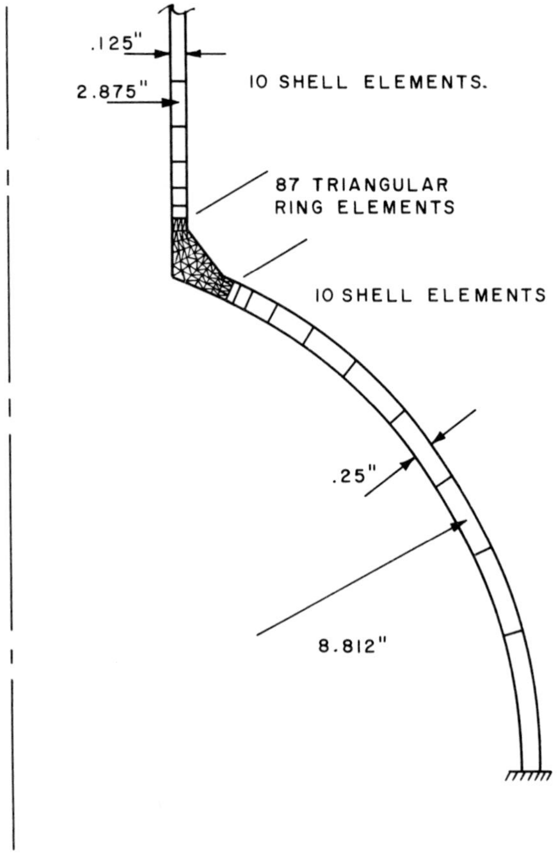

Fig. 6 Mesh for Shell-Nozzle Junction (Not to Scale)

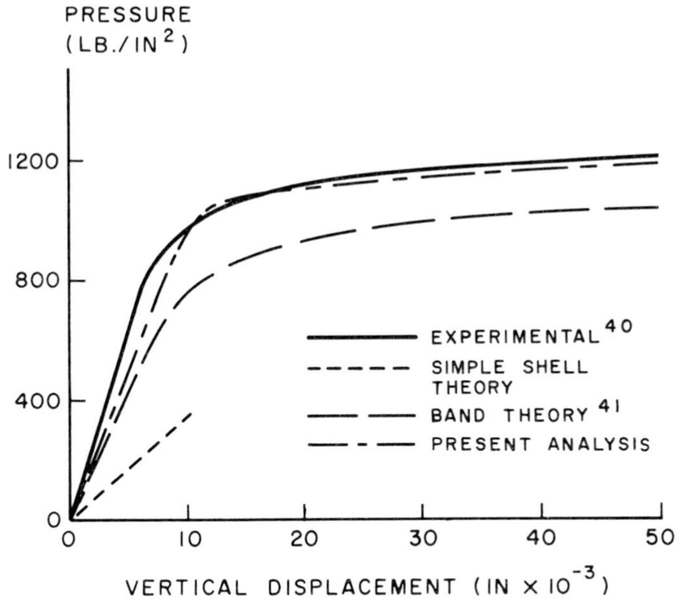

Fig. 7 Nozzle Displacement

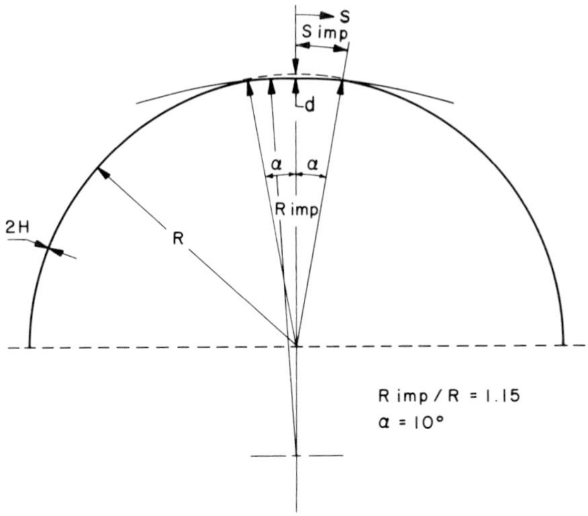

Fig. 8 Externally Pressurized Imperfect Hemisphere

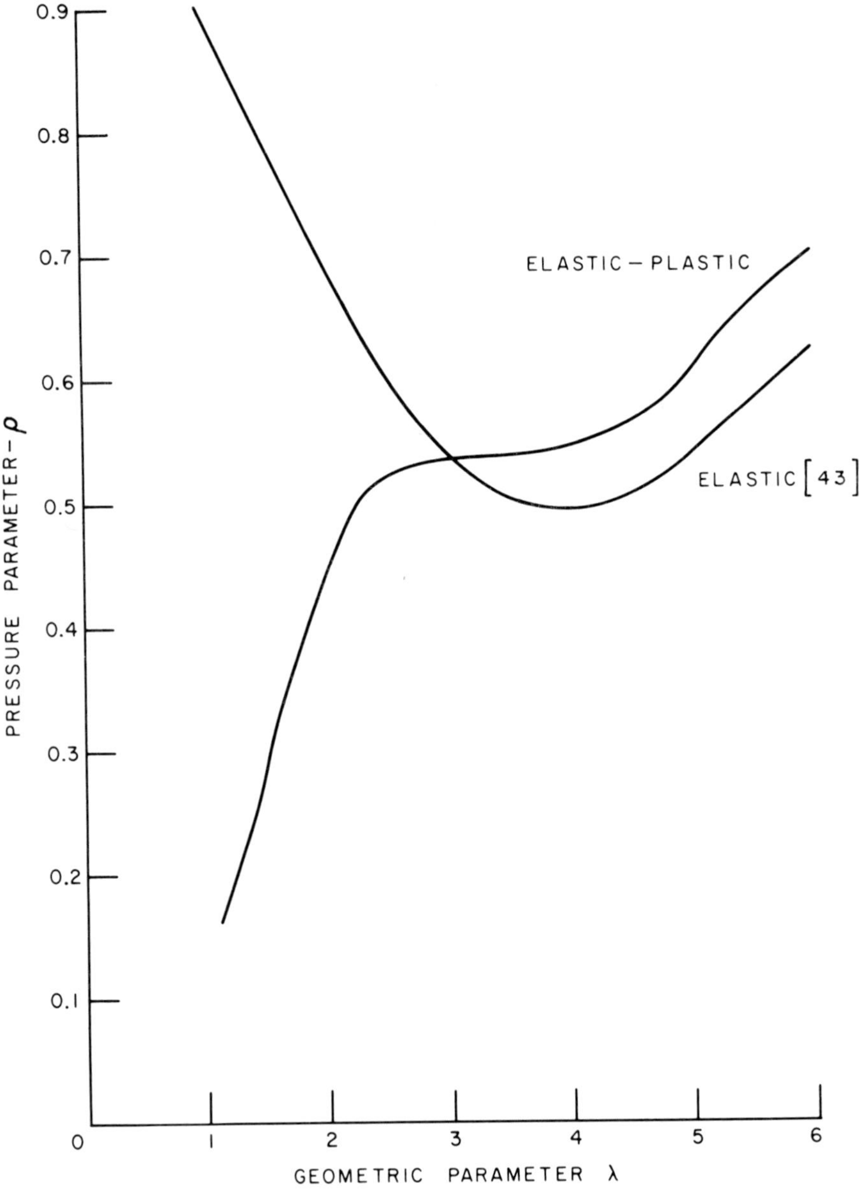

Fig. 9 Buckling Pressures for Oblate Shells, Rimp/R = 1.15

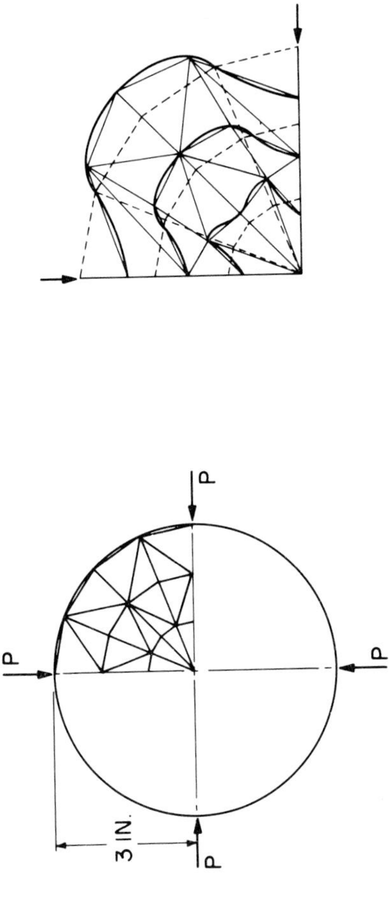

Fig. 10 Finite Incompressible Log

ON THE THEORY OF THE SPLITTING-UP METHOD

G. I. Marchuk*

The ideas of the reduction of complicated problems to a sequence of simpler problems were formed under a considerable influence of the alternating direction method developed by J. Douglas, D. Peaceman, and G. Rachford [1],[2],[3]. Our investigations were founded on the splitting up method, or the method of weak approximation, based on non-homogeneous approximations of the problems and developed by N. N. Yanenko [4],[5], E. Y. Diakonov [6], A. A. Samarsky [7], G. I. Marchuk [9], and others.

1. A Scheme of Splitting Up of Nonstationary Problems

The numerical solution of linear and quasi-linear equations is an important problem of computational mathematics, as the equations of this kind are the basic elements of algorithms for the solution of the problems in hydrodynamics, plasma physics, and other branches of science. Therefore the development of effective methods for the solution of such problems is of paramount importance.

Let us consider an abstract Cauchy problem

$$(1.1) \qquad \frac{\partial \Phi}{\partial t} + L\Phi = 0 , \qquad \Phi(0) = f$$

* Director of the Computer Center, Academy of Sciences of the U.S.S.R., Siberian Branch, Novosibirsk.

in a Hilbert space R with the positive semi-definite linear operator L

$$(L\Phi, \Phi) \geqslant 0 . \tag{1.2}$$

Assume that the operator L is dependent, in general, on the time and other variables and can be presented as a sum of simpler operators

$$L = \sum_{\alpha=1}^{n} L_\alpha , \tag{1.3}$$

each of which is also positive semi-definite. Assume a sufficient smoothness of the solution Φ, of the operators L_α and of the initial data f. Smoothness requires, each time, a special analysis.

In a sense, the best difference approximation of Eq. (1.1) in the interval $t_j \leqslant t \leqslant t_{j+1}$ is the Crank-Nicolson difference scheme

$$(E + \frac{\tau}{2} \Lambda^j)\phi^{j+1} = (E - \frac{\tau}{2} \Lambda^j)\phi^j . \tag{1.4}$$

Here ϕ^j is an approximate solution of (1.1) at the time t_j, and Λ^j is an approximation of the operator L in the interval $t_j \leqslant t \leqslant t_{j+1}$ accurate to the first or second order in τ, where $\tau = t_{j+1} - t_j$. The superscript denotes the number of the time interval.

If we solve Eq. (1.4) with respect to the unknown, we see

$$\phi^{j+1} = (E + \frac{\tau}{2} \Lambda^j)^{-1} (E - \frac{\tau}{2} \Lambda^j)\phi^j . \tag{1.5}$$

Investigation of the stability of (1.5) reduces, ultimately, to the evaluation of the norm of the operator

$$T^j = (E + \frac{\tau}{2}\Lambda^j)^{-1}(E - \frac{\tau}{2}\Lambda^j) .$$

Let us use for such an evaluation the important Kellogg's lemma (1963) which states that under the condition

$$(1.6) \qquad (\Lambda\phi,\phi) \geqslant 0$$

the following relation holds

$$(1.7) \qquad \|(E + \frac{\tau}{2}\Lambda)^{-1}(E - \frac{\tau}{2}\Lambda)\| \leqslant 1 .$$

Using (1.5) we get

$$(1.8) \qquad \|\phi^{j+1}\| \leqslant \|(E + \frac{\tau}{2}\Lambda^j)^{-1}(E - \frac{\tau}{2}\Lambda^j)\| \, \|\phi^j\| .$$

Then by Kellogg's lemma, we have

$$\|\phi^{j+1}\| \leqslant \|\phi^j\| \leqslant \cdots \leqslant \|f\| .$$

Consider now a very important case for hydrodynamics when the operator Λ is skew-Hermitian, that is

$$(1.9) \qquad (\Lambda\phi,\phi) = 0 .$$

According to Kellogg's lemma, we have

$$\|\phi^{j+1}\| \leqslant \|\phi^j\| .$$

To generalize the lemma for this case, we extend the theorem of stability. For this purpose, we consider the functional $\|\phi^{j+1}\|^2$ and, using Eq. (1.5), we obtain the identity

(1.10)
$$\|\phi^{j+1}\|^2 = \frac{((E+\frac{\tau}{2}\Lambda^j)^{-1}(E-\frac{\tau}{2}\Lambda^j)\phi^j,(E+\frac{\tau}{2}\Lambda^j)^{-1}(E-\frac{\tau}{2}\Lambda^j)\phi^j)}{(\phi^j,\phi^j)}\|\phi^j\|^2 .$$

Next rewrite Eq. (1.10) in the equivalent form

$$\|\phi^{j+1}\| = \frac{\|(E - \frac{\tau}{2}\Lambda^j)\xi^j\|}{\|(E + \frac{\tau}{2}\Lambda^j)\xi^j\|}\|\phi^j\| , \quad (1.11)$$

where

$$\xi^j = (E + \frac{\tau}{2}\Lambda^j)^{-1}\phi^j .$$

It is easy to verify that, if $(\Lambda^j\xi^j , \xi^j) = 0$, we have

$$\|(E - \frac{\tau}{2}\Lambda^j)\xi^j\| = \|(E + \frac{\tau}{2}\Lambda^j)\xi^j\| = \sqrt{\|\xi^j\|^2 + \frac{\tau^2}{4}\|\Lambda^j\xi^j\|^2} .$$

Taking the above equality into account, we write formula (1.11) in the form

$$\|\phi^{j+1}\| = \|\phi^j\| = \cdots = \|f\| . \quad (1.12)$$

Thus, in the case of a skew-Hermitian operator Λ , we come to the equality of the norms of the solution on the sequence of time intervals. We shall need this extension of Kellogg's lemma later for construction of non-dissipative difference schemes for hydrodynamic equations.

It should be stressed that the stability of scheme (1.4), under the condition of positive semi-definiteness, takes place for linear operators both with constant and with variable coefficients. Later these results will be extended to the quasi-linear hydrodynamic equations which can be linearized at each small time interval.

Let us consider the accuracy of approximation of Eq. (1.5). For this purpose, the operator T^j will be expanded in a series in τ . Then we get

$$T^j = E - \tau\Lambda^j + \frac{\tau^2}{2}(\Lambda^j)^2 - \cdots , \tag{1.13}$$

where the dots indicate the neglected terms of higher order smallness in τ . Using (1.13), we can write Eq. (1.5) as

$$H_\tau\phi \equiv \frac{1}{\tau}\{\phi^{j+1} - [E - \tau\Lambda^j + \frac{\tau^2}{2}(\Lambda^j)^2 - \cdots]\phi^j\} = 0 .$$

The operator of Eq. (1.1) will be represented in the form

$$H\Phi \equiv \frac{\partial\Phi}{\partial t} + L\Phi .$$

Now we determine the norm in the grid domain $c_\tau[0 \leqslant t_j \leqslant T]$,

$$\|H\Phi - H_\tau\Phi\|_{c_\tau} = \|H_\tau\Phi\|_{c_\tau} = \max_{t_j} |H_\tau\Phi| .$$

Next expand the solution of Eq. (1.1) in a Taylor's series in the neighborhood of $t = t_j$. Then we get

$$\Phi^{j+1} = \Phi^j + \tau\Phi_t^j + \frac{\tau^2}{2}\Phi_{tt}^j + \cdots .$$

Taking into consideration the obvious equalities

$$\Phi_t = -L\Phi \ , \quad \Phi_{tt} = L^2\Phi - \frac{\partial L}{\partial t}\Phi \ ,$$

we write Taylor's series for Φ in the form

$$\Phi^{j+1} = \Phi^j - \tau L^j\Phi^j + \frac{\tau^2}{2}\left[(L^j)^2\Phi^j - \left(\frac{\partial L}{\partial t}\right)^j\Phi^j\right] - \cdots \ . \tag{1.14}$$

Here we used the notation

$$L^j = L(t_j) \ .$$

Consider the simplest case when

$$\Lambda^j = L^j = L(t_j) \ .$$

Substituting (1.14) into (1.12), we get

$$\|H\Phi - H_\tau\Phi\|_{C_\tau} = \frac{\tau}{2}\max_{t_j}\left|\left(\frac{\partial L}{\partial t}\right)^j\Phi^j + O(\tau)\right| \ .$$

From this follows the first order approximation. Consider now the second order approximation of L, i.e.,

$$\Lambda^j = L^j + \left(\frac{\partial L}{\partial t}\right)^j \frac{\tau}{2} + \cdots \ .$$

Then we get the resulting second order approximation

$$\|H\Phi - H_\tau\Phi\|_{C_\tau} = O(\tau^2) \ .$$

In determining the order of approximation we used the

smoothness of the operator with respect to the variable t which is assumed a priori.

It will be noted that if the operator L is independent of time, the difference equation (1.5) always has the second order approximation.

Our problem now is to find such difference approximations of Eq. (1.1) which would remain unconditionally stable, would have second order approximation and could be efficiently realized on computers. It will be assumed, then, that

$$\Lambda^j = \sum_{\alpha=1}^{n} \Lambda_\alpha^j$$

and that all Λ_α^j are positive semi-definite operators. Consider a system of the following difference equations:

$$(E + \frac{\tau}{2}\Lambda_\alpha^j)\phi^{j+\frac{\alpha}{n}} = (E - \frac{\tau}{2}\Lambda_\alpha^j)\phi^{j+\frac{\alpha-1}{n}}, \quad (\alpha = 1,2,\ldots,n) . \tag{1.15}$$

The splitting up schemes of the above form have been considered in a number of papers [4],[10]. In the case of positive semi-definite and commutative operators Λ_α^j such a scheme is unconditionally stable and has the second order approximation. However, for the non-commutative operators Λ_α^j, as is easily verified, (1.15) is a scheme of the first order approximation, hence, it has no interest for applications, as long as the similar splitting up scheme

$$\phi^{j+\frac{\alpha}{2n}} = (E - \frac{\tau}{2}\Lambda_\alpha)\phi^{j+\frac{\alpha-1}{n}}, \quad (\alpha = 1,2,\dots,n),$$

(1.16)

$$(E + \frac{\tau}{2}\Lambda_\alpha)\phi^{j+\frac{\alpha}{2n}} = \phi^{j+\frac{\alpha-1}{2n}}, \quad (\alpha = n+1,\ n+2,\dots,\ 2n)$$

remains a scheme of the second order accuracy.

The present paper suggests a special cyclic construction of the splitting up schemes of type (1.15) which is used for the solution of the Cauchy problems with positive semi-definite non-commutative operators Λ_α^j. The central theoretical questions here concern the analysis of stability and approximation in the case of an arbitrary n-component representation of the operator Λ^j in terms of the elementary Λ_α^j. This analysis can be based on Kellogg's lemma. It will be noted that it seems impossible to obtain such general results for the scheme (1.16). Excluding the unknowns in (1.15) with fractional indices, we get

$$\phi^{j+1} = \prod_{\alpha=1}^{n} (E + \frac{\tau}{2}\Lambda_\alpha^j)^{-1}(E - \frac{\tau}{2}\Lambda_\alpha^j)\phi^j .$$

Considering Eq. (1.15) by the norm, we get

$$\|\phi^{j+1}\| \leqslant \prod_{\alpha=1}^{n} \|(E + \frac{\tau}{2}\Lambda_\alpha^j)^{-1}(E - \frac{\tau}{2}\Lambda_\alpha^j)\| \|\phi^j\| .$$

Then, applying Kellogg's lemma to each factor in the product, we come to the condition of unconditional stability

$$\|\phi^{j+1}\| \leqslant \|\phi^j\| \leqslant \cdots \leqslant \|f\| .$$

If the operators Λ_α are skew-Hermitian, we have

$$\|\phi^{j+1}\| = \|\phi^j\| = \cdots = \|f\| .$$

To determine the order of approximation we shall expand the operator

$$T^j = \prod_{\alpha=1}^{n} (E + \frac{\tau}{2} \Lambda_\alpha^j)^{-1} (E - \frac{\tau}{2} \Lambda_\alpha^j)$$

in powers of the parameters τ .

Since $T^j = \prod_{\alpha=1}^{n} T_\alpha^j$ and according to (1.13)

$$T_\alpha^j = E - \tau\Lambda_\alpha^j + \frac{\tau^2}{2} (\Lambda_\alpha^j)^2 - \cdots ,$$

then, developing the main part, we get

(1.17)

$$T^j = E - \tau\Lambda^j + \frac{\tau^2}{2} [(\Lambda^j)^2 + \sum_{\alpha=1}^{n} \sum_{\beta=\alpha+1}^{n} (\Lambda_\alpha^j \Lambda_\beta^j - \Lambda_\beta^j \Lambda_\alpha^j) \cdots] + O(\tau^3) .$$

In the case of the commutative operators the expression in square brackets from (1.17) is equal to $(\Lambda^j)^2$, therefore T^j , up to the second order, coincides with (1.13). This means that scheme (1.15) is equivalent in accuracy to the Crank-Nicolson difference scheme (1.4). If the operators Λ_α^j do not commute, then the splitting up scheme (1.15) has only first order accuracy. In order to build a splitting up scheme accurate to the second order in the case of non-commutative operators, the following two schemes are suggested:

$$(1.18) \qquad \phi^j = \prod_{\alpha=1}^{n} T_\alpha^j \phi^{j-1} , \qquad \phi^{j+1} = \prod_{\alpha=n}^{1} T_\alpha^j \phi^j .$$

These should be alternately used. Algorithmically it means that, first, in the interval $t_{j-1} \leqslant t \leqslant t_j$, system (1.15) is solved in the usual sequence $\alpha = 1,2,\ldots,n$, and in the interval $t_j \leqslant t \leqslant t_{j+1}$ in the reverse sequence $\alpha = n, n-1,\ldots, 1$. These two steps taken together will be termed a computation cycle. It is not difficult to see that the operator of the whole cycle S^i is equal to

$$(1.19) \qquad S^j = \prod_{\alpha=1}^{n} T_\alpha^j \prod_{\alpha=n}^{1} T_\alpha^j = E - 2\tau\Lambda^j + \frac{(2\tau)^2}{2} (\Lambda^j)^2 - \cdots .$$

Since the computation cycle is formed in the interval $t_{j-1} \leqslant t \leqslant t_{j+1}$, for such an interval we have a scheme of second order accuracy. This is easily seen if we compare (1.19) with the operator for the Crank-Nicolson difference scheme written in the interval $t_{j-1} \leqslant t \leqslant t_{j+1}$.

Above it has been assumed that the operators Λ_α^j are known at each step and have second order accuracy. As a rule this holds for linear equations. In the case of quasi-linear equations such operators depend on the solution itself. In order to have the second order accuracy in this case, it is appropriate to use the method of two-step correction which implies that, first, in the interval $t_{j-1} \leqslant t \leqslant t_j$ the problem is solved with the operator Λ^j depending on ϕ^{j-1} with first order accuracy in τ . After the problem is solved in this interval, one finds ϕ^{j+1} and the operators Λ_α^j are created through the use of the expression $\phi^j = \frac{1}{2}(\phi^{j+1}-\phi^{j-1})$. This gives second order approximation in all independent variables. If the computation is organized in this way, the amount of computing increases twofold.

2. Hydrodynamic Transfer Equations of Transfer Along Trajectories

Let us consider the equation of hydrodynamic transfer

$$(2.1) \qquad \frac{\partial \Phi}{\partial t} + \sum_{\alpha=1}^{n} v_\alpha \frac{\partial \Phi}{\partial x_\alpha} = 0 , \qquad (0) = f(\underline{x})$$

on the assumption that the coefficients of Eq. (2.1) v_α satisfy the equation of continuity

$$(2.2) \qquad \sum_{\alpha=1}^{n} \frac{\partial v_\alpha}{\partial x_\alpha} = 0 .$$

The problem (2.1), (2.2) can be reduced to the divergence form

$$(2.3) \qquad \frac{\partial \Phi}{\partial t} + \sum_{\alpha=1}^{n} \frac{\partial v_\alpha \Phi}{\partial x_\alpha} = 0 .$$

This equation will be taken as the basic element for developing a numerical algorithm for the solution of hydrodynamic equations by the splitting up method [11].

There are many different methods for the solution of Eq. (2.3). Among the the most popular, in recent years, are the explicity Lax-Wendroff method [13], the explicit variant of the "box-method" [13], [14], and others. The main problem in the numerical solution of the equation of transfer of substances is the construction of unconditionally stable difference schemes of at least second order approximation which can be efficiently realized on computers. All these conditions are, in a sense, inconsistent. In some investigations it was found that the local criterion of stability for the difference schemes of the quasi-linear equations of form (2.1) sometimes does not ensure the conditions of stability of the difference schemes [15]. Short wave perturbations, rapidly increasing with time, may arise during the solution by such schemes. They are caused by the non-linear interaction of perturbations of different wave lengths. One of the ways to suppress the non-linear instability is to introduce sufficient viscosity or to use dissipative difference schemes. Only in recent years have a number of new methods been formed for the construction of unconditionally stable difference schemes which are good approximations and which can be described from a common point of view.

To construct the difference equation corresponding to Eq. (2.3) we use the Crank-Nicolson difference scheme

$$\frac{\phi^{j+1} - \phi^{j}}{\tau} + \sum_{\alpha=1}^{n} \frac{\partial u_{\alpha}^{j} \phi^{j+\frac{1}{2}}}{\partial x_{\alpha}} = 0 \quad (t_j \leqslant t \leqslant t_{j+1}) \tag{2.4}$$

where

$$\phi^{j+\frac{1}{2}} = \frac{1}{2} (\phi^{j+1} + \phi^{j}) \quad .$$

Besides in (2.4) we use an approximation of the coefficients v_{α} . Naturally, this approximation of the coefficients can be either of first or of second order accuracy in τ . For example, it has first order accuracy for

$$u_\alpha^j = v_\alpha^j$$

and second order accuracy for

$$u_\alpha^j = \frac{v_\alpha^{j+1} + v_\alpha^j}{2} \quad .$$

Next we introduce the notation for the operator A^j , A_α^j

$$A^j\phi = \sum_{\alpha=1}^{n} \frac{\partial u_\alpha^j}{\partial x_\alpha} \phi \quad , \quad A_\alpha^j \phi = \frac{\partial u_\alpha^j}{\partial x_\alpha} \phi \ .$$

Then Eq. (2.4) can be written as

$$\frac{\phi^{j+1} - \phi^j}{\tau} + A^j \phi^{j+\frac{1}{2}} = 0$$

or

$$(E + \frac{\tau}{2} A^j)\phi^{j+1} = (E - \frac{\tau}{2} A^j)\phi^j \ . \tag{2.5}$$

For simplicity we assume that the solution of the problem under consideration is periodic in the n-dimensional parallelepiped D . It is easy to verify directly that the equality

$$(A^j\phi, \phi) = 0 \tag{2.6}$$

is valid.

Hence, by the extended Kellogg's lemma, from (2.6) it follows that

(2.7) $$\|\phi^{j+1}\| = \|\phi^j\| .$$

The unconditional stability of (2.4) follows from (2.7).

Let us consider now a difference approximation of Eq. (2.4) in the space variables. The corresponding difference equation will be written as

(2.8) $$\frac{\phi^{j+1} - \phi^j}{\tau} + \Lambda^j \phi^{j+\frac{1}{2}} = 0 ,$$

where

(2.9) $$\Lambda^j = \sum_{\alpha=1}^{n} \Lambda^j_\alpha ,$$

and

(2.10) $$\Lambda^j_\alpha \phi = \frac{1}{2\Delta x_\alpha} (u^j_{\alpha,k_\alpha+\frac{1}{2}} \phi_{k_\alpha+1} - u^j_{\alpha,k_\alpha-\frac{1}{2}} \phi_{k_\alpha-1})$$

Here k_α are the indices corresponding to the mesh points of the variables x_α. The approximation of the operator A^j_α by the difference operator Λ^j_α was used in [13], [14], and by others.

Now introduce into consideration the inner product

(2.11) $$(a,b) = \sum_{k_{\alpha_1} \cdots k_{\alpha_n}} a_{k_1 k_2 \cdots k_n} b_{k_1 k_2 \cdots k_n} \Delta x_1 \cdots \Delta x_n .$$

It is easy to verify that

$$(\Lambda^j\phi,\phi) = 0 \; , \tag{2.12}$$

hence, using (2.12), we obtain

$$\|\phi^{j+1}\| = \|\phi^j\| = \cdots = \|f\| \; .$$

This means that, if we use a difference approximation of the operators Λ^j in form (2.9), (2.10), we again come to the unconditionally stable schemes.

Now, consider the approximation of A^j by Λ^j. In this connection, consider first the elementary operators A^j_α and Λ^j_α. Next use the expression for $\Lambda^j_\alpha\phi$ in the form (2.10) and transform the latter using the identity

$$\begin{aligned} u^j_{\alpha,k_\alpha+\frac{1}{2}} &= u^j_{\alpha,k_\alpha+1} - \frac{1}{2}(u^j_{\alpha,k_\alpha+1} - u^j_{\alpha,k_\alpha}) \; , \\ u^j_{\alpha,k_\alpha-\frac{1}{2}} &= u^j_{\alpha,k_\alpha-1} + \frac{1}{2}(u^j_{\alpha,k_\alpha} - u^j_{\alpha,k_\alpha-1}) \; . \end{aligned} \tag{2.13}$$

Then we have

$$\begin{aligned} \Lambda^j_\alpha\phi = {} & \frac{u^j_{\alpha,k_\alpha+1}\,\phi_{k_\alpha+1} - u^j_{\alpha,k_\alpha-1}\,\phi_{k_\alpha-1}}{2\Delta x_\alpha} \\ & - \frac{\phi_{k_\alpha}}{2}\,\frac{u^j_{\alpha,k_\alpha+1} - u^j_{\alpha,k_\alpha-1}}{2\ x} - \frac{\Delta x^2_\alpha}{4}\,R^j_\alpha \; , \end{aligned} \tag{2.14}$$

where

$$\begin{aligned} R^j_\alpha = {} & \frac{1}{\Delta x^3_\alpha}\,[(u^j_{\alpha,k_\alpha+1} - u^j_{\alpha,k_\alpha})(\phi_{k_\alpha+1} - \phi_{k_\alpha}) \\ & - (u^j_{\alpha,u_\alpha} - u^j_{\alpha,k_\alpha-1})(\phi_{k_\alpha} - \phi_{k_\alpha-1})] \; . \end{aligned} \tag{2.15}$$

In the above formulas we write only essential indices, the others are omitted for simplicity.

If $\Delta x_\alpha \to 0$, then, taking into account a sufficient smoothness of ϕ and of the coefficients u^j, we have

$$R_\alpha^j \to \frac{\partial}{\partial x_\alpha}\left(\frac{\partial u^j}{\partial x_\alpha}\cdot\frac{\partial \phi}{\partial x_\alpha}\right).$$

From (2.14) it follows that in general the operator Λ_α^j does not approximate the operator A^j.

Consider now the total operator Λ^j and study the expression $\Lambda^j\phi$. We have

$$\Lambda^j\phi = \sum_{\alpha=1}^{n}\frac{u^j_{\alpha,k_\alpha+1}\phi_{k_\alpha+1}-u^j_{\alpha,k_\alpha-1}\phi_{k_\alpha-1}}{2\Delta x_\alpha}$$

(2.16)

$$-\frac{\phi_{\underline{k}}}{2}\sum_{\alpha=1}^{n}\frac{u^j_{\alpha,k_\alpha+1}-u^j_{\alpha,k_\alpha-1}}{2\Delta x_\alpha}-\frac{1}{4}\sum_{\alpha=1}^{n}\Delta x_\alpha^2 R_\alpha^j,$$

where $\underline{k}$ stands for $\{k_\alpha\}$.

Since we assumed that the coefficients u_α^j satisfy the equation of continuity, in the difference form we have

$$\sum_{\alpha=1}^{n}\frac{u^j_{\alpha,k_\alpha+1}-u^j_{\alpha,k_\alpha-1}}{2\Delta x_\alpha}=0. \tag{2.17}$$

Hence, the second sum in (2.16) is identically equal to zero and we get

$$\Lambda^j\phi = \sum_{\alpha=1}^{n}\frac{u^j_{\alpha,k_\alpha+1}\phi_{k_\alpha+1}-u^j_{\alpha,k_\alpha-1}\phi_{k_\alpha-1}}{2\Delta x_\alpha}+O(h^2), \tag{2.18}$$

where $h = \max\left\{\max\{\Delta x_\alpha\}\right\}$.

From here it immediately follows that the total operator Λ^j approximates the initial A^j to the second order in geometrical variables. Unfortunately, despite the pleasant qualities of the considered difference scheme -- i.e. good approximation and unconditional stability -- the solution of the equation is very complicated and this compels us to look for simpler algorithms using the splitting up method. We will try to find such difference approximations which would not affect, in principle, the good qualities of the Crank-Nicolson difference scheme. The author studied this problem together with V. V. Penenko.

Consider a system of difference equations (1.15)

$$(E + \frac{\tau}{2}\Lambda_\alpha^j)\phi^{j+\frac{\alpha}{n}} = (E - \frac{\tau}{2}\Lambda^j)\phi^{j+\frac{\alpha-1}{n}}, \tag{2.19}$$

$$(\alpha = 1,2, \ldots, n),$$

and rewrite it in the equivalent form

$$\phi^{j+\frac{\alpha}{n}} = (E + \frac{\tau}{2}\Lambda_\alpha^j)^{-1}(E - \frac{\tau}{2}\Lambda_\alpha)\phi^{j+\frac{\alpha-1}{n}} \tag{2.20}$$

$$(\alpha = 1,2, \ldots, n).$$

Since

$$(\Lambda_\alpha^j\phi, \phi) = 0, \tag{2.21}$$

the following equality holds

$$\|\phi^{j+1}\| = \|\phi^{j+\frac{n-1}{n}}\|. \tag{2.22}$$

Using the recurrence relation (2.22), we come to the equality

$$\|\phi^{j+1}\| = \|\phi^{j}\| = \dots = \|f\| .$$

Thus, the difference system of Eqs. (2.19) based on the splitting up method is unconditionally stable. So far as the operators Λ_α^j of the problem under consideration are noncommutative, one must introduce into consideration a sequence of alternating schemes, according to (1.18), in order to obtain schemes accurate to the second order in all variables.

Let us consider now a compressible fluid. In this case the equation of transfer of substances along trajectories and the equation of continuity take the form

$$\frac{\partial \rho \Phi}{\partial t} + \sum_{\alpha=1}^{n} \frac{\partial \rho v_\alpha \Phi}{\partial x_\alpha} = 0 , \tag{2.23}$$

$$\frac{\partial \rho}{\partial t} + \sum_{\alpha=1}^{n} \frac{\partial \rho v_\alpha}{\partial x_\alpha} = 0 . \tag{2.24}$$

For simplicity and uniformity of presentation it will be assumed that the time intervals, at which the system of Eqs. (2.23)-(2.24) is approximated, are chosen as follows: $t_{j-1} \leqslant t \leqslant t_{j+1}$ $(j = 1,2, \dots)$. Then the approximation of these equations will be chosen in the following form:

$$\frac{\rho^{j+\frac{1}{2}}\phi^{j+1} - \rho^{j-\frac{1}{2}}\phi^{j-1}}{2\tau} + \tag{2.25}$$

$$+ \sum_{\alpha=1}^{n} \frac{(\rho u_\alpha)^j_{k_\alpha+\frac{1}{2}}\phi^j_{k_\alpha+1} - (\rho u_\alpha)^j_{k_\alpha-\frac{1}{2}}\phi^j_{k_\alpha-1}}{2\Delta x_\alpha} = 0 ,$$

(2.26) $$\frac{\rho^{j+1} - \rho^{j-1}}{2\tau} + \sum_{\alpha=1}^{n} \frac{(\rho u_\alpha)^j_{k+1} - (\rho u_\alpha)^j_{k-1}}{2\Delta x_\alpha} = 0 .$$

Next introduce the notations

$$\rho^{i+\frac{1}{2}} = \rho^{j+1} - \frac{1}{2}(\rho^{j+1} - \rho^j), \quad \rho^{j-\frac{1}{2}} = \rho^{j-1} + \frac{1}{2}(\rho^j - \rho^{j-1}),$$

$$\phi^j = \frac{1}{2}(\phi^{j+1} + \phi^j)$$

and take (2.13) into account with respect to the value ρu_α. Then Eq. (2.25) becomes

$$\frac{\rho^{j+1}\phi^{j+1} - \rho^{j-1}\phi^{j-1}}{2\tau}$$

$$+ \sum_{\alpha=1}^{n} \frac{(\rho u_\alpha)^j_{k_\alpha+1}\phi^j_{k_\alpha+1} - (\rho u_\alpha)^j_{k_\alpha-1}\phi^j_{k_\alpha-1}}{2\Delta x_\alpha}$$

(2.27)
$$- \frac{\phi_k}{2}\left[\frac{\rho^{j+1} - \rho^{j-1}}{2\tau} + \sum_{\alpha=1}^{n} \frac{(\rho u_\alpha)^j_{k_\alpha+1} - (\rho u_\alpha)^j_{k_\alpha-1}}{2\Delta x_\alpha}\right]$$

$$= 0(\tau^2 + h^2) .$$

Using the equation of continuity (2.26), we come to a scheme of second order accuracy

$$(2.28)\qquad \frac{\rho^{j+1}\phi^{j+1} - \rho^{j-1}\phi^{j-1}}{2\tau} + \sum_{\alpha=1}^{n} \frac{(\rho u_\alpha)^j_{k_\alpha+1}\phi^j_{k_\alpha-1} - (\rho u_\alpha)^j_{k_\alpha-1}\phi^j_{k_\alpha-1}}{2\Delta x} = 0 \ .$$

Examine some other properties of (2.25). For this purpose let us multiply Eq. (2.25) by ϕ^j_k and sum the result in all basic mesh points of the domain of definition with regard for the periodicity of the solution. Then we get the relation of balance in the form

$$J^{j+1} = J^j \ ,$$

where

$$J^j = (\rho^{j-\frac{1}{2}}\phi^j,\ \phi^{j-1}) \ .$$

It is useful to note that the following equality

$$J^j = (\rho^{j-\frac{1}{2}}\phi^{j-\frac{1}{2}},\ \phi^{j-\frac{1}{2}}) + O(\tau^2) \ .$$

is valid. Using the results obtained for the incompressible fluid, one can come to the construction of different splitting up schemes for a compressible fluid which are analogous to (2.19).

In conclusion it will be noted that in a similar way one can construct difference analogues of Eq. (2.1) with periodic boundary conditions of higher order accuracy in geometrical variables. For this purpose we use the following approximations:

$$\frac{\partial u_\alpha \phi}{\partial x_\alpha} \Rightarrow \Lambda_\alpha \phi \ , \tag{2.29}$$

where

$$\Lambda_\alpha \phi = \sum_{m=1}^{p} \beta_m \frac{\frac{1}{2}(u_{k+m}+u_k)\phi_{k+m} - \frac{1}{2}(u_k-u_{k-m})\phi_{k-m}}{2(m\Delta x_\alpha)} \ ,$$

where the β_m satisfy the following system of equations:

$$\sum_{m=1}^{p} \beta_m = 1 \ , \qquad \sum_{m=1}^{p} m^2 \beta_m = 0 \ , \quad \cdots \quad \sum_{m=1}^{p} m^{2p} \beta_m = 0 \ .$$

Then, if $x_\alpha = x_{\alpha_k}$, we have

$$\sum_{\alpha=1}^{n} \Lambda_\alpha \phi = \left(\sum_{\alpha=1}^{n} \frac{\partial u_\alpha \phi}{\partial x_\alpha} \right)_{x_\alpha = x_{\alpha_k}} + O(h^{2p})$$

hence,

$$\frac{\phi^{j+1} - \phi^{j}}{\tau} + \sum_{\alpha=1}^{n} \Lambda_\alpha^j \phi^{j+\frac{1}{2}} = O(\tau^2 + h^{2p}) \quad . \tag{2.30}$$

If we use approximation of the form (2.29) analogously to the foregoing, it is easy to get a splitting up scheme of $O(\tau^2 + h^{2p})$ accuracy.

3. Solution of Time-Independent Problems by the Splitting Up Method with Variational Optimization

Let us find the solution of the problem

$$\Lambda\Phi = f , \tag{3.1}$$

where Λ is a linear operator which is a matrix when the differential problem is reduced to the difference problem. Let

$$\Lambda = \sum_{\alpha=1}^{n} \Lambda_\alpha$$

and $\Lambda_\alpha > 0$. Consider the iterative process

$$B_j \frac{\phi^{j+1} - \phi^j}{\tau_j} + \Lambda\phi^j = f, \quad \phi^0 = 0 , \tag{3.2}$$

where the B_j will be chosen in the form

$$B_j = \prod_{\alpha=1} (E + \frac{\sigma_j}{2} \Lambda_\alpha) . \tag{3.3}$$

The choice of the operator in such a form is made in practically all economical schemes of the solution of difference equations using the alternating direction method or the splitting up methods. Our problem now is to choose free relaxation parameters τ_j and σ_j so that the iterative method converge as fast as possible. Refs. [10], [16], [17], [18] and others deal with the problem of optimization methods for the choice of relaxation parameters in the alternating direction method. Most optimization methods assume that the spectrum of the operator AB_j^{-1} is known. We shall develop another method of algorithm optimization which will be based on a posteriori information obtained during the solution of problem [19].

In addition to the general assumption of the positive definiteness of the matrices Λ and B_j, it will be assumed (and this is the most probable situation) that the spectral properties of Λ and B_j are not known. The problem is to choose the parameters τ_j for a fixed choice of σ_j and, hence, B_j, so that the norm of discrepancy of the iterative process (3.2) approach zero the fastest possible way. For this purpose we shall go from the functions ϕ^j to the discrepancies

$$\xi^j = \Lambda\phi^j - f \, . \tag{3.4}$$

Let us apply the operator B_j^{-1} to Eq. (3.2). Then, using relation (3.4), we have

$$\frac{\phi^{j+1} - \phi^j}{\tau_j} + B_j^{-1}\xi^j = 0, \quad \xi^0 = - f \, . \tag{3.5}$$

Now, let us apply the operator Λ to Eq. (3.5) and add and subtract f. As a result we obtain the equation for discrepancy

$$\frac{\xi^{j+1} - \xi^j}{\tau_j} + AB_j^{-1}\xi^j = 0, \quad \xi^0 = - f \tag{3.6}$$

or

$$\xi^{j+1} = (E - \tau_j AB_j^{-1})\xi^j, \quad \xi^0 = - f \, . \tag{3.7}$$

Consider the inner produce (ξ^{j+1}, ξ^{j+1}). Using (3.7), we get

$$(3.8) \qquad (\xi^{j+1}, \xi^{j+1}) = q_j(\xi^j, \xi^j) \ ,$$

where

$$(3.9) \qquad q_1 = 1 - 2\tau_j \frac{(AB_j^{-1}\xi^j\xi^j)}{(\xi^j,\xi^j)} + \tau_j^2 \frac{(AB_j^{-1}\xi^j, AB_j^{-1}\xi^j)}{(\xi^j,\xi^j)}$$

Let the matrix B_j be fixed. Choose the parameter τ_j which minimizes the functional q_j. We get

$$(3.10) \qquad \frac{dq_j(\tau_j)}{d\tau_j} = 0, \qquad \frac{d^2q_j(\tau_j)}{d\tau_j^2} > 0 \ .$$

The second relation from (3.10) under the above assumptions with respect to matrices Λ and B_j is always fulfilled. From the first equation we find

$$(3.11) \qquad \tau_j = \frac{(\Lambda B_j^{-1}\xi^j, \xi^j)}{(\Lambda B_j^{-1}\xi^j, \ B^{-1}\xi^j)}$$

Now introduce the notations

$$(3.12) \qquad \begin{aligned} y^{j+1} &= B_j^{-1}\xi^j \ , \\ z^{j+1} &= y^{j+1} \ . \end{aligned}$$

The following is the scheme for the realization of the splitting up algorithm. Given ξ^j we find the auxiliary function y^{j+1} from the equation

$$(3.13) \qquad B_j y^{j+1} = \xi^j \ .$$

This equation reduces to the system

$$(E + \frac{\sigma_j}{2} \Lambda_1) y^{j+\frac{1}{n}} = \xi^j ,$$

$$(E + \frac{\sigma_j}{2} \Lambda_2) y^{j+\frac{2}{n}} = y^{j+\frac{1}{n}} ,$$

(3.14)

- - - - - - - - - - - - - -

$$(E + \frac{\sigma_j}{2} \Lambda_n) y^{j+1} = y^{j+\frac{n-1}{n}} .$$

After finding y^{j+1}, we determine another auxiliary function

$$Z^{j+1} = \Lambda y^{j+1}$$

and then

(3.15) $$\tau_j = \frac{(Z^{j+1}, \xi^j)}{(Z^{j+1}, Z^{j+1})} .$$

The approximation ξ^{j+1} is determined by

(3.16) $$\xi^{j+1} = \xi^j - \tau_j Z^{j+1}$$

for ϕ^{j+1} where

(3.17) $$\phi^{j+1} = \phi^j - \tau_j y^{j+1} .$$

Up to this point it has been assumed that σ_j is fixed and, using its value, we found τ_j ensuring the quickest suppression of the discrepancy. Thus, we have in principle one more free parameter which makes possible additional optimization of the process.

To understand this process in general we shall consider a trivial case when the Λ_α are positive and commutative and, hence, the operators Λ_α have a common basis. Under this assumption, the solution of problem (3.7) will be sought with the help of a Fourier series according to the eigenfunctions of the spectral problem

$$\Lambda\omega = \lambda\omega \tag{3.18}$$

in the form

$$\xi = \sum_k \xi_k \omega_k \ . \tag{3.19}$$

Let Λ be a positive and, in general, asymmetric operator. Introduce into consideration a conjugate system of functions as the solution of the problem

$$\Lambda^* \omega^* = \lambda \omega^* \ . \tag{3.20}$$

Substitute (3.19) into (3.7) and multiply the result by ω_k^* and use the commutativity of the operator Λ_α. Then we get the relation for the Fourier coefficients of the discrepancy

$$\xi_k^{j+1} = \left[1 - \frac{\tau\lambda_k}{\Pi\left(1 + \frac{\sigma}{2}\lambda_k^{(\alpha)}\right)}\right]\xi_n^j \ . \tag{3.21}$$

where λ_k is the eigenvalue of the spectral problem (3.18) and $\lambda_k^{(\alpha)}$ the eigenvalue of the operator Λ_α. Expression (3.21) will reduce to

$$\xi_k^{j+1} = \left(1 - \frac{\tau\lambda_k}{\mu_k + \frac{\sigma}{2}\lambda_k}\right)\xi_k^j , \tag{3.22}$$

where

$$\mu_k = \prod_\alpha (1 + \frac{\sigma}{2}\lambda_k^{(\alpha)}) - \frac{\sigma}{2}\lambda_k > 0 . \tag{3.23}$$

The positiveness of the coefficients μ_k follows immediately from that of $\lambda_k^{(\alpha)}$.

Next we analyze the recurrence relation (3.22). First, we determine the range of the parameter σ so that at every σ from this range and for every k there take place the uniform evaluation

$$|\xi_k^{j+1}| \leq |\xi_k^j| . \tag{3.24}$$

Taking account of (3.22) we get

$$\left| 1 - \frac{\tau\lambda_k}{\mu_k + \frac{\sigma}{2}\lambda_k} \right| \leq 1 . \tag{3.25}$$

From here it follows that the condition

$$\frac{\tau - \sigma}{2}\lambda_k \leq \mu_k \tag{3.26}$$

must be fulfilled. This relation defines the range of possible parameters σ depending on the spectrum of the operators. If one chooses

$$\sigma = \tau , \tag{3.27}$$

then (3.26) will be satisfied for any harmonics of the spectrum. In this case for each harmonic q_k

$$q_k = \frac{\mu_k - \frac{\tau}{2}\lambda_k}{\mu_k + \frac{\tau}{2}\lambda_k} \tag{3.28}$$

is the amplification factor ensuring a uniform approach to zero of all harmonics of the Fourier series of the discrepancy. In this way we come to the conclusion of the appropriate choice of the parameter σ by (3.27).

Thus, it is shown that, at least for the commutative and positive operators Λ_α, the optimal choice of the parameters σ_j is given in a sense by the formula

$$\sigma_j = \tau_j . \tag{3.29}$$

The solution of this recurrence equation is very complicated. However, there is a well-known fact that the parameters τ_j and σ_j for positive operators Λ_α are, in general, weakly changing functions of the index j. This suggests an idea that one can use an effective method of choice of σ_j by

$$\sigma_j = \tau_{j-1} \quad (j = 1,2,\ldots) . \tag{3.30}$$

It should be noted that the method of choice of σ_j by

(3.30) is effective also when commutativity of the operators is absent and there is only the condition of their positive definiteness. Unfortunately, we have no exact theorems on the optimal choice of σ by (3.29) in the absence of the assumption of commutativity of the operators Λ_α. Therefore the problem of the optimal choice of the parameter σ in a general case remains to be solved.

It will be shown that under the above assumptions concerning ΛB_j^{-1} the iterative process of the splitting up method converges to the exact solution of the problem. For this purpose let us consider relation (3.8) where, because of (3.11), the functional q_j has the form

$$(3.31) \qquad q_j = 1 - \frac{(\Lambda B_j^{-1}\xi^j)^2}{(\Lambda B_j^{-1}\xi^j, \Lambda B_j^{-1}\xi^j)\,(\xi^j, \xi^j)} .$$

From the property of the norm and (3.8) it follows that q_j is non-negative and, if ΛB_j^{-1} is assumed to be positive, we have

$$(3.32) \qquad q_j < 1 .$$

Thus, the sequence of the norms $\|\xi^j\|$ tends to zero not slower than the terms of the geometrical progression with the denominator

$$r = \sqrt{\max_j q_j} .$$

It is easy to see that from this fact there follows convergence of ϕ^j itself to the exact solution of the problem. Indeed, consider the square of the norm of the discrepancy

$$(\xi^j, \xi^j) = (\Lambda\phi^j + f, \Lambda\phi^j + f) .$$

Since $f = \Lambda\phi$, where ϕ is the exact solution of the problem, then

$$(\xi^j, \xi^j) = (\Lambda\varepsilon^j, \Lambda\varepsilon^j) ,$$

where

$$\varepsilon^j = \phi - \phi^j .$$

From the equality $\lim_{j\to\infty} (\xi^j, \xi^j) = 0$ it follows that

$$\Lambda(\phi - \phi^\infty) = 0 . \tag{3.33}$$

As Λ is positive matrix, Eq. (3.33) has only a trivial solution, hence

$$\phi^\infty = \phi .$$

Thus, under the above assumptions, the iterative process for the splitting up method converges to the exact solution of the problem.

In conclusion it will be stressed that for the method of minimal descrpancies one need not have information about the spectrum of matrices. In this sense the method is self-adjusting to the optimal process because of the use of the a posteriori information about the solution.

The splitting up method is effectively used in the solution of the problems of meteorology, oceanology, transport theory and other branches of science [9], [21], etc.

REFERENCES

1. Peaceman, D. W. and Rachford, H. H. Jr., "The Numerical Solution of Parabolic and Elliptic Differential Equations," J. Soc. Indust. Appl. Math 3 (1955), 28-41.
2. Douglas, Jim Jr., "On the Numerical Integration of $\frac{\partial^2 u}{\partial x^2} + \frac{\partial^2 u}{\partial y^2} = \frac{\partial u}{\partial t}$ by Implicit Methods," J. Soc. Indust. Appl. Math 3 (1955), 42-65.
3. Douglas, Jim Jr., and Rachford H. H. Jr., "On the Numerical Solution of Heat Conduction Problems in Two and Three Space Variables," Trans. Am. Math. Soc. 82 (1916), 421-439.
4. Yanenko, N. N., "A Difference Method of Solution in the Case of the Multidimensional Equation of Heat Conduction," Dokl. Akad Nank SSSR 125 (1959), 1207-1210 (Russian).
5. Yanenko, N. N., "The Method of Fractional Steps for Solving Multidimensional Problems of the Mathematical Physics," Novosibirsk, 1967.
6. D'yakonov, E. G., "The Method of Alternating Directions in the Solution of Finite-Difference Equations," Dokl. Ahad. Nank SSSR 138 (1961), 271-274 (Russian).
7. Samarskiǐ, A. A., and Andreev, V. B., "Iteration Alternating Direction Schemes for the Numerical Solution of the Dirichlet Problem," Z. Vycisl. Mat. i Mat. Fiz, 4 (1964), 1025-1036.
8. Samarskiǐ, A. A., Lectures on the Theory of Difference Schemes (Russian), M. 1969.
9. Marchuk, G. I., "Numerical Weather Forecasting on the Sphere," Doklady AN SSSR, 156 (1964), 810-813.
10. Marchuk, G. I. and Yanenko, N. N., "The Application of the Splitting Up Method for the Solution of Problems of the Mathematical Physics," Lecture on IFIP Congress, New York, 1965.
11. Kellogg, R. B., "Another Alternating-Direction-Implicit Method," J. Soc. Indust. Appl. Math. 11 (1963), 976-979.

12. Marchuk, G. I., Numerical Methods of Weather Forecasting (Russian) Gidrometeoizdot, 1967.
13. Lax, P. and Wendroff, B., "Systems of Conservation Laws," Comm. Pure Appl. Math. 13 (1960), 217-237.
14. Kurihara, Y. and Holloway, J. L., "Numerical Integration of a Nine-Level Global Primitive Equations Model Formulated by the Box Method," Monthly Weather Rev., Vol. 15 (1967), 509-530.
15. Bryan, K., "A Scheme for Numerical Integration of the Equations of Motion on an Irregular Grid Free of Nonlinear Instability," Monthly Weather Rev., 94 (1966), 39-40.
16. Kreiss, H. O., "Uber Implisite Differenzmethoden fur Partielle Differentialgluchungen," Num. Math. 5 (1963), 24-47.
17. Habetler, G. Y. and Wachspress, E. L., "Symmetric Successive Overrelaxation in Solving Diffusion Difference Equations," Math. Comp. 15 (1961), 356-362.
18. Birkhoff, G. and Varga, R. S., "Implicit Alternating Direction Method," Trans. Am. Math. Soc. 92 (1959), 13-24.
19. Vorob'ev. Ju. V., "A Random Iteration Process in the Method of Alternating Directions," (Russian) Z. Vycisl. Mat. i Mat. Fiz, 8 (1960), 663-670.
20. Marchuk, G. I., Lectures in IRIA, Paris, 1968.
21. Marchuk, G. I., Numerical Computation of Nuclear Reactors (Russian), Atomizdat, 1961.

ON CLASSES OF n-DIMENSIONAL NONLINEAR MAPPINGS GENERALIZING SEVERAL TYPES OF MATRICES*

Werner C. Rheinboldt**

1. Introduction

Consider a nonlinear mapping $F:D \subset R^n \to R^n$ and the system of equations $Fx = z$. Many of the well-known convergence results about iterative processes for solving this system place only very general analytic conditions upon F , such as differentiability, Lipschitz-continuity, etc. This provides, of course, for rather broad theorems which are often generalizable to infinite-dimensional spaces. But at the same time, when applied to particular mappings on R^n , such as, for example, discrete analogs of elliptic boundary value problems, or nonlinear network flow functions, these general convergence results tend to give only relatively limited or localized information.

* This work was supported in part by the National Science Foundation under Grant GJ-231 and the National Aeronautics and Space Administration under Grant NGL-21-002-008.

** Computer Science Center, University of Maryland, College Park, Maryland.

The situation is analogous to the one in which only a norm condition $\|B\| < 1$ is used to ensure the convergence of an iterative process $x^{k+1} = Bx^k + z$, $k = 0,1,\ldots,$ for solving the linear system $Ax = z$. It is well-known that stronger convergence results here require a much deeper knowledge of the spectral properties of the iteration matrix B and hence of the structural properties of A itself. Similarly, it appears to be beyond question that also in the nonlinear case stronger convergence theorems will have to be based, in general, on more specific assumptions about the inherent finite-dimensional structure of the mapping F, as, for instance, the specific dependence of the components f_j of F on the individual variables x_i. This in turn leads to the need for defining and analyzing appropriate classes of n-dimensional nonlinear mappings as they occur in various applications. So far only a few structurally different classes of such mappings have been considered, and a need for more work along this line certainly exists.

This article presents a survey - and some new results - on recent work about a group of related classes of n-dimensional functions. Following an unpublished suggestion of Ortega, Rheinboldt [1969b] investigated the so-called M-functions on R^n which represent a nonlinear generalization of the well-known M-matrices. In particular, it was shown that the discrete analogs of mildly nonlinear elliptic problems considered by Bers [1953], Greenspan and Parter [1965], Ortega and Rheinboldt [1967], [1970a] and others, as well as the network flow functions analyzed by Birkhoff and Kellogg [1966] and Porsching [1969], are specific cases of M-functions. Moreover, the global convergence of the (underrelaxed) nonlinear (point-) Jacobi-, and (point-) Gauss-Seidel processes was established for continuous, surjective M-functions, thereby generalizing the corresponding well-known results for M-matrices (see, e.g., Varga [1962]).

The latter result is typical for many surprising similarities between the behavior of M-functions and M-matrices, and in turn these similarities suggest the idea of looking for analogous nonlinear extensions of other types of matrices as well. In this connection, the P- and S-matrices and their weaker forms, the P_0- and S_0-matrices considered by Fiedler and Ptak [1962], [1966] are of some interest, especially since every M-matrix is also a P- , as well as an S-matrix. Recently, Moré and Rheinboldt [1970] introduced and studied such nonlinear n-dimensional generalizations of these four types of matrices - accordingly named P_0 , P , S_0 , and S-functions. As expected, M-functions are special cases of P-functions and the continuous P-functions are S-functions. It also turned out that earlier results of Gale and Nikaido [1965] and Karamardian [1968] have a natural place in this theory, and that certain mappings, considered by Willson [1968] and Sandberg and Willson [1969a/b] in connection with particular electronic circuit problems, are included among these new functions.

In Section 2 we present the basic definitions of the mentioned function-classes and of several related types of mappings. This is followed in Section 3 by a survey of the major properties of these functions and of their interrelations. For clarity the results are not always stated in their most general form, and for further details about the material in the first two sections, as well as for many of the proofs, reference is made to Rheinboldt [1969b], Moré and Rheinboldt [1970], and Moré [1970]. Section 4 concerns the problem of determining the surjectivity of certain of the mappings under consideration and presents some new generalizations of earlier results on M-functions. In Section 5 connections between n-dimensional nonlinear mappings and network flow problems are discussed, and, finally Section 6 concerns a general type of iterative process similar to the processes obtained by regular splittings in the linear case. In particular, a global convergence theorem is proved which covers as a special case the convergence of the block-Jacobi-, and block-Gauss-Seidel process for continuous surjective M-functions.

At this point, I would like to extend my special thanks to Jorge Moré for his helpful cooperation in preparing this article and to the Gesellschaft für Mathematik und Datenverarbeitung, m.b.H., Birlinghoven/Germany, where, in 1969, I began work on several of the new results reported here.

2. Basic Definitions

Throughout this paper, $x \leq y$ denotes the natural (component-wise) partial ordering on the n-dimensional real linear space R^n of column vectors, and $x < y$ stands for $x_i < y_i$, $i \in N = \{1,2,\ldots,n\}$. The corresponding notation is used on the space $L(R^n)$ of real $n \times n$ matrices.

We begin by recalling the following standard terminology:

Definition 2.1. (a) A mapping $F: D \subset R^n \to R^n$ is isotone (or antitone) on D if $x \leqslant y$, $x,y \in D$, implies that $Fx \leqslant Fy$ (or $Fx \geqslant Fy$) , and strictly isotone (or strictly antitone) if, in addition, it follows from $x < y$, $x,y \in D$, that also $Fx < Fy$ (or $Fx > Fy$).

(b) The function $F:D \subset R^n \to R^n$ is inverse isotone on D if $Fx \leqslant Fy$, $x,y \in D$, implies that $x \leqslant y$.

Note the self-evident fact that an affine mapping $Fx = Ax + b$ is isotone exactly if $A \geqslant 0$ and inverse isotone if and only if A is nonsingular and $A^{-1} \geqslant 0$.

There is a close connection between nonlinear network flows and several of the function classes to be discussed here. In fact, many of the results about these functions have inherent network-theoretical aspects and appear to be intuitively clearer if a network terminology is used to state them. Following Rheinboldt [1969b] - and in analogy with the connection between graphs and their incidence matrices - a particular network is associated with any function on R^n .

Definition 2.2. Consider $F:D \subset R^n \to R^n$ with the components $f_1,\dots,f_n$.

(a) For any fixed $x \in R^n$ the n^2 functions

$$\psi_{ij}: \{t \in R^1 \mid x+te^j \in D\} \to R^1 ,$$

$$\psi_{ij}(t) = f_i(x+te^j) , \quad i,j \in N$$

are the link-functions of F at x . Here e^j are the usual unit basis vectors in R^n .

(b) The associated network $\Omega_F = \{N,\Lambda_F\}$ of F

consists of the set of nodes $N = \{1,\ldots,n\}$ and the set of links

$$\Lambda_F = \{(i,j) \in N \times N \mid i \neq j \ , \ \psi_{ij} \text{ not constant for some } x \in R^n\} \ .$$

A link $(i,j) \in \Lambda_F$ is <u>permanent</u> if ψ_{ij} is not constant for any $x \in R^n$.

This notation can be interpreted as follows: The variables $x_1,\ldots,x_n$ are state variables associated with the n nodes of Ω_F, and the value $f_i(x_1,\ldots,x_n)$ is the (total) efflux from node i at state x. The nodes i and j of N are connected by a link, if there is at least one state x at which the link function ψ_{ij} is not constant; we might say that at state x the link $(i,j) \in \Lambda_F$ is conducting. A permanent link is then conducting at any state x.

We shall now place various conditions upon the behavior of the link functions, and in all cases these conditions will be assumed permanent, that is, they are to hold independently of the particular state.

<u>Definition 2.3</u>. A mapping $F:D \subset R^n \to R^n$ is <u>off-diagonally antitone</u> if for any state $x \in R^n$ the "off-diagonal" link functions ψ_{ij}, $i \neq j$, $i,j \in N$, are antitone. Similarly, F is <u>diagonally</u> (<u>strictly</u>) <u>isotone</u> if for any $x \in R^n$ the "diagonal" link functions $\psi_{11},\ldots,\psi_{nn}$ are (strictly) isotone.

Off-diagonal antitonicity states that for any linked nodes $i,j \in N$ a change of the state x_j of the receiving node produces a change with the opposite sign in the efflux f_i from the originating node i. This is, of course, the expected situation in a linear potential network where the flow from i to j is proportional to the potential difference $x_i - x_j$. The matrix $A = (a_{ij})$ describing such a linear network flow then satisfies $a_{ij} \leq 0$, $i \neq j$, which is one of the properties of an M-matrix. Since the other property, $A^{-1} \geq 0$, is equivalent with inverse isotonicity, we are led to the following nonlinear generalization of M-matrices.

Definition 2.4. An inverse isotone and off-diagonally antitone mapping $F:D \subset R^n \to R^n$ is an M-function.

It is now hardly surprising that an affine mapping $Fx = Ax + b$ is an M-function if and only if $A \in L(R^n)$ is an M-matrix.

Ky Fan [1958] has shown that all principal minor determinants of an M-matrix are necessarily positive. The same result, of course, holds for all symmetric, positive definite matrices. More generally, Fiedler and Ptak [1962] considered the class of all matrices in $L(R^n)$ with this property and called them P-matrices.

A different generalization of the M-matrices can be obtained from the following characterization result of Ky Fan [1958]: A matrix $A \in L(R^n)$ with $a_{ij} \leq 0$, $i \neq j$, $i,j \in N$, is an M-matrix if and only if $Au > 0$ for some $u > 0$. Following earlier work by Stiemke [1915], Fiedler and Ptak [1966] called any $A \in L(R^n)$ an S-matrix if $Au > 0$ for some $u > 0$. In the same article they also showed that any P-matrix is an S-matrix and proved a number of results about these and related matrices.

Stimulated by these linear results, as well as by some nonlinear results of Gale and Nikaido [1965], Karamardian [1968], and Sandberg and Willson [1969a], Moré and Rheinboldt [1970] introduced the following nonlinear generalizations of the P- and S-matrices and of their weaker forms:

Definition 2.5. (a) A mapping $F:D \subset R^n \to R^n$ is a P_0-function (or P-function) on D, if for any $x,y \in D$, $x \neq y$, there exists a $k \in N$ such that $(x_k - y_k)(f_k(x) - f_k(y)) \geq 0$, $x_k \neq y_k$, (or $(x_k - y_k)(f_k(x) - f_k(y)) > 0$).

(b) $F:D \subset R^n \to R^n$ is an S_0-function (or S-function) on D, if for any $x \in D$ there exists a $y \in D$ such that $y \geq x$, $y \neq x$, and $Fy \geq Fx$ (or $Fy > Fx$).

Again it is easily verified - using the results of Fiedler and Ptak [1962], [1966] - that an affine mapping $Fx = Ax + b$ belongs to one of these four classes of functions if and only of $A \in L(R^n)$ is a member of the corresponding class of matrices.

In network terminology, P-functions have the property that for any (non-zero) change of the state there is at least one node at which the change of the efflux has the same sign as the change of state. For many applications this appears to be a rather natural condition.

If $A \in L(R^n)$ is diagonally dominant and has a nonnegative diagonal, then A is a P_0-matrix, since, if $x \neq 0$ and $k \in N$ such that $|x_k| = \|x\|_\infty$, we have

$$x_k(Ax)_k \geq x_k^2(a_{kk} - \sum_{j \neq k} |a_{kj}|) \geq 0 \; , \quad x_k \neq 0 \; .$$

If A is even strictly diagonally dominant, it is a P-matrix. This suggests the question whether the concept of diagonal dominance can also be extended to nonlinear functions. Already a simple reflection shows that several natural direct generalizations are not entirely satisfactory; this makes the following ingenious definition of Moré [1970] rather interesting:

Definition 2.6. A mapping $F:D \subset R^n \to R^n$ is strictly diagonally dominant, if for any $x,y \in D$, $x \neq y$, it follows from $f_k(x) = f_k(y)$ that $|x_k - y_k| < \|x - y\|_\infty$.

Moré [1970] shows that, again, an affine mapping $Fx = Ax + b$ is strictly diagonally dominant if and only if A is a strictly diagonally dominant matrix. He also introduces an extension of the concept which includes the irreducibly diagonally dominant matrices. This generalization is based on the existence of certain paths in the associated network.

3. Properties of the Different Functions

In line with the survey nature of this article we summarize now without proof some of the major properties of the classes of functions introduced in the previous section. For the sake of simplicity, these results are not given in their most general form, and, in particular, it is always assumed that $F:R^n \to R^n$ is defined on all of R^n, although more restricted domains could also be admitted.

Theorem 3.1. Relations Between the Classes.

(a) Any P- or S-function $F:R^n \to R^n$ is also a P_0- or S_0-function, respectively.

(b) Any continuous P_0 or P-function $F:R^n \to R^n$ is also an S_0 or S-function, respectively.

(c) Any isotone mapping $F:R^n \to R^n$ is an S_0-function, and strict isotonicity implies that F is an S-function.

(d) A continuous, inverse-isotone mapping $F:R^n \to R^n$ is an S-function.

(e) If $F:R^n \to R^n$ is a P_0- or P-function, then F is diagonally isotone, or strictly diagonally isotone, respectively.

(f) Any continuous, diagonally isotone, and strictly diagonally dominant mapping $F:R^n \to R^n$ is a P-function.

(g) $F:R^n \to R^n$ is an M-function if and only if it is an off-diagonally antitone P-function.

The implications (a), (c), and (e) are rather straightforward consequences of the definitions; (b) represents a result of Karamardian [1968] phrased in this terminology; (d) and (g) are proved by Moré and Rheinboldt [1970] while (f) is a result of Moré [1970].

We indicate the general structure of these relations in the following diagram:

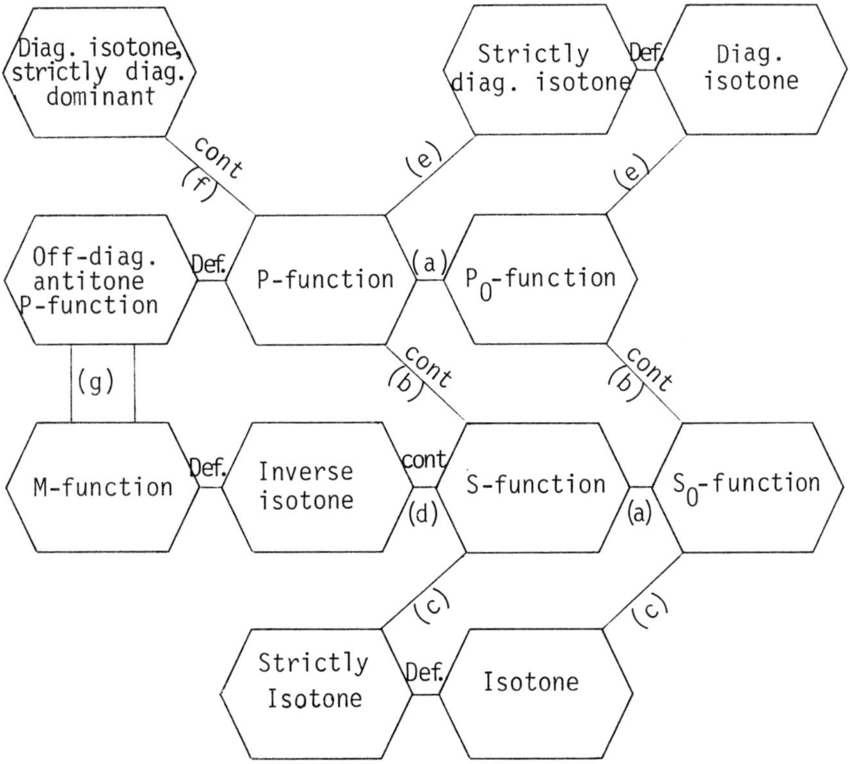

Note that, by definition, any isotone, or strictly isotone function is diagonally isotone, or strictly diagonally isotone, respectively. This relation is not shown. Note also that the diagram contains several derived implications, such as, for example, that any M-function is strictly diagonally isotone.

<u>Theorem 3.2. Inverses.</u>

(a) $F:R^n \to R^n$ is inverse isotone if and only if F is injective and $F^{-1}:FR^n \to R^n$ is isotone.

(b) If $F:R^n \to R^n$ is a P-function, then F is injective and $F^{-1}:FR^n \to R^n$ is again a P-function.

(c) If $F:R^n \to R^n$ is an F-differentiable, injective P_0-function, then $F^{-1}:FR^n \to R^n$ is again a P_0-function.

The proofs of (a) and (b) are straightforward consequences of the definitions. Part (c) is proved by Moré and Rheinboldt [1970]; it is conjectured that the result remains valid if F is only continuous.

In the case of P_0-, P-, or M-matrices also any principal submatrix belongs to the same class. In order to consider the nonlinear analog of this result, we formalize first the concept of a subfunction.

Definition 3.3. The subfunction of the mapping $F:R^n \to R^n$ corresponding to the index set $M = \{i_1,\ldots,i_m\} \subset N$, $0 < m \leq n$, and the constants c_j, $j \notin M$, is the mapping $G:R^m \to R^m$ with the components

$$g_k(y) = f_{i_k}\left(\sum_{j=1}^{m} y_j e^{i_j} + \sum_{j \notin M} c_j e^j\right), \quad k = 1,\ldots,m, \quad y \in R^m,$$

where e^j are the unit basis vectors in R^n.

We shall see in Section 5 that these subfunctions are of particular interest in connection with Dirichlet boundary value problems for network flows.

In generalization of the cited result for P_0-, P-, or M-matrices we have now:

Theorem 3.4. Subfunctions. For any P_0-, P-, or M-function $F:R^n \to R^n$, also any subfunction of F belongs to the same function class.

In the case of P_0- and P-functions, the proof follows directly from the definitions. For continuous, surjective M-functions, the result was given by Rheinboldt [1969b]. Its generalization to arbitrary M-functions is due to Moré and Rheinboldt [1970]; interestingly, the proof is based on the corresponding result for P-functions together with the relation (g) of Theorem 3.1 between the two function classes.

In the applications, characterization theorems for the functions of the various classes are of considerable importance. For M-functions Rheinboldt [1969b] gave four related theorems of this type, none of which required more than continuity. In order to illustrate the connection to other results given later, we prove here some modification of one of these theorems. As usual, a path from i to ℓ in Ω_F is a sequence of links of the form

$$(3.1) \quad (i_0,i_1),\ (i_1,i_2),\ldots,(i_m,i_{m+1}),\quad i_0 = i,\quad i_{m+1} = \ell,\quad m \geq 0 .$$

<u>Theorem 3.5</u>. Let $F:R^n \to R^n$ be off-diagonally antitone, and assume that for any $x \in R^n$ there exists a vector $u = u(x) > 0$ such that the mapping

$$p^x:R^1 \to R^n,\quad p_i^x(t) = f_i(x+tu(x)),\quad i \in N,$$

is isotone. Suppose further that for any $x \in R^n$ and

$i \in N$ there is a path (3.1) to a node $\ell = \ell(i,x) \in N$ such that p_ℓ^x is strictly isotone and that at any state the link functions $\psi_{i_j i_{j+1}}$, $j = 0,\ldots,m$, are strictly antitone. Then F is an M-function.

Proof. Suppose that $Fx \leqslant Fy$. Then, with $u = u(y)$,

$$+\infty > t_0 = \inf \{t \in R^1 \mid tu \geqslant y - x\} > -\infty$$

and $N_0 = \{i \in N \mid t_0 u_i = y_i - x_i\}$ is not empty. If $t_0 \leqslant 0$, then $x \leqslant y$, hence suppose that $t_0 > 0$. If $i \in N_0$, then also $j \in N_0$ for any $j \neq i$ such that ψ_{ij} is always strictly antitone. In fact, otherwise $y_i + t_0 u_i = x_i$ and $y_j + t_0 u_j > x_j$, and

$$\begin{aligned} & f_i(y+t_0u) \\ < \; & f_i(y_1+t_0u_1,\ldots,y_{j-1}+t_0u_{j-1},x_j,y_{j+1}+t_0u_{j+1},\ldots,y_n+t_0u_n) \\ \leqslant \; & f_i(x) \leqslant f_i(y) \leqslant f_i(y+t_0u) \end{aligned}$$

provides a contradiction. Hence, there exists a node $i \in N_0$ such that p_i^y is strictly isotone. But then

$$\begin{aligned} & f_i(y+t_0u) \\ = \; & f_i(y_1+t_0u_1,\ldots,y_{i-1}+t_0u_{i-1},x_i,y_{i+1}+t_0u_{i+1},\ldots,y_n+t_0u_n) \\ \leqslant \; & f_i(x) \leqslant f_i(y) < f_i(y+t_0u) \end{aligned}$$

is again a contradiction. Altogether, therefore, $t_0 > 0$

is impossible and the result is proved.

Note that when P^x is strictly isotone for any x, then F is already an M-function since we can take $\ell = \ell(i,x) = i$. This generalizes the sufficiency portion of the earlier cited characterization of M-matrices by Ky Fan, namely, that A is an M-matrix if and only if $a_{ij} \leq 0$, $i \neq j$, and $Au > 0$ for some $u > 0$. The necessity part is trivial, since we can take $u = A^{-1}e > 0$, with $e = (1,1,\ldots,1)^T$. It might be conjectured that similarly for M-functions the conditions of Theorem 3.5 are also necessary. This is not the case as the following example shows:

$$F:R^2 \to R^2 , \quad Fx = \begin{pmatrix} (\arctan x_1) - x_2 \\ \arctan x_2 \end{pmatrix} .$$

It is readily verified that F is an M-function, but for no $u > 0$ is $F(tu)$ an isotone function of t for all $t \in R^1$.

The above characterization result for M-functions does not even require F to be continuous. If F is assumed to be differentiable, simpler characterizations can be obtained in terms of properties of the derivative.

Theorem 3.6. Let $F:R^n \to R^n$ be F-differentiable on all of R^n.

(a) F is a P_0-function if and only if, for any $x \in R^n$, $F'(x)$ is a P_0-matrix.

(b) If, for any $x \in R^n$, $F'(x)$ is a P-matrix, then F is a P-function.

(c) If, for any $x \in R^n$, $F'(x)$ is an M-matrix, then F is an M-function.

(d) If F is an M-function, then $F'(x)$ is an M-matrix whenever it is nonsingular.

(e) If, for any $x \in R^n$, $F'(x)$ is a strictly diagonally dominant matrix, then F is a strictly diagonally dominant mapping.

Parts (a) and (d) are given by Moré and Rheinboldt [1970], and (e) is due to Moré [1970]. It may be noted that in some of these implications F-differentiability may be reduced to G-differentiability. Parts (b) and (c) represent results of Gale and Nikaido [1965] phrased in our terminology. It may be noted that (c) follows from (b) and Theorem 3.1(g). In fact, since $\partial_j f_i(x) \leqslant 0$ for $i \neq j$ and any $x \in R^n$, the mean value theorem applied to ψ_{ij} ensures that F is off diagonally antitone.

As a typical application of Theorem 3.6, consider the two-point boundary value problem

$$(3.2) \qquad u'' = (t,u,u') \ , \quad 0 \leqslant t \leqslant 1 \ , \quad u(0) = \alpha \ , \quad u(1) = \beta$$

where ϕ is F-differentiable on

$$S = \{(t,u,p)^T \in R^3 \mid 0 \leqslant t \leqslant 1 \ , \ u,p \in R^1\} \ ,$$

and

$$\partial_2\phi(t,u,p) \geqslant 0 \ , \quad |\partial_3\phi(t,u,p)| < \gamma \ , \quad \forall (t,u,p)^T \in S \ .$$

A simple discrete analog of (3.2) has the form

$$F : R^n \to R^n \ , \quad Fx = Ax + h^2\Phi x + b$$

where $h = (n+1)^{-1}$, $tj = jh$, $j = 0,1,\ldots,n+1$,

$$A = \begin{pmatrix} 2 & -1 & & & 0 \\ -1 & 2 & \ddots & & \\ & \ddots & \ddots & \ddots & \\ & & \ddots & \ddots & -1 \\ 0 & & & -1 & 2 \end{pmatrix} , \quad b = (-\alpha,0,\ldots,0,-\beta)^T ,$$

and $\Phi:R^n \to R^n$ has the components $\phi(t_i,x_i,(2h)^{-1}(x_{i+1}-x_{i-1}))$, $i = 1,\ldots,n$, with $x_0 = \alpha$, $x_{n+1} = \beta$. For $h < 2/\gamma$ it is readily verified that $F'(x) = A + h^2\Phi'(x)$ is always an M-matrix. In fact, $F'(x)$ is tridiagonal with positive diagonal and strictly negative first subdiagonals, and we have irreducible diagonal dominance. (See, e.g., Varga [1962].) Thus, F is an M-function. It turns out that F is also surjective; this can be shown in various ways; a simple proof follows from Theorem 4.7.

Since a linear mapping $A:R^n \to R^n$ is inverse isotone if and only if $A^{-1} \geq 0$, Theorem 3.6 suggests the conjecture that when $F:R^n \to R^n$ has, for any $x \in R^n$, a nonsingular F-derivative for which $F'(x)^{-1} \geq 0$, then F is inverse isotone. So far, this still represents an open problem, but there are several partial answers. We conclude this section with one of these; a second one is contained in the next section.

<u>Theorem 3.7</u>. Let $F:R^n \to R^n$ be convex and G-differentiable on R^n . Then F is inverse isotone if and only if, for any $x \in R^n$, $F'(x)$ is nonsingular and $F'(x)^{-1} \geq 0$.

The proof is given by Moré [1970].

4. Surjectivity

In this section we turn to the question when certain of the functions considered so far are surjective, that is, when the corresponding equation $Fx = z$ is solvable for any $z \in R^n$. The basic tool for our discussion will be the following "norm-coerciveness" theorem which appears to be due to Cacciopoli [1932] and which is also a special case of a more general result of Rheinboldt [1969a].

Theorem 4.1. Let $F:R^n \to R^n$ be a local homeomorphism. Then F is bijective if and only if F is norm-coercive in the sense that

$$\lim_{\|x\| \to \infty} \|Fx\| = +\infty \quad . \tag{4.1}$$

In addition, we shall frequently use the well-known domain invariance theorem which ensures that a continuous, injective mapping $F:R^n \to R^n$ has an open range FR^n and is a homeomorphism from R^n onto FR^n.

As a direct application of Theorem 4.1, we prove the following generalization of a result of Sandberg and Willson [1969a]. Following Ortega and Rheinboldt [1970b], a mapping $\Phi:R^n \to R^n$ is diagonal if the ith component ϕ_i of Φ is a function of only the ith variable x_i, or, in other words, if the link-set Λ_Φ of the associated network of Φ is empty.

Theorem 4.2. Let $F:R^n \to R^n$ be a continuous P_0-function such that, independent of x, the off-diagonal link-functions are Lipschitz continuous; that is

$$(4.2) \quad |f_i(x+se^j)-f_i(x+te^j)| \leq \gamma_{ij}|s-t| \ , \quad s,t \in R^1 \ , \quad x \in R^n \ , \quad i \neq j \ .$$

Then $\hat{F} = F + \Phi$ is a surjective P-function for any diagonal, strictly isotone, and surjective mapping $\Phi: R^n \to R^n$, ϕ_i, $i = 1,\ldots,n$.

<u>Proof:</u> From the definitions it follows readily that F is a P-function and, hence, injective. Moreover, each component of Φ is necessarily continuous on R^1 and thus also $\hat{F}$ is continuous. In order to apply Theorem 4.1, it remains to show only that $\hat{F}$ is norm-coercive. For this we proceed by induction with respect to the dimension n .

For $n = 1$, F is isotone and and the statement is trivial. Assume therefore that the theorem is valid for dimension $n - 1$, and that $\{x^k\} \subset R^n$ is any sequence such that $\{Fx^k\}$ is bounded. By the definition of P_0-functions, there exists for any $k \geq 0$ an index $i_k \in N$ such that

$$x^k_{i_k}(f_{i_k}(x^k)-f_{i_k}(0)) \geq 0 \ , \quad x^k_{i_k} \neq 0 \ ,$$

and hence that

$$(4.3) \qquad x^k_{i_k}(\hat{f}_{i_k}(x^k)-\hat{f}_{i_k}(0))$$

$$= x^k_{i_k}(f_{i_k}(x^k)-f_{i_k}(0)) + x^k_{i_k}(\phi_{i_k}(x^k_{i_k})-\phi_{i_k}(0))$$

$$\geq \ x_{i_k}^k(\phi_{i_k}(x_{i_k}^k)-\phi_{i_k}(0)) \ .$$

We can select a subsequence of $\{x^k\}$ - again denoted by $\{x^k\}$ - such that i_k is constant, and, for ease of notation, that $i_k = n$ for all $k \geq 0$. Then (4.3) assumes the form

$$x_n^k\phi_n(x_n^k) \leq x_n^k(\hat{f}_n(x^k)-f_n(0)) \ .$$

Since $\{\hat{F}x^k\}$ is bounded, and, say, $|\hat{f}_n(x^k)-f_n(0)| \leq c$, $k \geq 0$, it follows for $x_n^k \geq 0$ that $\phi_n(0) \leq \phi_n(x_n^k) \leq c$ or $0 \leq x_n^k \leq \phi_n^{-1}(c)$, while for $x_n^k < 0$ we obtain $\phi_n(0) \geq \phi_n(x_n^k) \geq -c$, and thus $0 \geq x_n^k \geq \phi_n^{-1}(-c)$. Altogether, therefore, $\{x_n^k\}$ is bounded.

Now consider the subfunction $\hat{G}:R^{n-1} \to R^{n-1}$ of $\hat{F}$ with the components

$$\hat{g}_i(x_1,\ldots,x_{n-1}) = \hat{f}_i(x_1,\ldots,x_{n-1},0)$$

$$= f_i(x_1,\ldots,x_{n-1},0) + \phi_i(x_i) \ ,$$

$$i = 1,\ldots,n-1 \ .$$

By Theorem 3.4, $\hat{G}$ satisfies again the conditions of the theorem, and it follows from the boundedness of $\{x_n^k\}$ and $\{Fx^k\}$ that

$$|\hat{g}_i(x_1^k,\ldots,x_{n-1}^k,0)|$$

$$\leq \; |\hat{f}_i(x^k)| + |f_i(x_1^k,\ldots,x_{n-1}^k,0) - f_i(x_1^k,\ldots,x_n^k)|$$

$$\leq \; |\hat{f}_i(x^k)| + \gamma_{in}|x_n^k| \leq \text{constant} \;, \quad i = 1,\ldots,n \;, \quad k \geq 0 \;.$$

Therefore, by induction hypothesis, $\{x_i^k\}$, $i = 1,\ldots,n-1$, must be bounded sequences, and this implies that $\hat{F}$ is indeed norm-coercive.

Note that the condition (4.2) certainly holds if F itself is uniformly Lipschitz-continuous. Thus the theorem applies, in particular, to the case $\hat{F} = A + \Phi$, where A is a P_0-matrix. This represents exactly the mentioned result of Sandberg and Willson [1969a]. At the same time, the one-dimensional example $Fx = e^x + x$ shows that (4.2) does not require F to be uniformly Lipschitzian.

In the case of inverse isotone, or M-functions, the rather stringent norm-coercivity assumption (4.1) can be replaced by the following condition of <u>order-coercivity</u>:

$$(4.4) \qquad \lim_{k\to\infty} \|Fx^k\| = +\infty \quad \text{whenever} \begin{cases} \lim_{k\to\infty} \|x^k\| = +\infty \\ \text{and either } x^k \leq x^{k+1} \text{ or} \\ \quad x^k \geq x^{k+1} \text{ for all } k \geq 0. \end{cases}$$

For M-functions this was proved by Rheinboldt [1969b]. For the proof of the corresponding more general result on inverse isotone mappings the next simple observation will be useful:

<u>Lemma 4.3</u>. A continuous, inverse isotone mapping $F:R^n \to R^n$ is surjective if and only if

(4.5) $$\{y \in R^n \mid y = z + tv \ , \quad -\infty < t < +\infty\} \subset FR^n$$

for some $v > 0$ and $z \in R^n$.

Proof: The necessity of the condition is trivial. For the proof of the sufficiency, observe first that F is a homeomorphism between R^n and FR^n and hence, by Theorem 4.1, that $FR^n = R^n$ if F is norm-coercive. Let $\{x^k\} \subset R^n$ be any sequence such that $\{Fx^k\}$ is bounded. Then, because of $v > 0$, we can choose constants α, β such that $\alpha v \leqslant Fx^k - z \leqslant \beta v$ for all $k \geqslant 0$, and hence, by (4.5), that

$$Fa = z + \alpha v \leqslant Fx^k \leqslant z + \beta v = Fb \ , \quad k = 0,1,\ldots$$

for certain $a,b \in R^n$. Therefore, it follows from the inverse isotonicity of F that $a \leqslant x^k \leqslant b$ for all k , which means that F is indeed norm-coercive.

The mentioned order-coercivity result has now the form:

Theorem 4.4. A continuous, inverse isotone mapping $F:R^n \to R^n$ is surjective if and only if it is order-coercive; that is, if and only if (4.4) holds.

Proof: If F is surjective, then Theorem 4.1 ensures that F is norm-coercive and hence order-coercive.

Conversely, let F be order-coercive and, with any fixed $z = Fx^0 \in FR^n$ and $\alpha > 0$, set $q(t)\colon [0,1] \to R^n$, $q(t) = z + t\alpha e$, $0 \leqslant t \leqslant 1$, where $e = (1,1,\ldots,1)^T$.

Since F is a homeomorphism between R^n and FR^n, we then have

$$\hat{t} = \sup \{t \in [0,1] \mid z + s\alpha e \in FR^n , \; s \in [0,t]\} > 0 ,$$

and $p(t) = F^{-1}q(t)$ is well-defined for $t \in [0,\hat{t}]$. Suppose that $\hat{t} < 1$ and let $\{t_k\} \subset R^1$ be such that $0 \leqslant t_k < t_{k+1} < \hat{t}$, $k = 0,1,\ldots$, and $\lim_{k\to\infty} t_k = \hat{t}$. Then $z \leqslant q(t_k) \leqslant q(t_{k+1}) \leqslant z + \hat{t}\alpha e$, $k = 0,1,\ldots$, and, by the inverse isotonicity of F, $p(t_k) \leqslant p(t_{k+1})$, $k = 0,1,\ldots$. But then the order-coercivity implies that $\{p(t_k)\}$ must be bounded, and, therefore, that $\lim_{k\to\infty} p(t_k) = \hat{x}$ exists. Now, by continuity, $F\hat{x} = q(\hat{t})$ and, since FR^n is open, $\hat{t}$ is clearly not maximal against assumption. Thus, necessarily, $\hat{t} = 1$ and, because $\alpha > 0$ was arbitrary, $z + te \in FR^n$ for all $t \geqslant 0$. Similarly it follows that $z + te \in FR^n$ for all $t \leqslant 0$ and now the result is a direct consequence of Lemma 4.3.

At the end of the previous section we mentioned the conjecture that when $F: R^n \to R^n$ is F-differentiable and $F'(x)^{-1} \geqslant 0$ for any $x \in R^n$, then F is inverse isotone. A partial answer to this question was given by Theorem 3.7. In the case of surjective mappings another partial result can be obtained with the help of a proof technique similar to that of the previous theorem.

Theorem 4.5. Suppose that $F: R^n \to R^n$ is continuously F-differentiable and that, for any $x \in R^n$, $F'(x)$ is nonsingular and satisfies $F'(x)^{-1} \geqslant 0$. Then F is inverse isotone and surjective if and only if it is order-coercive.

Proof: The necessity part was proved in Theorem 4.4. Suppose therefore that F is order-coercive. By the inverse-function theorem F is a local homeomorphism. More specifically, for any $x^0 \in R^n$ there exist open neighborhoods U of x^0 and V of Fx^0 such that the restriction F_U of F to U is a homeomorphism from U onto V and that the inverse $G = F_U^{-1}: V \to U$ is again continuously F-differentiable with $G'(y) = F'(Gy)^{-1} \geq 0$ for any $y \in V$. But then, for any $y \leq z$, $y,z \in V$, it follows from

$$Gz - Gy = \int_0^1 G'(y - t(z-y))(z - y)dt \geq 0$$

that G is isotone on V, and hence, by Theorem 3.2, that F_U is inverse isotone on U. In other words, for any $x^0 \in R^n$ there exists an open neighborhood U of x^0 in which F is inverse isotone.

Let now $y^0 = Fx^0$, $u \geq 0$, $u \neq 0$, be any vectors, and set $q:[0,1] \to R^n$, $q(t) = y^0 + tu$, $0 \leq t \leq 1$. By the local homeomorphism property, there is a $t_1 > 0$ and a continuous function $p:[0,t_1] \to R^1$ such that $p(0) = x^0$, and $Fp(t) = q(t)$, $t \in [0,t_1]$. If $t_1 < 1$, we can repeat this argument and continue p beyond t_1 to some $t_2 > t_1$, etc. This continuation process ensures the existence of a continuous $p:[0,\hat{t}) \to R^1$ such that $p(0) = x^0$, and $Fp(t) = q(t)$ for $t \in [0,\hat{t})$. Let $\hat{t} \in (0,1]$ be the maximal value up to which p can be extended. For any $r,s \in [0,\hat{t})$, $r < s$, the set $p([r,s])$ is compact and hence can be covered by finitely many open sets $U_1,\ldots,U_m$ in each of which F is inverse isotone. More specifically, we can select points $r = r_1 < r_2 < \cdots < r_{n+1} = s$, such that $\{p(t) \mid r_i \leq t \leq r_{i+1}\} \subset U_i$, $i = 1,\ldots,m$. Then $q(r_{i+1}) \geq q(r_i)$ and the inverse isotonicity of F in each U_i imply that $p(r_{i+1}) \geq p(r_i)$, $i = 1,\ldots,m$, and hence that

$$(4.6) \quad p(s) - p(r) = \sum_{i=1}^{m} (p(r_{i+1})-p(r_i)) \geq 0 \,, \quad 0 \leq r < s < \hat{t} \,.$$

Suppose now that $\hat{t} < 1$ and let $\{t_k\} \subset R^1$ be such that $0 \leq t_k < t_{k+1} < \hat{t}$, $k = 0,1,\ldots$, and $\lim_{k\to\infty} t_k = \hat{t}$. By (4.6) we have $p(t_k) \leq p(t_{k+1})$, $k = 0,1,\ldots$, and hence $y^0 \leq q(t_k) \leq y^0 + \hat{t}u$, $k \geq 0$, together with the order-coercivity, implies that $\{p(t_k)\}$ is bounded and, therefore, that $\lim_{k\to\infty} p(t_k) = \hat{x}$ exists. Because of the continuity of F, we now have $F\hat{x} = q(t)$ and the openness of FR^n contradicts the maximality of $\hat{t}$.

With this we have shown that $y^0 + u \in FR^n$ for any $y^0 \in FR^n$ and $u \geq 0$, $u \neq 0$. By the same argument it follows that also $y^0 - v \in FR^n$ whenever $y^0 \in FR^n$ and $v \geq 0$, $v \neq 0$. Since any point $y \in R^n$ can be written in the form $y = y^0 + u - v$, where $y^0 \in FR^n$ and $u \geq 0$, $v \geq 0$, we see that $y \in FR^n$ and hence that $FR^n = R^n$. Threefore, by Theorem 4.1, F is bijective and hence a homeomorphism from R^n onto itself. Thus, if $Fy \geq Fx$, then either $y = x$ or $u = Fy - Fx \geq 0$, $u \neq 0$. With $p(t) = F^{-1}(x+tu)$, $0 \leq t \leq 1$, the argument leading to (4.6) shows that $y = p(1) \geq p(0) = x$, and, hence, that

F is inverse isotone.

For continuous M-functions $F:R^n \to R^n$, Rheinboldt [1969b] has proved two surjectivity results which are based on Theorem 4.3; in other words, the assumptions placed upon F are sufficient to prove the order coercivity. For later reference we quote here without proof one of these results, and more specifically the one which represents a continuation of Theorem 3.5.

<u>Theorem 4.6</u>. Let $F:R^n \to R^n$ be continuous and off-diagonally antitone, and suppose that for some continuous, strictly isotone, surjective functions $h_j:R^1 \to R^1$, $j = 1,\ldots,n$, the mapping $P:R^n \to R^n$ with components $p_i(t) = f_i(h_1(x_1+t), \ldots, h_n(x_n+t))$, $i = 1,\ldots,n$, is isotone for any fixed $x \in R^n$. Assume further that for every node i of Ω_F there is a path (3.1) to some node ℓ such that, for any $x \in R^n$, the link functions $\psi_{i_j i_{j+1}}$, $j = 0,\ldots,m$, as well as the component p_ℓ are strictly isotone and surjective. Then F is a surjective M-function.

As a simple application to differentiable mappings, we present the following corollary:

<u>Theorem 4.7</u>. Let $F:R^n \to R^n$ be off-diagonally antitone and F-differentiable, and suppose that $F'(x)u \geq v \geq 0$ for all $x \in R^n$ and some fixed $u > 0$, $v \geq 0$. Assume further that for any node i of Ω_F there is a path (3.1) to some node ℓ such that $v_\ell > 0$ and

$$(4.7) \qquad \partial_{i_{j+1}} f_{i_j}(x) \leq a_j < 0 \ , \quad j = 0,\ldots,m \ ,$$

with constant $a_0,\dots,a_m$. Then F is a surjective M-function.

With $h_j(t) = t/u_j$, $j = 1,\dots,n$, this theorem reduces immediately to the previous one.

As mentioned earlier, Theorem 3.7 provides an easy means of verifying that the discrete analog of (3.2) - as considered in Section 3 - is a surjective M-function.

Clearly, there are various other corollaries of Theorem 4.6 along the lines of the previous result. Rather than to detail these possibilities, we end this section with a somewhat different observation. The discussion in this section was based on the norm-coerciveness theorem 4.1. As mentioned, Rheinboldt [1969a] obtained this theorem as a special case of a more general continuation theory; another special case of this theory is the well-known Hadamard theorem which states that a continuously F-differentiable mapping $F:R^n \to R^n$ is surjective if $F'(x)$ is nonsingular and $\|F'(x)^{-1}\| \leq \gamma$ for all $x \in R^n$. The rather simple and direct proof of the next theorem shows that also the Hadamard theorem can be used to obtain surjectivity results of the type considered here.

Theorem 4.8. Let $F:R^n \to R^n$ be off-diagonally antitone and continuously F-differentiable. Suppose further that for any $x \in R^n$ there is a vector $u(x) > 0$ such that $u(x) \leq u$ and $F'(x)u(x) \geq v > 0$ with fixed $u > 0$, $v > 0$. Then F is a surjective M-function.

Proof: By the earlier-cited characterization result for M-matrices, $F'(x)$ is an M-matrix for any $x \in R^n$, and hence Theorem 3.6 ensures that F is an M-function. Moreover, $F'(x)^{-1} \geq 0$ implies that $0 \leq F'(x)^{-1}v \leq u$. Let $F'(x)^{-1} = (b_{ij}(x))$, then

$$0 \leq b_{ik}(x)v_k \leq \sum_{j=1}^{n} b_{ij}(x)v_j \leq u_i \, , \quad i,k = 1,\ldots,n \, ,$$

and $v_k > 0$ shows that

$$\|F'(x)^{-1}\|_\infty \leq n(\max_i u_i)/(\min_k v_k) \, ,$$

and, therefore, that the Hadamard theorem applies.

5. Boundary Value Problems

For a mapping $F:R^n \to R^n$ representing an equilibrium flow on a given network, the following problem is basic: A state vector x is to be determined which satisfies certain specified conditions at the boundary nodes of the network and for which the efflux from all other nodes equals a prescribed value. In line with Birkhoff and Kellogg [1966], we consider here the case when the state at the boundary nodes is a given function of the efflux from that node. The system of equations to be solved has then the form

$$(5.1) \qquad \begin{cases} x_i - h_i(f_i(x)) = z_i \, , & i \in N_b \\ f_i(x) = z_i \, , & i \notin N_b \end{cases}$$

where $N_b \subset N$ is the set of boundary nodes. It is no restriction to assume always that $N_b = \{1,\ldots,m\}$, $1 \leq m \leq n$.

Any system of the form (5.1) is a (specific) boundary

value problem for F with respect to the boundary set N_b. We shall combine the given functions h_i into a mapping $H:R^1 \to R^m$ with components $h_1,\ldots,h_m$, and denote (5.1) by $\{H,z\}$ where $z \in R^n$. Correspondingly, $x = \text{sol}\,\{H,z\}$ designates any solution of (5.1).

The simplest type of boundary value problem is obtained when $H \equiv 0$, that is, when the boundary conditions reduce to $x_i = z_i$, $i \in N_b$; this will be called the Dirichlet boundary value problem. In this case, the system (5.1) is equivalent with

$$(5.2) \qquad f_i(z_1,\ldots,z_m,\, x_{m+1},\ldots,x_n) = z_i \;, \quad i = m+1,\ldots,n \;,$$

and hence we are again led to consider the properties of the subfunctions of F.

In many instances it is of interest to provide results for all boundary value problems $\{H,z\}$ obtained by letting $z \in R^n$ be any vector and H any function from a given class $\mathcal{H}$ of mappings. For abbreviation, we denote the collection of all problems $\{H,z\}$ of this type by $B(F,N_b,\mathcal{H})$, or $B(\mathcal{H})$ for short, if F and N_b are fixed. The two basic problems connected with such a class $B(\mathcal{H})$ are, of course, as in the case of differential equations, the existence and uniqueness of solutions for any $\{H,z\} \in B(\mathcal{H})$.

In view of the above indicated connection between the Dirichlet problems and the subfunctions of the mapping F, the following result is a direct consequence of Theorem 3.4:

Theorem 5.1. Let $F:R^n \to R^n$ be a P-, or M-function. Then the solution of any Dirichlet boundary value problem $\{0,z\}$, $z \in R^n$, of F is unique provided it exists.

For continuous, surjective M-functions $F:R^n \to R^n$, Rheinboldt [1969b] has shown that also any subfunction is again a surjective M-function. Thus, in that case any Dirichlet problem of F always has a unique solution. The proof of this result was based on a characterization of surjective M-functions in terms of the convergence of the Jacobi process. We give here a simple, direct proof based on the order coercivity theorem 4.3 and on the fact that the subfunctions of an M-function are again M-functions.

Theorem 5.2. Let $F:R^n \to R^n$ be a continuous surjective M-function. Then also any subfunction of F is again a surjective M-function.

Proof: We prove the result for the subfunction

$$G:R^{n-1} \to R^{n-1},\ g_i(x_1,\ldots,x_{n-1}) = f_i(x_1,\ldots,x_{n-1},c_n),\quad i = 1,\ldots,n-1\ ;$$

the general case then follows by repeated application of this result and by appropriate permutations of the variables and components. Because of Theorems 3.4 and 4.3 we need to show only that G is order-coercive. Let $\bar{x}^k = (x_1^k,\ldots,x_{n-1}^k)^T \in R^{n-1}$ be any monotonically increasing sequence such that $\{G\bar{x}^k\}$ is bounded. If, say, $b_i \geq g_i(\bar{x}^k)$, $i = 1,\ldots,n-1$, then

$$\left.\begin{aligned} b_i &\geq f_i(x_1^k,\ldots,x_{n-1}^k,c_n) \ , \quad i = 1,\ldots,n-1 \ , \\ b_n &= f_n(x_1^0,\ldots,x_{n-1}^0,c_n) \geq f_n(x_1^k,\ldots,x_{n-1}^k,c_n) \ , \end{aligned}\right\} k=0,1,\ldots ,$$

and hence it follows, with $v = F^{-1}(b_1,\ldots,b_n)$, that $v_i \geq x_i^k$, $i = 1,\ldots,n-1$, $k = 0,1,\ldots$. Therefore, $\{x^k\}$ is bounded, and the same result can be obtained if $\{x^k\}$ is monotonically decreasing. This shows that G is order coercive and thus surjective.

It may be interesting to note that the subfunction of surjective, inverse isotone mappings need neither be inverse isotone nor surjective. In fact, it is easily verified that

$$F:R^3 \to R^3, \ Fx = \begin{pmatrix} -x_1 & +x_2 & \\ x_1 & -x_2 & +x_3 \\ x_1 & & -x_3^3 \end{pmatrix}$$

is inverse isotone and surjective, while the subfunctions $G_1:R^1 \to R^1$, $g_1(x_1) = -x_1$, is not inverse isotone and the subfunction

$$G_2:R^2 \to R^2, \ G_2x = \begin{pmatrix} -x_1 & +x_2 \\ & \\ x_1 & -x_2 \end{pmatrix}$$

is not surjective.

In the case of elliptic partial differential equations, the validity of a maximum principle is an important tool in the study of the corresonding boundary value problems. In line with Rheinboldt [1969b] - where a somewhat different terminology was used - we define a maximum principle for network boundary value problems as follows:

Definition 5.3. Consider a class $B(F,N_b,\mathcal{H})$ of boundary value problems for a mapping $F:R^n \to R^n$ with respect to some set of boundary nodes $N_b \subset N$. The class $B(\mathcal{H})$ admits the maximum principle if for any $\{H^i,z^i\} \in B(\mathcal{H})$, $i = 1,2$, with $H^1(t) \leq H^2(t)$, $t \in R^1$, and $z^1 \leq z^2$, it follows that $x^1 \leq x^2$ for any $x^i = \text{sol}\ \{H^i,z^i\}$, $i = 1,2$.

For continuous, off-diagonally antitone F and the class $\mathcal{H} = \mathcal{A}$ of all continuous, antitone $H:R^1 \to R^m$, Rheinboldt [1969b] proved a theorem ensuring the validity of the maximum principle. This result was based on the observation that $B(F,N_b,\mathcal{A})$ admits the maximum principle if and only if, for any $H \in \mathcal{A}$, the mapping

$$(5.3) \quad F^h:R^n \to R^n, \quad f_i^H(x) = \begin{cases} x_i - h_i(f_i(x)), & i = 1,\dots,m \\ f_i(x), & i = m+1,\dots,n, \end{cases}$$

is inverse isotone.

In the differentiable case there is a simpler version of this result for which a proof can be obtained directly from Theorem 4.7. We denote by $\mathcal{A}^*$ the class of all F-differentiable, antitone mappings $H:R^1 \to R^m$.

Theorem 5.4. Let $F:R^n \to R^n$ be off-diagonally antitone and

F-differentiable and suppose that $F'(x)u \geqslant v \geqslant 0$ for any $x \in R^n$ and fixed $u > 0$, $v \geqslant 0$. Assume further that for any $i \notin N_b = \{1,\ldots,m\}$ there exists a path (3.1) from i to a boundary node ℓ for which (4.7) holds. Then every boundary value problem $\{H,z\} \in B(F,N_b,\mathcal{A}^*)$ has a unique solution and the class $B(\mathcal{A}^*)$ admits the maximum principle.

To apply Theorem 4.7 we need to note only that for any $H \in \mathcal{A}^*$ the function F^H of (5.3) is off-diagonally antitone and F-differentiable, and that $(F^H)'(x)u \geqslant \hat{v} \geqslant 0$ with $\hat{v}_i = u_i > 0$ for $i \in N_b$ and $\hat{v}_i = v_i$ for $i \notin N_b$.

6. Iterative Processes

This section is not intended to give a survey about the iterative solution of the equation $Fx = z$ when F belongs to any one of the function classes discussed here. There are many relevant results in the literature, and such a survey would, by necessity, be rather extensive (see, e.g., Ortega and Rheinboldt [1970b]). Even if we restrict ourselves to global convergence theorems, the list of results still remains surprisingly lengthy. For the Jacobi- or Gauss-Seidel processes it would include a well-known theorem of Schechter [1962], which in our terminology concerns a special type of P-function, new results of Moré [1970] for diagonally dominant functions, and also the result of Rheinboldt [1969b] for surjective M-functions, cited in the introduction. In addition, we would have to mention the global convergence theorem of Greenspan and Parter [1965] for the Newton-one-step Gauss-Seidel process, which applies to a simple type of M-function, and also the Newton-convergence theorem of Baluev [1952] which, as we shall see, assumes the inverse isotonicity of F .

Instead of going into further details about these and other related results, we shall consider here a particular class of implicit iterative processes of the form

$$G(x^{k+1}, x^k) = z \;, \quad k = 0,1,\ldots \;, \tag{6.1}$$

which represents a nonlinear generalization of the family of linear methods obtained from regular splittings.

For the iterative solution of a linear system $Ax = z$, Varga [1962] introduced a regular splitting of the matrix A as a decomposition $A = B - C$ in which B is nonsingular and $B^{-1} \geq 0$ as well as $C \geq 0$. He then showed that for such splittings the iterative process $x^{k+1} = B^{-1}Cx^k + B^{-1}z$, $k = 0,1,\ldots$, converges (for any $x^0 \in R^n$) to the solution of $Ax = z$ if $A^{-1} \geq 0$. This covers, in particular, the well-known result about the convergence of the Jacobi- and the Gauss-Seidel process when A is an M-matrix.

The mentioned similarity between the behavior of the M-matrices and M-functions suggests the idea of generalizing these regular splittings to nonlinear mappings. In direct analogy to the linear definition we introduce this generalization as follows:

Definition 6.1. A mapping $G: D_0 \times D_0 \subset R^n \times R^n \to R^n$ is a regular iteration function for $F: D \subset R^n \to R^n$ on the subset D_0 of D if

$$G(x,x) = Fx \;, \quad \text{for any } x \in D_0 \;, \tag{6.2a}$$

(6.2b) $G(\cdot,x):D_0 \to R^n$ is inverse isotone, for any fixed $x \in D_0$,

(6.2c) $G(y,\cdot):D_0 \to R^n$ is antitone, for any fixed $y \in D_0$.

Note that in the linear case this definition reduces exactly to that of a regular splitting. Note also that in the nonlinear case it is evidently necessary to add some assumption about the solvability of the equation $G(y,x) = z$, for given $x \in D_0$, before we can hope to establish general convergence results about the (implicit) iterative process (6.1).

In the remainder of this section we shall denote order intervals in R^n by $\langle u,v\rangle = \{x \in R^n \mid u \leq x \leq v\}$, and we write $x^k \uparrow x^*$, $k \to \infty$, as an abbreviation for $x^k \leq x^{k+1}$, $k = 0,1,\ldots$, $\lim_{k\to\infty} x^k = x^*$, and, similarly, $x^k \downarrow x^*$, $k \to \infty$, for $x^k \geq x^{k+1}$, $k = 0,1,\ldots$, $\lim_{k\to\infty} x^k = x^*$. Moreover, we will use the following well-known result of Kantorovich [1939]:

<u>Lemma 6.2</u>. Let $H:\langle x^0,y^0\rangle \subset R^n \to R^n$ be continuous and isotone, and assume that $x^0 \leq Hx^0$ and $y^0 \geq Hy^0$. Then the sequences $x^{k+1} = Hx^k$, $y^{k+1} = Hy^k$, $k = 0,1,\ldots$, satisfy $x^k \uparrow x^*$, $k \to \infty$, and $y^k \downarrow y^*$, $k \to \infty$, where $x^0 \leq x^* = Hx^* \leq y^* = Hy^* \leq y^0$.

The next theorem extends to processes of the form (6.1) a corresponding monotone convergence results for Jacobi and Gauss-Seidel processes proved by Rheinboldt [1969b].

Theorem 6.3. Given $F:D \subset R^n \to R^n$, suppose that for some $z \in R^n$ there are $x^0, y^0 \in D$ such that $x^0 \leq y^0$, $J = \langle x^0, y^0 \rangle \subset D$ and $Fx^0 \leq z \leq Fy^0$. Let $G:J \times J \to R^n$ be a continuous, regular iteration function for F on J with the property that $z \in G(J,x)$ for any fixed $x \in J$. Then the sequences $\{y^k\}$ and $\{x^k\}$ specified by (6.1) and starting from y^0 and x^0, respectively, are well-defined and satisfy $x^k \uparrow x^*$, $k \to \infty$, $y^k \downarrow y^*$, $k \to \infty$, where $x^0 \leq x^* \leq y^* \leq y^0$ and x^*, y^* are both solutions of $Fx = z$.

Proof: By assumption $G(\cdot,x) = z$ has a solution $y \in J$ and, because of (6.2b), this solution is unique. Hence the mapping $H:J \to J$ satisfying $G(Hx,x) = z$ is well-defined and clearly continuous. Moreover, by (6.2c) it follows from $x^0 \leq x \leq y \leq y^0$ that

$$G(Hy,y) = z = G(Hx,x) \geq G(Hx,y)$$

and hence, by (6.2b), that $Hy \geq Hx$. Therefore, H is isotone on J. Finally, again by (6.2b), $G(x^0,x^0) = Fx^0 \leq z = G(Hx^0,x^0)$ implies that $x^0 \leq Hx^0$, and, similarly, we obtain from $G(y^0,y^0) = Fy^0 \geq z = G(Hy^0,y^0)$ that $y^0 \geq Hy^0$. The convergence statement is now a direct consequence of Lemma 6.2, and $x^* = Hx^*$ and $y^* = Hy^*$ are equivalent with $Fx^* = G(x^*,x^*) = z$ and $Fy^* = G(y^*,y^*) = z$.

As a corollary we obtain the following global convergence result.

Theorem 6.4. Let $F:R^n \to R^n$ be continuous, inverse isotone, and surjective. Suppose, further, that $G:R^n \times R^n \to R^n$ is a regular iteration function for F on R^n with the property that $G(\cdot,x):R^n \to R^n$ is surjective for any fixed $x \in R^n$. Then, for any $z \in R^n$ and any initial point $x^0 \in R^n$, the process (6.1) converges to the unique solution $x^* \in R^n$ of $Fx = z$.

Proof: Note first that (6.2b) and (6.2c) together with the surjectivity of $G(\cdot,x)$ imply the existence of a continuous, isotone mapping $H:R^n \to R^n$ for which $G(Hx,x) = z$ for all $x \in R^n$. Moreover, for any $x^0 \in R^n$ the sequence $\{x^k\} \subset R^n$ given by (6.1) is uniquely defined by $x^{k+1} = Hx^k$, $k = 0,1,\ldots$.

Let $a,b \in R^n$ be the vectors with the components

$$a_j = \min\,(f_j(x^0),z_j)\,, \qquad b_j = \max\,(f_j(x^0),z_j)\,, \qquad j = 1,\ldots,n\,,$$

and set $u^0 = F^{-1}a$, $v^0 = F^{-1}b$. Then $Fu^0 \leqslant z \leqslant Fv^0$ as well as $Fu^0 \leqslant x^* \leqslant Fv^0$, and it follows from the inverse isotonicity of F that $u^0 \leqslant x^* \leqslant v^0$, as well as $u^0 \leqslant x^0 \leqslant v^0$. Consider now the sequences $\{u^k\}$, $\{x^k\}$, $\{v^k\}$ given by (6.1) and starting from u^0, x^0, and v^0, respectively. Then $u^{k+1} = Hu^k$, $x^{k+1} = Hx^k$, $v^{k+1} = Hv^k$, $k = 0,1,\ldots$, and, because of the isotonicity of H, we obtain from $u^0 \leqslant x^0 \leqslant v^0$, by induction, that $u^k \leqslant x^k \leqslant v^k$, $k = 0,1,\ldots$. Moreover, from Theorem 6.3 it follows that $u^k \uparrow u^*$, $k \to \infty$ and $v^k \downarrow v^*$, $k \to \infty$ where $u^0 \leqslant u^* \leqslant v^* \leqslant v^0$ and $Fu^* = z$, $Fv^* = z$. The injectivity of F then implies that $u^* = v^* = x^*$ and hence necessarily that, $\lim_{k\to\infty} x^k = x^*$.

This result covers as a special case the global convergence of the Jacobi and the Gauss-Seidel process for continuous surjective M-functions mentioned in the introduction. Instead of going into details of that case we extend it immediately to the corresponding block processes.

With $n_1 + n_2 + \cdots + n_p = n$, $n_j \geq 1$, $p \geq 1$, consider R^n as the product-space $R^{n_1} \times R^{n_2} \times \cdots \times R^{n_p}$ and let $P_i : R^n \to R^{n_i}$, $i = 1,\ldots,p$, denote the corresponding natural projections. Then any $x \in R^n$ may be partitioned in the form $x = (x^1,\ldots,x^p)$ where $x^i = P_i x$, $i = 1,\ldots,p$, and, similarly, we define the block-components $F^i : R^n \to R^{n_i}$ of any mapping $F : R^n \to R^n$ by $F^i x = P_i F x$, $i = 1,\ldots,p$.

For the solution of the equation $Fx = z$, the block-Gauss-Seidel process, with respect to the particular partition, has now the form:

$$(6.3)\quad \begin{cases} \text{Determine a solution } x^i \in R^{n_i} \text{ of} \\ F^i((x^1)^{k+1},\ldots,(x^{i-1})^{k+1}, x^i, (x^{i+1})^k,\ldots,(x^n)^k) = z^i \\ \text{and set } (x^i)^{k+1} = x^i , \quad i = 1,\ldots,p , \quad k = 0,1,\ldots . \end{cases}$$

Analogously, the block Jacobi process can be defined, and the form of the relaxation-versions of both processes should be self-evident. For reasons of space, we restrict ourselves here to (6.3), especially since in the mentioned other cases the discussion remains essentially the same.

Observe now that with

$$(6.4)\qquad G : R^n \times R^n \to R^n ,$$

$$P_i G(y,x) = F^i(y^1,\ldots,y^i,x^{i+1},\ldots,x^n)\ ,\quad i = 1,\ldots,p\ ,$$

the process (6.3) assumes the general form (6.1). For M-functions F, the following result then ensures the applicability of Theorems 6.3 and 6.4 to the block Gauss-Seidel process (6.3).

Theorem 6.5. Let $F:R^n \to R^n$ be an M-function; then the mapping $G:R^n \times R^n \to R^n$ defined by (6.4) is a regular iteration function for F on R^n. If, in addition, F is continuous and surjective, then $G(\cdot,x):R^n \to R^n$ is surjective for any fixed $x \in R^n$.

Proof: The conditions (6.2a) and (6.2c) are evidently satisfied for F, and hence, in order to complete the proof of the first statement, it remains to show only that $G(\cdot,x):R^n \to R^n$ is inverse isotone for any $x \in R^n$. Let $G(v,x) \geq F(u,x)$ for some $x,u,v \in R^n$. By Theorem 3.4, the subfunction $F^1(\cdot,x^2,\ldots,x^n):R^{n_1} \to R^{n_1}$ of F is again an M-function, and, hence, $P_1G(v,x) \geq P_1G(u,x)$ implies that $v^1 \geq u^1$. To proceed by induction, suppose that $v^i \geq u^i$ for $i = 1,\ldots,k-1$ and some k with $2 \leq k \leq n$. Then the off-diagonal antitonicity of F ensures that

$$\begin{aligned} &F^k(u^1,\ldots,u^{k-1},v^k,x^{k+1},\ldots,x^n) \\ &\geq F^k(v^1,\ldots,v^k,x^{k+1},\ldots,x^n) \\ &= P_kG(v,x) \geq P_kG(u,x) \\ &= F^k(u^1,\ldots,u^k,x^{k+1},\ldots,x^n)\ . \end{aligned}$$

Since $F^k(u^1,\ldots,u^{k-1},\cdot,x^{k+1},\ldots,x^n):R^{n_k} \to R^{n_k}$ is again an M-function, this shows that also $v^k \geq u^k$, and hence altogether, that $v \geq u$.

For the proof of the second part, assume that F is a continuous, surjective M-function. Then, by Theorem 5.2, also any subfunction of F has the same properties. Hence, for given $x,z \in R^n$, the equation

$$P_1G(y,x) = F^1(y^1,x^2,\ldots,x^n) = z^1$$

has a unique solution $y^1 \in R^n$. If for some k, with $2 \leq k \leq n-1$, vectors $y^i \in R^{n_i}$, $i = 1,\ldots,k-1$, have already been found with the property that

$$(6.5) \qquad P_iG(y,x) = F^i(y^1,\ldots,y^i,\ x^{i+1},\ldots,x^n) = z^i\ , \qquad i = 1,\ldots,k-1\ ,$$

then the surjectivity of $F^k(y^1,\ldots,y^{k-1},\cdot,x^{k+1},\ldots,x^n)$ ensures the existence of a $y^k \in R^{n_k}$ for which (6.5) holds with $i = k$. Hence, $G(y,x) = z$ has a unique solution $y \in R^n$ and the proof is complete.

Theorems 6.4 and 6.5 together state the global convergence of the block Gauss-Seidel process (6.3) for continuous, surjective M-functions. As mentioned above, this result also carries over to the block-Jacobi process and the underrelaxed versions of both these block methods, and the proofs for these cases are essentially the same as that for the Gauss-Seidel method.

These results raise the question whether some of the other global convergence theorems mentioned in the beginning of this section might also be subsumed under Theorem 6.4. This is not the case due to the fairly restrictive nature of the regular iteration functions. There appear to be many possible modifications of Definition 6.1, but it is doubtful whether any one of them actually covers a broader class of methods and not just again only a few specific convergence theorems. We end this section with some results about one such generalized form of the regular iteration functions. Since this discussion is primarily intended to be an example, no attempt was made to phrase these results in their most general form.

Theorem 6.6. Let $F:R^n \to R^n$ be inverse isotone and $G:R^n \times R^n \to R^n$ a continuous mapping with the properties

(6.6a) $G(x,x) = Fx$ for any $x \in R^n$

(6.6b) $G(\cdot,x):R^n \to R^n$ is inverse isotone and surjective for any $x \in R^n$

(6.6c) $G(x,x) \geq G(y,x)$, $x,y \in R^n$ implies that $G(y,y) \geq G(y,x)$.

If for some x^0, y^0, $z \in R^n$ we have $Fx^0 \leq z \leq Fy^0$, then the sequence $\{y^k\}$ given by (6.1) and starting from y^0 satisfies $y^k \downarrow x^*$, $k \to \infty$, where $x^* \in \langle x^0,y^0 \rangle$ is the (unique) solution of $Fx = z$ in R^n .

Proof: By (6.6b) the mapping $H:R^n \to R^n$ satisfying $G(Hx,x) = z$ is well-defined and certainly continuous.

Then $\{y^k\}$ is uniquely specified by $y^{k+1} = Hy^k$, $k = 0,1,\ldots$. Moreover, from (6.6c) it follows that when $G(x,x) = Fx \geq z = G(Hx,x)$, then also $FHx = G(Hx,Hx) \geq G(Hx,x) = z$. Thus we obtain from $Fy^0 \geq z$, by induction, that $Fy^k \geq z \geq Fx^0$, $k = 0,1,\ldots$. Now the inverse isotonicity of F implies that $y^k \geq x^0$, for all $k \geq 0$, which, in turn, ensures the existence of $\lim_{k\to\infty} y^k = x^* \varepsilon \langle x^0,y^0\rangle$. By the continuity of G we then have $Fx^* = G(x^*,x^*) = z$ and, clearly, x^* is unique.

Note that the analogous result holds for the lower sequence $\{x^k\}$ starting from x^0 provided that all inequalities in (6.6c) are reversed. Any regular iteration function satisfies (6.6c) as well as the corresponding reversed implication. In fact, if $G(x,x) \geq G(y,x)$, $x,y \varepsilon R^n$, then we obtain from (6.2b) that $x \geq y$ and hence from (6.2c) that $G(y,y) \geq G(y,x)$. The analogous argument applies when the inequalities are reversed.

As an application of this result we give the following theorem, proved in more generality by Ortega and Rheinboldt [1967].

Theorem 6.7. Let $F:R^n \to R^n$ be continuously differentiable, convex, and inverse isotone, and assume that $F'(x) = B(x) - C(x)$ is, for any $x \varepsilon R^n$, a regular splitting of $F'(x)$ with continuous $B:R^n \to L(R^n)$. If for given $x^0, y^0, z \varepsilon R^n$ we have $Fx^0 \leq z \leq Fy^0$ (and thus $x^0 \leq y^0$), then the sequence

$$(6.7) \qquad y^{k+1} = y^k - B(y^k)^{-1}(Fy^k - z), \quad k = 0,1,\ldots$$

satisfies $y^k \downarrow x^*$, $k \to \infty$, where $x^* \in \langle x^0, y^0 \rangle$ is the (unique) solution of $Fx = z$ in R^n.

In order to use Theorem 6.6, we note first that for

$$G: R^n \times R^n \to R^n , \quad G(y,x) = B(x)(y-x) + Fx$$

the process (6.7) is equivalent with (6.1). Clearly G satisfies (6.6a) and (6.6b); for the proof of (6.6c) let $G(x,x) \geqslant G(y,x) = w$ for some $x,y \in R^n$. This inequality is equivalent with $0 \geqslant B(x)(y-x)$, and hence, $B(x)^{-1} \geqslant 0$ implies that $x \geqslant y$. With the help of the convexity inequality

$$(6.8) \qquad Fy - Fx \geqslant F'(x)(y-x) , \quad x,y \in R^n ,$$

we find then that indeed

$$\begin{aligned} G(y,y) &= Fy \geqslant Fx + F'(x)(y-x) \\ &= Fx - (B(x)-C(x))B(x)^{-1}(Fx-w) \\ &= C(x)B(x)^{-1}(Fx-w) + w \geqslant w = G(y,x) . \end{aligned}$$

Note that G is, in general, not a regular iteration function.

Ortega and Rheinboldt [1967] have shown that this result contains as a corollary the global convergence theorem for Newton's method of Baluev [1952] mentioned in the beginning of this section. It may be interesting that Theorem 3.7 allows us to state this theorem in a slightly modified, and yet equivalent, form:

Theorem 6.8. Let $F:R^n \to R^n$ be continuously differentiable, convex, and inverse isotone. If $Fx = z$ has a solution x^*, then, for any $x^0 \in R^n$, the Newton iterates $x^{k+1} = x^k - F'(x^k)^{-1}(Fx^k - z)$, $k = 0,1,\ldots$, are well-defined and satisfy $x^k \geq x^{k+1}$, $k = 1,2,\ldots$, and $\lim_{k\to\infty} x^k = x^*$.

By Theorem 3.7, $F'(x)$ is always nonsingular and $F'(x)^{-1} \geq 0$. Thus the mapping $G:R^n \times R^n \to R^n$, $G(y,x) = F'(x)(y-x) + Fx$ clearly satisfies (6.6a) and (6.6b), while (6.6c) is simply (6.8). From (6.8) also follows that $Fx^1 \geq Fx^0 + F'(x^0)(x^1 - x^0) = z$, and hence, that Theorem 6.6 applies on the order interval $\langle x^*, x^1 \rangle$.

REFERENCES

1. Baluev, A. [1952], "On the Method of Chaplygin," Dokl. Akad. Nauk SSSR 83, 781-784.
2. Bers, L. [1953], "On Mildly Nonlinear Partial Difference Equations of Elliptic Type," J. Res. Nat. Bur. Stand. 51, 229-236.
3. Birkhoff, G., and Kellogg, B. [1966], "Solution of Equilibrium Equations in Thermal Networks," Proc. Symp. General. Networks, Brooklyn Polyt. Press.
4. Cacciopoli, R. [1932], "Sugli elementi uniti delle trasformazioni funzionali," Rend. Sem. Mat. Univ. Padova 3, 1-15.
5. Fan, Ky [1958], "Topological Proofs for Certain Theorems on Matrices with Nonnegative Elements," Monatsh. Math. 62, 219-237.
6. Fiedler, M., and Ptak, V. [1962], "On Matrices with Non-Positive Off-Diagonal Elements and Positive Principal Minors," Czech. Math. J. 12, 382-400.
7. Fiedler, M., and Ptak, V. [1966], "Some Generalizations of Positive Definiteness and Monotonicity," Num. Math. 9, 163-172.
8. Gale, D., and Nikaido, H. [1965], "The Jacobian Matrix and Global Univalence of Mappings," Mathem. Annalen 159, 81-93.

9. Greenspan, D., and Parter, S. [1965], "Mildly Nonlinear Elliptic Partial Differential Equations and Their Numerical Solution II," Num. Math. 7, 129-147.
10. Kantorovich, L. [1939], "The Method of Successive Approximations for Functional Equations, Acta Math. 71, 63-97.
11. Karamardian, S. [1968], "Existence of Solutions of Certain Systems of Nonlinear Inequalities, Num. Math. 12, 237-334.
12. Moré, J. [1970], "A Class of Nonlinear Functions and the Convergence of Gauss-Seidel and Newton-Gauss-Seidel Iterations," Ph.D. Dissertation, University of Maryland.
13. Moré, J., and Rheinboldt, W. [1970], "On P- and S-Functions and Related Classes of n-Dimensional, Nonlinear Mappings," Computer Science Center, University of Maryland, Technical Report 70-120.
14. Ortega, J., and Rheinboldt, W. [1967], "Monotone Iterations for Nonlinear Equations with Application to Gauss-Seidel Methods," SIAM J. Num. Anal. 4, 171-190.
15. Ortega, J., and Rheinboldt, W. [1970a], "Local and Global Convergence of Generalized Linear Iterations," Studies in Numerical Analysis 2, 122-143, SIAM Publications, Philadelphia, Pa.
16. Ortega, J., and Rheinboldt, W. [1970b], Iterative Solution of Nonlinear Equations in Several Variables, Academic Press, Inc., New York, N.Y.
17. Porsching, T. [1969], "Jacobi and Gauss-Seidel Methods for Nonlinear Network Problems," SIAM J. Num. Anal. 6, 437-449.
18. Rheinboldt, W. [1969a], "Local Mapping Relations and Global Implicit Function Theorems," Trans. Am. Math. Soc. 138, 183-198.
19. Rheinboldt, W. [1969b], "On M-Functions and their Application to Nonlinear Gauss-Seidel Iterations and Network Flows," Ges. f. Math. u. Datenverarb., Birlinghoven/Germany, Tech. Rep. BMwF-GMD-23; J. Math. Anal. Appl., 32,1, in press.
20. Sandberg, I. W., and Willson, Jr., A. N. [1969a], "Some Theorems on Properties of DC Equations of Nonlinear Networks," The Bell System Tech. J. 48, 1293-1311.
21. Sandberg, I. W., and Willson, Jr., A. N. [1969b], "Some Network-Theoretic Properties of Nonlinear DC Transistor Networks," The Bell System Tech. J. 48, 1-34.
22. Schechter, S. [1962], "Iteration Methods for Nonlinear Problems," Trans. Am. Math. Soc. 104, 179-189.

23. Stiemke, E. [1915], "Über positive Lösungen homogener linearer Gleichungen," Math. Annalen 76, 340-342.
24. Varga, R. [1962], Matrix Iterative Analysis, Prentice Hall, Englewood Cliffs, N.J.
25. Willson, Jr., A. N. [1968], "On the Solution of Equations for Nonlinear Resistive Networks," The Bell System Tech. J. 48, 1755-1773.

THE FINITE ELEMENT METHOD AND APPROXIMATION THEORY*

Gilbert Strang**

The essence of the finite element method appears to be this, that one seeks an approximate solution of the form

$$u^h(x) = \sum v_j^h \phi_j^h(x) \ , \tag{1}$$

and by a suitable choice of the trial functions ϕ_j^h, <u>the Galerkin equations for the coefficients</u> v_j^h <u>turn out to be difference equations</u>. It is to this combination of two fundamental techniques -- to the fact that it is a variational method and at the same time takes the form of a difference equation -- that the finite element method owes much of its success. The variational aspect is crucial to the <u>formulation</u> of the approximating scheme, allowing flexibility in the geometry and a physically intuitive derivation of accurate discrete analogues; in these respects it is superior to more conventional difference equations.

---

* The preparation of this paper was supported by the National Science Foundation (GP-13778) and by the Office of Naval Research.

** Professor of Mathematics, Massachusetts Institute of Technology.

In the solution of these discrete analogues, it is the finite difference aspect which dominates.

This paper describes a systematic approach to the choice of such trial functions ϕ_j^h, and then pursues the mathematical questions which arise in a model problem, linear and elliptic. The principal questions are these:

(i) The degree of approximation which can be achieved by combinations of the ϕ_j^h;

(ii) The numerical stability of the equations for the v_j^h, which hinges on the degree of linear independence of the ϕ_j^h;

(iii) The error estimates for $u^h - u$, where u is the true solution of the differential problem and u^h is its finite element approximation.

The parameter h measures the size of the finite elements into which the region is divided, so that we are contemplating a sequence of approximations with $h \to 0$. The error estimates give the rate h^r of decrease of the error $u^h - u$. These estimates, and in fact most of the main ideas in this paper, have been taken from the manuscript [1] written jointly with George Fix. A very early report on our work was published [2] in Studies in Applied Mathematics.

The ultimate question is whether or not the finite element method is more effective than its competitors, when everything is taken into account -- the derivation of approximate equations, the treatment of boundaries and material interfaces, the programming, the numerical solution, and so on. Our impression is that the weight of opinion -- which differs from the weight of evidence and may even be more important -- is increasingly on the side of the finite element method. (We are speaking here about steady-state problems; for evolution equations the challenge to difference methods is just beginning.) As for evidence, we hope it will be possible and useful to compare constants in the error estimates for competing methods, but the final decision will inevitably rest on a more empirical basis. In fact, many more unbiased comparisons on realistic problems are very much needed.

Our model problem will be governed by a linear elliptic equation of order $2m$ in n variables $x = (x_1, \ldots, x_n)$:

$$Lu \equiv \sum_{|\alpha| \leq 2m} q_\alpha(x)\, D^\alpha u = f(x) \;. \tag{2}$$

The equation (2) is itself very general; what we will restrict is the geometry of the problem, and the boundary conditions. We want to rule out misshapen regions, and to permit the use of the Fourier transform. This we can achieve most easily by investigating (2) either on the whole of n-dimensional space R^n, or in a cube with periodicity imposed as the boundary condition; we choose the former. The resulting analysis will surely apply to much more general domains and boundary conditions, provided that we look at an interior region where the mesh, or rather the shape of "elements" into which the region is subdivided, is regular. Of course this assertion requires proof; progress has been made in this direction by Aubin, Babuska, Bramble, Schatz, and many others.

To construct the trial functions ϕ_j^h, we first choose one or more functions $\phi_1(x), \ldots, \phi_N(x)$; N will be the number of basis functions associated with each mesh-point. Numerically this means that there will be N unknowns v to be computed for each point of the grid; the Galerkin equations for these unknowns will have the form of N coupled finite difference equations. We require the fixed functions $\phi_1, \ldots, \phi_N$ to have m derivatives (in the L_2 sense) so that they are admissible in the variational arguments, and we also require that they have compact support. In other words, they must vanish outside some sphere $|x| \leq R$; this property will lead in the next paragraph to a crucial feature of the finite element method, that each of the trial functions ϕ_j^h vanishes over all but a fixed finite number of elements. Such a basis ϕ_j^h for the trial space is called a local basis.

To form the trial functions associated with one particular gridpoint, the one at the origin, we simply rescale the independent variable, obtaining $\phi_1(x/h), \ldots, \phi_N(x/h)$. These vanish outside $|x| \leq Rh$, and therefore, as required, they are supported in only a fixed number of the grid cubes of edge h. To form the trial functions associated with an arbitrary gridpoint $jh = (j_1h, \ldots, j_nh)$, we translate the functions just constructed, replacing x by $x-jh$. Thus these new functions have the same shape, and to denote them we need a pair of subscripts i, j:

(3) $$\phi^h_{i,j}(x) = \phi_i\left(\frac{x}{h} - j\right), \quad i = 1, \ldots, N .$$

The totality of all these functions, as j runs over the set Z^n of all multi-integers $(j_1, \ldots, j_n)$, forms the local basis we want. For our problem on R^n, their number is admittedly infinite; the reader may prefer in this respect a periodic problem on the unit cube, for which a similar construction leads to Nh^{-n} trial functions. In fact, there is no difference in the analysis.

We emphasize that everything in our construction was determined by the original choice of $\phi_1, \ldots, \phi_N$; therefore the answers to questions (i) and (ii) above must depend exclusively on this choice. This turns the analysis of the finite element method into a very agreeable problem in function theory. For the error estimates of question (iii), the only added information we need is the ellipticity and the order $2m$ of the operator L; otherwise this question too is essentially function-theoretic.

To illustrate this pattern for the construction of trial functions, we describe several practical choices of $\phi_1, \ldots, \phi_N$, together with their associated trial spaces S^h:

(1) The most familiar example, in one dimension $(n = 1)$ and using one function $(N = 1)$, is the "roof function"

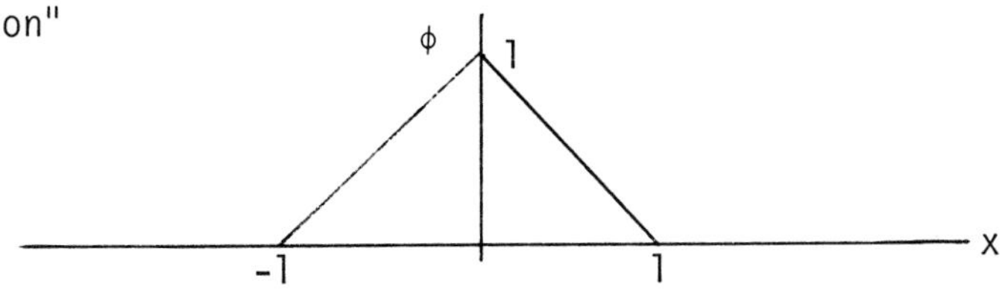

This function is admissible if the differential operator L is of second order $(m = 1)$. The trial space S^h spanned by the functions ϕ_j^h is made up of continuous functions which are linear in each of the intervals $[jh, (j+1)h]$.

(2) An increase in the complexity of ϕ leads to the spline functions, for example to the cubic spline

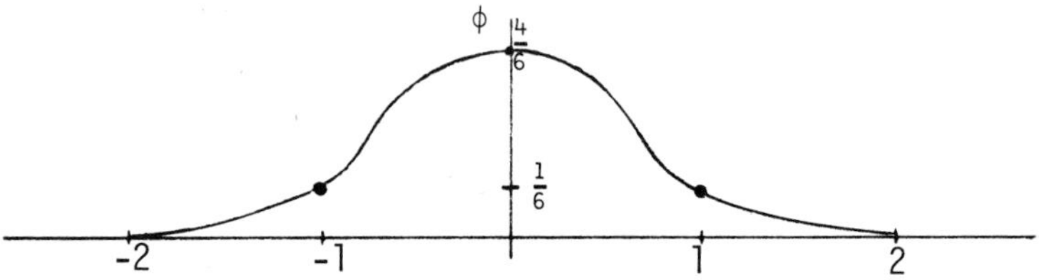

Now S^h is made up of functions with two continuous derivatives, agreeing with cubic polynomials in each of the intervals $[jh, (j+1)h]$. Thus at the nodes $x = jh$, only the third derivative is allowed to jump.

(3) A useful example with $N = 2$ is provided by the "cubic Hermite" basis functions

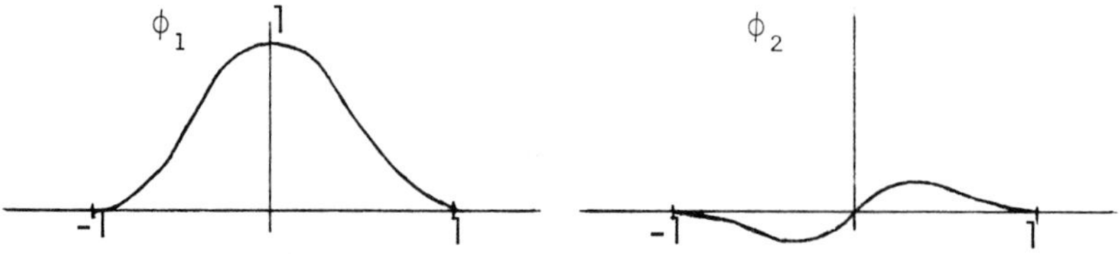

This S^h contains the previous one (so that the spline is a combination of these two and their translates). The combinations of $\phi_{1,j}$ and $\phi_{2,j}$ are again piecewise cubic, but with only one continuous derivative in general.

(4) To extend these examples to n dimensions, we may think either of a rectangular mesh or of one further

divided into right triangles (or right simplices, for $n > 2$). Apparently engineering computations favor the triangular choice, although a rectangular subdivision has had equal attention from numerical analysts. In extending the piecewise linear example to triangles, we get

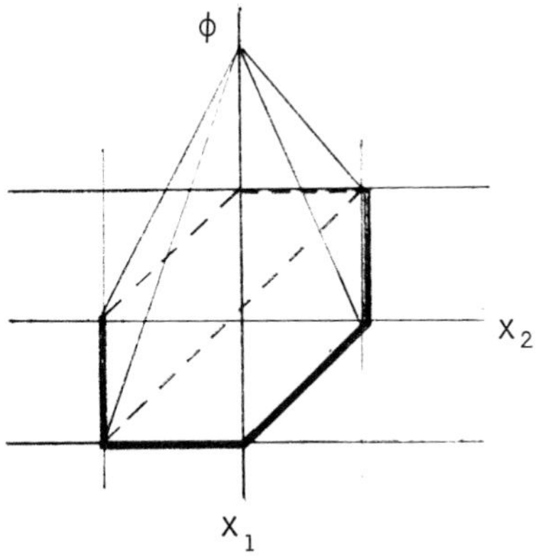

The associated trial space S^h is the one originally suggested by Courant, containing continuous functions which are linear in each triangle. (Buckminster Fuller had similar ideas, and actually carried out the constructions.)

The preceding examples can be generalized to piecewise polynomials of arbitrary degree p. N remains equal to one for the splines -- in either the rectangular case or di Guglielmo's simplicial case. For Hermite spaces, the support of the ϕ_i remains small, but in return the number N grows rapidly with p and n. A further class with large N is exemplified by

(5) Suppose S^h is made up of continuous functions which are bi-quadratic $(a + bx + cy + dx^2 + exy + fy^2)$ in

each of the right triangles of Courant's example (4). Then it turns out that $N = 4$ functions ϕ_i can be constructed which generate this space. Here the space is easier to describe, in terms of compatibility conditions across element boundaries, than the choice of the ϕ_i. Nevertheless our theory applies without difficulty to this example (discussed by Zlamal [3]), as well as to the higher-order piecewise polynomials for which $N >> 1$ treated by Bramble and Zlamal [4]. These appear throughout the engineering literature (for bibliography see [5]) and are therefore among the most important examples of the method.

Returning from examples to a general choice of $\phi_1, \ldots, \phi_N$ and their induced trial functions $\phi^h_{i,j}$, we want to derive variational equations for the coefficients $v^h_{i,j}$ in the finite element approximation

$$(4) \qquad u^h(x) = \sum_{\substack{1 \leqslant i \leqslant N \\ j \in Z^n}} v^h_{i,j} \, \phi^h_{i,j}(x) .$$

To do so we use Galerkin's idea: we cannot expect $Lu^h - f$ to be zero, but we <u>can</u> require that it be orthogonal to each of the $\phi^h_{i,j}$; in other words, the projection of $Lu^h - f$ onto S^h vanishes. The $v^h_{i,j}$ are therefore determined by

$$(5) \qquad (Lu^h - f, \phi^h_{i,k}) = 0 \quad \text{for} \quad 1 \leqslant i \leqslant N , \quad k \in Z^n .$$

This is simply a discrete linear system

(6) $$A^h v^h = f^h ,$$

where the unknown vector v^h is formed from the components $v^h_{i,j}$, and the entries of A^h and f^h are given by the familiar inner products. In case $N = 1$, so that we can ignore the index i, these entries are

(7) $$(A^h)_{kj} = (L\phi^h_j, \phi^h_k) , \qquad (f^h)_k = (f, \phi^h_k) .$$

In general these will be $N \times N$ blocks and N-vectors, respectively. We note that <u>A^h is a band matrix</u>: the k,j entry vanishes if $|k-j| \geq 2R$, given that the original ϕ_i were supported in $|x| \leq R$. Furthermore <u>A^h is of convolution type</u>: if L has constant coefficients, then

(8) $$(A^h)_{kj} = (A^h)_{k+\mu,j+\mu} \quad \text{for all} \quad \mu \in Z^n .$$

The proof is just by change of variables $x \to x + \mu h$ in the inner product

(9) $$(L\phi^h_j, \phi^h_k) = \int \sum q_\alpha D^\alpha \phi^h_j(x) \overline{\phi^h_k(x)} dx .$$

If the q_α depend smoothly on x, then in place of (8) there is slow variation with respect to μ; A^h is then a pseudo-convolution matrix. As usual, since the order $|\alpha|$ of the derivatives D^α reaches $2m$ while ϕ is only required to have m derivatives, one carries out m integrations by parts in the inner products (9).

We shall not discuss the numerical solution of the finite element equations $A^h v^h = f^h$, for which direct elimination methods are currently most popular. (Irons has described a variant which may be effective in production codes.) We do want to insert this conjecture, that even for non-selfadjoint problems pivoting is theoretically unnecessary; this we anticipate to be a consequence of the ellipticity condition

$$\text{(10)} \qquad \mathrm{Re}\,(Lu,u) \geq \rho \sum_{|\alpha|\leq m} \|D^\alpha u\|^2_{L_2} ,$$

which implies that Re L dominates Im L, and thus that Re A^h dominates Im A^h.

1. Approximation

We are ready to begin on the central mathematical problem of the subject, the degree of approximation which can be achieved by the spaces S^h. The goal is to find necessary and sufficient conditions on the $\phi_i(x)$ in order that an arbitrary smooth function u can be approximated by a combination of the $\phi^h_{i,j}$ with an error of order h^{p+1}.

Let us consider the simplest and most transparent case. Suppose we choose a function $\Omega(x)$ of compact support, and ask how closely a given $u(x)$ is approximated by the particular combination

$$\tilde{u}^h(x) = \sum_{j\in Z^n} u(jh)\Omega\left(\frac{x}{h} - j\right) .$$

We write $x = kh + th$, where k is the integral part of x/h and t lies in the unit cube $0 \leq t_\nu < 1$. Expanding in powers of h,

$$(11) \qquad u(x) - \tilde{u}^h(x) = \sum_{|\alpha| \leq p} \frac{(th)^\alpha}{\alpha!} D^\alpha u(kh)$$

$$- \sum_j \sum_{|\alpha| \leq p} \frac{((j-k)h)^\alpha}{\alpha!} D^\alpha u(kh)\Omega(k+t-j) + O(h^{p+1}|u|_{p+1})$$

$$= \sum_{|\alpha| \leq p} \frac{h^{|\alpha|} D^\alpha u(kh)}{\alpha!} [t^\alpha - \sum_j j^\alpha \Omega(t-j)] + O(h^{p+1}|u|_{p+1}) \; .$$

At the last step we wrote j for $j-k$. The compact support of Ω, by admitting only finitely many j into the sum, allowed the estimate given for the remainder term.

Obviously <u>the error $u - \tilde{u}^h$ is of order h^{p+1} for all smooth u if and only if Ω satisfies</u>

$$(12) \qquad \sum_{j \in Z^n} j^\alpha \Omega(t-j) = t^\alpha \quad \text{for} \quad |\alpha| \leq p \; .$$

These identities are easily verified, with $p = 1$, for our first and fourth examples: the expressions $\Sigma j^\alpha \phi(t-j)$ are piecewise linear, and agree with t^α at the meshpoints, so they agree everywhere. (Since a combination of piecewise linear functions can never be quadratic, these examples give exactly $p = 1$.) In these cases $n = 1$ and $n = 2$, respectively, and to treat both at once we have used the standard multi-integer conventions

$$\alpha = (\alpha_1, \ldots, \alpha_n) , \quad \alpha_\nu \geq 0 , \quad |\alpha| = \sum \alpha_\nu$$

$$t^\alpha = t_1^{\alpha_1} \ldots t_n^{\alpha_n} , \quad D^\alpha = (\partial/\partial x_1)^{\alpha_1} \ldots (\partial/\partial x_n)^{\alpha_n} .$$

We shall see in a moment how the other examples fit the identities (12).

These identities are fundamental not only to pointwise approximation, but also to interpolation and L_2 approximation. Suppose for example that the *interpolation problem*

$$(13) \qquad \sum \alpha_j^h \, \Omega(k-j) = u(kh)$$

is correctly posed, i.e., the coefficient matrix defined by $B_{kj} = \Omega(k-j)$ is invertible. (This will be true, over all the ℓ_p spaces, if and only if the symbol $b(\theta) = \Sigma\, \Omega(k)e^{ik\theta}$ never vanishes.) We claim that the error in interpolation is also of order

$$(14) \qquad u(x) - \sum \alpha_j^h \, \Omega\left(\frac{x}{h} - j\right) = O(h^{p+1})$$

whenever the identities (12) hold. The proof is nearly trivial: we have shown already that

$$(15) \qquad \sum u(jh) \, \Omega(k-j) = u(kh) + O(h^{p+1}) .$$

This right side differs from the one in (13) by $O(h^{p+1})$, so by the invertibility of B, α_j^h differs from $u(jh)$

only to the same order. But then the error in (14) will be of order h^{p+1} because the error in (11) was.

The question of L_2 approximation is a little deeper. If u is nice enough, then our pointwise error estimates for either $\tilde{u}^h$ or the interpolate can be converted to L_2 bounds. Properly, however, we should only require u to have $p + 1$ derivatives in the L_2 sense, and the error bound ought to look like

$$(16)\quad Ch^{p+1} \|u\|_{p+1} \equiv Ch^{p+1} \left(\sum_{|\alpha|=p+1} \|D^\alpha u\|^2_{L_2} \right)^{1/2} .$$

For large n, this expression may be finite even though u is discontinuous, in which case the values $u(jh)$ and the construction of $\tilde{u}^h$ may have no meaning. Therefore we first smooth the function u by annihilating its high frequency components. We define w^h to have Fourier transform

$$\hat{w}^h(\xi) = \begin{cases} \hat{u}(\xi) & \text{when each } |\xi_\nu| \leqslant \pi/h, \quad \nu = 1, \ldots, n \\ 0 & \text{otherwise} \end{cases}$$

From this function we form

$$(17)\qquad \tilde{w}^h(x) = \sum w^h(jh)\, \Omega\!\left(\frac{x}{h} - j\right) .$$

We claim that if the identities (12) hold, then

$$(18)\qquad \|u - \tilde{w}^h\|_{L_2} \leqslant Ch^{p+1} \|u\|_{p+1} .$$

For proof we have to translate these identities into a form appropriate for calculations in L_2. The link is provided by Poisson's summation formula $\Sigma\, f(j) = \Sigma\, \hat{f}(2\pi j)$, applied to $f(x) = x^{\alpha}\, \Omega(t-x)$, and the result is straightforward: the identities (12) hold if and only if the Fourier transform $\hat{\Omega}(\xi)$ and its derivatives up to order p vanish at the points $\xi = 2\pi j$, with the exception that $\hat{\Omega}(0) = 1$. Thus

$$(19) \qquad D^{\alpha}\hat{\Omega}(2\pi j) = \delta_{0\alpha}\, \delta_{0j} \quad \text{for} \quad |\alpha| \leq p \, , \qquad j \in Z^n \, .$$

We note, for example, that the roof function has transform $(\sin \xi/2)^2/(\xi/2)^2$, which satisfies (19) for $p = 1$.

The equivalence of (12) and (19) was first established in Schoenberg's 1946 paper [16], as the condition for a smoothing formula to be exact when applied to polynomials of degree p. In that paper he goes on to develop splines as the most important examples of functions which lead to such identities. Thus our work has (unconsciously) picked up the thread of a general function Ω, and related it to the approximation properties of the wide variety of choices of $\phi_1, \ldots, \phi_N$ which enter multi-dimensional computations. In cther words, we are trying to describe the underlying pattern which Schoenberg's splines and the finite element method (which was initially restricted to piecewise-linear approximations in solid mechanics) have in common. In particular, we want to show that the order of approximation can be decided very quickly, by looking for the degree of polynomials which are present (locally) in the trial spaces S^h.

To estimate $u - \tilde{w}^h$ in ℓ_2 we shall compare $\hat{u}$ with the Fourier transform

$$\hat{\tilde{w}}^h(\xi) = \Sigma\, w^h(jh)\, h^n\, e^{-i\xi jh}\, \hat{\Omega}(h\xi) \tag{20}$$

$$= \left[\Sigma\, \hat{w}^h \left(\frac{2\pi j}{h} + \xi \right) \right] \hat{\Omega}(h\xi)\ .$$

At the last step we appealed again to Poisson. The sum in brackets is a function of period $2\pi/h$, agreeing with $\hat{u}(\xi)$ in the fundamental period $|\xi_\nu| \leqslant \pi/h$.

The one remaining ingredient is the hypothesis that Ω has compact support, which is translated by the Paley-Wiener theorem into: $\hat{\Omega}$ is an entire function of exponential type. With this property, the required estimate of

$$\|u-\tilde{w}^h\|^2_{L_2} = \int \left|\hat{u}(\xi) - \Sigma\, \hat{w}^h \left(\frac{2\pi j}{h} + \xi\right)\hat{\Omega}(h\xi)\right|^2 d\xi$$

is based in [1] on a theorem of Duffin and Schaeffer for functions of exponential type. (This is the special case of [1, Theorem I] in which the weights are $q_\alpha = \delta_{o\alpha}$; there we were approximating with a general ϕ, and introduced Ω only after the approximation theorems.)

Our goal is now to transfer these approximation results from Ω to the functions $\phi_1, \ldots, \phi_N$ actually chosen for use in the finite element method. This we can do, <u>provided that some finite combination of the</u> ϕ_i <u>and their translates satisfies the required identities</u>. In other words, suppose we can form a finite sum

(21) $$\Omega(x) = \Sigma\ \rho_{ij}\phi_i(x-j)$$

for which the identities hold. Then Ω has compact support, and approximation of order $p+1$ is possible with some combination of the Ω_j^h, i.e., with some combination $\Sigma\ w_{i,j}^h\ \phi_{i,j}^h$. This establishes the direct part of our main theorem.

To prove the converse, that if approximation is possible then some translates of the ϕ_i must combine to yield a suitable Ω, we have to impose a condition on the weights w which appear in the approximating combination $\Sigma\ w_{i,j}^h\ \phi_{i,j}^h$. (Such a condition looks unnecessary in L_1 approximation; we are concerned here only with L_2.) The constraint is a natural one:

(22) $$h^n \sum_{i,j} |w_{i,j}^h|^2 \leqslant \text{constant as } h \to 0 .$$

Such a bound is easily verified for our constructions above; in (17), for example, this expression becomes the Riemann sum

$$h^n\ \Sigma\ |w^h(jh)|^2 \to \int |u(x)|^2\ dx .$$

We call the approximation <u>controlled</u> if (22) is satisfied.

Approximation Theorem. The trial spaces S^h spanned by the $\phi^h_{i,j}$ provided controlled L_2 approximation of order $p+1$,

$$\inf_{w} \|u-\Sigma\, w^h_{i,j}\phi^h_{i,j}\|_{L_2} \leq Ch^{p+1}\|u\|_{p+1} \tag{23}$$

$$\text{as} \quad h \to 0 \ ,$$

if and only if there exists a function Ω of the form (21) which satisfies the identities (12).

A number of remarks should be added at once. The estimate (23) is certainly not new for many of the piece-wise polynomial spaces used in practice. Without prejudice to the claims of others, including the reader, we want to mention the fundamental work of Birkhoff, de Boor, and Varga. For general ϕ, the direct part was first proved (in the context of $N = 1$) by di Guglielmo [6]. Babuska [7] independently proved a similar result, and we owe to him the suggestion to try to consolidate the N functions ϕ_i into a single Ω. The only steps we have seen toward the converse are in [7] and in Goël [8], who considered special choices of ϕ. Our paper with Fix [1] was again entirely independent of the others, and proves the result in full.

Approximation to the derivatives $D^{\alpha}u$ follows the familiar pattern: their order of approximation is $h^{p+1-|\alpha|}$.

It would be very useful to know the constant in (23), or rather the infimum C_{inf} of possible values of C. Since this constant enters the estimate of the finite element error u^h-u, it is crucial to the comparison of alternative choices of ϕ. Presumably the observed accuracy of finite elements, even for comparatively large h, arises from the appearance of a smaller constant than in some of the standard difference methods.

In fact there are at least three ways in which to define C_{inf}. In every case we ask that

$$\inf_{s^h \in S^h} \|u-s^h\| \leq (C_{inf}+\varepsilon)h^{p+1} \|u\|_{p+1} , \tag{24}$$

but apparently we can choose whether to require this for all h, for all $h \leq h_0(\varepsilon)$, or for all $h \leq h_0(\varepsilon,u)$. Surprisingly, the third definition leads to a simple and computable formula: C_{inf} <u>is the error constant in approximating polynomials of degree</u> $p+1$. This is suggested by the computation (11) of the pointwise error. If u is smooth, then the term of order h^{p+1} in $\tilde{u}^h-u$ comes exactly from a polynomial of degree $p+1$, and the rest is $O(h^{p+2})$. The extension to less smooth u needs the uniform boundedness theorem and is carried out for the L_2 case in [1].

This value for C_{inf} has a natural interpretation: <u>the more closely you look at smooth functions, the more they look like polynomials</u>. Thus as $h \to 0$, and the approximation to polynomials of degree p is exact, the error arises from the polynomial component of next higher degree.

To compute this error we depend on a simple consequence of our identities.

Lemma. If Ω satisfies the identities (12), and u is a polynomial of degree $p+1$, then

$$u(x) - \Sigma\, u(jh)\Omega\left(\frac{x}{h} - j\right)$$

has period h in each variable $x_1, \ldots, x_n$.

We give the proof in one dimension. Writing $x = th$, the lemma means that

$$E(t) = t^{p+1} - \Sigma\, j^{p+1}\, \Omega(t-j)$$

is 1-periodic. Therefore we compute

$$E(t+1) = (t+1)^{p+1} - \Sigma\, j^{p+1}\, \Omega(t+1-j)$$

$$= (t+1)^{p+1} - \Sigma\, (j+1)^{p+1}\, \Omega(t-j)\ .$$

Oh, it's obvious.

The Fourier series of E is interesting:

(25) $$E(t) = \Sigma\, e^{2\pi ijt}\, D^{p+1}\, \hat{\Omega}(2\pi j)\ .$$

The problem of approximating x^{p+1} by combinations of the Ω_j^h is reduced by the lemma to the approximation of the h-periodic function $E(x/h)$. If it were a question of interpolation, the answer would be clear: the constant function $E(0)$ interpolates at the meshpoints $x = kh$, and (since $\Sigma\, \Omega(t-j) = 1$) it lies in our space. Therefore the h^{p+1} term in the interpolation error is easy to compute.

Fortunately, the same is true of approximation both in L_2 and in the sup norm: the best approximation to E is a constant. We want to reserve the proof for a later paper, together with the discussion of a still more striking property of best approximation: the error constant C_{inf} for x^{p+1} is asymptotically correct for all functions, in the sense that for every u with $p+1$ derivatives,

(26)
$$\inf_{w_j^h} \| u - \Sigma\, w_j^h \Omega_j^h \| / h^{p+1} \to C_{inf} \|u\|_{p+1} \quad \text{as} \quad h \to 0 .$$

To complete the calculation of C_{inf} we now have to approximate the periodic function E by a constant. In L_2 the best constant is just the $j = 0$ term in the Fourier expansion (25). Therefore we are left with

$$C_{inf}^2 = \sum_{j \neq o} \left| \frac{D^{p+1}\hat{\Omega}(2\pi j)}{(p+1)!} \right|^2 \tag{27}$$

For splines these constants have a closed form. The one-dimensional splines of degree p lead (as we show below for $p = 3$) to

$$\hat{\Omega}(\xi) = \rho(\xi) \left(\frac{\sin \xi/2}{\xi/2} \right)^{p+1} , \tag{28}$$

where ρ is periodic, $\rho(0) = 1$, and the other factor is the transform of the B-spline ϕ. (The cases $p = 1$ and

$p = 3$ of ϕ were the first of our examples; ϕ is the unique spline of degree p with unit area which vanishes for $|x| \geqslant (p+1)/2.$) This second factor has zeros of order $p+1$ at the points $2\pi j$, $j \neq 0$, and we find

$$C_{\inf}^2 = \sum_{j\neq 0} (\pi j)^{-2(p+1)} .$$

The right side, modulo some normalization, is the Bernoulli number B_{2p+2}. Interpolation and sup norm approximation lead also from splines to Bernoulli numbers, and connect our theory to the constants discovered by Golomb [9].

It remains to see how, given an arbitrary set $\phi_1, \ldots, \phi_N$, the presence of a suitable combination Ω can be detected. With $N = 1$ this is especially simple, since the transform of $\Sigma\, \rho_j \phi(x-j)$ is just

$$\Sigma\, \rho_j e^{ij\xi} \hat{\phi}(\xi) = \rho(\xi)\hat{\phi}(\xi) .$$

If this is to equal one at $\xi = 0$, we must have

$$(29) \qquad \hat{\phi}(0) \neq 0 .$$

Furthermore, since also $\rho(0) \neq 0$ and ρ has period 2π, the product $\rho\hat{\phi}$ will have zeros at $\xi = 2\pi j$ just if $\hat{\phi}$ does; therefore we must require that

$$(30) \qquad D^{\alpha}\hat{\phi}(2\pi j) = 0 \quad \text{for} \quad |\alpha| \leqslant p, \quad j \neq 0 .$$

<u>These are the conditions on</u> ϕ <u>for approximation of order</u> h^{p+1}. If satisfied, we can choose ρ so that

$$(31) \qquad D^{\alpha}(\rho\hat{\phi})_{\xi=0} = 0 \text{ for } 1 \leq |\alpha| \leq p ,$$

and thus $\hat{\Omega} = \rho\hat{\phi}$ meets the conditions (19).

All this can be translated very simply out of the transform space and back to the function $\phi(x)$. The condition on ϕ is that for $|\alpha| \leq p$, the combination $\Sigma j^{\alpha}\phi(t-j)$ should be a polynomial with leading term Ct^{α}, $C \neq 0$. Then by taking a combination with weights ρ_j of the translates of ϕ, we can arrange that this polynomial is exactly t^{α}, i.e., we can produce an Ω satisfying the identities (12).

The best illustration of these ideas is provided by our second example, the cubic splines. In this case ϕ is the convolution of the roof function with itself, and

$$\hat{\phi}(\xi) = \left(\frac{\sin \xi/2}{\xi/2} \right)^4 .$$

This satisfies the conditions (29-30) with $p = 3$: $\hat{\phi}(0) = 1$, and there are fourth-order zeros at $\xi = 2\pi j \neq 0$. Around the origin,

$$\hat{\phi} = \left(\frac{\xi/2 - \xi^3/48 \ldots}{\xi/2} \right)^4 = 1 - \frac{\xi^2}{6} + 0(\xi^4) .$$

Therefore we choose

$$\rho(\xi) = \frac{4 - \cos \xi}{3} = 1 + \frac{\xi^2}{6} + 0(\xi^4) ,$$

which leads to $\rho\hat{\phi} = 1 + 0(\xi^4)$ and

$$\Omega(x) = -\frac{1}{6}\phi(x-1) + \frac{4}{3}\phi(x) - \frac{1}{6}\phi(x+1) .$$

This is exactly the spline associated with Jenkins' smoothing formula [16]. Perhaps we should christen the splines of degree p which satisfy (12) and have minimal support with the name <u>Ω-splines</u> (with the implied hope that they will be the last!). Pointwise approximation is now furnished by

$$\tilde{u}^h(x) = \Sigma\, u(jh)\Omega\left(\frac{x}{h} - j\right) .$$

Changing the index of summation,

$$\Sigma\, u(jh)\phi\left(\frac{x}{h} \pm 1 - j\right) = \Sigma\, u((j\pm 1)h)\phi\left(\frac{x}{h} - j\right) ,$$

this approximation can be rewritten as

$$\tilde{u}^h(x) = \Sigma\left[\frac{-u((j+1)h)}{6} + \frac{4u(jh)}{3} - \frac{u((j-1)h)}{6}\right]\phi\left(\frac{x}{h} - j\right) .$$

This form suggests a convenient new approximating function, the <u>quasi-interpolate</u> of u:

$$(32) \qquad \tilde{\tilde{u}}^h(x) = \Sigma\left[u(jh) - \frac{h^2 u''(jh)}{6}\right]\phi\left(\frac{x}{h} - j\right) .$$

Since this differs from its predecessor by $O(h^4|u|_4)$, it also approximates u to this order. In fact it can be derived directly from an arbitrary ϕ, in the same way that Ω was, by multiplying $\hat{\phi}$ by a polynomial ρ in order to achieve (19); the difference is that now ρ is an algebraic instead of a trigonometric polynomial in ξ. In the cubic spline example ρ is $1 + (\xi^2/6)$, and the inverse transform of $\rho(\xi h)$ is just the operator $1 - (h^2 D_x^2/6)$ which appears in $\tilde{\tilde{u}}^h$.

With $N > 1$, we can again give conditions (cf. [1]) on the transforms $\phi_1, \ldots, \phi_N$ which are equivalent to the existence of a suitable Ω, and we can again construct a quasi-interpolate. To repeat, the whole theory comes down to this: approximation of order h^{p+1} is possible if and only if all polynomials of degree p can be manufactured from the trial basis. We conjecture that this statement holds also for an irregular mesh; the constants must be allowed to depend on the geometry of the mesh, but when this geometry is non-degenerate, the order should depend precisely on the polynomials which are present. Furthermore a suitable $\tilde{u}^h$ will emerge when Taylor series are made to cancel through terms in h^p. This requirement on the trial basis singles out piecewise polynomials as the optimal form for the ϕ_i: they make the computation of the inner products $(L\phi_j^h, \phi_k^h)$ and (f, ϕ_k^h) about as simple as possible, and at the same time they are the most efficient units from which to make polynomials. Thus theory supports practice, even if the support comes a little late.

2. Stability

Ordinarily there are two kinds of stability to be considered. The first requires that the approximate solutions depend continuously on the data, and that this dependence is uniform in the parameter h. It is this kind of stability which is likely to be absent in approximating the backward heat equation. Together with a suitable consistency between the given equation and its approximations, this stability is equivalent to convergence -- the reader will instantly think of the celebrated Lax-Richtmyer theorem, the author of its less celebrated analogue for Hadamard's class of well-posed problems, and Aubin of his equivalence theorem [10-11] for variational approximations. As he shows, this first kind of stability is automatically present for elliptic problems. The ellipticity of L expressed in (10) holds a fortiori for u in the subspace S^h, with the same constant ρ, and this guarantees the existence of a unique u^h, depending continuously on the data f. (Such stability is not so automatic in non-dissipative initial-value problems, or even in elliptic problems approximated in the following way: find u^h in S_1^h so that $Lu^h - f$ is orthogonal to a second space S_2^h. Written operationally, this technique will be recognized as a familiar one:

$$(L \sum v_j^h \phi_j^h, \Psi_k^h) = (f, \Psi_k^h) \ .$$

This generalization of our case $\phi = \Psi$ seems to us rather dangerous; if the angle between the space S_1^h spanned by the ϕ_j^h and S_2^h spanned by the Ψ_k^h is not controlled, an instability will emerge which is unjustified in an elliptic problem.)

The second kind of instability is numerical. It is concerned with the growth of roundoff error, and is governed by the condition of the equations to be solved (such as $A^h v^h = f^h$) and by the precise algorithms which are used to solve them. In the finite element method as described above, the condition number of A^h will exceed

ch^{-2m}, except in very special problems. This reflects the order $2m$ of the differential operator L, and in fact this much ill-conditioning could be essentially removed, by noticing that it arises within A^h in a very systematic way. With $L = (\partial/\partial x)^{2m}$, for example, the elements in every row of A^h will have vanishing moments of all orders less than $2m$, and a clever algorithm (along the lines suggested by Babuska) can use these identities to find v^h with high accuracy. In many applications, of course, the perfection of such an algorithm will be too much trouble, and the condition of the given finite element matrix A^h will have direct significance.

The numerical instabilities which we should certainly avoid are those which arise from a poor choice of basis for the space S^h of trial functions. Such a choice will make the condition of A^h unnecessarily bad, either by increasing the constant or the exponent in ch^{-2m}, or by adding further identities to the rows of A^h which would then have to be removed. Thus, while the solution u^h and the error u^h-u depend only on the trial space S^h, numerical stability depends on the uniform linear independence of the basis elements ϕ_j^h.

To measure this independence we use the condition number $\kappa = \|G^h\| \; \|(G^h)^{-1}\|$ of the Gram matrix, formed from the inner products of the basis elements:

$$(G^h)_{kj} = (\phi_j^h, \phi_k^h) \; . \tag{33}$$

(We are supposing that $N = 1$; in general there will be a block of order N for each pair k,j.) The Gram matrix is always symmetric and nonnegative-definite; with finitely many functions ϕ_j^h it is positive definite if and only if they are linearly independent. Our model problem on R^n unfortunately involves infinite-dimensional subspaces S^h, since j ranges over all integers, and this may displease the reader. In fact, however, the model does accurately gauge the degree of independence of the basis, in the interior even of a bounded domain.

Thus the first requirement for numerical stability is that the infinite matrix G^h be positive-definite. If we adopt the ℓ^2 norm, then as in [1-2] the Fourier transform and Poisson's formula lead to the exact value

$$(34) \qquad \kappa(G^h) = \frac{\sup \Sigma \ |\hat{\phi}(2\pi j+\theta)|^2}{\inf \Sigma \ |\hat{\phi}(2\pi j+\theta)|^2}$$

The supremum and infimum are over all real θ, and (in case $N > 1$) over all combinations

$$\phi = c_1\phi_1 + \dots + c_N\phi_N \ , \quad \sum |c_i|^2 = 1 \ .$$

This leads immediately to our

Stability Theorem: Stability holds if and only if the denominator in (34) is positive, that is, there exist no $c = (c_1, \dots, c_N)$ and $\theta = (\theta_1, \dots, \theta_n)$ such that

$$(35) \qquad \sum c_i \ \hat{\phi}_i \ (2\pi j+\theta) = 0 \quad \text{for all} \quad j \in Z^n \ .$$

With stability, the earlier hypothesis that approximations be controlled is automatically fulfilled. None of the examples listed earlier are unstable, and in fact we would not expect to find any in the literature which are. The stability requirement would be violated at $\theta = \pi$ if we made the insane choice

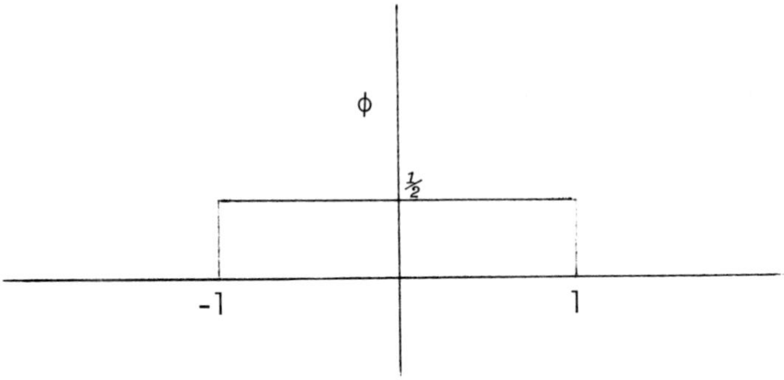

In this case $\hat{\phi}(\xi) = \sin \xi/\xi$. We note that the approximation identity is satisfied for $p = 0$.

A part of the value of constructing a precise stability theory lies, as it did for approximations, in the possibility of actually determining the crucial constants in the stable case. If ϕ is nonnegative, for example, then so are all the inner products (ϕ_j^h, ϕ_k^h), and the numerator in (34) is given by the value at $\theta = 0$. This is the case for splines, generated by the second factor in (28), for which the numerator equals one. The infimum in the denominator is attained (we conjecture) at $\theta = \pi$, and thus κ is the reciprocal of

$$2 \sum_0^\infty \left(\frac{2}{\pi(2j+1)}\right)^{2p+2} .$$

Again this constant is related to the Bernoulli numbers. A formula similar to (34) can be given for the stability constant in the sup norm.

We want to relate the Gram matrix G^h to the coefficient matrices which enter the interpolation problem. Notice that for the trivial elliptic equation $u = f$, in which L is the identity and the order $2m$ is zero, G^h arises from applying the least squares method (minimizing $\|Lu - f\|^2$) just as the interpolation matrices arise in collocation ($Lu = f$ at the meshpoints). In fact we may say that interpolation is to collocation as L_2 approximation is to the method of least squares.

The interpolation problem (if we take $h = 1$, $N = 1$) is to find a combination $\Sigma\, \alpha_j\, \phi(x-j)$ which interpolates given values y_k at the nodes $x = k = (k_1, \ldots, k_n)$. This leads to the linear system

(36) $$B\,\alpha = y\ , \quad \text{where} \quad B_{kj} = \phi(k-j)\ .$$

Suppose we consider the same problem on a shifted set of nodes $x = k + \gamma$, where $\gamma = (\gamma_1, \ldots, \gamma_n)$ is fixed in the unit cube C. (There are no canonical nodes.) Then again there is a linear system

(37) $$B_\gamma\,\alpha = y\ , \quad \text{where} \quad (B_\gamma)_{kj} = \phi(k+\gamma-j)\ .$$

Theorem. The Gram matrix G and the interpolation matrices B_γ satisfy

$$G = \int_C B_\gamma^* B_\gamma \, d\gamma$$

Proof. The entries of $B_\gamma^* B_\gamma$ are

$$(B_\gamma^* \; B_\gamma)_{kj} = \sum (B_\gamma^*)_{k\ell} (B_\gamma)_{\ell j}$$

$$= \sum \bar{\phi} \; (\ell + \gamma - k)\phi(\ell + \gamma - j) \; .$$

As γ runs over the unit cube C and ℓ runs over Z^n, the combinations $\ell + \gamma$ cover R^n. Therefore

$$\int_C (B_\gamma^* B_\gamma)_{kj} \, d\gamma = \int_{R^n} \bar{\phi} \; (x-k)\phi(x-j)dx$$

$$= \int \phi_j(x) \; \overline{\phi_k(x)} dx = G_{kj} \; .$$

We conclude from the theorem that if the family of interpolation problems is uniformly well-posed, i.e. the B_γ are invertible, then the Gram matrix is non-singular. The converse is false; stability (which is not in practice a serious threat to the finite element method in the interior of the domain) does not guarantee the solvability of the interpolation problem.

The relation given in the theorem also holds between the least squares matrix, with entries $(L \, \phi_j^h, \, L \, \phi_k^h)$, and the collocation matrices. Therefore collocation will again be numerically no more stable, and may very possibly be less so, than least squares.

3. Error Estimates

Suppose we apply the finite element method to an elliptic problem $Lu = f$ on R^n, using as a basis for the trial space S^h the functions

$$\phi^h_{i,j}(x) = \phi_i(\frac{x}{h} - j) \;,\quad 1 \leq i \leq N \;,\quad j \in Z^n \;.$$

The approximation theorem gave the conditions on $\phi_1, \ldots, \phi_N$ in order that there exist an element of S^h such that

$$(38) \qquad \|u - \sum w^h_{i,j}\, \phi^h_{i,j}\| \leq Ch^{p+1}\, \|u\|_{p+1}.$$

Our first question is, what about the weights $v^h_{i,j}$ produced by Galerkin's equation $A^h v^h = f^h$? Do these weights also give approximation of the optimal order $p+1$, or does $u^h = \Sigma\, v^k_{i,j}\, \phi^h_{i,j}$ differ from u by more than $O(h^{p+1})$?

In variational theory, the derivatives of order m have a special place. It is well known (cf. Varga [12]) that when L is elliptic of order $2m$, the error in these derivatives $D^\alpha(u^h-u)$, $|\alpha| = m$, will be of the optimal order h^{p+1-m}.

For other derivatives, and in particular for u^h-u itself, the order of the error is not so clear. In fact it is implausible that u^h should remain accurate to order $p+1$ as m imcreases, until suddenly, when m exceeds the differentiability of ϕ, the trial functions ϕ^h_j become inadmissible and the method fails. Thus we anticipate an error estimate which depends on m.

Since the finite element method is simultaneously a variational and a finite difference system, there are two ways to estimate the error. The variational estimates, when they work, are by far the simpler; in the (non-reflexive) sup norm there are some difficulties, and conceivably also with non-coercive operators. The finite difference device of substituting the true solution u (or its quasi-interpolate) into the equation and estimating truncation error becomes awkward with variable coefficients, or $N > 1$ generating functions ϕ_i, or non-periodic boundary conditions. Fortunately both techniques can be applied to the model elliptic problem $Lu = f$, and they led us in [1] to the following

Error Estimate. If the trial space S^h satisfies the conditions of the approximation theorem, then the error in the finite element method is precisely of order

(39)
$$\|u^h - u\| = O(h^r \|f\|_{p+1-2m}) , \quad r = \min(p+1, 2(p+1-m)) .$$

With stability the corresponding estimate holds in the sup norm. Thus the exponent r is positive -- the finite element method converges -- if and only if $p \geqslant m$.

We note that there is no loss of accuracy in the sup norm; Sobolev's lemma, which leads to more pessimistic estimates, ought not to be used. A valuable feature of the finite element method is that the estimates can also be made local: wherever u has the required smoothness, the error (in u^h or in approximation) is of the optimal order.

As far as we know the estimate (39) is new. For derivatives of order s, the exponent in the error becomes the smaller of $p+1-s$ and $2(p+1-m)$; with $s = m$ the first is smaller, and the order $p+1-m$ coincides with that of optimal approximation. Since the least squares method for $Lu = f$ coincides with Galerkin's method for $L^*Lu = L^*f$, the same estimates apply to this case -- but with m replaced by $2m$. These estimates extend also to parabolic problems [1].

One very useful result which comes from the difference equation aspect of the finite element method is an asymptotic expansion for the error. By substitution into $A^h v^h = f^h$, we find functions $V_{i,s}(x)$ such that asymptotically as $h \to 0$,

$$v^h_{i,j} = \sum h^s V_{i,s}(jh) . \tag{40}$$

In the standard finite difference situation, the discrete unknowns u^h_j possess such an expansion; here, at least in the first instance, it is the intermediate unknowns v. These are related to the values u^h_k which the (continuous) approximation $u^h(x)$ assumes at the meshpoints $x = kh$ by

$$u^h_k = \sum_{i,j} v^h_{i,j} \phi_i(k-j) . \tag{41}$$

Substituting (40) we are led to functions $U_s(x)$ such that asymptotically

$$u^h_k = \sum h^s U_s(kh) . \tag{42}$$

These functions U will undergo a simple change if we look at a mesh other than the one centered at the origin, say the set $x = (k+\gamma)h$, with γ fixed.

Supposing that $p \geqslant m$, so that u^h converges to u, the expansion (42) will behave in the familiar way: U_0 will coincide with the true solution u, the next $r-1$ functions U_s will vanish identically, and U_r will be the principal error function, with structure as in Henrici's book. We plan to discuss in our next paper the constants involved in U_r, which provide a basis for comparing the accuracy of different finite element methods. This asymptotic behavior has been observed in computation, and it justifies the use of various inexpensive tricks for improving the accuracy.

We want finally to note that, through the work of others, an extension of the sort of theory described here is already partially complete for several more difficult classes of equations:

(i) elliptic boundary-value problems

(ii) initial-value and initial-boundary-value problems

(iii) nonlinear problems

In connection with (i), the most effective treatment of unnatural boundary conditions, e.g. of Dirichlet data, has been an outstanding question. The present alternatives are either to modify the ϕ_j^h so as to satisfy these conditions, or to add to the variational problem a boundary integral divided by a power of h. For two-point boundary value problems and for rectangular polygons a modification of the ϕ_j^h is normally convenient, and Birkhoff, Schultz, and Varga [14] have given a comprehensive discussion of Hermite methods for these problems. For more general regions, Babuska has described elsewhere in this volume how to combine a "penalty integral" around the boundary with Galerkin's method in the interior. At the same time Bramble and Schatz (cf. these proceedings, and [17]) have discovered that if <u>least squares</u> is used in the interior, then there is a <u>natural</u> boundary integral leading to the best estimates for the error $u - u^h$. Thus the Dirichlet problem can be solved without finding trial functions which satisfy the essential boundary conditions; an abstract discussion is given by Aubin in this volume. This means that curved boundaries are infinitely simpler to handle in variational than in standard finite difference approximations, where Schaeffer [13] has for the first time penetrated to the essential mathematics; but even variationally, the appearance of small elements introduces new questions of numerical stability as well as approximation.

Just by mentioning (ii), we risk excommunication from the von Neumann-Lax-Kreiss tradition of stability theory. The reason is that their problems become almost trivial; the Galerkin method has the supreme virtue (to a lazy analyst) that the evolution in time of the approximations $u^h(t,x)$ is automatically stable if the given differential problem is dissipative. We expect the order of the error to coincide, at least for a large class of these problems, with that given above for elliptic equations on R^n. This is confirmed in the estimates reported by Douglas to this conference.

For nonlinear equations in one dimension the series of papers [15] is a valuable reference, and we understand that Schultz has given a constructive form of the Leray-Schauder theory. In certain eigenvalue (bifurcation) problems the convergence of u^h has been established -- but not always with useful error bounds. Thus in all three of the classes listed above, and above all for non-linear equations, important and challenging problems remain. We hope that their solution will be accomplished, not solely in the classical existence -- uniqueness -- *a priori* estimate framework, but with the aid also of ideas from boundary layer analysis and asymptotic expansions. We are convinced that, given global control of the problems by classical techniques, such ideas will yield by far the best local information.

REFERENCES

1. Strang, Gilbert and Fix, George, "The Finite Element Variational Method, 150 pp. (to appear, probably in the series Notes on Math. and its Appl., Gordon-Breach).
2. Fix, George and Strang, Gilbert, "Fourier Analysis of the Finite Element Method in Ritz-Galerkin Theory," Studies in Appl. Math. 48 (1969) 265-273.
3. Zlamal, M., "On the Finite Element Method," Numerische Math. 12 (1968) 394-409.
4. Bramble, J. and Zlamal, M., "Triangular Elements in the Finite Element Method," Math. of Comp., to appear.
5. Zienkiewicz, O. C., "The Finite Element Method: from Intuition to Generality," Appl. Mech. Rev. 23 #3 (1970) 249-256.
6. di Guglielmo, F., "Construction d'approximations des espaces de Sobolev sur des réseaux en simplexes," Calcolo 6 (1969) 279-331.
7. Babuška, I., "Approximation by Hill Functions," Tech. Note BN-648, Univ. of Md., 1970.

8. Goël, J.-J., "Construction of Basic Functions for Numerical Utilization of Ritz's Method," Numerische Math. 12 (1969).
9. Golomb, Michael, "Approximation by Periodic Spline Interpolants on Uniform Meshes, J. Approx. Th. 1. (1968) 26-65.
10. Aubin, J. -P., "Approximation des espaces de distributions et des opérateurs differentiels," Bull. Soc. Math. France, Memoire 12 (1967).
11. Aubin, J. -P., Manuscript on numerical methods for variational problems.
12. Varga, R. S., "Hermite Interpolation-type Ritz Methods for Two-point Boundary Value Problems. Numerical Solution of Partial Differential Equations, ed. by J. Bramble, Academic Press 1966.
13. Schaeffer, David, "Wiener-Hopf Factorization of the Symbol of an Elliptic Difference Operator," J. Fcnl. Anal., to appear in 1970.
14. Birkhoff, G., Schultz, M. H., and Varga, R. S., "Piecewise Hermite Interpolation in One and Two Variables with Applications to Partial Differential Equations," Numerische Math. 11 (1968) 232-256.
15. Ciarlet, P. G., Schultz, M. H., and Varga, R. S., "Numerical Methods of High-order Accuracy for Nonlinear Boundary Value Problems." I. (See also II-V) Numerische Math. 9 (1967) 394-430.
16. Schoenberg, I. J., "Contributions to the Problem of Approximation of Equidistant Data by Analytic Functions," Quart. Appl. Math. 4 (1946) 45-99, 112-141.
17. Bramble, J., and Schatz, A. H., "Rayleigh-Ritz-Galerkin Methods for Dirichlet's Problem Using Subspaces Without Boundary Conditions," to appear.

ON THE RATE OF CONVERGENCE OF DIFFERENCE SCHEMES FOR HYPERBOLIC EQUATIONS

Vidar Thomée*

1. Introduction

The purpose of this paper is to present a survey of recent results on the rate of convergence of finite difference schemes applied to initial-value problems for hyperbolic equations. In particular, we shall consider the dependence of this rate of convergence upon the smoothness of the initial values. The main sources of our presentation are the following papers, namely Peetre and Thomée [17], Hedstrom [8], Brenner and Thomée [3], [4], and Apelkrans [1]. At certain points we have extended and improved the results in these papers.

In Section 2 we introduce the spaces of functions in which we shall express our results and also state some interpolation properties of these spaces which are essential in our approach. In Section 3 we give a general convergence result [17] which applies to the case when the hyperbolic system is correctly posed and the difference operator is stable in the L_p space under consideration. Since these assumptions are in general satisfied only in L_2 their

* Department of Mathematics, Chalmers Institute of Technology and The University of Göteborg, Göteborg, Sweden.

applicability to other L_p spaces is limited to special cases. Therefore we give in Section 4 a maximum-norm convergence result based on the L_2-theory and on an embedding lemma of the Sobolev type. In the rest of the paper we then consider the scalar one-dimensional case. In this case the correctness assumption for the initial-value problem is satisfied even in the maximum-norm but in important cases we still have stability only in L_2. In Section 5 we therefore give a very precise result in the constant coefficient scalar one-dimensional case on the rate of convergence in L_p for L_2-stable schemes [3]. Those results are best possible in a certain sense but in the case of isolated singularities they can be improved as is shown in a new result in Section 6. In the case of isolated singularities in the initial data it is also of interest to describe the propagation properties of these singularities. Some estimates in this situation [4] are given in Section 7. In Section 8 finally we present analogues of these estimates in the variables coefficient case based on [1].

The analogous problem for parabolic systems has been treated in Peetre and Thomée [17], Hedstrom [8], [9], Löfström [15], Widlund [22], [23], and Kreiss, Thomée, and Widlund [12].

Except for new results most of the material in this paper is presented without proofs and the reader is referred to the original papers for more precise information. We have chosen for simplicity to discuss only explicit schemes. In most cases the results carry over to implicit schemes and in some cases the references contain such generalizations.

For general information on stability, correctness, and convergence theory, the reader is referred to Richtmyer and Morton [18] and Thomée [21].

2. Function Spaces and Interpolation

In this section we shall introduce some spaces of functions, namely the Sobolev spaces W_p^m and the Besov spaces $B_p^{s,q}$. We shall also collect some known properties of these spaces which will be of special importance for us, in particular the property of the spaces $B_p^{s,q}$ of being interpolation spaces between the spaces W_p^m. We shall not give any proofs of the results in this section. The reader is referred to Butzer and Berens [5], Lions and Peetre [14], Peetre and Thomée [17], and references for more information on this subject.

For $1 \leq p \leq \infty$ let L_p denote the space of measurable functions on euclidean d-space R^d with

$$\|v\|_p = \begin{cases} \left(\int_{R^d} |v(x)|^p dx\right)^{1/p} < \infty \quad , \quad 1 \leq p < \infty \quad , \\ \operatorname{ess\,sup}_x |v(x)| < \infty \quad , \quad p = \infty \, , \end{cases}$$

and let $\mathcal{C}$ denote the subspace of L_∞ consisting of uniformly continuous bounded functions. Set $W_p = L_p$ for $1 \leq p < \infty$ and $W_\infty = \mathcal{C}$. Further let W_p^m be the Sobolev space of order m, the space of $v \in W_p$ such that $D^\alpha v \in W_p$ for $|\alpha| \leq m$ $(D^\alpha = (\partial/\partial x_1)^{\alpha_1} \ldots (\partial/\partial x_d)^{\alpha_d}$, $\alpha = (\alpha_1, \ldots, \alpha_d)$, $|\alpha| = \Sigma_j \alpha_j)$ and set for $v \in W_p^m$,

$$\|v\|_{W_p^m} = \sum_{|\alpha|\le m} \|D^\alpha v\|_p \ .$$

We shall write $\mathcal{C}^\infty = \bigcap_m W_\infty^m$. Notice that $v \in \mathcal{C}^\infty$ then means that $D^\alpha v$ is bounded for any α .

We shall now define the Besov space $B_p^{s,q}$, $s > 0$, $1 \le p$, $q \le \infty$. For $v \in W_p$, $t > 0$, we introduce the modulus of continuity in W_p of order N ,

$$\omega_{p,N}(t,v) = \sup_{|y|\le t} \|(T_y - I)^N v\|_p \ ,$$

where $T_y v(x) = v(x+y)$. Let $s = S+s_0$ where S is a non-negative integer and $0 < s_0 \le 1$. Then $B_p^{s,q}$ is the subspace of W_p^S defined for $1 \le q < \infty$ by

$$\|v\|_{B_p^{s,q}} = \|v\|_p + \begin{cases} \sum_{|\alpha|=S} \left[\int_0^\infty (t^{-s_0}\omega_{p,1}(t,D^\alpha v))^q \frac{dt}{t} \right]^{1/q} , \\ \qquad\qquad 0 < s_0 < 1 , \\ \sum_{|\alpha|=S} \left[\int_0^\infty (t^{-1}\omega_{p,2}(t,D^\alpha v))^q \frac{dt}{t} \right]^{1/q} , \\ \qquad\qquad s_0 = 1 , \end{cases}$$

and for $q = \infty$ by

$$\|v\|_{B_p^{s,\infty}} = \|v\|_p + \begin{cases} \sum_{|\alpha|=S} \sup_{t>0} t^{-s_0}\omega_{p,1}(t,D^\alpha v), \quad 0 < s_0 < 1 , \\ \\ \sum_{|\alpha|=S} \sup_{t>0} t^{-1}\omega_{p,2}(t,D^\alpha v) \ , \quad s_0 = 1 . \end{cases}$$

In particular, $B_p^{s,\infty}$ is defined by a type of Lipschitz condition for the derivatives of order S (for $s_o = 1$ a Zygmund type condition). For $p = \infty$ we recognize the classical Lipschitz condition. It turns out that for any $N > s$ the following is an equivalent norm for $B_p^{s,q}$, namely

$$\|v\|_{B_p^{s,q},N} = \|v\|_p + \begin{cases} \left[\displaystyle\int_0^\infty (t^{-s}\omega_{p,N}(t,v))^q \frac{dt}{t}\right]^{1/q}, & 1 \leq q < \infty, \\ \sup\limits_{t>0} t^{-s}\omega_{p,N}(t,v), & q = \infty . \end{cases}$$

The following embedding results hold, where the inclusion stands for continuous embedding so that there is in each case a corresponding inequality between the respective norms; one has

$$B_p^{s_1,q_1} \subset B_p^{s_2,q_2} \quad \text{if} \quad s_1 > s_2 \quad \text{or} \quad s_1 = s_2 , \quad q_1 \leq q_2 ,$$

and for s a natural number,

$$B_p^{s,1} \subset W_p^s \subset B_p^{s,\infty} .$$

Let R^- denote the interval $(-\infty,0] \subset R = R^1$. In some of the results below in one dimension we shall have use for the space $B_\infty^{s,\infty}(R^-)$ of functions on R^- corresponding for instance to the norm $(N > s)$

$$\|v\|_{B_\infty^{s,\infty}(R^-)} = \sup_{x\leq 0} |v(x)| + \sup \{t^{-s}|(T_h-I)^N v(x)| ; \; x\leq 0,\ x+Nh\leq 0,\ t>0\} .$$

We now turn to the interpolation properties of these spaces. In order to avoid excessive generality we shall only collect in three lemmas the precise results we shall need.

Let F_1 and F_2 be normed linear spaces. We denote by $\mathcal{L}(F_1,F_2)$ the set of bounded linear operators from F_1 into F_2. For $A \in \mathcal{L}(F_1,F_2)$ we set

$$\|A\|_{F_1,F_2} = \sup_{0\neq v\in F_1} \frac{\|Av\|_{F_2}}{\|v\|_{F_1}} .$$

We then have the following lemmas.

<u>Lemma 2.1</u>. Let F be a normed linear space, $1 \leqslant p \leqslant \infty$, m a positive integer, and $0 < s < m$. Then there is a positive constant C such that if $A \in \mathcal{L}(W_p^m,F)$ and $\theta = s/m$ then

$$\|A\|_{B_p^{s,\infty},F} \leqslant C\|A\|^{1-\theta}_{W_p,F} \|A\|^{\theta}_{W_p^m,F} .$$

<u>Lemma 2.2</u>. Let F be a normed linear space and let $0 < s_1 < s_2$, $0 < \theta < 1$. Then there is a positive constant C such that if $A \in \mathcal{L}(B_2^{s_j,1},F)$, $j = 1,2$, then $A \in \mathcal{L}(B_2^{s,\infty},F)$ where $s = s_1 + \theta(s_2-s_1)$ and

$$\|A\|_{B_2^{s,\infty},F} \leqslant C\, \|A\|^{1-\theta}_{B_2^{s_1,1},F} \|A\|^{\theta}_{B_2^{s_2,1},F} .$$

Lemma 2.3. For any non-negative integer m and $1 \leq p \leq \infty$ there is a positive constant C such that if $A \in \mathcal{L}(W_p^m, W_p) \cap \mathcal{L}(W_p^{m+d}, W_p^d)$, then $A \in \mathcal{L}(B_p^{m+d/2,1}, B_p^{d/2,1})$ and

$$\|A\|_{B_p^{m+\frac{d}{2},1},B_p^{\frac{d}{2},1}} \leq C\,\|A\|^{\frac{1}{2}}_{W_p^m,W_p}\;\|A\|^{\frac{1}{2}}_{W_p^{m+d},W_p^d}\;.$$

In the proofs below C will denote a positive constant, in general not the same at different occurrences.

3. The Rate of Convergence in the Stable Cases

Consider the initial-value problem

$$\text{(3.1)}\qquad \frac{\partial u}{\partial t} = P(x,D)u \equiv \sum_{j=1}^{d} A_j(x)\,\frac{\partial u}{\partial x_j} + B(x)u\;,$$

$$t \geq 0\;,\quad x \in R^d\;,$$

$$\text{(3.2)}\qquad u(x,0) = v(x)\;,$$

where A_j and B are C^∞ $N \times N$ matrices and $u = u(x,t)$ and $v = v(x)$ are N-vectors. We shall assume that the initial-value problem is correctly posed in W_p so that $P(x,D)$ generates a strongly continuous semi-group $E(t)$ for $t \geq 0$ for which for any $T > 0$ there is a positive constant C such that

$$\text{(3.3)}\qquad \|E(t)v\|_{W_p} \leq C\|v\|_{W_p}\;,\quad 0 \leq t \leq T\;.$$

It is well-known that for $p = 2$ sufficient conditions for

correctness are

a) that the A_j are hermitian,

b) that the A_j are constant and $\Sigma_j A_j \xi_j$ is uniformly equivalent to a real diagonal matrix when $\xi \in R^d$.

For $p \neq 2$ and A_j constant and hermitian, the problem is correctly posed if and only if the A_j commute (cf. [2]).

In the case of a first order system, (3.3) implies the following condition of strong correctness, namely for any non-negative integer m there is a constant C such that

$$\|E(t)v\|_{W_p^m} \leqslant C\|v\|_{W_p^m} , \quad 0 \leqslant t \leqslant T .$$

Consider now a finite difference operator which for simplicity we shall take to be explicit,

$$E_k\, v(x) = \sum_{\gamma} a_{\gamma}(x,h)\, v\,(x-\gamma h) \ , \quad k/h = \lambda = \text{constant} \ ,$$

where the summation is over a finite set of $\gamma = (\gamma_1, \ldots, \gamma_d)$, γ_j integer, and the a_γ are C^∞ $N \times N$ matrices in x and h . The operator shall be assumed to be accurate of order μ so that for smooth solutions u of (3.1) ,

$$u(x,t+k) = E_k\, u(x,t) + k\, O(h^{\mu}) \text{ as } \quad h \to 0 .$$

One can prove the following more precise estimate.

Lemma 3.1. Under the above assumptions, there is a constant

C such that for $v \in W_p^{\mu+1}$,

$$\|(E_k - E(k))v\|_{W_p} \leq C\, h^{\mu+1} \|v\|_{W_p^{\mu+1}} .$$

Proof. See [17], Theorem 4.1.

The operator E_k is said to be stable in W_p if for $T > 0$ there is a C such that

$$\|E_k^n v\|_{W_p} \leq C\|v\|_{W_p} , \quad nk \leq T , \quad v \in W_p .$$

Related to the above mentioned fact that correctness is exceptional in W_p for $p \neq 2$ is the fact that most of the criteria for stability concerns L_2. However, for the scalar one-dimensional case the question of stability is thoroughly investigated also for general p (cf. Section 5 below). For references on stability, see [18] and [21].

We shall want to approximate the solution $E(nk)v$ of (3.1), (3.2) by $E_k^n v$ and to estimate the difference

$$F_{nk}v = E_k^n v - E(nk)v .$$

By the well-known Lax equivalence theorem we have

$$\lim_{h\to 0} \|F_{nk}v\|_{W_p} = 0 , \quad \text{for} \quad v \in W_p , \tag{3.4}$$

uniformly in $0 \leq nk \leq T$ if and only if E_k is stable in W_p. Our purpose is to give estimates for the rate of convergence in (3.4) under more precise assumptions on the initial function v.

By Lemma 3.1 we can easily obtain the following convergence estimate.

Theorem 3.1. Under the above assumptions with E_k stable in W_p there is a constant C such that for $v \in W_p^{\mu+1}$,

$$\|F_{nk}v\|_{W_p} \leq C\, h^{\mu} \|v\|_{W_p^{\mu+1}} , \quad nk \leq T . \tag{3.5}$$

Proof. We have

$$F_{nk}v = (E_k^n - E(nk))v$$

$$= \sum_{j=0}^{n-1} E_k^{n-1-j}(E_k - E(k))\, E(jk)v ,$$

and so by the stability of E_k, Lemma 3.1, and the strong correctness

$$\|F_{nk}v\|_{W_p} \leq C \sum_{j=0}^{n-1} k\, h^{\mu}\, \|E(jk)v\|_{W_p^{\mu+1}} \leq C\, nk\, h^{\mu}\, \|v\|_{W_p^{\mu+1}} ,$$

which proves the theorem.

We now consider the case that the initial-values are less smooth.

Theorem 3.2. Under the above assumptions, with E_k stable in W_p, then if $0 < s < \mu + 1$ there is a C such that for $v \in B_p^{s,\infty}$ we have

$$\|F_{nk}v\|_{W_p} \leq C\, h^{\frac{s\mu}{\mu+1}}\, \|v\|_{B_p^{s,\infty}} , \quad nk \leq T .$$

Proof. The proof follows immediately by interpolation (Lemma 2.1) between (3.5) and the inequality

$$\|F_{nk}v\|_{W_p} \leq C \, \|v\|_{W_p} \, , \quad nk \leq T \, ,$$

which is an obvious consequence of stability and correctness.

4. Maximum-Norm Estimates for L_2-Stable Operators

For numerical purposes it is desirable to have uniform rather than L_2 bounds for the error. We shall see how such estimates can be derived from the results in the preceding section even for operators E_k which are stable only in L_2. The main tool which makes the transition possible is the following embedding result.

Lemma 4.1. There is a constant C such that $B_2^{\frac{d}{2},1} \subset \mathcal{C}$ and

$$\|v\|_\infty \leq C\|v\|_{B_2^{\frac{d}{2},1}} \, .$$

Proof. See [16]. This inequality is somewhat stronger than the classical Sobolev inequality,

$$\|v\|_\infty \leq C \, \|v\|_{B_2^{s,\infty}} \, , \quad s > \frac{d}{2} \, ,$$

and gives slightly sharper estimates below.

We shall need the following sharpening of Theorem 3.1.

Lemma 4.2. Under the assumptions of Theorem 3.1 there is for any non-negative integer m and $T > 0$ a constant C such that

$$(4.1) \qquad \|F_{nk}v\|_{W_p^m} \leq C\, h^{\mu}\, \|v\|_{W_p^{\mu+m+1}} , \quad nk \leq T .$$

Proof. The proof is identical with that of Theorem 3.1 under the assumption that the following two statements have been proved, namely

$$(4.2) \qquad \|(E_k - E(k))v\|_{W_p^m} \leq C\, h^{\mu+1}\, \|v\|_{W_p^{\mu+m+1}} ,$$

$$(4.3) \qquad \|E_k^n v\|_{W_p^m} \leq C\, \|v\|_{W_p^m} , \quad 0 \leq nk \leq T .$$

The proof of (4.2) is only a trivial modification of that of Lemma 3.1. In order to prove (4.3), let α be such that with $\tilde{E}_k = e^{-\alpha k} E_k$ we have for all $n \geq 0$, $h \leq 1$ (with no restriction on nk),

$$(4.4) \qquad \|\tilde{E}_k^n\|_{W_p} \leq C\, \|v\|_{W_p} .$$

We set

$$|||v|||_{W_p} = \sup_{n\geq 0} \|\tilde{E}_k^n v\|_{W_p} ,$$

$$|||v|||_{W_p^m} = \sum_{|\alpha|\leq m} |||D^{\alpha}v|||_{W_p} .$$

By (4.4) this norm is then equivalent with the ordinary norm on W_p^m and we have

$$(4.5) \qquad |||\tilde{E}_k v|||_{W_p} \leq |||v|||_{W_p} .$$

We easily obtain for $|\alpha| \leq m$,

$$\|(D^\alpha \tilde{E}_k - \tilde{E}_k D^\alpha) v\|_{W_p} \leq Ck \|v\|_{W_p^m} ,$$

and consequently by (4.5),

$$|||D^\alpha \tilde{E}_k v|||_{W_p} \leq |||D^\alpha v|||_{W_p} + Ck |||v|||_{W_p^m} .$$

Hence by summation over $|\alpha| \leq m$,

$$|||\tilde{E}_k v|||_{W_p^m} \leq (1 + Ck) |||v|||_{W_p^m} ,$$

which clearly proves (4.3).

We can now prove the following maximum-norm estimates.

Theorem 4.1. Under the general assumptions of Section 3, with E_k stable in L_2 and for $0 \leq s \leq \mu + 1$ there is a positive constant $C = C_s$ such that for v in the respective space ,

$$
(4.6) \qquad \|F_{nk}v\|_\infty \leq \begin{cases} C\, h^{\mu} \|v\|_{B_2^{\mu+\frac{d}{2}+1,1}} , \\ C\, h^{\frac{s\mu}{\mu+1}} \|v\|_{B_2^{s+\frac{d}{2},\infty}} , \quad 0 < s < \mu + 1 , \\ C\, \|v\|_{B_2^{\frac{d}{2},1}} . \end{cases}
$$

Proof. We obtain by interpolation (Lemma 2.3) between the estimates (3.5) and (4.1) with $m = d$,

$$
\|F_{nk}v\|_{B_2^{\frac{d}{2},1}} \leq C\, h^{\mu} \|v\|_{B_2^{\mu+\frac{d}{2}+1,1}} .
$$

By strong correctness, stability, and (4.3) we have for $m = 0,d$,

$$
\|F_{nk}v\|_{W_2^m} \leq C\, \|v\|_{W_2^m} ,
$$

and again by interpolation

$$
\|F_{nk}v\|_{B_2^{\frac{d}{2},1}} \leq C\, \|v\|_{B_2^{\frac{d}{2},1}} .
$$

This implies as above the latter of the inequalities (4.6). The middle inequality is now a consequence of Lemma 2.2.

5. L_p-Estimates in the Non-Stable Scalar One-Dimensional Constant Coefficient Case

We recall that for $p \neq 2$ the initial-value problem (3.1), (3.2) with constant hermitian A_j is only correctly posed in W_p when the A_j commute and hence may be simultaneously diagonalized. But in that case, after the corresponding change of dependent variables the equations are only coupled in the lower order terms so that the system consists essentially of independent scalar equations. It is therefore natural to consider separately the case of the initial-value problem for the scalar one-dimensional equation

$$\frac{\partial u}{\partial t} = \rho \frac{\partial u}{\partial x} , \quad \rho \text{ real constant} .$$

In this case the solution operator is $(E(t)v)(x) = v(x+\rho t) = T_{\rho t}v(x)$.

We shall analyze corresponding difference schemes with constant coefficients,

$$(5.1) \qquad E_k v(x) = \sum_{|j|\leq J} a_j \, v(x-jh) , \quad k/h = \lambda ,$$

in terms of its symbol or characteristic trigonometric polynomial,

$$a(\xi) = \Sigma_j \, a_j \, e^{ij\xi} .$$

It is well known that E_k is stable in L_2 if and only if

$$(5.2) \qquad |a(\xi)| \leqslant 1 \ , \quad \xi \text{ real} \ ,$$

and we shall consider only the case when this is satisfied. The results in Section 3 give then error estimates in L_2 and in the case that E_k is stable in W_p for other p we get analogous convergence results in W_p . However, many operators E_k are stable only in L_2 and our purpose in this section is to find precise error estimates also in W_p .

We shall assume for simplicity that rather than (5.2),

$$(5.3) \qquad |a(\xi)| < 1 \quad \text{for} \quad 0 < |\xi| \leqslant \pi \ .$$

For small ξ we can then write

$$a(\xi) = \exp(-i\lambda\rho\xi + \psi(\xi)) \ ,$$

where as $\xi \to 0$,

$$(5.4) \qquad \psi(\xi) = \beta\xi^{\nu}(1 + o(1)) \ , \quad \beta \neq 0 \ .$$

Here $\nu = \mu + 1$ where μ is the order of accuracy of E_k . By (5.3) there is a smallest (even) number σ , the order of dissipation of E_k , such that

$$\text{Re } \psi(\xi) \leqslant -\gamma\xi^{\sigma} \ , \quad |\xi| \leqslant \pi \ .$$

Clearly $\nu \leqslant \sigma$ and if $\nu < \sigma$, β is purely imaginary.

It is known that E_k is stable in W_p , $p \neq 2$, if and only if $\nu = \sigma$. The following more precise result holds. Here and below we shall denote the operator norm $\|\cdot\|_{W_p, W_p}$ by $\|\cdot\|_p$.

Theorem 5.1. Under the above assumption on the operator E_k there are positive constant c and C such that

$$cn^{|\frac{1}{2}-\frac{1}{p}|(1-\frac{\nu}{\sigma})} \leq \|E_k^n\|_p \leq Cn^{|\frac{1}{2}-\frac{1}{p}|(1-\frac{\nu}{\sigma})} .$$

Proof. See [3].

We now state the result on the rate of convergence.

Theorem 5.2. Under the above assumptions on the operator E_k , then for $s \geq 0$, $s \neq \nu$ and $\nu|\frac{1}{2} - \frac{1}{p}|$, there is a constant C such that for $v \in B_p^{s,\infty}$,

$$(5.5) \qquad \|F_{nk}v\|_p \leq Ch^{g(s)}\|v\|_{B_p^{s,\infty}} , \quad nk \leq T ,$$

where

$$(5.6) \qquad g(s) = \min\{\mu, s(1 - \frac{1}{\nu}) , s(1 - \frac{1}{\sigma}) - |\frac{1}{2} - \frac{1}{p}| (1 - \frac{\nu}{\sigma})\} .$$

Proof. See [3].

It can also be proved that this result is best possible in the sense that the function $g(s)$ defined in (5.6) is the largest for which an estimate of the form (5.5) holds for all $v \in B_p^{s,\infty}$. In the stable cases, i.e. when $\nu = \sigma$ or $p = 2$, the order of convergence is $s(1-1/\nu) = s\mu/(\mu+1)$

when $0 < s < \nu$, in agreement with Theorem 3.2. In the opposite case the error is larger for $s < \nu|\frac{1}{2} - \frac{1}{p}|$. For small s, $g(s)$ is then negative and for $s = 0$ we recognize the exponent in Theorem 5.1.

For $p = \infty$ one finds

$$\|E_k^n\|_\infty = \sum_j |a_{nj}| ,$$

where a_{nj} are the Fourier coefficients of $a(\xi)^n$. Using the saddle-point method for estimating a_{nj}, Serdjukova [19], [20] and Hedstrom [6], [7] were able to prove Theorem 5.1 for this case, and in [8] Hedstrom succeeded in obtaining the corresponding special case of Theorem 5.2. In the proof in [3] which works for general p one notices that

$$\|E_k^n\|_p = M_p(a^n) ,$$

where $M_p(\cdot)$ denotes the Fourier multiplier norm ($\hat{\phi}$ is the Fourier transform of ϕ),

$$M_p(\phi) = \sup\{\|\hat{\phi} * v\|_p , \|v\|_p \leq 1\}$$

$$= \sup\{\|\phi\hat{v}\|_p , \|v\|_p \leq 1\} .$$

For basic material on Fourier multipliers, see Hörmander [10]. The central tool in our proof of the estimates is the following form of the Carlson-Beurling inequality: Assume that $\phi \in L_2$, $\phi' \in L_2$. Then there is a positive C such that

$$M_p(\phi) \leqslant C \, \|\phi\|_\infty^{\frac{2}{p}} \, \|\phi\|_2^{\frac{1}{2}-\frac{1}{p}} \, \|\phi'\|_2^{\frac{1}{2}-\frac{1}{p}} \quad ,$$

and for $p^{-1} + q^{-1} = 1$, $M_p(\phi) = M_q(\phi)$.

6. Isolated Singularities

In this section we shall consider the scalar one-dimensional variable coefficient equation

$$(6.1) \qquad \frac{\partial u}{\partial t} = \rho(x) \frac{\partial u}{\partial x} ,$$

where ρ is a real $\mathcal{C}^\infty$ function. We shall treat the corresponding initial-value problem when the initial-function v is smooth for $x \neq 0$ but has a singularity at $x = 0$ which makes v belong to a $B_\infty^{s,\infty}$ space with a lower s . In spite of the results in Section 5 we shall see that essentially the same results hold in the maximum-norm for operators which are only L_2-stable as if the operator had been stable in $\mathcal{C}$. For our purpose it is clearly no restriction of the generality to assume that v vanishes for positive x and has compact support.

We introduce the set of functions

$$\check{B}_M^s = \{v;\ v \in B_\infty^{s+\frac{1}{2},\infty}(R^-) ,$$
$$\operatorname{supp} v \subseteq [-M,o], v(x) = O(x^s) \text{ as } x \to 0\} .$$

This is a Banach space under the norm

$$\|v\|_{\check{B}_M^s} = \|v\|_{B^{s+\frac{1}{2},\infty}(R^-)} + \Lambda_s(v) ,$$

where

$$\Lambda_s(v) = \sup_x |x^{-s}v(x)| \quad .$$

The following embedding result holds.

Lemma 6.1. For any positive M and s there is a constant C such that $\check{B}_M^s \subseteq B_2^{s+\frac{1}{2},\infty}$ and

$$\|v\|_{B_2^{s+\frac{1}{2},\infty}} \leq C \|v\|_{\check{B}_M^s} \quad \text{for} \quad v \in \check{B}_M^s \quad .$$

Proof. A simple calculation with $N > s + \frac{1}{2}$ shows for $h > 0$,

$$|(T_h - I)^N v(x)| \leq \begin{cases} C\, h^s\, \Lambda_s(v) \ , & -Nh < x \leq 0 \ , \\ C\, h^{s+\frac{1}{2}} \|v\|_{B^{s+\frac{1}{2},\infty}(R^-)} \ , & -M-Nh \leq x \leq -Nh \ , \\ 0 \ , \quad x > 0 \ \text{and} \ x < -M-Nh \ , \end{cases}$$

and hence

$$\omega_{2,N}(t,v) \leq C\, t^{s+\frac{1}{2}} \|v\|_{\check{B}_M^s} \ ,$$

which proves the result.

We can now state our convergence result.

Theorem 6.1. Consider a L_2-stable operator E_k for the equation (6.1). Then for given positive $s \neq \mu + 1$ there is a constant C such that for $v \in \check{B}^s_M$,

$$\|F_{nk}v\|_\infty \leq C\, h^{\min(\mu,\frac{s}{\mu+1})} \|v\|_{\check{B}^s_M} .$$

Proof. For $0 < s < \mu + 1$ the result is an immediate consequence of Theorem 4.1 and Lemma 6.1. For $s > \mu + 1$ the result follows in the same way after noticing that $B_2^{\mu+\frac{3}{2},1} \supseteq B_2^{s+\frac{1}{2},\infty}$.

We complete the result by some information on the case when v has a discontinuity at $x = 0$. For simplicity we shall assume that the initial-function has the following simple form, namely,

$$v_0(x) = \begin{cases} 0 , & x \notin [-M,0] , \\ 1 , & x \in [-M+1,0] , \\ x+M , & x \in [-M,-M+1] , \end{cases} \tag{6.2}$$

where $M > 1$. Clearly, by Theorem 6.1 the result can be generalized by considering any linear combination of v_0 and a function in $\check{B}^s_M$ for $s > \frac{1}{2}$.

Theorem 6.2. Assume again that E_k is a L_2-stable operator consistent with (6.1). Then there is a constant C such that for the function v_0 defined in (6.2) and small h ,

$$\|E^n_k v_0\|_\infty \leq C \log \frac{1}{h} .$$

Proof. Given x_0 we want to estimate $E_k^n v(x_0)$. This number only depends on the values of v_0 at $x_0 - jh$, $j = 0, \pm 1, \pm 2, \ldots$. Let j_0 be such that $-\frac{1}{2}h < x_0 - j_0 h \leq \frac{1}{2}h$. We shall define a function v_h which coincides with v_0 at all points $x_0 - jh$. If $x_0 - j_0 h \leq 0$ we set $v_h = v_0$ except in $(0, \frac{1}{2}h)$ where we interpolate linearly between $v_h(0) = 1$ and $v_h(\frac{1}{2}h) = 0$. If $x_0 - j_0 h > 0$ we set similarly $v_h = v_0$ except in $(-\frac{1}{2}h, 0)$ where we interpolate linearly between $v_h(-\frac{1}{2}h) = 1$ and $v_h(0) = 0$. We obtain by Theorem 4.1,

$$|E_k^n v_0(x_0)| = |E_k^n v_h(x_0)| \leq C \, \|v_h\|_{B_2^{\frac{1}{2},1}},$$

and we want to estimate the quantity on the right. A simple calculation shows

$$\omega_{2,1}(t, v_h) \leq \begin{cases} C\, t\, h^{-\frac{1}{2}}, & 0 \leq t \leq h, \\ C\, t^{\frac{1}{2}}, & h \leq t \leq 1, \\ C, & t > 1, \end{cases}$$

and hence for small h,

$$\|v_h\|_{B_2^{\frac{1}{2},1}} \leq \sqrt{M} + \int_0^\infty t^{-\frac{3}{2}} \, \omega_{2,1}(t, v_h)\,dt \leq C \log \frac{1}{h},$$

which completes the proof.

7. Propagation of Discontinuities in the Constant Coefficient Case

We shall now study the initial-value problem

$$\frac{\partial u}{\partial t} = \rho \frac{\partial u}{\partial x} , \quad t > 0 , \quad \rho \text{ real} , \tag{7.1}$$

$$u(x,0) = v(x) , \tag{7.2}$$

where $v(x)$ vanishes for positive x but may have a discontinuity at $x = 0$. As above, the general case of an isolated discontinuity at $x = 0$ may be reduced to this case by subtracting a smooth initial function.

The discontinuity propagates along the line $x + \rho t = 0$ and for $x + \rho t > 0$ we have $u(x,t) = 0$ if u is the solution of (7.1), (7.2). If E_k is a stable consistent difference operator of the form (5.1) the general convergence results show that $E_k^n v(x)$ tends to zero with h for $x + \rho t > 0$ where $t = nk$. Our purpose here is to give stronger results on the rate of decay of $E_k^n v$.

Our main source in this section is Brenner and Thomée [4]. The same problem has been treated previously by Hedstrom [8] and Apelkrans [1] when $v = 1 - \chi_0$. Hedstrom gave very precise estimates based on his estimates for the Fourier coefficients of $a(\xi)^n$ (cf. Section 5). His approach would, although quite technical, permit more general conclusions. Apelkrans used a technique developed by Kreiss and Lundqvist [11] to obtain somewhat less precise estimates. The contribution in [4] is to show that this latter technique can be used to obtain the sharper results.

Following Apelkrans [1] (cf. also Kreiss and Lundqvist [11]) we say that E_k is contractive of order τ if there is a positive constant C such that the symbol of E_k satisfies

$$(7.3) \qquad |a(\xi - i\eta)| \leqslant \exp(-\lambda\rho\eta + C\eta^{\tau}) \ , \quad \xi \text{ real} \ , \quad \eta \geqslant 0 \ .$$

The following lemma holds.

<u>Lemma 7.1</u>. Assume that E_k is dissipative of order σ and accurate of order $\mu = \nu - 1$. Then E_k is contractive of order τ where with β defined by (5.4),

$$(7.4) \qquad \tau = \begin{cases} \nu & \text{if } \nu \text{ is odd and } \operatorname{Im}\beta < 0 \ , \\ \sigma(\sigma-\mu)^{-1} & \text{otherwise} \ . \end{cases}$$

<u>Proof</u>. See [4] or Lemma 8.1 below.

Let χ_y be the characteristic function of the interval $[y,\infty)$. We have the following result in L_2.

<u>Theorem 7.1</u>. Assume that E_k is dissipative of order σ and accurate of order $\mu = \nu - 1$, and let v vanish for positive x. Then there is a positive constant c such that with $nk = t$, $\delta_t = \delta t^{-1}$, $v \in L_2$,

$$\|\chi_{\delta-\rho t} E_k^n v\|_2 \leqslant \exp(-cn\delta_t^{\kappa}) \, \|v\|_2 \ ,$$

where

$$\kappa = \begin{cases} \nu/\mu & \text{if } \nu \text{ is odd and } \operatorname{Im}\beta < 0\ , \\ \sigma/\mu & \text{otherwise}\ . \end{cases} \tag{7.5}$$

Proof. The main idea in the proof is to introduce the operator

$$E_{k,\eta} v = e^{\eta x} E_k (e^{-\eta x} v)\ ,$$

which has the symbol $a_\eta(\xi) = a(\xi - ih\eta)$. We set

$$\varepsilon_{\delta,n} = \| \chi_{\delta-\rho t}\ E_k^n\ v \|_2\ .$$

For η positive, $\chi_{\delta-\rho t}(x) \leq \exp(\eta(x+\rho t-\delta))$ and hence

$$\begin{aligned} \varepsilon_{\delta,n} &\leq \| \exp(\eta(x+\rho t-\delta)) E_k^n v \|_2 \\ &\leq \exp(\eta(\rho t-\delta)) \| E_{k,\eta}^n (e^{\eta x} v) \|_2 \\ &\leq \exp(\eta(\rho t-\delta)) \| E_{k,\eta}^n \|_2\ \| v \|_2\ , \end{aligned}$$

where we have used that $\exp(\eta x) \leq 1$ for all x where $v(x)$ is non-zero. By Lemma 7.1 we have for $\eta > 0$,

$$\| E_{k,\eta} \|_2 \leq \exp(-\lambda\rho h\eta + C(h\eta)^\tau)\ ,$$

where τ is defined by (7.4). Hence for such η ,

$$\varepsilon_{\delta,n} \leq \exp(-\delta\eta + Cn(h\eta)^\tau)\ \| v \|_2\ .$$

The result now follows by choosing

$$h\eta = \left(\frac{1}{2}\lambda\delta_t C^{-1}\right)^{\frac{1}{(\tau-1)}}$$

and noticing that $\kappa = \tau(\tau-1)^{-1}$.

We now turn to results in the maximum-norm. The analogue of Theorem 7.1 is then the following.

<u>Theorem 7.2</u>. Under the same assumptions as in Theorem 7.1 there are positive constants c and C such that for $v \in L_\infty$,

$$\|\chi_{\delta-\rho t} E_k^n v\|_\infty \leq C\, n^\omega \exp(-cn\, \delta_t^\kappa)\, \|v\|_\infty ,$$

where κ is defined by (7.5) and

$$\omega = \begin{cases} 0 & \text{if } \nu \text{ is odd and } \operatorname{Im}\beta < 0 , \\ \frac{1}{2}\left(1 - \frac{\nu}{\sigma}\right) & \text{otherwise.} \end{cases}$$

<u>Proof</u>. See [4]. We recognize in the value of ω the exponent in Theorem 5.1.

If v has a discontinuity at $x = 0$ but is smooth for negative x , the factor n^ω may be suppressed as is stated in the following theorem.

<u>Theorem 7.3</u>. Under the same assumptions as in Theorem 7.1 there are positive constants c and C such that if $v \in B_\infty^{s,\infty}(R^-)$ for some $s > \frac{\nu}{2}$,

$$\|\chi_{\delta-\rho t}E_k^n v\|_\infty \leq C\,\exp(-cn\delta_t^\kappa)\,\|v\|_{B_\infty^{s,\infty}(R^-)}\ ,$$

where κ is again defined by (7.5) .

Proof. See [4]. The proofs of the last two theorems are considerably more complicated than in the L_2 case and use the same techniques as the results in Section 5.

In the above discussion we have only considered the behavior to the right of the discontinuity. Clearly a corresponding analysis holds to the left. This case may be reduced to the one treated by a change of sign in x . Notice that since the symbol of the corresponding operator is $a(-\xi)$ the coefficient of $i\xi^\nu$ for ν odd then changes sign so that the values of τ , κ , and ω above are altered accordingly. Consequently, for odd ν the estimates will be different on the two sides of the discontinuity.

Let us remark finally that the results in this (and also in the next) section can be interpreted as investigations on the domain of influence of the initial-values at a point.

8. Propagation of Discontinuities in the Variable Coefficient Case

We shall consider generalizations to variable coefficients of the results in Section 7. In doing so we shall use the approach in Apelkrans [1] to obtain somewhat sharper and more general results than in that paper. As in Section 6, let the equation be

(8.1) $$\frac{\partial u}{\partial t} = \rho(x)\frac{\partial u}{\partial x} ,$$

where $\rho \in \mathcal{C}^\infty$ is real, and consider a consistent difference operator

(8.2) $$E_k v(x) = \sum_{|j|\leq J} a_j(x) v(x-jh) ,$$

where $a_j \in \mathcal{C}^\infty$. Let the symbol of E_k have the form

$$a(x,\xi) = \sum_j a_j(x) e^{ij\xi} = \exp(-i\lambda\rho(x)\xi + \psi(x,\xi))$$

where (cf.(5.3))

(8.3) $$\text{Re}\,\psi(x,\xi) < 0 \quad \text{for} \quad 0 < |\xi| \leq \pi , \quad x \quad \text{real} .$$

We say that E_k is uniformly accurate of order $\mu = \nu - 1$ if, uniformly in x,

(8.4) $$\psi(x,\xi) = \beta(x)\xi^\nu(1+o(1)) \quad \text{as} \quad \xi \to 0 ,$$

where $\beta(x)$ is bounded away from zero and infinity. For ν odd it follows by (8.3) that $\beta(x)$ is necessarily purely imaginary. Further, we say that E_k is dissipative of order σ if σ is the smallest (even) number such that for some positive γ,

(8.5) $$\text{Re}\,\psi(x,\xi) \leq -\gamma\xi^\sigma, \quad |\xi| \leq \pi , \quad x \text{ real} .$$

As in (7.3), E_k will be said to be contractive of order

τ if there is a positive constant C such that

$$\text{(8.6)} \qquad \operatorname{Re} \psi(x,\xi-i\eta) \leq C\eta^{\tau} , \quad \eta \geq 0 .$$

We have the following analogue of Lemma 7.1.

Lemma 8.1. Under the above general assumptions, if E_k is dissipative of order σ and uniformly accurate of order $\mu = \nu - 1$, then E_k is contractive of order τ where τ is defined in (7.4) with $\beta = \beta(x)$ as in (8.4).

Proof. Since for $\eta \geq 0$,

$$|a(\xi-i\eta)| \leq Ce^{J\eta} ,$$

it is sufficient to prove (8.6) for small η . For any $\varepsilon_0 > 0$, it follows by (8.5) that

$$\sup\{\operatorname{Re} \psi(x,\xi) \; ; \; \varepsilon_0 \leq |\xi| \leq \pi , \quad x \in R\} < 0 ,$$

and hence by continuity for small η ,

$$\sup\{\operatorname{Re} \; (x,\xi-i\eta) \; ; \; \varepsilon_0 \leq |\xi| \leq \pi , \quad x \in R\} < 0 .$$

It therefore suffices to prove (8.6) when both ξ and η are small. Consider first the case ν odd, $\operatorname{Im} \beta(x) = \tilde{\beta}(x) < 0$. By our assumptions, we then have

$$\begin{aligned} \operatorname{Re} \psi(x,\xi-i\eta) &\leq \operatorname{Re} [\psi(x,\xi-i\eta) - \psi(x,\xi)] \\ &\leq \frac{1}{2} \nu\beta(x)\eta\xi^{\nu-1} + C(\eta^2|\xi|^{\nu-2} + \eta^{\nu}) . \end{aligned}$$

Since the coefficient of the first term on the right is negative and bounded away from zero we can use the obvious inequality

$$\eta^2|\xi|^{\nu-2} \leqslant \varepsilon\eta\,\xi^{\nu-1} + C_\varepsilon\eta^\nu ,$$

to obtain the result in this case. In the opposite case we obtain by (8.5),

$$\mathrm{Re}\,\psi(x,\xi-i\eta) \leqslant -\gamma\xi^\sigma + C(\eta|\xi|^{\nu-1} + \eta^\nu) ,$$

and the result follows as above by the inequality

$$\eta|\xi|^{\nu-1} + \eta^\nu \leqslant \varepsilon\xi^\sigma + C_\varepsilon\eta^\tau .$$

Let Γ be the characteristic of (8.1) through the origin. In order to obtain a measure of the distance $d(x,t)$ from (x,t) to Γ parallel to the x-axis we introduce the solution $g(x,t)$ of (8.1) with initial-values $g(x,0) = x$. Then Γ has the equation $g(x,t) = 0$ and $g(x,t)$ is positive to the right and negative to the left of Γ. Since $w = \frac{\partial}{\partial x} g(x,t)$ is a solution of the initial-value problem

$$\frac{\partial w}{\partial t} = \rho\frac{\partial w}{\partial x} + \rho' w , \quad t \geqslant 0 ,$$

$$w(x,0) = 1 ,$$

we easily find that for any finite t-interval, $\frac{\partial g}{\partial x}$ is bounded away from zero and infinity. Hence Γ can be written in the form $x = G(t)$ and there are positive constants c_1 and C_1 such that

(8.7) $$c_1 d(x,t) \leqslant |g(x,t)| \leqslant C_1 d(x,t) .$$

Notice that for the equation (7.1), $G(t) = -\rho t$ and that in general, $d(x,t) = x-G(t)$ to the right of Γ .

We now introduce for $\eta \geqslant 0$ two operators, namely

$$E_{k,\eta,t}v(x) = e^{\eta g(x,t+k)} E_k(e^{-\eta g(x,t)} v)$$

$$= \sum_j a_j(x)\exp(\eta[g(x,t+k)-g(x-jh,t)])v(x-jh) ,$$

$$\tilde{E}_{k,\eta,t}v(x) = e^{k\eta g_t'} \sum_j a_j(x)e^{jh\eta g_x'} v(x-jh) .$$

We have the following estimate for their difference.

<u>Lemma 8.2</u>. For $h\eta$ and t bounded there is a constant C such that

$$\|E_{k,\eta,t} - \tilde{E}_{k,\eta,t}\|_2 \leqslant Ck .$$

<u>Proof</u>. Since for $|j| \leqslant J$,

$$|g(x,t+k) - g(x-jh,t) - kg_t'(x,t) - jhg_x'(x,t)| \leqslant Chk ,$$

it follows that for t , $h\eta$ bounded,

$$\|E_{k,\eta,t} - \tilde{E}_{k,\eta,t}\|_2$$

$$\leq \sum_j \sup_x\{|a_j(x)|\, |\exp(\eta[g(x,t+k) - g(x-jh,t)])$$

$$- \exp(\eta[kg_t'(x,t) + jhg_x'(x,t)])|\} \leq Ck ,$$

which proves the lemma.

We shall need the following result by Lax and Nirenberg.

Lemma 8.3. If E_k is an operator of the form (8.2) with symbol satisfying $|a(x,\xi)| \leq 1$, then there is a constant C depending only on an upper bound for $\sum_j |a_j''(x)|$ such that

$$\|E_k\|_2 \leq 1 + Ck .$$

Proof. See [13].

We set for $\eta \geq 0$,

$$E_{k,\eta}^{n,0} = E_{k,\eta,(n-1)k}\, E_{k,\eta,(n-2)k} \cdots E_{k,\eta,0} .$$

The following is the central technical lemma in this section.

Lemma 8.4. Assume that E_k is contractive of order τ . Then for bounded t and $h\eta$ there is a constant C such that

$$\|E_{k,\eta,t}\|_2 \leq (1 + Ck) \exp(C(h\eta)^{\tau}) , \tag{8.8}$$

and

(8.9) $$\|E^{n,0}_{k,\eta}\|_2 \leq C \exp(Cn(h\eta)^{\tau}) \ .$$

Proof. By Lemma 8.2, in order to prove (8.8) it is sufficient to prove the corresponding statement for the operator $\tilde{E}_{k,\eta,t}$. This operator has the symbol

$$\tilde{a}_{k,\eta,t}(x,\xi) = \exp(k\eta g_t') \sum_j a_j(x) \exp(jh\eta g_x' + ij\xi)$$

$$= \exp(k\eta\rho g_x')a(x,\xi - ih\eta g_x') \ ,$$

and consequently by the definition of contractivity,

$$|\tilde{a}_{k,\eta,t}(x,\xi)| \leq \exp(C(h\eta)^{\tau}) \ .$$

Hence the operator $\exp(-C(h\eta)^{\tau})\tilde{E}_{k,\eta,t}$ has a symbol which satisfies the assumptions of Lemma 8.3. This proves (8.8), and (8.9) is now a trivial consequence.

We can now give the variable coefficient analogue of Theorem 7.1.

Theorem 8.1. Assume that E_k is dissipative of order σ and accurate of order $\mu = \nu - 1$ and let v vanish for positive x . Then for t bounded there are positive constants c and C such that with κ as in (7.5) , $nk = t$, $\delta_t = \delta t^{-1}$, $v \in L_2$,

(8.10) $$\|\chi_{\delta+G(t)} E^n_k v\|_2 \leq C \exp(-cn\delta_t^{\kappa}) \|v\|_2 \ .$$

Proof. We easily find that with $\tilde{\rho} = \|\rho\|_\infty$ we have the estimate $|G(t)| \leqslant \tilde{\rho}t$. Since $E_k^n v$ vanishes for $x > \frac{Jt}{\lambda}$ this implies that $\chi_{\delta+G(t)} E_k^n v$ vanishes for $\delta_t > J\lambda^{-1} + \tilde{\rho} = \tilde{\delta}$. It remains to prove (8.10) for $\delta_t \leqslant \tilde{\delta}$. By (8.7) we may replace $\chi_{\delta+G(t)}$ in (8.10) by the characteristic function $\tilde{\chi}_{\delta,t}$ of the set $\{x ; g(x,t) > \delta\}$. We have for t and $h\eta$ bounded, $\eta \geqslant 0$,

$$(8.11) \qquad \|\tilde{\chi}_{\delta,t} E_k^n v\|_2 \leqslant \|\exp(\eta(g(\cdot,nk) - \delta)) E_k^n v\|_2$$

$$\leqslant e^{-\eta\delta} \|E_{k,\eta}^{n,0}\, e^{\eta x} v\|_2 \leqslant e^{-\eta\delta} \|E_{k,\eta}^{n,0}\|_2 \, \|v\|_2 ,$$

and hence using Lemmas 8.1 and 8.4,

$$\|\tilde{\chi}_{\delta,t} E_k^n v\|_2 \leqslant C \exp(-\eta\delta + Cn(h\eta)^\tau) \, \|v\|_2 .$$

The result now follows as in the proof of Theorem 7.1 by choosing for $\delta_t \leqslant \tilde{\delta}$,

$$h\eta = (\tfrac{1}{2} \lambda \delta_t C^{-1})^{\frac{1}{(\tau-1)}} .$$

In order to prove the analogous maximum-norm estimate we shall need the following lemma.

Lemma 8.5. Let A_h be an operator of the form

$$A_h v(x) = \sum_j a_j(x,h) v(x-jh) ,$$

where the summation is over J points and the coefficients are continuous in x . Then

$$\|A_h\|_\infty \leq \sqrt{J}\,\|A_h\|_2 \ .$$

Proof. Let $\|A_h\|_2 = M$. We shall prove that for any $x_0 \in R$,

(8.12) $$\sum_j |a_j(x_0,h)| \leq \sqrt{J}\, M \ .$$

Let $0 < \varepsilon < \frac{h}{2}$ and set with $\phi_j = \arg a_j(x_0,h)$,

$$v_\varepsilon(x) = \begin{cases} \exp(-i\phi_j) & \text{for } |x - x_0 + jh| < \varepsilon \\ & \qquad \text{if } a_j(x_0,h) \neq 0 \ , \\ 0 & \text{for other } x \ . \end{cases}$$

We then have $\|v_\varepsilon\|_2 \leq \sqrt{2\varepsilon J}$. On the other hand,

$$\|A_h v_\varepsilon\|_2^2 \geq \int_{|x-x_0|<\varepsilon} |A_h v_\varepsilon(x)|^2 dx$$

$$= \int_{|x-x_0|<\varepsilon} |\sum_j a_j(x,h) e^{-i\phi_j}|^2 dx$$

$$= 2\varepsilon\{(\sum_j |a_j(x_0,h)|)^2 + o(1)\} \quad \text{as } \varepsilon \to 0 \ ,$$

and hence for small ε ,

$$\sqrt{2\varepsilon}\,\{\sum_j |a_j(x_0,h)| + o(1)\} \leq M\sqrt{2\varepsilon J} \ ,$$

which clearly implies (8.12) for $\varepsilon \to 0$.

The following is a weaker analogue of Theorem 7.2.

Theorem 8.2. Under the assumptions of Theorem 8.1 there are positive constants c and C such that for $v \in L_\infty$,

$$\|\chi_{\delta+G(t)} E_k^n v\|_\infty \leq C\, n^{\frac{1}{2}} \exp(-cn\delta_t^\kappa)\, \|v\|_\infty .$$

Proof. The result follows in the same way as in Theorem 8.1 if we replace the subscript 2 in (8.11) by ∞ and notice that application of Lemma 8.5 to $E_{k,\eta}^{n,0}$ gives

$$\|E_{k,\eta}^{n,0}\|_\infty \leq C\sqrt{n}\, \|E_{k,\eta}^{n,0}\|_2 .$$

In the case of an isolated singularity at $x = 0$ we have finally the following result.

Theorem 8.3. Under the assumptions of Theorem 8.1 and for any positive ε, M, and s with $s > \frac{1}{2}$ there are constants c and C such that for $v \in \check{B}_M^s$,

$$\|\chi_{\delta+G(t)} E_k^n v\|_\infty \leq C\, n^{\varepsilon} \exp(-cn\delta_t^\kappa)\, \|v\|_{\check{B}_M^s} .$$

Proof. By Theorem 8.2 we have

$$\|\chi_{\delta+G(t)} E_k^n v\|_\infty \leq C\, n^{\frac{1}{2}} \exp(-cn\delta_t^\kappa)\, \|v\|_\infty \tag{8.13}$$

and by Theorem 6.2,

$$\|\chi_{\delta+G(t)} E_k^n v\|_\infty \leq C\, n^{\frac{\varepsilon}{2}}\, \|v\|_{\check{B}_M^s} . \tag{8.14}$$

The result now follows by taking a geometric mean of the bounds in (8.13) and (8.14).

REFERENCES

1. Apelkrans, M. Y. T., "On Difference Schemes for Hyperbolic Equations with Discontinuous Initial Values," Math. Comp. 22 (1968), 525-539.
2. Brenner, Ph., "The Cauchy Problem for Symmetric Hyperbolic Systems in L_p," Math. Scand. 19 (1966), 27-37.
3. Brenner, Ph., and Thomée, V.,"Stability and Convergence Rates in L_p for Certain Difference Schemes," Math. Scand, to appear.
4. Brenner, Ph., and Thomée, V., "Estimates Near Discontinuities for Some Difference Schemes," to appear.
5. Butzer, P. L., and Berens, H., Semi-Groups of Operators and Approximation, Springer-Verlag, Berlin, Heidelberg, New York, 1967.
6. Hedstrom, G. W., "The Near-Stability of the Lax-Wendroff Method," Numer. Math. 7 (1965), 73-77.
7. Hedstrom, G. W., "Norms of Powers of Absolutely Convergent Fourier Series," Michigan Math. J. 13 (1966), 393-416.
8. Hedstrom, G. W., "The Rate of Convergence of Some Difference Schemes," SIAM J. Numer. Anal. 5 (1968), 363-406.
9. Hedstrom, G. W., "The Rate of Convergence of Difference Schemes with Constant Coefficients," BIT 9 (1969), 1-17.
10. Hörmander, L., "Estimates for Translation Invariant Operators in L_p Spaces," Acta. Math. 104 (1960), 93-140.
11. Kreiss, H. O., and Lundqvist, E., "On Difference Approximations with Wrong Boundary Values," Math. Comp. 22 (1968), 1-12.
12. Kreiss, H. O., Thomée, V., and Widlund, O. B., "Smoothing of Initial Data and Rates of Convergence for Parabolic Difference Equations," Comm. Pure Appl. Math, 23 (1970), 159-176.
13. Lax, P. D., and Nirenberg, L., "On Stability for Difference Schemes: A Sharp Form of Gårding's Inequality," Comm. Pure Appl. Math. 19 (1966), 473-492.
14. Lions, J. L., and Peetre, J., "Sur une classe d'espaces d'interpolation," Inst. Hautes Études Sci. Publ. Math. No. 19 (1964), 5-68.

15. Löfström, J., "Besov Spaces in Theory of Approximation," Ann. Math. Pura Appl. 85 (1970), 93-184.
16. Peetre, J., "Applications de la théorie des espaces d'interpolation dans l'analyse harmonique," Ricerche Mat. 15 (1966), 1-36.
17. Peetre, J., and Thomée, V., "On the Rate of Convergence for Discrete Initial-Value Problems," Math. Scand. 21 (1967), 159-176.
18. Richtmyer, R. D., and Morton, K. W., Difference Methods for Initial-Value Problems, 2nd ed., Interscience, New York, 1967.
19. Serdjukova, S. I., "A Study of Stability of Explicit Schemes with Constant Real Coefficients," Ž. Vyčisl. Mat. i Mat. Fiz. 3 (1963), 365-370.
20. Serdjukova, S. I., "On the Stability in C of Linear Difference Schemes with Constant Real Coefficients," Ž. Vyčisl. Mat. i Mat. Fiz. 6 (1966), 477-486.
21. Thomée, V., "Stability Theory for Partial Difference Operators," SIAM Rev. 11 (1969), 152-195.
22. Widlund, O. B., "On the Rate of Convergence for Parabolic Difference Schemes, I," Proc. Amer. Math. Symp. for Appl. Math. vol. 21, Durham, N. C., 1968, 60-73.
23. Widlund, O. B., "On the Rate of Convergence for Parabolic Difference Schemes, II," Comm. Pure Appl. Math. 23 (1970), 79-96.

SOME RESULTS IN APPROXIMATION THEORY WITH APPLICATIONS TO NUMERICAL ANALYSIS*

Richard S. Varga**

1. Introduction

The object of this paper is to present results in two rather different areas of approximation theory, and to sketch their applications to numerical analysis. In Sections 2-4, we discuss results concerning Chebyshev rational approximations of reciprocals of certain entire functions (such as $f(z) = e^z$) on $[0, +\infty)$, and we show how these approximations can be used numerically in the solution of semi-discrete parabolic partial difference equations. We also discuss in Section 4 results of numerical experiments testing such Chebyshev semi-discrete approximations.

In Section 5, we discuss improved error bounds for spline and L-spline interpolation. These improved error bounds for spline interpolation are then used to

---

*This research was supported in part by AEC Grant AT(11-1)-2075.

**Kent State University.

deduce improved error bounds for Galerkin approximations of solutions of particular two-point nonlinear boundary value problems in Section 6.

2. Chebyshev Semi-Discrete Approximations of Parabolic Partial Difference Equations

As described in Cody, Meinardus, and Varga [3], consider any linear system of N coupled ordinary differential equations of the form

$$(2.1)\qquad \begin{cases} \dfrac{dc(t)}{dt} = -Ac(t) + g, & \forall t > 0, \\ c(0) = \tilde{c}, \end{cases}$$

where A is assumed to be an $N \times N$ time-independent Hermitian positive definite matrix, and $c(t)$, g, and $\tilde{c}$ are N-vectors. Typically, such coupled equations can arise from semi-discrete approximations to linear parabolic partial differential equations, in which all spatial variables are differenced, but the time variable, t, is left continuous (cf. [14, Chapter 8]). The solution $c(t)$ of (2.1) is given explicitly by

$$(2.2)\qquad c(t) = A^{-1}g + \exp(-tA)\{\tilde{c} - A^{-1}g\}, \quad \forall t \geq 0,$$

where as usual $\exp(-tA) \equiv \sum_{k=0}^{\infty} (-tA)^k/k!$.

To define the Chebeyshev semi-discrete approximations of (2.1), we turn to the following approximation problem. If π_m denotes all real polynomials $p(n)$ of degree at most m, and $\pi_{m,n}$ analogously denotes all real rational functions $r_{m,n}(x) = p(x)/q(x)$ with $p \in \pi_m$, $q \in \pi_n$, then let

$$(2.3) \qquad \lambda_{m,n} \equiv \inf_{\pi_{m,n}} \| e^{-x} - r_{m,n}(x) \|_{L_\infty[0,+\infty]}$$

denote the minimum error in approximating e^{-x} on $[0, +\infty)$ in the uniform norm over $\pi_{m,n}$. These constants $\lambda_{m,n}$ are called the <u>Chebyshev constants</u> for e^{-x} with respect to $[0, +\infty)$. It is obvious that $\lambda_{m,n}$ is finite if and only if $0 \leqslant m \leqslant n$, and moreover, given any pair (m,n) of nonnegative integers with $0 \leqslant m \leqslant n$, it is known (cf. Meinardus [6]) that there exists a unique $\hat{r}_{m,n}(x) \in \pi_{m,n}$ (after dividing out possible common factors) with

$$(2.4) \qquad \hat{r}_{m,n}(x) \equiv \hat{p}_{m,n}(x)/\hat{q}_{m,n}(x) \ ,$$

and with $\hat{q}_{m,n}(x) > 0$ on $[0, +\infty)$, such that

$$(2.5) \qquad \lambda_{m,n} = \| e^{-x} - \hat{r}_{m,n}(x) \|_{L_\infty[0,+\infty)} \ .$$

Since $\hat{q}_{m,n}(tA)$ is a polynomial in the matrix A, it is evident that $\hat{q}_{m,n}(tA)$ is a Hermitian positive definite $N \times N$ matrix for any finite $t \geqslant 0$. Thus, we can define the <u>$(m,n)^{th}$ Chebyshev approximation</u> $c_{m,n}(t)$ of $c(t)$ of (2.2) as

$$(2.6) \qquad c_{m,n}(t) = A^{-1}g + \hat{r}_{m,n}(tA)\{\tilde{c} - A^{-1}g\} \ , \qquad \forall t \geqslant 0 \ ,$$

$$= A^{-1}g + (\hat{q}_{m,n}(tA))^{-1}[\hat{p}_{m,n}(tA)\{\tilde{c} - A^{-1}g\}], \quad \forall t \geqslant 0 \ .$$

In other words, $c_{m,n}(t)$ is defined for each finite $t \geq 0$ as the solution v of the system of linear equations

$$(2.7) \qquad \hat{q}_{m,n}(tA) \cdot v = k(t) \equiv \hat{q}_{m,n}(tA) \cdot A^{-1}g + \hat{p}_{m,n}(tA)\{\tilde{c} - A^{-1}g\} .$$

To estimate the error in $c(t) - c_{m,n}(t)$, we use ℓ_2-vector norms, i.e., $\|v\|_2 = (v^{*}v)^{\frac{1}{2}}$. If $\{\lambda_i\}_{i=1}^{N}$ denote the (positive) eigenvalues of A, then the Hermitian character of A gives us for any $t \geq 0$ that

$$\|\exp(-tA) - \hat{r}_{m,n}(tA)\|_2 = \max_{1 \leq i \leq N} \left| e^{-t\lambda_i} - \hat{r}_{m,n}(t\lambda_i) \right| ,$$

where $\|\cdot\|_2$ denotes the induced operator norm (or spectral norm) relative to the ℓ_2-vector norm. But as $t\lambda_i \geq 0$ for all $1 \leq i \leq N$, it follows from the definition of $\lambda_{m,n}$ in (2.3) that

$$(2.8) \qquad \begin{aligned} \|c(t) - c_{m,n}(t)\|_2 &\leq \|\exp(-tA) - \hat{r}_{m,n}(tA)\|_2 \cdot \|\tilde{c} - A^{-1}g\|_2 \\ &\leq \lambda_{m,n} \|c - A^{-1}g\|_2 \qquad \forall t \geq 0 . \end{aligned}$$

In general, the error of the spatial discretization leading to (2.1) must also be bounded to give the total error (i.e., space and time) of the Chebyshev semi-discrete approximations. This is discussed, for example, in [3] in a particular case.

Unlike usual methods of time-discretization, such as Crank-Nicolson, which depended upon repeatedly taking small time steps $\Delta t = T/M$ to achieve precision at a time T, the Chebyshev semi-discrete approximation directly (in one step) gives an approximation at time T by simply setting $t = T$ in (2.5). The accuracy of this method is clearly dependent from (2.8) on how the Chebyshev constants $\lambda_{m,n}$ behave as $n \to \infty$. First, from (2.3), it is evident that

$$(2.9) \qquad 0 < \lambda_{n,n} \leqslant \lambda_{n-1,n} \leqslant \cdots \leqslant \lambda_{o,n}, \quad \forall n \geqslant 0 .$$

In [3], it was shown in particular that

$$(2.10) \qquad \lambda_{o,n} \leqslant (2e^{\alpha})^{-n} \qquad \forall n \geqslant 0 ,$$

where $\alpha = 0.13923\cdots$ is the real solution of $2\alpha e^{2\alpha+1} = 1$. Thus, (2.10) shows us that the Chebyshev constants $\lambda_{o,n}$ converge <u>geometrically to zero</u> as $n \to \infty$. In [3], it was also shown that this convergence is <u>not</u> faster than geometric, in that

$$\overline{\lim_{n\to\infty}} (\lambda_{o,n})^{1/n} \geqslant \frac{1}{6} .$$

Because of (2.9) and (2.10), we can state these results in the following form.

<u>Theorem 1</u>. Let $\{m(n)\}_{n=o}^{\infty}$ be any sequence of nonnegative integers with $0 \leqslant m(n) \leqslant n$ for each $n \geqslant 0$. Then,

$$(2.11) \qquad \overline{\lim_{n\to\infty}} (\lambda_{m(n),n})^{1/n} \leq \frac{e^{-\alpha}}{2} = 0.43501\cdots ,$$

and

$$(2.12) \qquad \overline{\lim_{n\to\infty}} (\lambda_{o,n})^{1/n} \geq \frac{1}{6} .$$

3. Theoretical Extensions.

It is natural to ask if the geometric convergence to zero of the Chebyshev constants $\lambda_{m,n}$ for $\frac{1}{e^x}$ in (2.11) and (2.12) hold for a wider class of entire functions than just $f(x) = e^x$. A generalization of these results has recently been given by Meinardus and Varga [8], and can be described as follows.

Let $f(z) = \sum_{k=o} a_k z^k$ be an entire function with $M_f(r) \equiv \sup_{|z|\leq r} |f(z)|$ as its maximum modulus function. Then, f is said to be of perfectly regular growth (ρ,B) (cf. Boas [2, p. 8], Valiron [13, p. 45]) if and only if there exist two (finite) positive numbers ρ and B such that

$$(3.1) \qquad \lim_{r\to\infty} \frac{\ln M_f(r)}{r^\rho} = B .$$

We then state (cf. [4])

Theorem 2. Let $f(z) = \sum_{k=o}^{\infty} a_k z^k$ be an entire function of perfectly regular growth (ρ,B) with $a_k \geq 0$ $\forall k \geq 0$, and for any pair (m,n) of nonnegative integers with $0 \leq m \leq n$, let

$$(3.2) \qquad \lambda_{m,n} \equiv \inf_{\pi_{m,n}} \left\| \frac{1}{f(x)} - r_{m,n}(x) \right\|_{L_\infty[0,+\infty]}$$

be its associated Chebyshev constants. Then, for any sequence $\{m(n)\}_{n=0}^{\infty}$ of nonnegative integers with $0 \leq m(n) \leq n$ for each $n \geq 0$,

$$(3.3) \qquad \overline{\lim_{n\to\infty}}(\lambda_{m(n),n})^{1/n} \leq 2^{-1/\rho} < 1 .$$

Moreover,

$$(3.4) \qquad \overline{\lim_{n\to\infty}} (\lambda_{0,n})^{1/n} \geq 2^{-2-1/\rho} .$$

Thus, Theorem 2 establishes the geometric convergence to zero of the Chebyshev constants for $1/f(x)$ on $[0,+\infty)$ for all entire functions of perfectly regular growth with nonnegative Taylor coefficients. As special cases of Theorem 2, we have of course $f(z) = e^z$, $f(z) = \sinh(z^n)$, and $f(z) = J_n(iz)$, where J_n denotes the nth Bessel function of the first kind. Note that for $f(z) = e^z$, for which $\rho = B = 1$ from (3.1), the results of (3.3)-(3.4) of Theorem 2, are slightly weaker than those of (2.11)-(2.12) of Theorem 1.

It should be mentioned that the proofs of Theorems 1 and 2 depend upon estimating

$$\frac{1}{s_n(x)} - \frac{1}{f(x)}$$

where $s_n(z) = \sum_{k=0}^{n} a_k z^k$ is the nth partial sum of $f(z)$.

It is shown in fact in [8] that, under the hypotheses of Theorem 2,

$$\lim_{n\to\infty}\left(\left\|\frac{1}{s_n}-\frac{1}{f}\right\|_{L_\infty[0,+\infty]}\right)^{1/n}=2^{-1/\rho},$$

so that the upper bound of (3.3) cannot be improved using this specific technique.

Upon examining the conclusions of Theorem 2, we see that the bounds of (3.3)-(3.4) depend upon ρ, but not upon B, and this would suggest the possibility of extensions of Theorem 1 to entire functions which are not of perfectly regular growth. Such extensions have recently been considered in Meinardus, Taylor, Reddy and Varga [7], and we state below a representative result.

Theorem 3. Let $f(z) = \sum_{k=0}^{\infty} a_k z^k$ be an entire function of order ρ, i.e., $\overline{\lim_{r\to\infty}} \dfrac{\ln\ln M_f(r)}{\ln r} \equiv \rho$, and assume that

$$(3.5)\qquad \overline{\lim_{r\to\infty}}\frac{\ln M_f(r)}{r^\rho}\equiv B,\quad \underline{\lim_{r\to\infty}}\frac{\ln M_f(r)}{r^\rho}\equiv b$$

satisfy $0 < b \leqslant B < \infty$, and that $a_k \geqslant 0$ $\forall k \geqslant 0$ with

$$(3.6)\qquad \frac{a_n}{a_{n+1}}$$

nondecreasing and unbounded for all n sufficiently large. Then, for any sequence $\{m(n)\}_{n=0}^{\infty}$ of nonnegative integers with $0 \leqslant m(n) \leqslant n$ for each $n \geqslant 0$,

(3.7) $$\overline{\lim_{n\to\infty}} (\lambda_{m(n),n})^{1/n} \leq \left(\frac{B}{2b} \right)^{1/\rho} ,$$

where the $\lambda_{m,n}$ are the Chebyshev constants of $1/f$, defined in (3.2). Moreover,

(3.8) $$\overline{\lim_{n\to\infty}} (\lambda_{o,n})^{1/n} \geq \frac{1}{4} \left(\frac{b}{2B} \right)^{1/\rho} .$$

The results of Theorems 2 and 3 give then sufficient conditions on $f(z)$ so that its associated Chebyshev constants $\lambda_{m(n),n}$, $0 \leq m(n) \leq n$, converge geometrically to zero as $n \to \infty$. In the spirit of Bernstein's classical inverse-type theorems for polynomial and trigonometric polynomial approximation on finite intervals, the following result of [7] which we sketch, gives necessary conditions for this geometric convergence.

Theorem 4. Let $f(x) > 0$ be a real continuous function on $[0, +\infty)$, such that there exist a sequence of real polynomials $\{p_n(x)\}_{n=o}^{\infty}$ with $p_n \in \pi_n$ $\forall n \geq 0$ and a real number $q > 1$ such that

(3.9) $$\overline{\lim_{n\to\infty}} \left(\left\| \frac{1}{p_n(x)} - \frac{1}{f(x)} \right\|_{L_\infty[o,+\infty]} \right)^{1/n} = \frac{1}{q} < 1 .$$

Then, there exists a function $F(z)$ with $F(x) = f(x)$ $\forall x \geq 0$ such that F is analytic in the whole complex plane, i.e., F is entire. Moreover, F is of finite order, i.e.,

$$\overline{\lim_{r\to\infty}} \frac{\ell n \ell n M_F(r)}{\ell n r} = \rho < \infty .$$

Proof. For any q_1 with $q > q_1 > 1$, it follows from (3.9) that there exists a positive integer $n_1(q_1)$ such that

$$\left\| \frac{1}{p_n} - \frac{1}{f} \right\|_{L_\infty[0,+\infty]} \leqslant \frac{1}{q_1^n}, \quad \forall n \geqslant n_1(q_1),$$

or equivalently,

$$(3.10) \qquad -\frac{1}{q_1^n} \leqslant \frac{1}{p_n(x)} - \frac{1}{f(x)} \leqslant \frac{1}{q_1^n}, \quad \forall n \geqslant n_1(q_1), \quad \forall x \geqslant 0 .$$

Next, define

$$(3.11) \qquad m_f(r) = \|f\|_{L_\infty[0,r]}, \quad \text{where} \quad 0 \leqslant r < +\infty .$$

Fixing $r > 0$, the fact that q_1 exceeds unity implies that there exists a positive integer $n_2(r)$ such that

$$(3.12) \qquad q_1^n \geqslant q_1^n - m_f(r) \geqslant \frac{q_1^n}{2}, \quad \forall n \geqslant n_2(r) .$$

With $n_3 \equiv \max(n_1(q_1), n_2(r))$, a simple manipulation of (3.10) gives

$$\frac{-f^2(x)}{q_1^n+f(x)} \leqslant p_n(x) - f(x) \leqslant \frac{f^2(x)}{q_1^n-f(x)}, \quad \forall 0 \leqslant x \leqslant r, \quad \forall n \geqslant n_3 .$$

From these inequalities and the inequalities of (3.12), it

follows that

$$(3.13)\qquad \|p_n - f\|_{L_\infty[0,r]} \leq \frac{m_f^2(r)}{q_1^n - m_f(r)} \leq \frac{2m_f^2(r)}{q_1^n}\ , \quad \forall n \geq n_3\ .$$

We now make the change of variables

$$\frac{r}{2}(1+t) = x\ ;\quad 0 \leq x \leq r\ ,\quad -1 \leq t \leq +1\ ,$$

and define

$$(3.14)\qquad h(t;r) = f\left(\frac{r}{2}(1+t)\right)\ .$$

If $E_n\{h(\cdot,r)\} \equiv \inf_{\sigma_n \in \pi_n} \|\sigma_n - h(\cdot,r)\|_{L_\infty[-1,+1]}$ denotes the error in the best Chebyshev polynomial approximation in π_n to $h(x,r)$ on $[-1,+1]$, the inequality of (3.13) immediately gives us that

$$(3.15)\qquad E_n\{h(\cdot,r)\} \leq \frac{2m_f^2(r)}{q_1^n}\ , \quad \forall n \geq n_3\ .$$

Since $r > 0$ is fixed and q_1 is an arbitrary number with $q > q_1 > 1$, we evidently have from (3.15) that

$$(3.16)\qquad \overline{\lim_{n\to\infty}}\,(E_n\{h(\cdot,r)\})^{1/n} \leq \frac{1}{q}\ , \quad \forall r > 0\ .$$

Using Bernstein's Theorem (cf. [6, p. 86]), it follows that $h(t;r)$ can, for each $r > 0$, be extended to a function analytic in the open ellipse $\mathcal{E}_q$ with foci at ± 1 and semi-major and semi-minor axes a and b such that

$a + b = q > 1$. In terms of f , this means that f can be extended to a function $F(z)$ analytic in the region $\Omega_r = \{z: z = \frac{r}{2}(1+t)$ where $t \in \mathcal{E}_q\}$. But, r is an arbitrary positive real number, and it is easily seen that for any complex number w , $w \in \Omega_r$ for r sufficiently large. Hence, $F(z)$ is analytic in the whole complex plane, i.e., F is an entire function, which proves the first assertion of Theorem 4. The proof that F has finite order, which depends on comparing $\sup_{\Omega_r} |F(z)|$ with $m_f(r)$ as $r \to \infty$, is only slightly more difficult, and can be found in [7].

Not all entire functions f which are real and positive on $[0,+\infty)$ satisfy the geometric convergence of (3.9). As shown in [7], the particular entire function

$$f(z) = (z + 1)\{2 + \cos z\}$$

which is real and positive on $[0,+\infty)$, fails to satisfy (3.9).

4. Numerical Results

Dr. W. E. Culham of the Gulf Research and Development Co. (Pittsburgh) has been numerically testing the Chebyshev semi-discrete method of Sections 2-3 for solving linear parabolic problems with one spatial variable. Though the results are not yet complete, some interesting conclusions can already be made.

First, for very small $T > 0$, it is generally preferable to use the Crank-Nicolson method rather than the Chebyshev semi-discrete method. The reason for this is almost obvious. The Crank-Nicolson method for one spatial variable problems can be viewed as giving a matrix Padé approximation $M(t)$ of $\exp(-tA)$ for which (cf. [14, p. 266])

$$\exp(-tA) = M(t) + O(t^3), \quad t \downarrow 0 .$$

In other words, $M(t)$ is a third-order approximation for $\exp(-tA)$ for t close to zero. The Chebyshev semi-discrete method, on the other hand, gives a matrix approximation of $\exp(-tA)$ for which the maximum error occurs at $t = 0$. This is a consequence of the Chebyshev equi-oscillation of the error curve. More precisely, we necessarily have (cf. (2.5)) that

$$\lambda_{m,n} = |e^{-\hat{x}} - \hat{r}_{m,n}(\hat{x})| \quad \text{for} \quad \hat{x} = 0 ,$$

and consequently,

$$\|\exp(-tA) - \hat{r}_{m,n}(tA)\|_2 = \lambda_{m,n} \quad \text{for} \quad t = 0 .$$

This short-coming of the Chebyshev semi-discrete method to small t can be partially off-set by using the following suggestion of Professor R. B. Kellogg. From the error curve $e^{-x} - \hat{r}_{m,n}(x)$, which necessarily has $m + n + 1$ distinct positive zeroes, let $\sigma_{m,n} > 0$ be the smallest such positive zero. Then, with $x = \sigma_{m,n} + t$, it follows from $|e^{-x} - \hat{r}_{m,n}(x)| \leq \lambda_{m,n}$, $\forall x \geq 0$ that

$$|e^{-t} - e^{\sigma_{m,n}} \hat{r}_{m,n}(\sigma_{m,n} + t)| \leq e^{\sigma_{m,n}} \lambda_{m,n} , \quad \forall t \geq 0 ,$$

and the rational approximation $e^{\sigma_{m,n}} \hat{r}_{m,n}(\sigma_{m,n} + t)$ of e^{-t} has zero error at $t = 0$. This is equivalent with solving the Chebyshev minimization problem over $\pi_{m,n}$ for e^{-t} on $[0,+\infty)$ with the linear constraint of zero error at $t = 0$. The numbers $\sigma_{n,n}$ have been computed by W. J. Cody, Jr., and they decrease very rapidly to zero as $n \to \infty$.

Because of the above-mentioned error behavior at $t = 0$, the numerical experiments comparing the Crank-Nicolson method with the Chebyshev semi-discrete method have centered about comparing total work on a computer for T large. Physically speaking, T in these experiments is selected to be about the half-life of the transient term. As previously mentioned, though the results are not complete, several typical cases have arisen where the Chebyshev semi-discrete method with $m = n = 3$ is about 100 times faster than the Crank-Nicolson method. More will be reported on this at a later time.

It should be stated that these Chebyshev semi-discrete methods as described are rather severely limited to linear problems for which the natural semi-group property for such parabolic problems holds. This is the essence of approximating e^{-x} on $[0,+\infty)$ by rational functionals in the uniform norm. It is not known if such techniques can be extended to the numerical solution of strongly nonlinear parabolic partial differential equations.

Finally, we wish to comment on the practical solution of the matrix equation of (2.7). Because $\hat{q}_{m,n}(x)$ is positive on $[0,+\infty)$, it can be factored in the form

$$\hat{q}_{m,n}(x) = \prod_{\ell=1}^{k_1} j_\ell(x) \cdot \prod_{\ell=1}^{k_2} h_\ell(x) , \quad 2k_2+k_1 = n ,$$

where each $j_\ell(x)$ is real and linear in x , and each $h_\ell(x)$ is real and quadratic in x . Thus, solving the matrix problem (2.7) amounts to solving a succession of simpler matrix problems of the form

$$j_\ell(tA)\cdot v = k \quad , \qquad h_\ell(tA)\cdot v = k \; ,$$

for given vectors k . For example, if A is a tridiagonal positive definite Hermitian $N \times N$ matrix, then solving (2.7) reduces to solving a succession of matrix problems for which the matrix involved is either a tri-diagonal or five-diagonal positive definite Hermitian $N \times N$ matrix. For problems arising from a two-dimensional (spatial) parabolic partial differential equation, this factorization allows one to use either a direct inversion procedure, or a multi-line successive overrelaxation iterative method.

5. Improved Error Bounds for Spline and L-Spline Interpolation

We now switch to another topic, concerned with approximation by spline and L-spline interpolation. The results of this section are from Swartz and Varga [12]. Fuller details can be found in [12].

We now introduce some standard notation. For $-\infty < a < b < +\infty$, for each integer m and for each extended real number with $1 \leqslant q \leqslant \infty$, $W^m_q[a,b]$ denotes the Sobolev space of all real-valued functions $w(x)$ defined on the interval $[a,b]$ such that $w \in C^{m-1}[a,b]$, $D^{m-1}w$

is absolutely continuous with $D^m w \in L_q[a,b]$, where $D \equiv \frac{d}{dx}$. It is well known that $W_q^m[a,b]$ is a Banach space, with its norm defined by

$$(5.1) \qquad \|w\|_{W_q^m[a,b]} = \sum_{j=0}^{m} \|D^j w\|_{L_q[a,b]} \; .$$

Next, for N a positive integer, let $\Delta: a = x_0 < x_1 < \cdots < x_N = b$ denote a partition Δ of $[a,b]$. The collection of all such partitions Δ of $[a,b]$ is denoted by $\mathcal{P}(a,b)$. We further define $\pi = \max_i (x_{i+1}-x_i)$ and $\underline{\pi} = \min_i (x_{i+1}-x_i)$ for each partition $\Delta \in \mathcal{P}(a,b)$. For any real number B with $B \geq 1$, $\mathcal{P}_B(a,b)$ then denotes the subset of all partitions Δ in $\mathcal{P}(a,b)$ for which $\pi \leq B\underline{\pi}$. In particular, $\mathcal{P}_1(a,b)$ is the collection of all uniform partitions of $[a,b]$, and its elements are denoted by Δ_u.

Since we shall make extensive use of L-splines, we now briefly describe them. Given the differential operator L of order m,

$$(5.2) \qquad Lu(x) \equiv \sum_{j=0}^{m} c_j(x) D^j u(x) \; , \quad m \geq 1 \; ,$$

where $c_j \in C^j[a,b]$, $0 \leq j \leq m$, with $c_m(x) \geq \delta > 0$ for all x in $[a,b]$, and given the partition $\Delta: a = x_0 < x_1 < \cdots < x_N = b$, for $N > 1$ let $z = (z_1, \ldots, z_{N-1})$, the <u>incidence vector</u>, be an $(N-1)$-triple of positive integers with $1 \leq z_i \leq m$, $1 \leq i \leq N-1$. Then, $Sp(L,\Delta,z)$, the L-spline space, is the collection of all real-valued functions on $[a,b]$ such that (cf. Ahlberg, Nilson, and Walsh [1, ch. 6] and Schultz and Varga [11])

$$(5.3)\quad \begin{cases} L^*Lw(x) = 0\ , & x \in [a,b] - \{x_i\}_{i=1}^{N-1}\ , \\ D^k w(x_i -) = D^k w(x_i +) & \text{for all } 0 \leq k \leq 2m-1-z_i\ , \\ & 0 < i < N\ , \end{cases}$$

where L^* is the formal adjoint of L. From (5.3), we see that $Sp(L,\Delta,z) \subset C^{2m-\sigma-1}[a,b]$ where $\sigma \equiv \max_{1 \leq i \leq N-1} z_i$.

In the special case $L = D^m$, the elements of $Sp(L,\Delta,z)$ are, from (5.3), polynomials of degree $2m-1$ on each subinterval of Δ, and, as such, are called <u>polynomial splines</u>. More specifically, when $L = D^m$ and $z_i = m$, $0 < i < N$, the associated L-spline space is called the <u>Hermite space</u>, and is denoted by $H^{(m)}(\Delta)$. Similarly, when $L = D^m$ and $z_i = 1$, $0 < i < N$, the associated L-spline space is called the <u>spline space</u>, and is denoted by $Sp^{(m)}(\Delta)$.

Finally, if f is any bounded function defined on $[a,b]$, then

$$\omega(f,\delta) \equiv \sup\{|f(x+t) - f(x)| : x,\ x+t \text{ are in } [a,b] \text{ and } |t| \leq \delta\}$$

denotes the usual modulus of continuity of f. In general, K will denote below any generic constant which is independent of the functions considered, and is independent of π.

The following interpolation error bounds are typical (cf. Hedstrom and Varga [4]), and follow from results of Jerome and Varga [5], Schultz and Varga [11], and Perrin [9].

Theorem 5. Given $f \in W_2^{2m}[a,b]$ and given $\Delta \in \mathcal{P}_B(a,b)$, let s be the unique element in $Sp(L,\Delta,z)$ which interpolates f in the sense that

$$(5.4)\qquad D^j(f-s)(x_i) = 0 , \quad 0 \leqslant j \leqslant z_i - 1 , \quad 0 \leqslant i \leqslant N \quad (z_0 = z_N = m) .$$

Then, for $2 \leqslant q \leqslant \infty$,

$$(5.5)\qquad \|D^j(f-s)\|_{L_q[a,b]} \leqslant K\pi^{2m-j-\frac{1}{2}+\frac{1}{q}} \|f\|_{W_2^{2m}[a,b]} , \quad 0 \leqslant j \leqslant 2m-1 .$$

For polynomial splines (i.e., $L = D^m$), $\|f\|_{W_2^2[a,b]}$ can be replaced by $\|D^{2m}f\|_{L_2[a,b]}$ in (5.5).

The above result, based on the second intregal relation (cf. [1, p. 205]), is for rather smooth functions. While the exponent of π in (5.5) is sharp in that it cannot in general be increased for the function spaces considered, our goal is to obtain sharp interpolation errors for less smooth functions f. We next state a result of [12, Lemma 3.2] based on the Peavo Kernel Theorem, which is useful in achieving this goal.

Theorem 6. Given $f \in C^k[a,b]$ with $0 \leqslant k < 2m$ and given

$\Delta \in \mathcal{P}_B(a,b)$, let g be the unique element in $H^{(2m+1)}(\Delta)$ such that

$$(5.6)\qquad \begin{cases} D^j(f-g)(x_i) = 0 , & 0 \leq j \leq k , \quad 0 \leq i \leq N , \\ D^j g(x_i) = 0 , & k < j \leq 2m , \quad 0 \leq i \leq N . \end{cases}$$

Then,

$$(5.7)\quad K\pi^{k-j}\omega(D^k f,\pi) \geq \begin{cases} \| D^j(f-g)\|_{L_\infty[a,b]} , & 0 \leq j \leq k , \\ \| D^j g\|_{L_\infty[a,b]} , & k < j \leq 2m . \end{cases}$$

With Theorem 6, we now prove an analogue of Theorem 5 for less smooth functions.

Theorem 7. Given $f \in C^k[a,b]$ with $0 \leq k < 2m$ and given $\Delta \in \mathcal{P}_B(a,b)$, let s be the unique element in $Sp(L,\Delta,z)$ such that for $z_0 = z_N = m$,

$$(5.8)\quad \begin{cases} D^j(f-s)(x_i) = 0 , & 0 \leq j \leq \min(k,z_i-1), \quad 0 \leq i \leq N, \\ D^j s(x_i) = 0 , & \text{if } \min(k,z_i-1) < j \leq z_i-1 , \\ & \qquad 0 \leq i \leq N . \end{cases}$$

Then, for $2 \leq q \leq \infty$,

$$(5.9)\qquad K\left\{\pi^{k-j-\frac{1}{2}+\frac{1}{q}}\,\omega(D^k f,\pi) + \pi^{2m-j-\frac{1}{2}+\frac{1}{q}}\,\|f\|_{W_2^k[a,b]}\right\}$$

$$\geq \begin{cases} \|D^j(f-s)\|_{L_q[a,b]} , & 0 \leq j \leq k , \\ \|D^j s\|_{L_q[a,b]} , & \text{if } k < j \leq 2m-1 . \end{cases}$$

For polynomial splines $(L = D^m)$, the term involving $\|f\|_{W_2^k[a,b]}$ can be deleted in (5.9).

Proof. Given $f \in C^k[a,b]$, let g be its interpolant in $H^{(2m+1)}(\Delta)$, in the sense of (5.6) of Theorem 6. For any $2 \leq q \leq \infty$, the triange inequality gives us that

$$(5.10) \qquad \|D^j(f-s)\|_{L_q[a,b]} \leq \|D^j(f-g)\|_{L_q[a,b]} + \|D^j(g-s)\|_{L_q[a,b]}, \quad 0 \leq j \leq k,$$

where s is the interpolant of f is $Sp(L,\Delta,z)$ in the sense of (5.8). Note from (5.8) that s is also the interpolant of g in $Sp(L,\Delta,z)$ in the sense of (5.4). Hence, applying Theorem 5 yields

$$(5.11) \quad \|D^j(g-s)\|_{L_q[a,b]} \leq K\pi^{2m-j-\frac{1}{2}+\frac{1}{q}} \|g\|_{W_2^{2m}[a,b]}, \quad 0 \leq j \leq 2m-1.$$

We now bound $\|g\|_{W_2^{2m}[a,b]}$. For any ℓ with $k < \ell \leq 2m$, (5.7) of Theorem 6 gives

$$(5.12) \quad \|D^\ell g\|_{L_2[a,b]} \leq K\pi^{k-\ell}\,\omega(D^k f,\pi), \quad k < \ell \leq 2m.$$

For any ℓ with $0 \leq \ell \leq k$, we evidently have

$$\|D^\ell g\|_{L_2[a,b]} \leq \|D^\ell(f-g)\|_{L_2[a,b]} + \|D^\ell f\|_{L_2[a,b]}, \quad 0 \leq \ell \leq k,$$

and, using the first inequality of (5.7) of Theorem 6, this can be bounded above by

$$(5.13)\quad \|D^{\ell}g\|_{L_2[a,b]} \leq K\pi^{k-\ell}\,\omega(D^k f,\pi) + \|D^{\ell}f\|_{L_2[a,b]}\,, \qquad 0 \leq \ell \leq k\,.$$

Summing the inequalities of (5.12) and (5.13) and using the norm definition of (5.1) gives

$$\|g\|_{W_2^{2m}[a,b]2} \leq K\left\{\pi^{k-2m}\,\omega(D^k f,\pi) + \|f\|_{W_2^k[a,b]}\right\}\,.$$

This bound, when substituted in (5.11), gives

$$\|D^j(g-s)\|_{L_q[a,b]} \leq K\left\{\pi^{k-j-\frac{1}{2}+\frac{1}{q}}\,\omega(D^k f,\pi) + \pi^{2m-j-\frac{1}{2}+\frac{1}{q}}\,\|f\|_{W_2^k[a,b]}\right\}$$

for $0 \leq j \leq 2m-1$, thus suitably bounding the last term of (5.10). If polynomial splines are used, the term involving $\|f\|_{W_2^k[a,b]}$ can be deleted (cf. Theorem 5). Finally, the first term of the right side of (5.10) can be bounded above from (5.7) of Theorem 6, and the combined upper bounds, when inserted in (5.10), give the desired result of the first inequality of (5.9) for $0 \leq j \leq k$. If $k < j \leq 2m-1$, the same technique can be used to bound the terms on the right-hand side of

$$\|D^j s\|_{L_q[a,b]} \leq \|D^j(g-s)\|_{L_q[a,b]} + \|D^j g\|_{L_q[a,b]}\,, \qquad k < j \leq 2m-1,$$

which then establishes the second inequality of (5.9) for $k < j \leqslant 2m-1$. Q.E.D.

As an easy extension of Theorem 7, we include (cf. [12])

Corollary 1. With the hypotheses of Theorem 7, if $f \in W_r^{k+1}[a,b]$ with $1 \leqslant r \leqslant \infty$ and $0 \leqslant k < 2m$, then for $\max(r,2) \leqslant q \leqslant \infty$,

$$(5.14)\qquad K\pi^{k+1-j+\frac{1}{q}+\min(-\frac{1}{r},-\frac{1}{2})}\|f\|_{W_r^{k+1}[a,b]} \geqslant \begin{cases} \|D^j(f-s)\|_{L_q[a,b]}, & 0 \leqslant j \leqslant k, \\ \|D^j s\|_{L_q[a,b]}, & \text{if } k < j \leqslant 2m-1. \end{cases}$$

One difficulty with the result of Theorem 7 is that one needs to know the explicit continuity class of f to define its interpolant s in $Sp(L,\Delta,z)$ in the sense of (5.8). Often, this continuity class is difficult to determine from raw data in a routine setting in, say, a computer center. However, this can be avoided through the use of Lagrange interpolation. We now state (cf. [12])

Theorem 8. Given $f \in C^k[a,b]$ with $0 \leqslant k < 2m$ and given $\Delta \in \mathcal{P}_B(a,b)$ with at least $2m$ knots, let s be the unique element in $Sp(L,\Delta,z)$ such that for $z_0 = z_N = m$,

$$(5.15)\qquad D^j s(x_i) = D^j(\mathcal{L}_{2m-1,i}f)(x_i), \quad 0 \leqslant j \leqslant z_i - 1, \quad 0 \leqslant i \leqslant N,$$

where $\mathcal{L}_{2m-1,i}f$ is any Lagrange polynomial (of degree $2m-1$) interpolation of f in $2m$ consecutive knots $x_j, x_{j+1}, \ldots, x_{j+2m-1}$ with $x_i \in [x_j, x_{j+2m-1}]$. Then, for

$2 \leq q \leq \infty$, the bounds of (5.9) are valid, where again, for polynomial splines, the term in (5.9) involving $\|f\|_{W_2^k[a,b]}$ can be deleted.

We remark that the inequalities of (5.14) of Corollary 1 also apply to the interpolant s in $Sp(L,\Delta,z)$, defined by (5.15).

Although the result of Theorem 8 lifts the objection raised concerning the application of Theorem 7, we note that, for polynomial splines and $q = \infty$, the inequalities of (5.9) become

$$K\pi^{k-j-\frac{1}{2}}\,\omega(D^k f,\pi) \geq \begin{cases} \|D^j(f-s)\|_{L_\infty[a,b]}, & 0 \leq j \leq k, \\ \|D^j s\|_{L_\infty[a,b]}, & \text{if } k < j \leq 2m-1, \end{cases}$$

and one naturally expects that the exponent of π in the above expression is too small by a factor ½ . To improve the inequalities above, we next state another result of [12].

<u>Theorem 9</u>. Given $f \in C^k[a,b]$ with $0 \leq k < 2m$ and given $\Delta_u \in \mathcal{P}_1(a,b)$ with at least $2m$ knots, let s be the unique element in $Sp^{(m)}(\Delta_u)$ (i.e., $z_1 = z_2 = \cdots = z_{N-1} = 1$) such that

$$(5.16)\qquad \begin{cases} (f-s)(x_i) = 0, & 0 \leq i \leq N, \\ D^j(f-s)(a) = D^j(f-s)(b) = 0, & 0 \leq j \leq \min(k,m-1), \\ D^j s(a) = D^j s(b) = 0, & \text{if } \min(k,m-1) < j \leq m-1, \end{cases}$$

or

$$(5.17)\qquad \begin{cases} (f-s)(x_i) = 0\ , & 0 \leq i \leq N\ , \\ D^j(f-s)(a) = D^j(\mathcal{L}_{2m-1,0}f)(a)\ , & 0 \leq j \leq m-1\ , \end{cases}$$

where $\mathcal{L}_{2m-1,0}f$ is, as in Theorem 8, the Lagrange polynomial interpolation of f in the knots $x_0, x_1, \ldots, x_{2m-1}$, with a similar definition at $x = b$. Then,

$$(5.18)\quad K\pi^{k-j}\ \omega(D^k f, \pi) \geq \begin{cases} \|D^j(f-s)\|_{L_\infty[a,b]}\ , & 0 \leq j \leq k\ , \\ \|D^j s\|_{L_\infty[a,b]}\ , & \text{if } k < j \leq 2m-1\ . \end{cases}$$

<u>Corollary 2</u>. With the hypotheses of Theorem 9, if $f \in W_r^{k+1}[a,b]$ with $0 \leq k < 2m$ and $1 \leq r \leq \infty$, then for $\max(r,2) \leq q \leq \infty$,

$$(5.19)\qquad K\pi^{k+1-j-\frac{1}{r}+\frac{1}{q}}\ \|D^{k+1}f\|_{L_r[a,b]}$$

$$\geq \begin{cases} \|D^j(f-s)\|_{L_q[a,b]}\ , & 0 \leq j \leq k\ , \\ \|D^j s\|_{L_q[a,b]}\ , & \text{if } k < j \leq 2m-1\ . \end{cases}$$

To give an explicit example of Theorem 9, consider the particular case $m = 2$ of cubic splines. The cubic spline $s \in Sp^{(2)}(\Delta_u)$ of (5.17) is defined then by

$$(f-s)(x_i) = 0\ , \quad 0 \leq i \leq N\ ,$$

$$Ds(a) = \frac{1}{6h}\left\{-11f(a) + 18f(a+h) - 9f(a+2h) + 2f(a+3h)\right\}\ ,$$

$$Ds(b) = \frac{1}{6h}\left\{11f(b) - 18f(b-h) + 9f(b-2h) - 2f(b-3h)\right\}\ ,$$

where $h = \pi$ measures the uniform mesh of Δ_u . In this special case, (5.19) becomes, for $f \in C^k[a,b]$ with $0 \leq k < 4$,

$$Kh^{k-j}\ \omega(D^k f,h) \geq \begin{cases} \|D^j(f-s)\|_{L_\infty[a,b]} , & 0 \leq j \leq k , \\ \|D^j s\|_{L_\infty[a,b]} , & \text{if } k < j \leq 3 . \end{cases}$$

More complete results can be found in [12] for Hermite L-splines. In addition, certain stability theorems are established in [12] concerning the use of Lagrange interpolation of f to define interpolants s in $Sp(L,\Delta,z)$.

6. Application to Two-Point Boundary Value Problems

The interpolation error bounds of Section 5 can be applied to the numerical solution of two-point nonlinear boundary value problems in the following way. Consider, as a very special case, the Galerkin approximation of the solution of

$$(6.1)\qquad \begin{cases} (-1)^{m+1}\ D^{2m}\ u(x) = f(x,u(x)) , & a < x < b , \\ D^j u(a) = D^j u(b) = 0 , & 0 \leq j \leq m-1 , \end{cases}$$

where it is assumed that $f(x,u)$ is a real-valued function defined on $[a,b] \times R$ such that $f(x,u)$ and $f_u(x,u)$ are in $C^\circ([a,b] \times R)$, and there exists a constant γ such that

(6.2) $f_u(x,u) \geq \gamma > -\Lambda$ for all $x \in [a,b]$, and all real u , where $\Lambda \equiv \inf\left\{\int_a^b (D^m w(x))^2 dx / \int_a^b (w(x))^2 dx : w \in \overset{\circ}{W}{}_2^m[a,b]\right\}$,

and where $\overset{\circ}{W}{}_2^m[a,b]$ denotes the subspace of functions of $W_2^m[a,b]$ which satisfy the boundary conditions of (6.1). Given a finite dimensional subspace S_M of $\overset{\circ}{W}{}_2^m[a,b]$ with $\{w_i(x)\}_{i=1}^M$ a basis for S_M, the Galerkin approximation $\hat{w}(x) \in S_M$ of the solution of (6.1) is characterized (cf. [10]) by

$$(6.3) \qquad \int_a^b \left\{D^m\hat{w}(t)\cdot D^m w_i(t) + f(t,\hat{w}(t))\cdot w_i(t)\right\}dt = 0 , \quad 1 \leqslant i \leqslant M .$$

Next, let $\overset{\circ}{Sp}{}^{(m)}(\Delta_u)$ denote the subspace of functions in $Sp^{(m)}(\Delta_u)$ which satisfy the boundary conditions of (6.1). Then, based on results of Perrin, Price, and Varga [10, Theorem 3] and Theorem 9 of Section 5, we have

Theorem 10. If $u(x)$, the generalized solution of (6.1) is in $C^{2m}[a,b]$, and $\hat{w}(x)$ is its unique Galerkin approximation in $\overset{\circ}{Sp}{}^{(m)}(\Delta_u)$, then

$$(6.4) \qquad \|D^j(u-w)\|_{L_\infty[a,b]} \leqslant K\pi^{2m-j} \|D^{2m}u\|_{L_\infty[a,b]} , \quad 0 \leqslant j \leqslant m-1 .$$

REFERENCES

1. Ahlberg, J. H., Nilson, E. N., and Walsh, J. L., The Theory of Splines and their Applications, Academic Press, New York, 1967.

2. Boas, R. P. Jr., Entire Functions, Academic Press, New York, 1954.
3. Cody, W. J., Meinardus, G., and Varga, R. S., "Chebyshev Rational Approximations to e^{-x} in $[0,+\infty)$ and Applications to Heat-Conduction Problems," J. Approx. Theory, 2, 50-65 (1969).
4. Hedstrom, Gerald W., and Varga, Richard S., "Application of Besov Spaces to Spline Approximation," to appear.
5. Jerome, J. W., and Varga, R. S., "Generalizations of Spline Functions and Applications to Nonlinear Boundary Value and Eigenvalue Problems," Theory and Applications of Spline Functions (edited by T. N. E. Greville), Academic Press, New York, 1969, 103-155.
6. Meinardus, Günter, Approximation of Functions: Theory and Numerical Methods, Springer-Verlag, New York, 1967. Translated by L. L. Schumaker.
7. Meinardus, G., Taylor, G., Reddy, A. R., and Varga, R. S., "Converse Theorems and Extensions in Chebyshev Rational Approximation in $[0,+\infty)$," to appear.
8. Meinardus, Günter, and Varga, Richard S., "Chebyshev Rational Approximation to Certain Entire Functions in $[0,+\infty)$," to appear.
9. Perrin, F. M., "An Application of Monotone Operators to Differential and Partial Differential Equations on Infinite Domains," Doctoral Thesis, Case Institute of Technology, 1967.
10. Perrin, F. M., Price, H. S., and Varga, R. S., "On Higher-Order Numerical Methods for Nonlinear Two-Point Boundary Value Problems," Numer. Math. 13 (1969), 180-198.
11. Schultz, M. H., and Varga, R. S., "L-Splines," Numer. Math. 10 (1967), 345-369.
12. Swartz, Blair, and Varga, Richard S., "Error Bounds for Spline and L-Spline Interpolation," to appear.
13. Valiron, G., Lectures on the General Theory of Integral Functions, Chelsea, New York, 1949.
14. Varga, Richard S., Matrix Iterative Analysis, Prentice-Hall, Englewood Cliffs, New Jersey, 1962.